in√ 015698
18/2/83

ernal Readers | S

RET

Methods in Enzymology

Volume 90
CARBOHYDRATE METABOLISM
Part E

METHODS IN ENZYMOLOGY

EDITORS-IN-CHIEF

Sidney P. Colowick   Nathan O. Kaplan

Methods in Enzymology

Volume 90

Carbohydrate Metabolism

Part E

EDITED BY

Willis A. Wood

DEPARTMENT OF BIOCHEMISTRY
MICHIGAN STATE UNIVERSITY
EAST LANSING, MICHIGAN

1982

ACADEMIC PRESS

A Subsidiary of Harcourt Brace Jovanovich, Publishers

New York   London
Paris   San Diego   San Francisco   São Paulo   Sydney   Tokyo   Toronto

ACADEMIC PRESS, INC.
111 Fifth Avenue, New York, New York 10003

United Kingdom Edition published by
ACADEMIC PRESS, INC. (LONDON) LTD.
24/28 Oval Road, London NW1 7DX

Library of Congress Cataloging in Publication Data
Main entry under title:

Carbohydrate metabolism.

    Includes bibliographical references and idexes.
    1. Carbohydrate metabolism. 2. Enzymes. I. Wood,
Willis A., Date. ed. II. Series: Methods in
enzymology, v. 90 [etc.] [DNLM: 1. Carbohydrates--
Metabolism. W1ME9615K v. 90]
QP601.M49 vol. 90., etc. 574.1'925s 72-26891
[QP701] [574.1'33]
ISBN 0-12-181990-6 (v. 90)

PRINTED IN THE UNITED STATES OF AMERICA

82 83 84 85   9 8 7 6 5 4 3 2 1

Table of Contents

Section I. Kinases

Section II. Aldolases and Transketolases

Section III. Dehydratases

Section IV. Synthases

Section V. Phosphatases

Section VI. Phosphoenolpyruvate : Glycose Phosphotransferase System

Contributors to Volume 90

Article numbers are in parentheses following the names of contributors.
Affiliations listed are current.

CLARENCE AHLEM (78), *Department of Chemistry and Molecular Biology Institute, San Diego State University, San Diego, California 92182*

JAMES MICHAEL ANCHORS (66), *Department of Medicine, Duke University Medical Center, Durham, North Carolina 27710*

RICHARD L. ANDERSON (15, 16, 34, 35, 48), *Department of Biochemistry, Michigan State University, East Lansing, Michigan 48824*

A. E. ANNAMALAI (56), *Department of Chemistry, University of Delaware, Newark, Delaware 19711*

ANNE-CHARLOTTE ARONSSON (85), *Department of Biochemistry, Arrhenius Laboratory, University of Stockholm, S-10691 Stockholm, Sweden*

N. JAYANTHI BAI (37), *Department of Biochemistry, Medical College, Trivandrum 695 011, Kerala, India*

RUDOLF BENDER (45), *Farbwerke Hoechst, 6230 Frankfurt-Hoechst, Federal Republic of Germany*

STEPHEN J. BENKOVIC (53), *Department of Chemistry, The Pennsylvania State University, University Park, Pennsylvania 16802*

ROBERT W. BERNLOHR (12, 64), *Biochemistry Program, The Pennsylvania State University, University Park, Pennsylvania 16802*

DONALD L. BISSETT (15, 34), *Department of Biochemistry, Michigan State University, East Lansing, Michigan 48824*

W. J. BLACK[1] (56), *Department of Biochemical Cytology, Rockefeller University, New York, New York 10021*

MARINA BOJANOVSKI (20), *Zentrum Biochemie, Medizinische Hochschule, 3000 Hannover 61, Federal Republic of Germany*

HENRI BUC (11), *Département de Biologie Moléculaire, Institut Pasteur, 75724 Paris Cedex 15, France*

JANINA BUCZŁKO (40), *Division of Biochemistry, Institute of Organic and Physical Chemistry, Technical University of Wroclaw, 50-370 Wroclaw, Poland*

CAROL A. CAPERELLI (53), *Department of Chemistry, New York University, New York, New York 10003*

JANET M. CARDENAS (24), *National Eye Institute, National Institutes of Health, Bethesda, Maryland 20205*

BRUCE M. CHASSY (5, 89), *Laboratory of Microbiology and Immunology, National Institute of Dental Research, National Institutes of Health, Bethesda, Maryland 20205*

TAPATI CHATTERJEE (59), *Department of Biochemistry, University of Health Sciences, The Chicago Medical School, North Chicago, Illinois 60064*

B. R. COPELAND (33), *Department of Genetics and Medicine (Division of Medical Genetics, Center for Inherited Diseases), University of Washington, Seattle, Washington 98195*

V. L. CROW (26), *New Zealand Dairy Research Institute, Palmerston North, New Zealand*

MARINA DACHÀ (1), *Istituto di Chimica Biologica, Università degli Studi di Urbino, 61029 Urbino, Italy*

STANLEY DAGLEY (43, 44), *Department of Biochemistry, College of Biological Sciences, University of Minnesota, St. Paul, Minnesota 55108*

A. STEPHEN DAHMS (42, 47, 49, 78), *Department of Chemistry and Molecular Biology Institute, San Diego State University, San Diego, California 92182*

MARGARET M. DEMAINE (53), *Department of Biology, University of Alabama, Huntsville, Alabama 35899*

[1] Deceased.

ALAN DONALD (42, 47, 49), *Department of Chemistry and Molecular Biology Institute, San Diego State University, San Diego, California 92182*

JOHN S. EASTERBY (2), *Department of Biochemistry, The University of Liverpool, Liverpool L69 3BX, England*

WILLIAM B. EUBANK (17), *School of Medicine, Tulane Medical Center, New Orleans, Louisiana 70112*

THOMAS FERENCI (75), *Department of Microbiology, University of Sydney, Sydney, New South Wales 2006, Australia*

HARALD FOELLMER (81), *Department of Pathology, School of Medicine, Yale University, New Haven, Connecticut 06510*

ALISON M. FORDYCE (13), *New Zealand Dairy Research Institute, Palmerston North, New Zealand*

GIORGIO FORNAINI (1), *Istituto di Chimica Biologica, Università degli Studi di Urbino, 61029 Urbino, Italy*

JOACHIM FUCHS (81), *Zentrum Biochemie Medizinische Hochschule Hannover, 3000 Hannover 61, Federal Republic of Germany*

CLEMENT E. FURLONG (33, 77), *Department of Genetics and Medicine (Division of Medical Genetics, Center for Inherited Diseases), University of Washington, Seattle, Washington 98195*

ROBERT GEE (82), *Department of Biochemistry, Michigan State University, East Lansing, Michigan 48824*

GERHARD GOTTSCHALK (14, 45), *Institut für Mikrobiologie, Universität Göttingen, D-3400 Göttingen, Federal Republic of Germany*

PETER F. HAN (55), *Atlanta University Center Science Research Institute, Atlanta, Georgia 30310*

PAUL A. HARGRAVE (40), *Department of Medical Biochemistry and Department of Chemistry and Biochemistry, Southern Illinois University, Carbondale, Illinois 62901*

BEN G. HARRIS (4, 9, 38), *Departments of Biochemistry, Biological Sciences, and Chemistry, North Texas State University, Texas College of Osteopathic Medicine, Denton, Texas 76203*

FRED C. HARTMAN (83), *Biology Division, Oak Ridge National Laboratory, Oak Ridge, Tennessee 37830*

CHARLES L. HAUSWALD (48), *Department of Biochemistry, Michigan State University, East Lansing, Michigan 48824*

JOHN B. HAYS (74), *Department of Chemistry, University of Maryland, Catonsville, Maryland 21228*

H. WERNER HOFER (9), *Department of Biology, University of Konstanz, D-7750 Konstanz, Federal Republic of Germany*

EBERHARD HOFMANN (10), *Institute of Physiological Chemistry, Karl Marx University, DDR-7010 Leipzig, German Democratic Republic*

ROBERT W. HOGG (76), *Department of Microbiology, School of Medicine, Case Western Reserve University, Cleveland, Ohio 44106*

B. L. HORECKER (56, 57), *Roche Institute of Molecular Biology, Nutley, New Jersey 07110*

M. MARLENE HOSEY (59), *Department of Biochemistry, University of Health Sciences, The Chicago Medical School, North Chicago, Illinois 60064*

WILLIAM HUISMAN (47, 78), *Department of Chemistry and Molecular Biology Institute, San Diego State University, San Diego, California 92182*

KAZUTOMO IMAHORI (22, 28), *Tokyo Metropolitan Institute of Gerontology, Itabashi-ku, Tokyo 173, Japan*

KIICHI IMAMURA (25), *Department of Nutrition and Physiological Chemistry, Osaka University Medical School, Kitaku, Osaka 530, Japan*

J. JIMENO-ABENDANO (46), *Laboratoire de Microbiologie, Institut National des Sciences Appliquées, F-69621 Villeurbanne Cedex, France*

JOE JOHNSON, JR. (55), *Atlanta University Center Science Research Institute, Atlanta, Georgia 30310*

AXEL KAHN (23), *Institut de Pathologie Moléculaire, INSERM U129, C. H. U. Cochin, 75674 Paris Cedex 14, France*

MANFRED L. KARNOVSKY (66), *Department of Biological Chemistry, Harvard Medical School, Boston, Massachusetts 02115*

ODILE KELLERMANN (75), Département de Biologie Moléculaire, Unité de Génétique Moléculaire, Institut Pasteur, 75724 Paris Cedex 15, France

GRAHAME J. KELLY (62), Department of Biochemistry and Nutrition, University of New England, Armidale, New South Wales 2351, Australia

ROBERT B. KILLION (80), Department of Chemistry, University of California, Santa Cruz, California 95064

KINUKO KIMURA (36), Laboratory of Biochemistry, Rikkyo (St. Paul's) University, Toshima-ku, Tokyo 171, Japan

G. A. KOCHETOV (32), Laboratory of Molecular Biology and Bioorganic Chemistry, Moscow State University, Moscow 117234, USSR

MARIAN KOCHMAN (39, 40), Division of Biochemistry, Institute of Organic and Physical Chemistry, Technical University of Wroclaw, 50-370 Wroclaw, Poland

GERHARD KOPPERSCHLÄGER (10), Institute of Physiological Chemistry, Karl Marx University, DDR-7010 Leipzig, German Democratic Republic

DENISE KOTLARZ (11), Département de Biologie Moléculaire, Institut Pasteur, 75724 Paris Cedex 15, France

WOLFGANG K. G. KRIETSCH (18, 19), Institut für Physiologische Chemie, Physikalische Biochemie und Zellbiologie, Universität München, 8000 München 2, Federal Republic of Germany

MARIA A. KUKURUZINSKA (69, 71), Department of Biology and the McCollum-Pratt Institute, The Johns Hopkins University, Baltimore, Maryland 21218

MARIA-REGINA KULA (52), Gesellschaft für Biotechnologische Forschung, D-3300 Braunschweig, Federal Republic of Germany

KLAUS D. KULBE (20, 81), Fraunhofer-Institut für Grenzflächen und Bioverfahrenstechnik, 7000 Stuttgart 80, Federal Republic of Germany

GÜNTER W. K. KUNTZ (18, 19), Institut für Physiologische Chemie, Physikalische Biochemie und Zellbiologie, Universität München, 8000 München 2, Federal Republic of Germany

DANUTA KWIATKOWSKA (39), Department of Biochemistry, School of Medicine, 50-368 Wroclaw, Poland

ERWIN LATZKO (62), Bontanisches Institut, Westfälische Wilhelms-Universität, D-4400 Münster, Federal Republic of Germany

HERBERT G. LEBHERZ (80), Department of Chemistry and Molecular Biology Institute, San Diego State University, San Diego, California 92182

CHI-YU LEE (3, 21), Andrology Laboratory, Obstetrics and Gynecology, Acute Care Hospital, The University of British Columbia, Vancouver, British Columbia V6T 2B5, Canada

FREDRICH H. LEIBACH (31, 68), Department of Cell and Molecular Biology, Medical College of Georgia, Augusta, Georgia 30912

PATRICIA LOBELLE-RICH (17, 88), Department of Biochemistry, Louisiana State University Medical Center, New Orleans, Louisiana 70112

MARY V. LONG (83), Biology Division, Oak Ridge National Laboratory, Oak Ridge, Tennessee 37830

STEPHEN D. MCCURRY (82), Research Department, Kellogg Company, Battle Creek, Michigan 49016

JOHN S. MACGREGOR (56), Cornell University Medical College, New York, New York 10021

W. C. MCGREGOR (57), Biopolymer Research Department, Hoffmann-La Roche, Inc., Nutley, New Jersey 07110

MAURO MAGNANI (1), Istituto di Chimica Biologica, Università degli Studi di Urbino, 61029 Urbino, Italy

MASSIMO MALCOVATI (27), Istituto di Biologia Generale, Università di Milano, 20133 Milano, Italy

BENGT MANNERVIK (85), Department of Biochemistry, Arrhenius Laboratory, University of Stockholm, S-10691 Stockholm, Sweden

FRANK MARCUS (59), Department of Biochemistry, University of Health Sciences, The Chicago Medical School, North Chicago, Illinois 60064

JOËLLE MARIE (23), Institut de Pathologie

Moléculaire, INSERM U129, C.H.U. Cochin, 75674 Paris Cedex 14, France

JOHN P. MARKWELL (35), Department of Biochemistry, Michigan State University, East Lansing, Michigan 48824

CHARLES K. MARSCHKE (12), The Upjohn Company, Kalamazoo, Michigan 49001

LINDA A. MAUCK (50), Eastman Kodak Health Safety and Human Factors Laboratory, Rochester, New York 14650

NORMAN D. MEADOW (69, 70, 73), Department of Biology and the McCollum-Pratt Institute, The Johns Hopkins University, Baltimore, Maryland 21218

E. MELLONI (57), Instituto Policattedra di Chimica Biologica, Università degli Studi di Genova, 16132 Genova, Italy

H. PAUL MELOCHE (41), Papanicolaou Cancer Research Institute, Miami, Florida 33136

JOSEPH MENDICINO (31, 68), Department of Biochemistry, University of Georgia, Athens, Georgia 30602

HIDEO MIZUNUMA (60), Department of Biochemistry, Akita University School of Medicine, Akita 010, Japan

C. H. MOORE (13), Department of Chemistry, Biochemistry, and Biophysics, Massey University, Palmerston North, New Zealand

K. MUNIYAPPA (31, 68), Department of Biochemistry, University of Georgia, Athens, Georgia 30602

P. SURYANARAYANA MURTHY (37), Department of Biochemistry, University College of Medical Sciences, New Delhi 110 029, India

HIROSHI NAKAJIMA (28), Research and Development Center, Unitika Limited Company, Uji, Kyoto 611, Japan

ATSUSHI NAKAZAWA (72), Department of Biochemistry, Yamaguchi University School of Medicine, Kogushi, Ube 755, Japan

GERALD NESLUND (78), Department of Chemistry and Molecular Biology Institute, San Diego State University, San Diego, Californa 92182

H. G. NIMMO (54), Department of Biochemistry, University of Glasgow, Glasgow G12 8QQ, United Kingdom

I. LUCILE NORTON (83), Biology Division, Oak Ridge National Laboratory, Oak Ridge, Tennessee 37830

SCOTT J. NORTON (86, 87), Departments of Chemistry and Biochemistry, North Texas State University, Texas College of Osteopathic Medicine, Denton, Texas 76203

KAZUKO ÔBA (84), Laboratory of Biochemistry, Faculty of Agriculture, Nagoya University, Nagoya 464, Japan

EDWARD L. O'CONNELL (41), Papanicolaou Cancer Research Institute, Miami, Florida 33136

DENNIS J. OPHEIM (64), Department of Medical Laboratory Sciences, Quinnipiac College, Hamden, Connecticut 06518

BEDII ORAY (86, 87), Departments of Chemistry and Biochemistry, North Texas State University, Texas College of Osteopathic Medicine, Denton, Texas 76203

M. RAMACHANDRA PAI (37), Department of Chemistry, College for Women, Trivandrum 695 014, Kerala, India

JAMES K. PETELL (80), Department of Cell and Tumor Biology, Roswell Park Memorial Institute, Buffalo, New York 14263

FLORA H. PETTIT (30, 67), Clayton Foundation Biochemical Institute and Department of Chemistry, The University of Texas, Austin, Texas 78712

E. F. PHARES (83), Biology Division, Oak Ridge National Laboratory, Oak Ridge, Tennessee 37830

S. PONTREMOLI (57), Instituto Policattedra di Chimica Biologica, Università degli Studi di Genova, 16132 Genova, Italy

E. VICTORIA PORTER (5, 89), Laboratory of Microbiology and Immunology, National Institute of Dental Research, National Institutes of Health, Bethesda, Maryland 20205

G. G. PRITCHARD (13, 26), Department of Chemistry, Biochemistry, and Biophysics, Massey University, Palmerston North, New Zealand

S. SALEHEEN QADRI (2), Department of Biochemistry, The University of Dacca, Dacca 2, Bangladesh

J. RODNEY QUAYLE (51), *Department of Microbiology, University of Sheffield, Sheffield S10 2TN, United Kingdom*

LESTER J. REED (30, 67), *Clayton Foundation Biochemical Institute and Department of Chemistry, The University of Texas, Austin, Texas 78712*

RICHARD E. REEVES (17, 88), *Department of Biochemistry, Louisiana State University Medical Center, New Orleans, Louisiana 70112*

JUDITH RITTENHOUSE (59), *Department of Biochemistry, University of Health Sciences, The Chicago Medical School, North Chicago, Illinois 60064*

J. ROBERT-BAUDOUY (46), *Laboratoire de Microbiologie, Institut National des Sciences Appliquées, F-69621 Villeurbanne Cedex, France*

ROBERT ROBERTS (29), *Cardiac Care Unit, Cardiovascular Division, Washington University School of Medicine, St. Louis, Missouri 63110*

SAUL ROSEMAN (69), *Department of Biology and the McCollum-Pratt Institute, The Johns Hopkins University, Baltimore, Maryland 21218*

DARIO C. SABULARSE (16), *Department of Biochemistry, Michigan State University, East Lansing, Michigan 48824*

HERMANN SAHM (52), *Institut für Biotechnologie der KFA Jülich, D-5170 Jülich 1, Federal Republic of Germany*

MARK A. SCHELL (6), *Genex Corporation, Rockville, Maryland 20852*

G. D. SCHELLENBERG (33), *Department of Medicine, Division of Neurology, University of Washington, Seattle, Washington 98195*

JOHN V. SCHLOSS (83), *Central Research and Development Department, E. I. du Pont de Nemours & Company, Inc., Wilmington, Delaware 19801*

HORST SCHÜTTE (52), *Gesellschaft für Biotechnologische Forschung, D-3300 Braunschweig, Federal Republic of Germany*

ROBERT K. SCOPES (79), *Department of Biochemistry, La Trobe University, Bundoora, Victoria 3083, Australia*

WILLIAM R. SHERMAN (50), *Department of Psychiatry, Washington University School of Medicine, St. Louis, Missouri 63110*

JIAN-PING SHI (58), *Department of Enzymology, Institute of Biochemistry, Chinese Academy of Sciences, Shanghai 200031, People's Republic of China*

DAVID SIBLEY (47), *Department of Chemistry and Molecular Biology Institute, San Diego State University, San Diego, California 92182*

ROBERT D. SIMONI (74), *Department of Biological Sciences, Stanford University, Stanford, California 94305*

CLARK F. SPRINGGATE (63), *Department of Microbiology and Immunology, Tulane University, School of Medicine, New Orleans, Louisiana 70112*

CHESTER S. STACHOW (63), *Department of Biology, Boston College, Chestnut Hill, Massachusetts 02167*

JANE A. STARLING (9), *Department of Biology, University of Missouri, St. Louis, Missouri 63121*

VILBERTO STOCCHI (1), *Istituto di Chimica Biologica, Università degli Studi di Urbino, 61029 Urbino, Italy*

F. STOEBER (46), *Laboratoire de Microbiologie, Institut National des Sciences Appliquées, F-69621 Villeurbanne Cedex, France*

KENNETH B. STOREY (8, 61), *Institute of Biochemistry, Carleton University, Ottawa, Ontario K1S 5B6, Canada*

ANN STOTER (79), *Department of Biochemistry, La Trobe University, Bundoora, Victoria 3083, Australia*

CLAUDE D. STRINGER (83), *Biology Division, Oak Ridge National Laboratory, Oak Ridge, Tennessee 37830*

TAKAHIKO SUMI (7), *Division of Psychopharmacology, Psychiatric Research Institute of Tokyo, Tokyo 156, Japan*

SCOTT C. SUPOWIT (4), *Department of Medicine, University of California at San Diego, La Jolla, California 92093*

KOICHI SUZUKI (22, 28), *Department of Biochemistry, Faculty of Medicine, University of Tokyo, Bunkyo-ku, Tokyo 113, Japan*

TAKEHIKO TANAKA (25), *Department of Nutrition and Physiological Chemistry, Osaka University Medical School, Kitaku, Osaka 530, Japan*

YOHTALOU TASHIMA (60), *Department of Biochemistry, Akita University School of Medicine, Akita 010, Japan*

W. MARTIN TEAGUE (67), *CPC International, Inc., Argo, Illinois 60501*

GUDRUN TIBBELIN (85), *Department of Biochemistry, Arrhenius Laboratory, University of Stockholm, S-10691 Stockholm, Sweden*

K. F. TIPTON (54), *Department of Biochemistry, Trinity College, Dublin 2, Ireland*

N. E. TOLBERT (82), *Department of Biochemistry, Michigan State University, East Lansing, Michigan 48824*

S. TRANIELLO (57), *Istituto di Chimica Biologica, Università degli Studi di Ferrara, 44100 Ferrara, Italy*

MICHIO UI (7), *Department of Physiological Chemistry, Faculty of Pharmaceutical Sciences, Hokkaido University, Sapporo 060, Japan*

SUSUMU UJITA (36), *Laboratory of Biochemistry, Rikkyo (St. Paul's) University, Toshima-ku, Tokyo 171, Japan*

IKUZO URITANI (84), *Laboratory of Biochemistry, Faculty of Agriculture, Nagoya University, Nagoya 464, Japan*

GIOVANNA VALENTINI (27), *Istituto di Chimica Biologica, Facoltà di Medicina e Chirurgia, Università di Pavia, 27100 Pavia, Italy*

A. VAN TOL (56), *Department of Biochemistry I, Erasmus Universiteit, 300 DR Rotterdam, The Netherlands*

T. A. VENKITASUBRAMANIAN (37), *Department of Biochemistry, V. P. Chest Institute, University of Delhi, New Delhi 110 007, India*

HASSO V. HUGO (14), *Bayer AG, D-5600 Wuppertal 1, Federal Republic of Germany*

YING-LAI WANG (58), *Department of Enzymology, Institute of Biochemistry, Chinese Academy of Sciences, Shanghai 200031, People's Republic of China*

E. BRUCE WAYGOOD (70, 71), *Department of Biochemistry, University of Saskatchewan, Saskatoon, Saskatchewan S7N OWO, Canada*

NANCY WEIGEL (71, 72), *Department of Cell Biology, Baylor College of Medicine, Houston, Texas 77030*

DAVID B. WILSON (6), *Section of Biochemistry, Molecular and Cell Biology, Cornell University, Ithaca, New York 14853*

N. M. WILSON (33), *Department of Pharmacology, University of Wisconsin, Madison, Wisconsin 53706*

YUN-HUA H. WONG (50), *Department of Psychiatry, Washington University School of Medicine, St. Louis, Missouri 63110*

IAN E. WOODROW (65), *Max Volmer Institut für Biophysikalische und Physikalische Chemie, Technische Universität Berlin, 1000 Berlin 12, Federal Republic of Germany*

GEN-JUN XU (58), *Department of Enzymology, Institute of Biochemistry, Chinese Academy of Sciences, Shanghai 200031, People's Republic of China*

STEPHEN J. YEAMAN (30), *Department of Biochemistry, The University, Newcastle upon Tyne NE1 7RU, England.*

DON R. YELTMAN (38), *Departments of Biochemistry, Biological Sciences, and Chemistry, North Texas State University, Texas College of Osteopathic Medicine, Denton, Texas 76203*

GERHARD ZIMMERMANN (62), *Bayerische Landesanstalt für Bodenkultur und Pflanzenbau, D-8050 Freising, Federal Republic of Germany*

MICHAEL A. ZOCCOLI (66), *SYVA Research Institute, Palo Alto, California 94303*

Preface

Volumes 89 and 90 of *Methods in Enzymology* contain new procedures that have been published since 1974 for the enzymes involved in the conversion of monosaccharides to pyruvate or to the breakdown products of pyruvate. During the seven years since the last volumes on this subject were published, the development of new chromatographic separations utilizing affinity and hydrophobic properties of enzymes have resulted in improved procedures differing radically from those available previously. Hence, many of the enzyme purifications now involve fewer steps, are easier to carry out, and often allow separation of isozymes. As with Volumes 41 and 42, preparation of the same enzyme from a number of sources has been included. This is in recognition of the increasing spectrum of interest of investigators in, for instance, the comparison properties from a variety of sources and phylogenetic relationships.

Because of the large amount of material to be presented in this well-defined region of metabolism, division of the chapters into two volumes was necessary. The placement of sections in each volume is arbitrary and is for the most part the pattern found in Volumes 41 and 42.

WILLIS A. WOOD

METHODS IN ENZYMOLOGY

EDITED BY

Sidney P. Colowick and Nathan O. Kaplan

VANDERBILT UNIVERSITY
SCHOOL OF MEDICINE
NASHVILLE, TENNESSEE

DEPARTMENT OF CHEMISTRY
UNIVERSITY OF CALIFORNIA
AT SAN DIEGO
LA JOLLA, CALIFORNIA

METHODS IN ENZYMOLOGY

EDITORS-IN-CHIEF

Sidney P. Colowick   Nathan O. Kaplan

VOLUME L. Complex Carbohydrates (Part C)
Edited by VICTOR GINSBURG

VOLUME LI. Purine and Pyrimidine Nucleotide Metabolism
Edited by PATRICIA A. HOFFEE AND MARY ELLEN JONES

VOLUME LII. Biomembranes (Part C: Biological Oxidations)
Edited by SIDNEY FLEISCHER AND LESTER PACKER

VOLUME LIII. Biomembranes (Part D: Biological Oxidations)
Edited by SIDNEY FLEISCHER AND LESTER PACKER

VOLUME LIV. Biomembranes (Part E: Biological Oxidations)
Edited by SIDNEY FLEISCHER AND LESTER PACKER

VOLUME LV. Biomembranes (Part F: Bioenergetics)
Edited by SIDNEY FLEISCHER AND LESTER PACKER

VOLUME LVI. Biomembranes (Part G: Bioenergetics)
Edited by SIDNEY FLEISCHER AND LESTER PACKER

VOLUME LVII. Bioluminescence and Chemiluminescence
Edited by MARLENE A. DeLUCA

VOLUME LVIII. Cell Culture
Edited by WILLIAM B. JAKOBY AND IRA H. PASTAN

VOLUME LIX. Nucleic Acids and Protein Synthesis (Part G)
Edited by KIVIE MOLDAVE AND LAWRENCE GROSSMAN

VOLUME LX. Nucleic Acids and Protein Synthesis (Part H)
Edited by KIVIE MOLDAVE AND LAWRENCE GROSSMAN

VOLUME 61. Enzyme Structure (Part H)
Edited by C. H. W. HIRS AND SERGE N. TIMASHEFF

VOLUME 62. Vitamins and Coenzymes (Part D)
Edited by DONALD B. MCCORMICK AND LEMUEL D. WRIGHT

VOLUME 63. Enzyme Kinetics and Mechanism (Part A: Initial Rate and Inhibitor Methods)
Edited by DANIEL L. PURICH

VOLUME 64. Enzyme Kinetics and Mechanism (Part B: Isotopic Probes and Complex Enzyme Systems)
Edited by DANIEL L. PURICH

Section I

Kinases

[1] Hexokinase from Rabbit Red Blood Cells

By Giorgio Fornaini, Marina Dachà,
Mauro Magnani, and Vilberto Stocchi

$$\text{Hexose} + \text{MgATP}^{2-} \rightarrow \text{hexose 6-phosphate} + \text{MgADP}^{-}$$

The hexokinase (EC 2.7.1.1) reaction in red blood cells has the lowest turnover with respect to all the other enzymes[1] and in rabbit erythrocytes represents only 0.0003% (w/w) of the total proteins.[2]

This enzyme shows a marked decrease during cell aging,[3] and two distinct molecular forms (designated hexokinase Ia and Ib) have been found in reticulocytes,[4] but only one of these was consistently present in the mature rabbit red cell.[5] Hexokinase Ia corresponds to type I in the nomenclature of Katzen and Schimke[6] whereas hexokinase Ib differs from every other previously reported hexokinase isozyme.[7]

In this chapter we describe first the purification and the properties of hexokinase from mature red cells and subsequently the purification and the properties of the two hexokinases found in rabbit reticulocytes.

Assay Method

Principle. The continuous spectrophotometric assay is based on the following sequence of reactions:

$$\text{Glucose} + \text{ATP} \xrightarrow{\text{hexokinase}} \text{glucose 6-phosphate} + \text{ADP}$$
$$\text{Glucose 6-phosphate} + \text{NADP}^{+} \xrightarrow{\text{G6PDH}} \text{gluconate 6-phosphate} + \text{NADPH} + \text{H}^{+}$$
$$\text{Gluconate 6-phosphate} + \text{NADP}^{+} \xrightarrow{\text{6-PGDH}} \text{D-ribose 5-phosphate} + \text{CO}_2 + \text{NADPH} + \text{H}^{+}$$

[1] G. J. Brewer, *in* "The Red Blood Cell" (D. MacN. Surgenor, ed.), Vol. 1, p. 387. Academic Press, New York, 1974.

[2] V. Stocchi, M. Magnani, P. Ninfali, and M. Dachà, *J. Solid-Phase Biochem.* **5**, 11 (1980).

[3] M. Magnani, V. Stocchi, M. Bossù, M. Dachà, and G. Fornaini, *Mech. Ageing Dev.* **11**, 209 (1978).

[4] M. Magnani, V. Stocchi, M. Dachà, F. Canestrari, and G. Fornaini, *FEBS Lett.* **120**, 264 (1980).

[5] M. Magnani, M. Dachà, V. Stocchi, P. Ninfali, and G. Fornaini, *J. Biol. Chem.* **255**, 1752 (1980).

[6] H. M. Katzen and R. T. Schimke, *Proc. Natl. Acad. Sci. U.S.A.* **54**, 1218 (1965).

[7] V. Stocchi, M. Magnani, F. Canestrari, M. Dachà, and G. Fornaini, *J. Biol. Chem.* **256**, 7856 (1981).

METHODS IN ENZYMOLOGY, VOL. 90

With glucose-6-phosphate dehydrogenase and 6-phosphogluconate dehydrogenase in excess, the rate of glucose phosphorylation is equivalent to the rate of $NADP^+$ reduction, which is measured by the absorbance increase at 340 nm. For each molecule of glucose utilized, 2 molecules of $NADP^+$ are reduced.

Reagents

Glycylglycine-NaOH buffer, 0.25 M, pH 8.1
$MgCl_2$, 100 mM
ATP, 50 mM, pH 7.0
Glucose, 50 mM
$NADP^+$, 5 mM
Glucose-6-phosphate dehydrogenase from yeast, Boehringer grade I, 200 units/ml in saturated $(NH_4)_2SO_4$ solution
6-Phosphogluconate dehydrogenase from yeast, 120 units/ml in saturated $(NH_4)_2SO_4$ solution

Procedure. In a cuvette, add 0.55 ml of glycylglycine buffer, 0.1 ml of $MgCl_2$, 0.1 ml of ATP, 0.1 ml of glucose, 0.05 ml of NADP, 0.005 ml of glucose-6-phosphate dehydrogenase, 0.01 ml of 6-phosphogluconate dehydrogenase, distilled water, and enzyme to give a final volume of 1.0 ml.

Definition of Units and Specific Activity. One unit of hexokinase activity is defined as the amount of enzyme that catalyzes the formation of 1 μmol of glucose 6-phosphate per minute at 37°. Specific activity is expressed in units per milligram of protein. In the hemolysate, hemoglobin concentration was determined spectrophotometrically at 540 nm with Drabkin's solution as described by Beutler.[8] Protein is routinely determined spectrophotometrically by the procedure of Lowry *et al.*[9] Crystalline bovine serum albumin is used as a standard.

Alternative Assay Procedure. According to Sapico and Anderson,[10] for substrate specificity and inhibition studies a pyruvate kinase–lactate dehydrogenase-linked assay has been used.[5]

Purification Procedure for Hexokinase from Rabbit Red Blood Cells

All operations are performed at 4°. All buffers contain 3 mM 2-mercaptoethanol and 3 mM KF.

Step 1. Preparation of Hemolysate. Rabbit red blood cells are collected using EDTA as anticoagulant. The red cells are washed twice with

---

[8] E. Beutler, *in* "Red Cell Metabolism," 2nd ed., p. 11. Grune & Stratton, New York, 1975.
[9] O. H. Lowry, N. J. Rosebrough, A. L. Farr, and R. J. Randall, *J. Biol. Chem.* **193**, 265 (1951).
[10] V. Sapico and R. L. Anderson, *J. Biol. Chem.* **242**, 5086 (1967).

isotonic sodium chloride, and the buffy coat is removed by suction. The washed cells are hemolyzed by adding an equal volume of 0.4% (w/w) saponin solution and incubating for 2 hr.

Step 2. Batch Treatment with DEAE-Sephadex A-50. The hemolysate is mixed with 2 volumes of DEAE-Sephadex A-50 suspension, equilibrated in 3 mM sodium–potassium phosphate buffer, pH 7.3, containing 5 mM glucose, and mechanically stirred for 60 min. The suspension is rinsed on a Büchner funnel with the same buffer until the eluate is colorless. This procedure removes the bulk of the hemoglobin while the hexokinase remains bound. The enzyme is eluted with 0.5 M KCl in phosphate buffer. For a complete hexokinase recovery, the elution procedure is repeated twice.

Step 3. Ammonium Sulfate Fractionation. Solid ammonium sulfate, 19.4 g/100 ml, is slowly added to the stirred enzyme solution from step 2. The suspension is gently stirred for 60 min and then centrifuged at 16,000 g for 30 min. The supernatant solution is removed, and 19.9 g per 100 ml of ammonium sulfate are added. After 60 min, the suspension is centrifuged as before. The second ammonium sulfate precipitate is dissolved in 20–30 ml of 5 mM sodium–potassium phosphate buffer, pH 7.5, containing 5 mM glucose and then dialyzed overnight against 200 volumes of 5 mM sodium–potassium phosphate buffer, pH 7.5, containing 5 mM glucose and 9% (v/v) glycerol. The dialyzed enzyme solution is then centrifuged at 16,000 g for 20 min. At this stage the enzyme can be stored at $-20°$ for 10 weeks, with no significant loss of activity.

Steps 4 and 5. Affinity Chromatography on Sepharose–N-amino-hexanoylglucosamine Column. The activated CH-Sepharose 4B (from Pharmacia) is swollen in 1 mM HCl. The gel is washed with distilled water on a sintered-glass filter using approximately 300 ml per gram of dry powder. Four milligrams of D(+)-glucosamine hydrochloride per milligram of dry powder are dissolved in the coupling solution (0.1 M NaHCO$_3$ containing 0.5 M NaCl), using 5 ml per gram of dry powder, and mixed with the gel. Excess ligand is removed by washing with the coupling solution, and the remaining active groups are blocked with 1 M ethanolamine, pH 9.

The product is washed with three cycles of alternating pH, consisting of a wash at pH 4 (0.1 M acetate buffer and 1 M NaCl) followed by a wash at pH 8 (0.1 M Tris-HCl and 1 M NaCl). The product is stored at $+4°$ in 5 mM sodium potassium phosphate buffer, pH 7.5, containing 3 mM KF and 3 mM 2-mercaptoethanol. Before affinity chromatography, the glucose present in the enzyme solution is completely removed by enzymic phosphorylation. By adding 7.5 mM ATP · MgCl$_2$ to the hexokinase solution (30 min at room temperature), the glucose present is phosphorylated

TABLE I
PURIFICATION OF HEXOKINASE FROM RABBIT RED BLOOD CELL

Fraction	Volume (ml)	Protein (mg/ml)	Activity (units/ml)	Specific activity (units/mg protein)	Yield (%)
Hemolysate	900	163	0.0806	0.00049	100
DEAE-Sephadex A-50 eluate	2280	1.4	0.0318	0.0227	99.9
Ammonium sulfate precipitate	18	134	3.82	0.0285	95
First immobilized N-acetylglucosamine eluate	120	0.104	0.458	4.40	76
Second immobilized N-acetylglucosamine eluate	94	0.005	0.378	75.6	49
Preparative gel electrophoresis	5	0.0348	5.07	145.6	35

and removed by passage through a Sephadex G-25 column (2.1 × 12 cm) equilibrated with 2.5 mM sodium potassium phosphate buffer, pH 7.5, containing 9% (v/v) glycerol.

The fractions containing hexokinase activity are applied to a Sepharose–N-aminohexanoylglucosamine column (1 × 4 cm) equilibrated with the same buffer. After the sample application, the affinity column is washed with buffer phosphate until the protein absorbance of the eluate at 280 nm, is lower than 0.1 A. Hexokinase is eluted by adding 5 mM glucose to the equilibrating buffer. The preparation obtained at this stage is 8000- to 10,000-fold purified. The pooled fractions containing hexokinase activity are concentrated by ultrafiltration through an Amicon PM-30 membrane. A 150,000-fold purification is obtained by a second similar affinity chromatography step in 10 mM sodium potassium phosphate buffer, pH 7.5, containing 9% (v/v) glycerol. The enzyme is eluted by adding 10 mM glucose to the developing buffer. At this stage, the enzyme can be stored at −20° over a period of months with no significant loss of activity.

Step 6. Preparative Polyacrylamide Gel Electrophoresis. The enzyme solution from step 5 is electrophoresed on 7.5% preparative polyacrylamide gel (1 × 8 cm). The gels are sliced into 0.2-cm sections and separately mashed in 1 ml of 10 mM sodium–potassium phosphate buffer, pH 7.5, containing 10 mM glucose and 9% (v/v) glycerol. The eluted fractions are assayed for hexokinase activity and pooled. The preparation obtained at this stage gave a single band in nondenaturing polyacrylamide disc gel electrophoresis and in the presence of sodium dodecyl sulfate.

A typical purification is summarized in Table I. The specific activity of the pure hexokinase was 145 units per milligram of protein, corresponding to a 300,000-fold purification over the starting material, with an overall recovery of approximately 35%. In the final stage of purification, the presence of 9% (v/v) glycerol in the buffer for elution is essential for good recovery of hexokinase activity.

Hexokinase recoveries indicated in Table I, also strictly depend upon the use of buffers containing 5 mM glucose. As reported above, removal of glucose is absolutely necessary for binding hexokinase to Sepharose–N-aminohexanoylglucosamine; in fact, N-acetylglucosamine is a competitive inhibitor (K_i = 0.7 mM) with respect to the glucose of red blood cell hexokinase.

Properties of Hexokinase from Rabbit Red Blood Cells

Stability. Purified hexokinase can be stabilized in a specific way by glucose or fructose at 5 mM concentrations or in a nonspecific way by high concentrations of glycerol (9–15% v/v). Hexokinase stored at $-20°$ is more stable in 5 mM sodium–potassium phosphate buffer, pH 7.5, than in other buffer systems.

Molecular Weight. The molecular weight of erythrocyte hexokinase has been found to be 110,000 by gel filtration, 112,000 by sedimentation velocity, and 110,000 by sodium dodecyl sulfate gel electrophoresis.

Isoelectric Point. The enzyme had a pI of 6.20–6.30 pH units under native conditions.

Kinetic Properties and Specificity.[5,11] The specificity of this hexokinase for nucleotide triphosphate and hexoses is reported in Table II. The enzyme is inhibited competitively with respect to MgATP^{2-} by uncomplexed ATP (K_i = 1.5 mM), which is also a noncompetitive inhibitor for glucose (K_i = 4 mM). N-Acetylglucosamine, a competitive inhibitor of glucose (K_i = 0.1 mM), is a noncompetitive inhibitor for MgATP^{2-} (K_i = 0.45 mM). These data, as for hexokinase I from other sources, are consistent with a rapid random kinetic mechanism.

Product inhibition: Glucose 6-phosphate causes a linear competitive inhibition when MgATP is the varied substrate (K_i = 25 μM); MgADP$^-$ is a mixed inhibitor versus MgATP^{2-} (K_i = 3.0 mM) and glucose (K_i = 7.8 mM). Uncomplexed Mg^{2+} inhibits competitively with respect to MgATP^{2-} (K_i = 9 mM). Reduced glutathione at 1 mM concentration can maintain red blood cell hexokinase in the reduced state with fully catalytic activity. At higher concentrations a marked inhibition is observed. In contrast, oxidized glutathione is a strong inhibitor of reduced erythrocyte hexokinase at all concentrations.[12]

[11] P. Ninfali, M. Magnani, M. Dachà, V. Stocchi, and G. Fornaini, *Biochem. Int.* **1,** 574 (1980).

[12] M. Magnani, V. Stocchi, P. Ninfali, M. Dachà, and G. Fornaini, *Biochim. Biophys. Acta* **615,** 113 (1980).

TABLE II
SPECIFICITY OF PURIFIED RED BLOOD CELL HEXOKINASE

Compounds	$K_m{}^a$ (mM)	Relative velocity[b] (%)
MgATP	0.6	100
MgGTP	—	—
MgITP	1.9	36
MgUTP	—	—
MgCTP	—	—
D(+)-Glucose	0.06	100
D(+)-Mannose	0.71	109
D(−)-Fructose	17.8	70
2-Deoxy-D-glucose	1.33	65.7
D(+)-Glucosamine	2.0	43.4
N-Acetyl-D-glucosamine	41.6	32.6
D(+)-Galactose	—	—

[a] K_m values were determined using a Lineweaver–Burk plot.
[b] Maximum velocities are expressed relative to the V for glucose (100%) when the enzyme is saturated by the substrate.

Purification Procedure for Hexokinases Ia and Ib from Rabbit Reticulocytes

All operations are performed at 4°. All buffers contain 3 mM 2-mercaptoethanol and 3 mM KF.

Step 1. Preparation of Hemolysate. Reticulocytosis is induced in rabbits by administration of phenylhydrazine (5 mg/kg body weight each day for 3 days). Blood containing 75–80% reticulocytes is collected using EDTA as anticoagulant. The reticulocytes are washed twice with isotonic sodium chloride. The buffy coat is removed by suction. After addition of an equal volume of 0.4% (w/v) saponin solution, the washed cells are hemolyzed for 2 hr. The red cell stroma is then removed by centrifuging the lysate at 13,000 g for 30 min.

Step 2. Ammonium Sulfate Fractionation. Four volumes of 5 mM sodium potassium phosphate, pH 7.5, containing 5 mM glucose and 9% (v/v) glycerol are added to the resulting supernatant from step 1. Solid ammonium sulfate (19.4 g/100 ml) is slowly added to the stirred enzyme solution. The suspension is gently stirred for 60 min and then centrifuged at 16,000 g for 30 min. The supernatant solution is removed, and 19.9 g per 100 ml of ammonium sulfate are added. After 60 min the suspension is centrifuged as before. This ammonium sulfate precipitate is dissolved in

40–50 ml of 5 mM sodium potassium phosphate buffer, pH 7.5, containing 5 mM glucose, and then dialyzed overnight against 200 volumes of the 5 mM sodium–potassium phosphate buffer, pH 7.5, containing 5 mM glucose and 9% (v/v) glycerol. The dialyzed enzyme solution is then centrifuged at 16,000 g for 20 min.

Step 3. DEAE-Cellulose Chromatography. The enzyme solution from step 2 is concentrated to 10 ml by ultrafiltration through an Amicon PM-30 membrane. An aliquot of 2.5 ml enzyme solution is applied to a column (20 × 1 cm) of DEAE-cellulose (Whatman DE-52) equilibrated with 5 mM sodium potassium phosphate, pH 7.5. The column is developed with a 280-ml linear gradient, 0 to 0.4 M KCl, in the same buffer. The peaks of hexokinase Ia and Ib are eluted at about 0.120 and 0.150 M KCl, respectively.

Step 4. Affinity Chromatography on a Sepharose–N-aminohexanoyl-glucosamine Column. The affinity column is prepared as described above for the purification of erythrocyte hexokinase. The pooled DEAE-cellulose fraction of hexokinase Ia and Ib are concentrated by ultrafiltration through an Amicon PM-30 membrane. Before affinity chromatography, the glucose present in the enzyme solutions is completely removed by enzymic phosphorylation and by passage through a Sephadex G-25 column as described above.

The fractions containing hexokinase Ia and Ib activity are applied to Sepharose–N-aminohexanoylglucosamine columns (4 × 1 cm) equilibrated with 10 mM sodium–potassium phosphate buffer, pH 7.5 containing 9% (v/v) glycerol. After application of the sample, the affinity columns are washed with buffer phosphate until the protein absorbance at 280 nm in the eluates is zero. The hexokinases are eluted by adding 10 mM glucose to the equilibrating buffer. The preparations obtained at this stage are 28,000-fold purified for hexokinase Ia and 26,000-fold for hexokinase Ib.

Step 5. Preparative Polyacrylamide Gel Electrophoresis. The enzyme solutions from step 4 are concentrated by ultrafiltration and electrophoresed on 7.5% preparative polyacrylamide gel (1 × 8 cm). The gels are sliced into 0.2-cm sections and separately mashed in 1 ml of 5 mM sodium–potassium phosphate buffer, pH 7.5, containing 10 mM glucose, 9% (v/v) glycerol, and 1 mM dithiothreitol. The eluted fractions are assayed for hexokinase activity and pooled. The preparations obtained at this stage gave a single band in nondenaturating polyacrylamide disc gel electrophoresis and in the presence of sodium dodecyl sulfate, both for hexokinase Ia and hexokinase Ib.

A typical purification is summarized in Table III. The specific activity of pure hexokinase Ia was 155 units per milligram of protein, and that of pure hexokinase Ib was 144 units per milligram of protein. In the final

TABLE III
PURIFICATION OF TWO DISTINCT FORMS OF HEXOKINASE, Ia AND Ib, FROM
RABBIT RETICULOCYTES[a]

Step and fraction	Volume (ml)	Protein (mg/ml)	Activity (units/ml)	Specific activity (units/mg protein)	Yield (%)
1. Hemolysate	650	137	0.387	0.0028	100
2. Ammonium sulfate precipitate	49	10.89	4.87	0.451	95
3. DE-52 cellulose eluate					
Hexokinase Ia	120	0.123	0.660	5.37	31 (100)[b]
Hexokinase Ib	150	0.195	0.978	5.01	58 (100)[b]
4. Immobilized N-acetyl-glucosamine eluate					
Hexokinase Ia	135	0.0066	0.528	80	28 (90)
Hexokinase Ib	195	0.0093	0.677	72.8	52 (90)
5. Preparative gel electrophoresis					
Hexokinase Ia	6.0	0.034	5.28	155	12.5 (40)
Hexokinase Ib	9.0	0.068	9.76	144	34.8 (60)

[a] The two forms of hexokinase were copurified to step 3, separated on DEAE-cellulose (DE-52), and further purified in the subsequent steps.
[b] Values in parentheses refer to the assumed 100% recovery of the separated hexokinase Ia and Ib.

stage of the purification recovery of hexokinase Ia is much more affected by separation on polyacrylamide gel than that of hexokinase Ib.

Properties of Hexokinase Ia and Ib from Rabbit Reticulocytes

Stability. Both hexokinase Ia and Ib can be stabilized by glucose, fructose, glycerol, and sulfhydryl protecting agents, as can hexokinase from mature rabbit red cells.

Molecular Weight. The molecular weight of both hexokinase Ia and Ib were estimated to be 105,000 by gel filtration or sedimentation velocity and 104,000 by sodium dodecyl sulfate gel electrophoresis.

Isoelectric Point. The isoelectric point of the native enzymes was 6.2–6.3 pH units for hexokinase Ia and 5.7–5.8 for the hexokinase Ib.

Kinetic Properties and Specificity. The K_m value of glucose was 0.04 mM for hexokinase Ia and 0.125 mM for hexokinase Ib. Both enzymes had the same K_m value for MgATP (0.5 mM).

Several hexoses could be phosphorylated by hexokinase Ia and Ib with different affinities.[7]

[2] Hexokinase Type II from Rat Skeletal Muscle

By JOHN S. EASTERBY and S. SALEHEEN QADRI

Hexose + ATP → hexose 6-phosphate + ADP

Hexokinase type II (EC 2.7.1.1) is the principal isozyme of rat skeletal muscle,[1] and this tissue is the best source of the enzyme. Unlike the type I hexokinase of brain[2] and heart,[3] it is largely soluble, and its extraction from the tissue is therefore simple. It is extremely unstable and is rapidly inactivated in the absence of hexose or thiol, which together with lack of abundance makes purification difficult. An affinity elution procedure has been adopted in this purification as it is highly specific, but, unlike affinity chromatography, does not require the synthesis of a matrix containing immobilized ligand. The free solution behavior of the enzyme is used to effect its specific elution by glucose 6-phosphate from an ion exchanger.

Assay Method

Principle. The glucose 6-phosphate produced in the hexokinase reaction is coupled to glucose-6-phosphate dehydrogenase. The reaction is followed by measuring the increase in absorbance at 340 nm due to NADPH formation. A sufficient excess of glucose-6-phosphate dehydrogenase is provided to minimize the transit time and to avoid inhibition of hexokinase by glucose 6-phosphate.[4]

Reagents

Assay buffer: 50 mM glucose, 30 mM MgCl$_2$, 30 mM Tris-HCl (0.1 I), pH 7.5
ATP, 100 mM, pH 7.5
NADP, 1 mM
Glucose-6-phosphate dehydrogenase, 2 units/ml in 30 mM Tris-HCl, pH 7.5

Procedure. This procedure is essentially the same as that described previously for heart hexokinase assay.[3] Assay buffer, 0.7 ml, is mixed in a

---

[1] H. M. Katzen and R. T. Schimke, *Proc. Natl. Acad. Sci. U.S.A.* **54**, 1218 (1965).
[2] A. C. Chou and J. E. Wilson, *Arch. Biochem. Biophys.* **151**, 48 (1972).
[3] J. S. Easterby and M. J. O'Brien, *Eur. J. Biochem.* **38**, 201 (1973).
[4] J. S. Easterby, *Biochim. Biophys. Acta* **293**, 552 (1973).

cuvette with 0.1 ml each of ATP, NADP, and glucose-6-phosphate dehydrogenase. The mixture is incubated at 30° for 5 min before starting the reaction by addition of 10 μl of hexokinase. The increase in absorbance at 340 nm is recorded in a spectrophotometer at 30°. Rates of change of absorbance of 0.1/min or less are monitored. When glucose 6-phosphate is present in the enzyme sample, hexokinase is incubated in the presence of NADP and glucose-6-phosphate dehydrogenase until no further absorbance increase occurs; the reaction is then started by addition of ATP. Contaminating enzymes have not been a problem with this assay system.

Alternative Procedures. When inhibition of hexokinase by glucose 6-phosphate is studied or when hexose substrates other than glucose are used, the enzyme is coupled to pyruvate kinase and lactate dehydrogenase through the ADP produced in the reaction. In this case the decrease in absorbance at 340 nm due to NADH oxidation is followed. The reaction cuvette contains 60 μmol of Tris-HCl, pH 7.5, 1.5 μmol of phosphoenolpyruvate, 50 μmol of KCl, 0.2 μmol of NADH, 7.5 units of pyruvate kinase, 4.5 units of lactate dehydrogenase, and appropriate concentrations of hexose, ATP, and $MgCl_2$ in a total volume of 1 ml. The reaction is started by the addition of enzyme.

Units. A unit of hexokinase is defined as the amount of enzyme catalyzing the formation of 1 μmol of glucose 6-phosphate per minute at 30°. For specific activity measurements, protein was determined by the method of Mejbaum–Katzenellenbogen and Dobryszycka.[5]

Purification Procedure

Buffers

Buffer 1: maleate, pH 7.0, I = 0.02, containing 0.86 g of maleic acid titrated to pH 7.0 with KOH in 1 liter

Buffer 2: maleate, pH 7.0, I = 0.05, containing 2.15 g of maleic acid titrated to pH 7.0 with KOH in 1 liter

Buffer 3: maleate, pH 6.5, I = 0.02, containing 1.04 g of maleic acid titrated to pH 6.5 with KOH in 1 liter

Buffer 4: phosphate, pH 7.0, I = 0.05, containing 2.23 g of K_2HPO_4, 1.58 g of KH_2PO_4, and 0.154 g dithiothreitol in 1 liter

All buffers additionally contain 0.1 M glucose and 1 mM disodium EDTA. Maleate buffers contain 5 mM 2-mercaptoethanol added immediately before use.

Phosphocellulose Preparation. Phosphocellulose P-11 is obtained from Whatman (Maidstone, England) and precycled by the following procedure. The exchanger is suspended in 15 volumes of 0.5 M KOH for 30

---

[5] W. Mejbaum-Katzenellenbogen and W. W. Dobryszycka, *Clin. Chim. Acta* **4**, 515 (1959).

min, washed on a Büchner funnel until the filtrate is neutral, and then suspended in 15 volumes of $0.5\,M$ HCl for 30 min. It is then washed again and resuspended in $0.5\,M$ KOH as before and washed to neutrality. Equilibration is performed by suspending the exchanger in buffer and titrating with the acid component to the correct pH. This is followed by several changes of buffer to obtain the correct pH and conductivity. The ion-exchange column once poured is further equilibrated by passage of buffer to stable pH and conductivity.

Animals. Wistar rats, 150–200 g, fed on laboratory chow are used as a source of skeletal muscle. It is possible to store whole rat carcasses deep-frozen without significant alteration of properties or yield of enzyme.

Extraction. The limbs from approximately 30 rats are removed, finely chopped, and minced. They are then homogenized for 2 min in a Waring blender in 2 volumes of buffer 1 (v/w). The homogenate is centrifuged at 25,000 g for 30 min, and the residue is reextracted. The extracts are combined, and the pH is carefully adjusted to 7.0 with $0.2\,M$ KOH.

Chromatography on DEAE-Cellulose. The extracted enzyme is adsorbed batchwise onto 600 g of preswollen DEAE-cellulose equilibrated against buffer 2, washed with 6 volumes of the buffer, and poured into a column (5.5 × 35 cm). A further 600 ml of buffer are then passed through the column, and it is developed with a 2-liter linear gradient, 0 to 0.4 M in

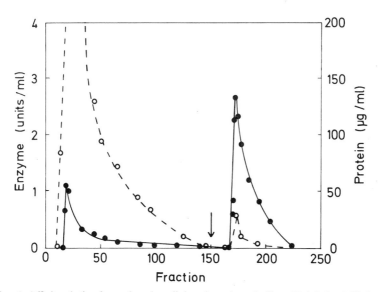

FIG. 1. Affinity elution from phosphocellulose in maleate buffer, pH 6.5, $I = 0.02$ (9 mM). Enzyme (●——●) and protein (○---○) are shown. The flow rate was 65 ml/hr, and 10-ml fractions were collected. Elution of hexokinase was effected by the inclusion of 100 μM glucose 6-phosphate in the buffer from fraction 150 onward (indicated by arrow).

KCl. The active fractions (100–155) are combined, dialyzed against buffer 3, and concentrated by ultrafiltration. The flow rate is 100 ml/hr, and 10-ml fractions are collected.

Affinity Elution from Phosphocellulose. The concentrated, dialyzed enzyme (15 ml) is pumped onto a column of phosphocellulose (7 × 8 cm) equilibrated in buffer 3. When the enzyme is absorbed, the column is washed with the buffer (approximately 5 column volumes) until unbound proteins are removed. The buffer is then made 100 μM in glucose 6-phosphate, and this effects elution of the enzyme (Fig. 1). The active fractions (163–195) are collected, and glucose 6-phosphate is removed by vacuum dialysis against buffer 4.

Gel Filtration on Sephadex G-200. The concentrated enzyme sample (1.5 ml) is applied to a column of Sephadex G-200 (2.1 × 140 cm) equilibrated with buffer 4 containing 1 mM dithiothreitol. The enzyme eluting between 205 and 240 ml is collected and vacuum dialyzed against the same buffer for storage. The flow rate of the column is 8 ml/hr, and 3-ml fractions are collected. The void volume is 146 ml.

Comments on the Procedure. The course of the procedure is outlined in Table I. The enzyme is essentially homogeneous after the affinity elution stage, but aggregated enzyme can be removed on Sephadex G-200. The enzyme is homogeneous by criteria of polyacrylamide gel electrophoresis, sodium dodecyl sulfate (SDS)–polyacrylamide electrophoresis, and sedimentation velocity analysis. Glucose-6-phosphate dehydrogenase, phosphoglucose isomerase, phosphoglucomutase, and 6-phosphogluconate dehydrogenase could not be detected in the preparation.

The glucose 6-phosphate used in the affinity elution step can be replaced by 20 μM glucose 1,6-bisphosphate, and this provides sharper elution. The maleate buffer used at this stage suffers one disadvantage. On prolonged storage in the presence of thiol the pH drops, presumably

TABLE I
PURIFICATION OF TYPE II HEXOKINASE

Fraction	Protein (mg)	Enzyme (units)	Specific activity (units/mg protein)	Purification (fold)	Yield (%)
>25,000 g supernatant from homogenate[a]	45,600	1030	0.0225	1	100
DEAE-cellulose	2,830	747	0.264	11.7	73
Phosphocellulose	3.8	410	108	4800	40
Sephadex G-200	3.25	390	120	5330	38

[a] Skeletal muscle from the limbs of 30 rats was used as a source of enzyme.

TABLE II
KINETIC PARAMETERS OF TYPE II HEXOKINASE

Substrate	Inhibitor	$K_m{}^a$ (μM)	K_i (μM)
D-Glucose	—	50	—
D-Mannose	—	60	—
D-Fructose	—	900	—
D-Glucosamine	—	60	—
ATP	—	500	—
—	Glucose 6-phosphate	—	15 (ATP)
			26 (hexose[b])
—	ADP	—	3500
—	Glucose 1,6-bisphosphate	—	3
—	P_i	—	5300

[a] The K_m values were mutually independent of the concentration of the second substrate

[b] This represents the dissociation constant of the ternary enzyme–hexose–hexose 6-phosphate complex. The dissociation constant from the binary enzyme–hexose 6-phosphate complex was 43 μM.

owing to an addition reaction of the thiol across the double bond. For this reason 2-mercaptoethanol is added only immediately prior to use. Phosphate buffer must not be used for this stage as it blocks the binding of glucose 6-phosphate to the enzyme and prevents elution. The equilibration of the phosphocellulose is critical, and for this reason the buffer system employed should not be altered. Owing to the polyprotic nature of the phosphate ionization and the extensive substitution of the ion exchanger, quite small changes in ionic strength can cause considerable pH fluctuations within the matrix. This can lead either to leakage of the enzyme or denaturation if a substantial drop in pH occurs.

Properties

The molecular weight of hexokinase determined by sedimentation equilibrium was 100,000. A value of 108,000 was determined by SDS–polyacrylamide gel electrophoresis, and 106,000 by gel filtration on Sephadex G-200. The isoelectric point was measured as pH 5.7 by analytical isoelectric focusing[6] and $E_{280\,nm}^{1\%,\,1\,cm}$ was 6.2 as determined by the method of Babul and Stellwagen.[7]

Kinetic data for substrate and inhibitor are given in Table II.

[6] C. W. Wrigley, this series, Vol. 22, p. 559.
[7] J. Babul and E. Stellwagen, *Anal. Biochem.* **28**, 216 (1969).

[3] Hexokinases from *Drosophila melanogaster*

By CHI-YU LEE

Hexose + ATP $\xrightleftharpoons{\text{Mg}^{2+}}$ hexose 6-phosphate + ADP

Three major forms of hexokinase (EC 2.7.1.1; ATP:hexose 6-phosphotransferase) are generally found in adult *Drosophila*.[1] They are designated as Hex A, Hex B, and Hex C, respectively, starting from the most anodal zone on starch gels.[2] In the larval extract, Hex B is the predominant form of hexokinase, whereas Hex A is completely absent.[2] Genetic studies of multiple forms of hexokinase in *Drosophila melanogaster* have revealed that Hex A and Hex B are the products of the same structural gene mapped on the X chromosome owing to the nature of their covariation and cosegregation.[3] Hex C, on the other hand, was mapped on the second chromosome. Hexokinase catalyzes the phosphorylation of glucose and other hexoses in the presence of ATP. Since the binding of glucose to *Drosophila* hexokinases does not depend on ATP, and vice versa, the affinity columns utilizing either coenzyme or glucose derivatives as ligands should be suitable for the purification of *Drosophila* enzymes.

Assay Method

Hexokinase activity is assayed spectrophotometrically at 25° by following either glucose 6-phosphate or ADP formed in the reaction. When the formation of glucose 6-phosphate is followed, a modification of DiPietro and Weinhouse method is used.[4] The assay mixture in a total volume of 1 ml contains 0.1 M Tris-HCl, pH 8.0, 25 mM glucose, 10 mM ATP, 25 mM MgCl$_2$, 1.1 mM NAD$^+$, and 0.6 unit of glucose-6-phosphate dehydrogenase from *Leuconostoc mesenteroides*. ADP formation is followed by method of Kornberg and Pricer[5] in a reaction mixture containing 25 mM glucose, 10 mM ATP, 25 mM MgCl$_2$, 10 mM KCl, 0.18 mM NADH, 0.1 M Tris-HCl, pH 8.0, 2.2 mM phosphoenolpyruvate, and 1 unit each of pyruvate kinase and lactate dehydrogenase in a total volume of 1

[1] K. Madhavan, D. J. Fox, and H. Ursprung, *J. Insect Physiol.* **18**, 1523 (1972).
[2] C. Knutsen, C. F. Sing, and G. J. Brewer, *Biochem. Genet.* **3**, 475 (1969).
[3] R. Voelker, C. H. Langley, A. Leigh-Brown, and S. Ohnishi, *Drosophila Inf. Serv.* **53**, 200 (1978).
[4] D. L. DiPietro and S. Weinhouse, *J. Biol. Chem.* **235**, 2542 (1960).
[5] A. Kornberg and W. E. Pricer, Jr., *J. Biol. Chem.* **193**, 481 (1951).

ml. Sufficient hexokinase is added to the reaction mixture to cause a change in absorbance at 340 nm of 0.05 to 0.1 A per minute. One unit of activity catalyzes the formation of 1 μmol of glucose 6-phosphate per minute under the described assay conditions.

Drosophila Culture

Culture conditions for adult and larval *Drosophila melanogaster* have been described elsewhere.[6]

Purification

Affinity Columns

Three different affinity gels are used for the selective purification of Hex A and Hex C of adult *Drosophila melanogaster*.

3-Aminopyridine Adenine Dinucleotide–Sepharose (AAD–Sepharose). 3-Aminopyridine adenine dinucleotide (AAD) is prepared according to the procedure of Fisher *et al.*[7] ϵ-Aminocaproic acid is coupled to CNBr-activated Sepharose by mixing 100 ml of 1 M ϵ-aminocaproic acid with 100 ml of the activated Sepharose at pH 10 and 4° for 5 hr. One gram of AAD is coupled to the Sepharose-bound spacer in the presence of 1 g of 1-ethyl-3-(3-dimethylaminopropyl)carbodiimide by incubation at pH 4.75 for 24 hr.[8] To block the unreacted carboxyl group on AAD–Sepharose, the gel is further incubated with 100 ml of 1 M ethanolamine at pH 4.75 for an additional 2 hr at room temperature in the presence of 1 g of 1-ethyl-3-(3-dimethylaminopropyl)carbodiimide.

8-(6-Aminohexyl)amino-ATP–Sepharose. The preparation, maintenance, and commercial sources of this affinity gel are described elsewhere.[9] The estimation of ligand density is also described elsewhere.[6]

The preparation of this affinity gel is identical to that of AAD–Sepharose, except that glucosamine is used as the ligand.

Purification Procedure

This procedure allows a large-scale purification of Hex A and Hex C from crude homogenates of adult *Drosophila melanogaster*. Typical results are summarized in Table I. Unless otherwise specified, all the phosphate buffers used are the potassium salt, and sodium dodecyl sulfate–acryl-

---

[6] D. W. Niesel, G. C. Bewley, C.-Y. Lee, and F. B. Armstrong, this series, Vol. 89 [51].
[7] T. L. Fisher, S. V. Vercellotti, and B. M. Anderson, *J. Biol. Chem.* **248**, 4293 (1973).
[8] C.-Y. Lee and N. O. Kaplan, *J. Macromol. Sci.* **A10**, 15 (1976).
[9] C.-Y. Lee, J. H. Yuan, and E. Goldberg, this series, Vol. 89 [61].

TABLE I
PURIFICATION OF HEX A AND HEX C₂ FROM ADULT *Drosophila melanogaster*

Step	Total activity (units)	Specific activity (unit/mg protein)	Purification (fold)
Hex A			
Crude extract	1580	0.1	—
AAD–Sepharose column	750	1.2	12
DE-52 cellulose column	350	7.2	72
Sephadex G-75 column	270	9.4	94
Glucosamine–Sepharose column	220[a]	137[a]	1370
Hex C₂			
Crude extract	1580	0.1 (0.02)[b]	—
Post-AAD–Sepharose column	280	0.28	2.8
Isoelectric focusing	145	3.1	31
8-(6-Aminohexyl)amino-ATP–Sepharose column	105	60	600 (3000)[c]

[a] The total activity presented is the sum of four individual experiments, and the final specific activity of Hex A is the average from these experiments.

[b] The value in parentheses refers to the actual specific activity of Hex Cs in the homogenate.

[c] The value in parentheses refers to actual fold of protein purification.

amide gel electrophoresis is used to determine the protein purity (10% acrylamide gels).

Purification of Hex A

AAD–Sepharose Affinity Chromatography. Two hundred grams of frozen adult *Drosophila* (*Samarkand*) are homogenized with a Polytron (Brinkmann Co., Westbury, New York) in about 750 ml of 10 mM phosphate, pH 6.5, containing 1 mM each of EDTA and α-thioglycerol (buffer A). The homogenate is centrifuged at 27,000 g for 20 min. The supernatant is divided into two portions and passed through two separate AAD–Sepharose columns (2.5 × 20 cm each) equilibrated with buffer A. About 20% of the total hexokinase activity leaks continuously through the column during the loading and washing. By starch gel electrophoresis, the nonadsorbed enzyme was identified as exclusively Hex C. Therefore, the unadsorbed enzyme fractions are pooled and stored at 4° in ammonium sulfate (52 g/100 ml) for the purification of Hex C as described later.

After a wash with 1 liter of buffer A, which removes all the Hex C, the adsorbed mixture of Hex A and Hex B is eluted from the column as a sharp peak by buffer A containing 10 mM ATP. Owing to the specificity of the AAD–Sepharose column, no Hex C is adsorbed by the affinity column and eluted with ATP as judged by starch gel electrophoresis.[2] Peak enzyme fractions eluted by ATP are pooled and precipitated with 52 g/100 ml of ammonium sulfate. After centrifugation at 27,000 g for 20 min, the pellet is redissolved in 20 ml of 5 mM phosphate, pH 7.0 (buffer B) and dialyzed overnight against 2 liters of buffer B.

DE-52 Column Chromatography. After removal of the precipitated protein from the dialyzed enzyme solution by centrifugation, the supernatant containing hexokinase activity is loaded on a DE-52 cellulose column (3 × 40 cm) equilibrated with buffer B. After adsorption and extensive washing with buffer B, the hexokinase activity is eluted in two main fractions by a linear salt gradient of 0 to 0.3 M NaCl in buffer B (1 liter by 1 liter). The first minor peak eluted at low salt concentration is mainly Hex B, whereas the major peak eluted at higher salt concentration (0.15 M) is exclusively Hex A. Fractions containing only Hex A or Hex B are pooled separately. No attempt is made for further purification of Hex B owing to its low activity (~50 units with a specific activity of ~10 units/mg).

Sephadex G-75 Gel Filtration Chromatography. Fractions containing Hex A activity are pooled and concentrated to 5 ml. The enzyme preparation is loaded on a Sephadex G-75 column (2 × 70 cm) and eluted with 0.1 M phosphate, pH 7.0. Peak enzyme fractions are pooled and dialyzed against 2 liters of buffer A.

Glucosamine–Sepharose Affinity Chromatography. The solution containing 85 units of the dialyzed enzyme is loaded on a glucosamine–Sepharose column (2 × 5 cm) that has been equilibrated with buffer A. More than 95% of the total enzyme is adsorbed on the column. The column is subsequently washed with 45 ml of buffer A followed by 90 ml of 50 mM phosphate, pH 7.0. A slowly increasing leakage of hexokinase activity is observed during the high-salt wash. The adsorbed hexokinase is finally eluted from the column with 150 ml of 1 M glucose in 50 mM phosphate, pH 7.0, containing 1 mM ATP and 2 mM MgCl$_2$. The peak fractions, which contains 60 units of enzyme, are pooled. At this stage, the enzyme is more than 90% pure and has a specific activity of 137 units/mg (determined by fluorescamine protein assay using bovine serum albumin as the standard).[10]

Purification of Hex C

Preparative Isoelectric Focusing. Hex C is not adsorbed on an AAD–Sepharose affinity column during the initial step of Hex A purification. After precipitation with 52 g of ammonium sulfate per 100 ml followed by

[10] P. Böhlen, S. Stein, W. Dairman, and S. Udenfriend, *Arch. Biochem. Biophys.* **155**, 213 (1973).

TABLE II
BIOCHEMICAL PROPERTIES OF *Drosophila* HEXOKINASES

	Relative activity (K_m)		
Substrate or coenzyme	Hex A	Hex B_1	Hex C_2
Glucose[a]	100 (0.30)[b]	100 (0.18)	100 (0.09)
Fructose	50 (50)	41 (50)	134 (2.5)
Mannose	100 (0.19)	98 (0.29)	44 (0.33)
2-Deoxyglucose	93	51	48
Glucosamine	19	0	44
ATP[c]	100 (0.5)	100 (0.35)	100 (0.25)
2'-Deoxy-ATP	73	76	80
UTP	0	0	0
GTP	0	0	0
CTP	0	5	11

[a] The concentration of the sugar substrates was 10 mM. The following hexoses and pentoses were found to be inactive for *Drosophila* hexokinases: N-acetylglucosamine, galactose, sorbitol, ribose, and arabinose.
[b] The values in parentheses refer to K_m values.
[c] Concentration of the coenzymes was 5 mM.

centrifugation, the pellet containing Hex C activity is redissolved in 50 ml of H_2O and dialyzed overnight against 4 liters of 1% glycine, pH 7.0. After centrifugation to remove the denatured protein, the enzyme solution is subjected to preparative column isoelectric focusing. The column (Model 8100-2, 440 ml, from LKB, Sweden) is routinely run for 48 hr at 1200 V in the presence of 2% ampholyte (pH 5 to 7). Two hexokinase activity peaks, Hex C_1 and Hex C_2, are resolved with isoelectric points of 6.1 and 6.4, respectively.

8-(6-Aminohexyl)amino-ATP–Sepharose Affinity Chromatography. Fractions containing 150 units of Hex C_2 are diluted 10-fold with buffer A and loaded directly on an 8-(6-aminohexyl)amino-ATP–Sepharose column (2.5 × 15 cm) equilibrated with buffer A. Hex C_2 is initially retarded, but begins to leak during the wash with buffer A. As a result, almost all of the loaded enzyme is recovered in the next two column volumes. Active enzyme fractions are pooled, concentrated to 10 ml by ultrafiltration, and loaded on another 8-(6-aminohexyl)amino-ATP–Sepharose column (2.5 × 15 cm) equilibrated with the same buffer. In this instance, the enzyme is quantitatively adsorbed on the column during the loading and the subsequent wash. The enzyme is readily eluted with 10 mM ATP in buffer A. The purified Hex C_2 has a specific activity of about 60 units/mg. Hex C_1, a genetic variant of Hex C_2 is purified to homogeneity according to the same procedure with about the same final specific activity (proteins are determined by fluorescamine assay).[10]

No cross contamination among purified Hex A, Hex B, and Hex C is observed by starch gel electrophoresis.[2]

Properties

All three *Drosophila* hexokinases (Hex A, Hex B, and Hex C) are monomers with molecular weights in the range of 40,000 to 50,000. Immunologically, Hex A and Hex B were shown to have complete identity, whereas Hex C is distinct from Hex A or Hex B.[11]

General biochemical properties of these three forms of *Drosophila* hexokinases are summarized in Table II. Consistent with the genetic evidence, biochemical and immunological studies suggest that Hex A and Hex B are the products of the same structural gene, but differ by epigenetic or posttranslational modifications.[11]

[11] D. Moser, L. Johnson, and C.-Y. Lee, *J. Biol. Chem.* **255**, 4673 (1980).

[4] Hexokinase from *Ascaris suum* Muscle

By SCOTT C. SUPOWIT and BEN G. HARRIS

$$\text{Hexose} + \text{ATP} \rightarrow \text{hexose 6-phosphate} + \text{ADP}$$

Assay Method

Principle. The hexokinase[1] reaction is monitored by coupling to NADPH formation (followed by absorbance at 340 nm) with glucose-6-phosphate dehydrogenase.[2]

Reagents

Assay mixture: 50 mM triethanolamine-HCl, pH 7.4, 8 mM MgCl$_2$, and 50 mM glucose

NADP, 25 mg/ml

ATP, 0.4 M, pH 7.0

Glucose-6-phosphate dehydrogenase, 100 units/ml, in 0.02% bovine serum albumin–25 mM triethanolamine-HCl, pH 7.4

[1] ATP: D-hexose 6-phosphotransferase, EC 2.7.1.1.
[2] D. DiPietro and S. Weinhouse, *J. Biol. Chem.* **235**, 2542 (1960).

Procedure. The assay is carried out in a total volume of 1 ml. To a 1-cm pathlength cuvette, add 0.9 ml of assay mixture, 0.01 ml of ATP, 0.01 ml of NADP, 0.01 ml of glucose-6-phosphate dehydrogenase, and H_2O. The reaction is initiated by the addition of enzyme. The increase in absorbance at 340 nm is measured in a spectrophotometer at 30°. One unit of hexokinase activity is defined as the amount of enzyme catalyzing the formation of 1 μmol of glucose-6-phosphate per minute at 30°. Specific activity is expressed as units per milligram of protein.

Purification Procedure

Source of Ascaris suum. This helminth resides in the upper small intestine of swine, and the female grows up to 30 cm in length with a weight of 10–12 g. The parasite can be obtained from most slaughterhouses that process swine. The initial contact with the slaughterhouse can be made through the local office of the United States Department of Agriculture (USDA), who can aid the investigator in contacting the USDA Inspector in the packing house, and through him the necessary permission and collecting permits can be obtained. In most cases, the USDA Inspectors are very familiar with the parasitic infestation and can provide valuable information on collection procedures. The number of worms obtained varies with the size of the slaughterhouse and number and type of animals processed, but several kilograms can usually be collected in just a few hours. The parasites are transported to the laboratory in a saline solution[3] maintained at 37–38°, which consists of the following components: 111 mM NaCl, 10 mM NaHCO$_3$, 24 mM KCl, 1 mM CaCl$_2$, 5 mM MgSO$_4$, 14 mM NH$_4$Cl, 0.5 mM KH$_2$PO$_4$ (pH 7.0).

The muscle, which comprises 50–60% of the weight of the worm, is obtained by dissecting the worm longitudinally and removing the white reproductive tissue and the yellow-to-green intestine. The remaining pink-to-red muscle cuticle is used as the source of enzyme. In the following procedures, fresh muscle is utilized, since freezing of the tissue results in a substantial loss of hexokinase activity.

Purification Steps

All steps, unless otherwise indicated, are performed at 4°. Glucose (0.1 M) is included in all buffers to stabilize the hexokinase. *Ascaris suum* muscle cuticle (100 g) is homogenized in a Waring blender with a buffer (4 : 1, v/w) containing 10 mM Tris-HCl, pH 7.6, 1 mM EDTA, 20 mM 2-mercaptoethanol, and 0.1 M glucose. After centrifugation at 20,000 g for 30 min, the supernatant solution is incubated with 0.5% Triton X-100 and

[3] M. J. Donahue, N. J. Yacoub, M. R. Kaeini, R. A. Masaracchia, and B. G. Harris, *J. Parasitol.* **67**, 505 (1981).

0.5 M KCl (prepared by the direct addition of the compounds to the supernatant solution) for a period of 1 hr. The level of Triton X-100 added to solubilize the enzyme during the purification has no detectable effects on the activity of the hexokinase or the glucose-6-phosphate dehydrogenase. The pellet from the centrifugation is rehomogenized in the same buffer as above, with the addition of 0.5 M KCl. After homogenization of the pellet, Triton X-100 is added to 0.5% and incubated for a period of 1 hr. Both solutions are combined and centrifuged for 1 hr at 105,000 g. The supernatant solution is dialyzed overnight against several changes of a buffer containing 4 mM Tris-HCl, pH 7.8, 1 mM EDTA, 20 mM 2-mercaptoethanol, and 0.1 M glucose.

DEAE-BioGel Chromatography. The dialyzate is centrifuged at 10,000 g for 10 min, and the supernatant solution is applied to a 2.5 cm × 50 cm column of DEAE-BioGel that has been equilibrated with the above buffer. The column is washed with buffer until the eluent is essentially protein-free by absorbance at 280 nm. The bound hexokinase is eluted by the application of a linear salt gradient. The reservoir chamber contains 1 liter of buffer plus 0.5 M KCl, and the mixing chamber contains 1 liter of buffer. Hexokinase elutes as a single peak at a KCl concentration of 0.26 M. Fractions containing maximum hexokinase activity are pooled and dialyzed overnight against several changes of a buffer containing 50 mM potassium phosphate, pH 7.4, 1 mM EDTA, 10 mM 2-mercaptoethanol, and 0.1 M glucose.

Ammonium Sulfate Fractionation. The dialyzate is made 0.39 M by the addition of solid ammonium sulfate and slowly stirred for 30 min. This fraction is centrifuged at 10,000 g for 20 min, and the pellet is discarded. Ammonium sulfate is again added to the supernatant solution to bring the sale concentration to 1.58 M. This fraction is centrifuged, and the supernatant solution is discarded. The pellet containing the hexokinase activity is dissolved in a minimum amount of the phosphate buffer and dialyzed for 12 hr against several changes of the buffer. The dialyzate is brought to 1.58 M ammonium sulfate by the slow addition of saturated ammonium sulfate (3.94 M at 4°) in the phosphate buffer (pH 7.4). The extract is centrifuged at 16,000 g for 20 min and the supernatant fraction is discarded. The resulting pellet is dissolved in 5 ml of the phosphate buffer. Buffer containing saturated ammonium sulfate (3.94 M) is added to the redissolved pellet fraction to bring the ammonium sulfate concentration to 0.94 M. The above steps of centrifugation and resuspension are then repeated several times, each time decreasing ammonium sulfate concentration by decrements of 0.12 M, until a concentration of 0.39 M is reached. Supernatant fractions between 0.45 and 0.85 M ammonium sulfate concentration have the highest specific activity and are pooled and dialyzed for 2 hr against 2.76 M ammonium sulfate to precipitate the protein. The dialy-

Purification of Hexokinase from *Ascaris suum*

Fraction	Volume (ml)	Total protein (mg)	Total activity (units)	Specific activity (units/mg protein)	Purification overall (fold)	Yield overall (%)
Crude homogenate	395	4120	177	0.041	—	(100)
Postdialysis	420	3521	162	0.046	1.1	92.0
DEAE-BioGel	230	222	120	0.540	13.0	68.0
Ammonium sulfate I	15	101	83	0.810	20.0	47.0
Ammonium sulfate II	5	41	46	1.100	27.0	26.0
Sephadex G-200 I	1	11	32	2.900	70.0	18.0
Sephadex G-200 II	1	1	6	6.000	146.0	3.3

zate is centrifuged at 16,000 g for 20 min, and the supernatant solution is discarded.

Sephadex G-200 Chromatography. The resulting pellet is redissolved in 2 ml of 50 mM phosphate (pH 7.2), 1 mM EDTA, 14 mM 2-mercaptoethanol, and 0.1 M glucose. The sample is applied to a 2.5 cm × 90 cm Sephadex G-200 column that has been equilibrated in the above buffer. Two-milliliter fractions are collected, and those containing maximum hexokinase activity are pooled. The enzyme is precipitated by dialysis for 2 hr against saturated ammonium sulfate. The precipitated protein is centrifuged at 16,000 g for 20 min and redissolved in 1 ml of the buffer. The protein sample is again applied to the Sephadex G-200 column used in the preceding step, and 1-ml fractions are collected. Fractions having maximum hexokinase activity are pooled and precipitated by dialysis for 2 hr in 2.76 M ammonium sulfate. At this stage the enzyme is essentially homogeneous.

A summary of the purification procedure is presented in the table. The procedure resulted in a 146-fold purification of hexokinase over the crude extract. As the protein is purified, it becomes increasingly labile and is extremely unstable in the absence of glucose. Dialysis to remove the glucose results in a total loss of activity. Storage of the enzyme in 2.76 M ammonium sulfate, 0.1 M glucose, and 10 mM 2-mercaptoethanol at 4° results in a 10% loss of activity per month.

Properties of *Ascaris suum* Hexokinase

The purified enzyme represents the only form of hexokinase in the muscle of *Ascaris suum* [4] and exhibits an apparent isoelectric point of 5.9.

---

[4] S. C. Supowit and B. G. Harris, *Biochim. Biophys. Acta* **422**, 48 (1976).

Both sodium dodecyl sulfate gel electrophoresis and Sephadex G-200 column chromatography give molecular weights in the range of 97,000–100,000. Most mammalian hexokinases are monomers with molecular weights of 100,000,[5] and it is probable that the ascarid enzyme is also a monomer. The Stokes' radius of *Ascaris* hexokinase is 3.75 nm, the diffusion coefficient ($D_{20,w}$) is 5.7×10^{-7} cm^2 sec^{-1}, and the frictional ratio is 1.3. The apparent K_m for glucose (4.7 mM) is reminiscent of the value for rat liver glucokinase[6] (10 mM), whereas the apparent K_m for ATP (0.2 mM) is similar to that of hexokinase from various rat tissues.[7] The enzyme is capable of phosphorylating fructose, but the apparent K_m value for fructose is 80 mM. The inhibition constants of glucose 6-phosphate versus glucose and ATP[4] are at least one order of magnitude higher than those for hexokinase from mammalian tissues.[7] The product inhibition pattern of the ascarid hexokinase suggests that the enzyme has a random kinetic mechanism[4] similar to that found in other hexokinases.[8,9]

[5] A. Chou and J. Wilson, *Arch. Biochem. Biophys.* **151**, 48 (1972).
[6] S. Grossman, C. Dorn, and V. Potter, *J. Biol. Chem.* **249**, 3055 (1974).
[7] L. Grossbard and R. T. Schimke, *J. Biol. Chem.* **241**, 3546 (1966).
[8] D. Purich, H. Fromm, and F. Rudolph, *Adv. Enzymol.* **39**, 249 (1973).
[9] K. Danenberg and W. W. Cleland, *Biochemistry* **14**, 28 (1975).

[5] Glucokinase from *Streptococcus mutans*

By E. VICTORIA PORTER and BRUCE M. CHASSY

ATP + glucose → ADP + glucose 6-phosphate + H$^+$

Streptococcus mutans OMZ70 (ATCC 33535) possesses a highly specific glucokinase (ATP: D-glucose 6-phosphotransferase, EC 2.7.1.2) that is synthesized during growth on a variety of substrates. The enzyme proved to be significantly more stable than the glucokinase described in another strain of *S. mutans*.[1] A method for purifying the enzyme to homogeneity from *S. mutans* OMZ70 and a characterization of this enzyme are described here.

Assay Method

Principle. Two spectrophotometric methods are used. One method measures glucose 6-phosphate formation by following the reduction of

[1] E. V. Porter, B. M. Chassy, and C. E. Holmlund, *Biochim. Biophys. Acta* **611**, 289 (1980).

NADP at 340 nm in the presence of added glucose-6-phosphate dehydrogenase (EC 1.1.1.49).[2] The other employs a lactate dehydrogenase (EC 1.1.1.27)–pyruvate kinase (EC 2.7.1.40) couple to measure ADP formation by following the reduction of phosphoenolpyruvate (PEP) to lactate in the presence of NADH at 340 nm.[3]

Reagents

HEPES (*N*-2-hydroxyethylpiperazine-*N'*-2-ethanesulfonic acid) buffer, 1 *M*, pH 7.5
ATP, 0.1 *M*, pH 7
MgCl$_2$, 0.1 *M*
NADP, 0.01 *M*, pH 7
D-Glucose, 0.5 *M*
Glucose-6-phosphate dehydrogenase, Sigma type XV
PEP (monopotassium salt, Sigma Chemical Co., St. Louis, Missouri), 0.1 *M*
KCl, 0.1 *M*
NADH, 0.01 *M*, pH 7
L-Lactate dehydrogenase, Sigma type XI
Pyruvate kinase, Sigma type III

Procedure. For the method based on glucose 6-phosphate formation, the following are added to a semimicrocuvette (0.4 × 1 × 3 cm internal dimensions) with a 1-cm light path: 0.1 ml of buffer, 0.05 ml of ATP, 0.05 ml of MgCl$_2$, 0.05 ml of glucose, 0.05 ml of NADP, excess glucose-6-phosphate dehydrogenase (5 IU), glucokinase, and water to a volume of 0.5 ml. For the alternative method, based on ADP formation, the following are added to a semimicrocuvette with a 1-cm light path: 0.1 ml of buffer, 2.5 μl of ATP, 0.05 ml of MgCl$_2$, 0.05 ml of glucose, 0.05 ml of PEP, 0.05 ml of KCl, 0.025 ml of NADH, excess lactate dehydrogenase (20–30 IU), excess pyruvate kinase (20–30 IU), glucokinase, and water to a volume of 0.5 ml. With both methods, the reaction is initiated by the addition of glucokinase. The changes in absorbance at 340 nm can be monitored with a Gilford multiple-sample absorbance spectrophotometer. Assays are conducted at ambient temperature.

Evaluation of the Assay. It is necessary to adjust the pH of the reagents used in the assays to pH 7 in order to minimize errors in rate determinations due to pH effects. With both methods, the rate of change in absorbance at 340 nm is linear with time and proportional to the amount of enzyme present under conditions where the enzymes of the couple employed are present in excess.

[2] M. W. Slein, G. T. Cori, and C. F. Cori, *J. Biol. Chem.* **186**, 763 (1950).
[3] A. Kornberg and W. E. Pricer, *J. Biol. Chem.* **193**, 481 (1951).

Definition of Unit and Specific Activity. A unit of enzyme activity is defined as the amount of glucokinase that catalyzes the formation of 1 μmol of glucose 6-phosphate per minute in the assay systems described. Specific activity is expressed as units per milligram of protein. Protein is determined spectrophotometrically.[4]

Purification Procedure

A seed culture (1.75 liters) of *S. mutans* OMZ70 (deposited with American Type Culture Collection) is grown at 37° overnight in still culture in a medium containing 0.5% each of yeast extract, Trypticase (BBL), and $K_2HPO_4 \cdot 3 H_2O$, 0.005% Na_2CO_3, 0.09% glucose, and 0.05 ml per liter of salt solution [containing 0.8% $MgSO_4 \cdot 7 H_2O$, 0.04% $FeSO_4 \cdot 7 H_2O$, and 0.019% $MnSO_4 \cdot 4 H_2O$ (w/v) in water], which is adjusted to pH 7.4 with 12 N HCl. A bottle containing 35 liters of the same medium is inoculated with the 1.75 liters of seed culture and allowed to stand overnight at 37°. Batches of cells for isolation of enzyme are grown by inoculating a 350-liter fermentor with the 35-liter subculture. The medium used for the 350-liter fermentation is the same as that described above except that the medium is buffered at pH 7 with 0.78% K_2HPO_4 and 0.26% KH_2PO_4, and contains 0.03% Tween 80 (Difco) and 1% sucrose (autoclaved separately) in place of glucose. The culture is grown (National Institute of Arthritis, Diabetes, and Digestive and Kidney Diseases, NIH, Bethesda, Maryland) at 37° with a gas flow rate of 10 liters of 95% N_2–5% CO_2 per minute. The agitation speed is increased gradually during the fermentation from 110 rpm to 320 rpm in order to minimize clumping that occurs during growth of *S. mutans* in sucrose-containing media. The cells are harvested with a Sharples continuous-flow centrifuge 7 hr after inoculation. Cell yield is approximately 11 g (wet weight) of cells per liter. Wet pastes are stored at −20°.

Extracts are prepared by sonic disruption (Branson W350 Sonifier) of a 20% (w/v) slurry of the cells suspended in 0.1 M HEPES buffer (pH 7.5) containing 5 mM $MgCl_2$ and 5 mM dithiothreitol (DTT). The supernatant fluid obtained by centrifugation of the broken-cell suspension at 48,000 g for 30 min is dialyzed overnight against 4 liters of 5 mM HEPES buffer (pH 7.5) containing 5 mM $MgCl_2$ and 5 mM DTT (two changes). The dialyzed extract is centrifuged at 105,000 g for 45 min, and the supernatant fluid is concentrated 10-fold in a cell fitted with a PM-10 membrane (Amicon Corp., Lexington, Massachusetts). The crude extract from 100 g of cells is used for the purification of glucokinase described below. All operations are carried out at 0–4°.

[4] M. M. Bradford, *Anal. Biochem.* **72,** 248 (1976).

Diethylaminoethyl (DEAE)-Cellulose Chromatography. The crude extract is applied to a 2.5 × 40 cm DEAE-cellulose (Whatman) column at a flow rate of 1.2 ml/min. The column is eluted using a 16-hr linear gradient from 0.1 M HEPES buffer (pH 7.5) containing both 5 mM MgCl$_2$ and 5 mM DTT as the initial buffer to 0.35 M KCl in the same buffer. The fraction size is 7.2 ml, and the fractions containing most of the glucokinase activity (fractions 89–100) are pooled and concentrated in an Amicon cell fitted with a PM-10 membrane.

Ultrogel AcA-54 Chromatography. The sample obtained from the DEAE-cellulose column is applied to a 1.5 × 110 cm Ultrogel AcA-54 (LKB) column that has been previously equilibrated with 0.1 M HEPES buffer (pH 7.5) containing 5 mM MgCl$_2$ and 5 mM DTT. The column is eluted at a flow rate of 30 ml/hr, and the fraction size is 3 ml. The peak fraction of glucokinase activity corresponds to an M_r of 41,000 for the enzyme based on standardization of the column with proteins of known molecular weights. The fractions containing glucokinase activity (fractions 86–95) are pooled and concentrated in dialysis bags using Aquacid III (Calbiochem) as a dehydrating agent.

Agarose–Hexane–Adenosine 5'-Triphosphate (AGATP) Affinity Gel Chromatography. The glucokinase sample recovered from Ultrogel AcA-54 chromatography is applied to a 1 × 10 cm AGATP affinity gel (P-L Biochemicals, type 2) column at a flow rate of 12 ml/hr. The starting buffer for the 16-hr gradient elution is 0.05 M HEPES buffer (pH 7.5) containing 5 mM MgCl$_2$ and 5 mM DTT ascending to a final concentration of 0.4 M KCl in the same buffer. The final preparation contains 198 IU of glucokinase per milligram of protein.

Results of a typical purification are summarized in the table.

Properties

Purity. The final preparation obtained after the purification described here was homogeneous based on the observation of a single band on both disc gel and SDS gel electrophoresis. Since the purified enzyme exhibited a single band on SDS gel electrophoresis that corresponded to an M_r value approximately half that of the nondenatured enzyme, the enzyme is probably composed of two subunits of equal size.

Effect of Growth Substrate. Glucokinase is present in *S. mutans* OMZ70 cells whether the cells are cultured in glucose-, sucrose-, or fructose-containing media. However, growth in 30 mM sucrose-containing media results in a 30-fold increase in the amount of glucokinase (specific activity) compared to the amount found in cells grown in 60 mM glucose-containing media.

PURIFICATION OF GLUCOKINASE FROM SUCROSE-GROWN
Streptococcus mutans OMZ70 CELLS

Preparation	Volume (ml)	Activity (units/ ml)	Total activity (units)	Protein (mg/ml)	Specific activity (units/mg protein)	Yield (%)	Purification (fold)
Crude extract	118	0.96	113	0.696	1.38	100	1.00
DEAE-cellulose	25	3.38	84.5	0.460	7.34	75	5.32
Ultrogel AcA-54	5.8	12.2	70.5	0.575	21.1	62	15.3
AGATP affinity gel	3	20.0	60.0	0.101	198	53	144

Substrate Specificity. Of the 20 compounds tested as potential carbohydrate substrates for the glucokinase reaction, only D-glucose (K_m = 0.61 mM) was phosphorylated. The following compounds were not phosphorylated: D-ribose, L-arabinose, D-mannose, D-sorbitol, L-sorbose, D-galactose, D-fructose, D-mannitol, maltose, sucrose, lactose, L-rhamnose, D-xylose, 3-O-methyl-D-glucose, 6-deoxy-D-glucose, 2-deoxy-D-glucose, N-acetyl-D-glucosamine, D-glucosamine-HCl, and α-methyl-D-glucoside.

Phosphoryl Donor Specificity. Adenosine 5'-triphosphate was by far the most effective phosphoryl donor tested. The nucleoside triphosphates tested had relative rates as follows: ATP, 100; ITP, 9; CTP, 8; TTP, 7; UTP, 6; GTP, 5; and XTP, 4. The K_m for ATP is 0.21 mM.

Metal Ion Specificity. A divalent metal ion is required for the glucokinase reaction. The following rates of D-glucose phosphorylation were observed in the presence of 10 mM metal salts: Co^{2+}, 264; Mn^{2+}, 173; Mg^{2+}, 100; Cd^{2+}, 91; Zn^{2+}, 56; Fe^{2+}, 47; Ca^{2+}, 34; Sr^{2+}, 11; and Cu^{2+}, <1.

Inhibition Studies. At saturating concentrations of glucose or ATP, ADP was shown to be a noncompetitive inhibitor of ATP (K_i = 0.67 mM) and an uncompetitive inhibitor of glucose (K_i = 0.71 mM), respectively. Glucose 6-phosphate was a competitive inhibitor of glucose (K_i = 0.31 mM) at saturating ATP. In the presence of 2 mM glucose or ATP, β, γ-methyleneadenosine 5'-triphosphate was shown to be a competitive inhibitor (K_i = 2.09 mM) of ATP utilization and a noncompetitive inhibitor (K_i = 5.01 mM) of glucose utilization, respectively. With the same substrate concentrations, 6-amino-6-deoxy-D-glucose was shown to be a noncompetitive inhibitor (K_i = 2.2 mM) of ATP utilization and a competitive inhibitor (K_i = 0.89 mM) of glucose utilization.

pH Optimum. Phosphorylation of D-glucose was tested over the range from pH 5.5 to 9.5 in 0.2 M MES (pH 5.5–7.1), HEPES (pH 6.7–8.1), and

Tris (pH 7.9–9.5) buffers with overlapping at the transition pH values. The pH optimum of the phosphorylation reaction is broad; it is maximal from pH 7.5 to 9.5.

Stability. When stored at $-20°$ in the presence of 1 mM DTT and 20–30% glycerol, glucokinase preparations are stable indefinitely. Storage at room temperature in the presence of 1 mM DTT and 5 mM MgCl$_2$ stabilizes enzyme preparations for 1 week. Enzyme activity is lost rapidly in the absence of DTT.

[6] Galactokinase from *Saccharomyces cerevisiae*

By David B. Wilson and Mark A. Schell

$$\text{D-Galactose} + \text{ATP} \rightarrow \text{galactose 1-phosphate} + \text{ADP}$$

Galactokinase (EC 2.7.1.6) catalyzes the first step in the Leloir pathway for galactose metabolism.[1] This enzyme has been purified and characterized from several organisms including *Escherichia coli,*[2,3] pig liver,[4] human red cells,[5] and yeast.[6,7]

Assay Method

Reagents

[^{14}C]Galactose, 0.01 M (4×10^5 cpm/μmol)
Sodium fluoride, 0.032 M
Dithioerythritol, 0.01 M
ATP, 0.1 M, pH 7.4
MgCl$_2$, 0.80 M
Triethanolamine–acetate, pH 8.0, 1.0 M
Glucose, 1.0 M
Solution 1: 1.0 ml of sodium fluoride and 1.0 ml of dithioerythritol
Solution 2: 0.15 ml of ATP, 0.05 ml of MgCl$_2$, 1.0 ml of triethanolamine, 0.01 ml of glucose, and 3.7 ml of distilled water

[1] C. E. Cardini and L. F. Leloir, *in* "Biochemists' Handbook" (E. King and W. Sperry, eds.), pp. 404 and 439. Van Nostrand-Reinhold, Princeton, New Jersey, 1961.
[2] D. B. Wilson and D. S. Hogness, this series, Vol. 8, p. 229.
[3] D. B. Wilson and D. S. Hogness, *J. Biol. Chem.* **244**, 2137 (1969).
[4] F. J. Ballard, *Biochem. J.* **98**, 347 (1966).
[5] K.-G. Blume and E. Beutler, *J. Biol. Chem.* **246**, 6507 (1971).
[6] M. A. Schell and D. B. Wilson, *J. Biol. Chem.* **252**, 1162 (1977).
[7] M. R. Henrich, *J. Biol. Chem.* **239**, 50 (1964).

Enzyme dilution buffer: 20 mM triethanolamine–acetate (pH 8.0) containing 1 mM EDTA, 1 mM dithioerythritol, and 0.1 mg of bovine serum albumin per milliliter

Procedure. The assay solution consists of 10 μl of [^{14}C]galactose, 20 μl of solution 1, 50 μl of solution 2, and 20 μl of the sample to be assayed. After 20 min at 30°, a 50-μl aliquot is spotted 3 cm from the end of a 1.5 by 15 cm strip of DEAE paper. The strip is eluted with water by descending chromatography until the front is 2 cm from the bottom. The bottom 7 cm is discarded and the rest of the strip is dried in an 80° oven for 10 min. The area containing the original spot is cut out and counted in a scintillation counter using a toluene-based fluid. This assay works well in crude extracts and is very sensitive and specific for galactokinase. The assay values are directly proportional to enzyme concentration as long as the enzyme concentration in the sample to be assayed is below 0.020 unit/ml.

Definition of Unit and Specific Activity. One unit of enzyme phosphorylates 1 μmol of galactose per minute in this assay. Specific activity is expressed in units per milligram of protein, and protein is determined by the procedure of Lowry *et al.*[8]

Purification

All procedures are carried out at 0–5° unless otherwise stated.

Preparation of Yeast. A diploid strain of *S. cerevisiae* X108D is used in this procedure. Cells are grown in a medium containing 2% bactotryptone, 1% yeast extract, and 2% galactose at 30° with moderate aeration in a 15-liter New Brunswick fermentor. Cells are grown from a density of $\sim$10^7 cells/ml to 10^8 cells/ml before harvest; the generation time is 2.5 hr, and the yield of cells is 10 g/liter.

Preparation of Extract. Fifty grams of frozen or fresh cells (the initial specific activity of kinase in frozen cells is unchanged after 3 months of storage at $-$20°) are resuspended in 100 ml of galactokinase buffer [20 mM triethanolamine–acetate (pH 8.0), 1 mM EDTA, 1 mM dithioerythritol]. The suspension is passed through a French pressure cell (Aminco Corp., Lexington, Massachusetts) at 10,000 to 15,000 psi. This extract is centrifuged at 20,000 g for 15 min, and the supernatant solution is decanted and saved. The pellet is resuspended in 100 ml of fresh galactokinase buffer and pressed and centrifuged as before. The supernatants from the two breakages are combined and diluted to 500 ml with galactokinase buffer (fraction I). This procedure results in 80–90% breakage of the yeast.

[8] O. H. Lowry, N. J. Rosebrough, A. L. Farr, and R. J. Randall, *J. Biol. Chem.* **193,** 265 (1951).

Ammonium Sulfate Fractionation. Ammonium sulfate, 133 g (0.266 g/ml) is added to fraction I over a period of 7 min with thorough mixing. After addition of the final aliquot of ammonium sulfate, the mixture is stirred an additional 15 min, allowed to sit for 20 min, and then centrifuged at 12,000 g for 20 min. The supernatant solution is decanted; an additional 73 g (0.146 g/ml of original extract volume) of ammonium sulfate is added, and it is allowed to sit as before. The suspension is centrifuged, and the supernatant fraction is decanted and discarded; the pellet is redissolved in 100 ml of galactokinase buffer (fraction II).

Streptomycin Sulfate Precipitation. Streptomycin sulfate (25 ml of 10% w/v) is added to fraction II while it is stirred slowly. The solution is then dialyzed against 2.0 liters of buffer A [10 mM Tris-HCl (pH 7.4), 1 mM dithioerythritol, 1 mM EDTA, 10% (v/v) glycerol] for 6 hr. The dialysis buffer is changed, and dialysis is continued for at least another 5 hr. The buffer is changed once more, and after 5 hr the solution is removed from the dialysis bag. The turbid solution is centrifuged at 20,000 g for 15 min to remove the precipitate. The supernatant fraction is decanted and saved (fraction III).

DEAE-Cellulose Chromatography. A column of DEAE-cellulose (Whatman, DE-52) (4 × 30 cm) is prepared and equilibrated with buffer A containing 7 mM ammonium sulfate. Fraction III is made 7 mM in ammonium sulfate by the addition of 0.90 ml of 1 M ammonium sulfate and then loaded onto the DEAE-cellulose column at a flow rate of 100 ml/hr; 15-ml fractions are collected. After the sample has entered the column, it is washed through with buffer A containing 7 mM ammonium sulfate until the $A_{280 nm}$ of the effluent is less than 0.04. Under these conditions galactokinase binds weakly to the column and is found in the wash-through fractions, whereas 85% of the protein in the extract remains tightly bound to the column. Fractions of specific activity greater than 7.5 units/$A_{280 nm}$ are pooled and concentrated to 50 ml in an Amicon ultrafiltration apparatus equipped with a UM-10 membrane. The concentrated fractions are then dialyzed overnight against 1.0 liter of buffer B [5 mM sodium phosphate (pH 6.8), 1 mM EDTA, 1 mM dithioerythritol, and 10% (v/v) glycerol]. The dialysis buffer was changed and dialysis continued for an additional 6 hr (fraction IV).

Hydroxyapatite Chromatography. A column (2.5 × 17 cm) of hydroxyapatite (Hypatite C, Clarkson Chemical Co.) is prepared and washed with 2-bed volumes of buffer B. Fraction IV is loaded onto the column at a flow rate of 100 ml/hr. After the sample has entered the column, it is washed with buffer B until the $A_{280 nm}$ is constant. The bound galactokinase is eluted with a linear gradient consisting of 600 ml of buffer B and 600 ml of buffer B containing 0.12 M sodium phosphate, pH 6.8.

TABLE I
PURIFICATION OF GALACTOKINASE

| Fraction | Step or fraction | Activity | | Recovery (%) | Protein (mg/ml) | Specific activity (units/mg protein) |
		Units/ml	Total units × 10⁻³			
I	Extract	7.5	3.76	100	10.0	0.75
II	(NH$_4$)$_2$SO$_4$ precipitate	22	2.62	70	18.7	1.17
III	Streptomycin supernatant	19	2.47	66	9.6	2.0
IV	DEAE-cellulose eluate	45	2.35	63	3.9	11.6
V	Hydroxyapatite chromatography	184	1.97	53	6.0	30.7
VI	BioGel chromatography	111	1.84	50	2.9	38.3

Fractions (20 ml) are collected at a flow rate of 100 ml/hr. The fractions with the highest specific activity are pooled and concentrated in an Amicon ultrafiltration apparatus to a final volume of 10 ml. The concentrated fractions are centrifuged at 20,000 g for 10 min, and the supernatant is saved and frozen at $-80°$ (fraction V).

BioGel A-0.5m Gel Filtration. Fraction V is chromatographed on a (1.6 × 90 cm) BioGel A-0.5m column equilibrated with buffer A + 10 mM galactose. The column is eluted with the same buffer at a flow rate of 6 ml/hr, and 2.5-ml fractions are collected. The most efficient purification is obtained with samples smaller than 5 ml. The galactokinase eluted from the column as a sharp peak in the area of molecular weight 50,000 to 70,000 (V_e = 100–110 ml). The fractions of highest specific activity are pooled and stored at $-80°$ (fraction VI).

The results of a typical purification are summarized in Table I.

Properties

Purity. The contaminating proteins in fraction VI appeared to represent less than 5% of the total protein when measured by the following techniques: BioGel A-0.5m chromatography, tube and slab gel SDS–polyacrylamide gel electrophoresis, and analytical ultracentrifugation.

This enzyme gave anomalous results when run on polyacrylamide gels in the absence of SDS. A very diffuse broad band was observed under nondenaturing conditions, and the addition of 7 M urea or 0.1% Triton X-100 did not alter this behavior.

Stability. Fraction VI lost less than 20% of its activity after 6 months at $-80°$ whereas fraction V lost less than 10% activity after 1 year at $-80°$.

TABLE II
AMINO ACID ANALYSIS OF YEAST
GALACTOKINASE

Amino acid	Residues/molecule	Mole %
Lysine	45.9	8.7
Histidine	11.8	2.2
Arginine	10.9	2.1
Tryptophan	5.3	1.0
Aspartic acid	51.8	9.8
Threonine	26.7	5.0
Serine	43.4	8.2
Glutamic acid	59.2	11.2
Proline	30.8	5.8
Glycine	26.0	4.9
Alanine	46.1	8.7
Cysteine	11.1	2.1
Valine	40.1	7.6
Methionine	8.8	1.7
Isoleucine	20.3	3.8
Leucine	46.6	8.8
Tyrosine	19.5	3.7
Phenylalanine	24.6	4.6

Yeast galactokinase is also stable (i.e., <20% loss of activity) when stored for 24 hr at pH values between 5.4 and 8.8. Furthermore, protease degradation of yeast galactokinase during purification was not a major problem. Inclusion of 1 mM phenylmethylsulfonyl fluoride or diisopropyl fluorophosphate in early stages of purification did not significantly affect the yield of galactokinase.

Structure. The molecular weight of native galactokinase was 58,000 as determined by sedimentation equilibrium using a value of $\bar{v}$ of 0.736 calculated from the amino acid composition. The molecular weight determined by SDS–polyacrylamide gel electrophoresis was also 58,000, showing that the enzyme contains a single polypeptide chain. The results of zone sedimentation on a sucrose gradient and molecular sieve chromatography confirmed the 58,000 M_r for the catalytically active galactokinase.

The amino acid composition is given in Table II, and the N-terminal sequence is NH$_2$-Thr-Lys-Ser-His-Arg-Glu-Arg-Val-Ile-Val-Pro. . . .[9]

|
Glu

Catalysis. The enzyme has a pH optimum at 8.0 with a sharp drop in activity above pH 9.0 and a more gradual drop below pH 7.0. The enzyme

[9] D. H. Schlesinger, M. A. Schell, and D. B. Wilson, *FEBS Lett.* **83**, 45 (1977).

is extremely specific for galactose, as no sugar tested (glucose, mannose, galactitol, arabinose, 2-deoxygalactose, D-fucose, and lactose) inhibited galactose phosphorylation even when present at 70 times the galactose concentration.

The K_m for galactose was 0.6 mM, and for ATP it was 0.15 mM. The turnover number of the purified enzyme was 55 molecules/sec per enzyme molecule (V_{max} = 3.35 mmol/mg/hr).

Immunology. Antisera prepared against the purified enzyme did not cross-react with *E. coli* galactokinse, and antisera against the *E. coli* enzyme did not cross-react with the yeast enzyme.

[7] Phosphofructokinase[1] from Ehrlich Ascites Tumor

By Michio Ui and Takahiko Sumi

Fructose 6-phosphate + ATP → fructose 1,6-bisphosphate + ADP + H[+]

Assay Method[1a]

Principle. A spectrophotometric assay is employed based on the coupling of fructose 1,6-bisphosphate formation to the oxidation of NADH with the use of a sequential reaction catalyzed by aldolase, triosephosphate isomerase, and α-glycerophosphate dehydrogenase.

Reagents

Magnesium buffer mixture: 75 mM Tris-HCl, 3 mM MgCl$_2$, and 15 mM (NH$_4$)$_2$SO$_4$, all adjusted to pH 7.5
Fructose 6-phosphate, 30 mM
ATP, 15 mM
NADH, 1.5 mM
Mercaptoethanol, 0.1 M
Auxiliary enzyme solution: aldolase (10 units/ml), triosephosphate isomerase (8 units/ml), and α-glycerophosphate dehydrogenase (4 units/ml) in 25 mM Tris-HCl (pH 7.5)

Procedure. A reaction mixture of 3.0 ml is made up as follows: 2.0 ml of magnesium buffer mixture, 0.1 ml of fructose 6-phosphate, 0.1 ml of ATP, 0.1 ml of NADH, 0.3 ml of mercaptoethanol, 0.3 ml of the auxiliary enzyme solution. After 5 min for temperature equilibration at 28°, 0.1 ml

---

[1] ATP: D-fructose-6-phosphate 1-phosphotransferase, EC 2.7.1.11.
[1a] T. Sumi and M. Ui, *Biochim. Biophys. Acta* **268**, 354 (1972).

METHODS IN ENZYMOLOGY, VOL. 90

of the test solution for phosphofructokinase is added. A change in the absorbance at 340 nm is recorded in a 10-mm light path cuvette in a thermostatted spectrophotometer. One unit of phosphofructokinase activity is defined as the amount of enzyme that catalyzes the formation of 1 μmol of fructose 1,6-bisphosphate per minute under the above conditions.

Kinetics of the purified enzyme are analyzed by incubating the dialyzed enzyme solution dissolved in 0.5% bovine serum albumin under the same conditions as above, but with the omission of $(NH_4)_2SO_4$ from the reaction mixture containing 50 mM imidazole-HCl (pH 7.5) instead of Tris-HCl. Auxiliary enzymes are also freed of $(NH_4)_2SO_4$ by dialysis in 0.5% albumin overnight against a large volume of 50 mM Tris buffer, pH 8.0.

Purification[1a]

Ehrlich diploid ascites tumor cells are harvested 7–8 days after inoculation in albino mice (SWJ/mk). Cells from ascitic fluids of 50 mice are combined and washed once by suspension and centrifugation in 500 ml of ice-cold 0.9% NaCl. All manipulations are performed at 4°.

Step 1. Preparation of Crude Extract. The once-washed packed cells, approximately 100 ml, are suspended in 500 ml of 0.1 M sodium and potassium phosphate buffer (pH 8.0) containing 1 mM EDTA (this solution is henceforth referred to as buffer A) and centrifuged at 1600 g for 10 min. The precipitated cells are disrupted by grinding with 100 g of sea sand and 30 ml of buffer A in a mortar. The mixture, suspended in 500 ml of buffer A, is centrifuged at 4500 g for 30 min, and the precipitate is reextracted with 200 ml of buffer A followed by centrifugation. The volume of the combined extract is about 800 ml.

Step 2. Fractionation with $(NH_4)_2SO_4$. To the crude extract from step 1, finely powdered $(NH_4)_2SO_4$ is added slowly with stirring to give a final concentration of 23 g/ml, and the solution, brought to pH 7.5 with 2.0 M NaOH, is allowed to stand for 4–6 hr. The precipitate is removed by centrifugation at 13,000 g for 30 min and discarded. Additional solid $(NH_4)_2SO_4$ in an amount sufficient to achieve a concentration of 45 g/ml is added to the supernatant, which is then readjusted to pH 7.5 and kept for 4–6 hr before the precipitate containing phosphofructokinase activity is collected by centrifugation at 13,000 g for 30 min.

Step 3. DEAE-Sephadex A-50 Column Chromatography. The precipitate obtained from step 2 is dissolved in 50 ml of 20 mM phosphate buffer (pH 7.8) containing 80 mM $(NH_4)_2SO_4$, 25 mM mercaptoethanol, and 1 mM EDTA (buffer B) and is dialyzed for 12 hr against 2 liters of the same buffer with change of buffer after 6 hr. The dialyzed enzyme solution is clarified by centrifugation at 12,000 g for 30 min and applied to a column (3

PURIFICATION OF ASCITES TUMOR PHOSPHOFRUCTOKINASE

Step and fraction	Activity (units)	Protein (mg)	Specific activity (units/mg protein)	Yield (%)
1. Crude extract	3620	3800	1.0	100
2. Ammonium sulfate	3160	803	3.9	87
3. DEAE-Sephadex	1160	15	80	32
4. Sephadex G-200	1030	7	150	29

cm × 25 cm) of DEAE-Sephadex A-50 that has been equilibrated with buffer B. The column is first washed with the same buffer (approximately 400 ml) until unadsorbed protein having no phosphofructokinase activity is washed away. Elution of the enzyme is then achieved with a linear gradient from 0 to 0.2 M KCl in buffer B at a flow rate of 60 ml/hr. Phosphofructokinase activity is eluted in two peaks, the bulk of the activity emerging in the second peak. Further purification is performed on the pooled eluate between fractions 80 and 90 (215 ml) corresponding to the second peak. The active protein is precipitated by dialysis overnight against a solution prepared by adding 1100 g of $(NH_4)_2SO_4$ to 1 liter of 0.1 M phosphate buffer (pH 7.8) containing 1 mM EDTA. $(NH_4)_2SO_4$ remaining undissolved when prepared is dissolved during dialysis.

Step 4. Fractionation on Sephadex G-200 Column. The precipitate from step 3 is collected by centrifugation at 25,000 g for 30 min and dissolved in 4 ml of 0.1 M phosphate buffer (pH 7.8) containing 0.5 M $(NH_4)_2SO_4$, 25 mM mercaptoethanol, and 1 mM EDTA (buffer C). The enzyme solution is applied to a Sephadex G-200 column (3 cm × 95 cm) that has been equilibrated with buffer C. Upon elution with the same buffer, phosphofructokinase activity emerges as the first of two protein boundaries. Fractions containing phosphofructokinase activity (fractions 26–36, 110 ml) are combined and dialyzed overnight against 1 liter of 0.1 M phosphate buffer (pH 7.8) containing 0.1 mM ATP, 1 mM EDTA, and 55 g of $(NH_4)_2SO_4$ per milliliter to precipitate the enzyme. The volume of this final enzyme suspension is reduced to about one-third during dialysis. A typical purification is summarized in the table.

Properties[1a,2]

Stability. The final enzyme suspension (step 4) in a concentrated ammonium sulfate solution can be stored at 3° for at least 1 month without

[2] T. Sumi and M. Ui, *Biochim. Biophys. Acta* **276**, 19 (1972).

loss of activity. The presence of ATP is essential; 50% of the activity is lost after 1 month if ATP is omitted from the stock solution.

Homogeneity and Molecular Weight. The enzyme preparation yields a single band without noticeable contamination when subjected to disc electrophoresis in a polyacrylamide gel at pH 7.5. The molecular weight as determined by gel filtration through a Sephadex G-200 column is approximately 300,000.

pH Optimum. The pH optimum of ascites tumor phosphofructokinase is in the vicinity of 7.1 regardless of the concentration of ATP in the reaction mixture. This is in sharp contrast to the muscle enzyme,[3] the optimum pH of which undergoes striking changes dependent on the ATP concentration. SO_4^{2-} and K^+ have been found to shift the optimum pH slightly toward alkaline side.

Dual Effects of K^+. Similar to phosphofructokinase from other sources, K^+ is essentially required for the tumor phosphofructokinase. Upon further increasing the concentration of K^+, however, an inhibition instead of activation is observed. The K^+-induced inhibition is reversed by increasing the concentration of ATP. This dual response to K^+ appears to be one of the unique properties that characterize the tumor phosphofructokinase among the mammalian enzymes.

Kinetics. The ascites tumor phosphofructokinase is a typical allosteric enzyme in the sense that the rate-concentration curves with respect to fructose 6-phosphate and ATP show highly cooperative patterns of activation and inhibition, respectively. The inhibition by ATP is reduced by increasing the concentration of fructose 6-phosphate or by the addition of the positive effectors (activators) such as AMP, P_i, ADP, and SO_4^{2-}, while cooperative binding of fructose 6-phosphate and the effectors is exaggerated by increasing the concentration of ATP. The cooperative interactions displayed by this enzyme are completely lost in the presence of a saturating concentration of the positive effectors. The resultant hyperbolic saturation curve for fructose 6-phosphate and ATP is consistent with a "Ping-Pong Bi-Bi" mechanism named by Cleland.[4] Thus, the kinetic properties are readily interpreted in terms of the reaction models of "allosteric Ping-Pong II," which is one the reaction models proposed by Sumi and Ui.[5] The model suggests that the conversion of free enzyme to another stable enzyme form is an essential step for allosteric transition.

[3] M. Ui, *Biochim. Biophys. Acta* **124**, 310 (1966).
[4] W. W. Cleland, *Biochim. Biophys. Acta* **67**, 104 (1963).
[5] T. Sumi and M. Ui, *Biochim. Biophys. Acta* **276**, 12 (1972).

[8] Phosphofructokinase from Oyster Adductor Muscle

By KENNETH B. STOREY

Fructose 6-phosphate + ATP → fructose 1,6-bisphosphate + ADP + H$^+$

In tissues that rely on carbohydrate as an energy source, phosphofructokinase (PFK; EC 2.7.1.11), the first committed step of the glycolytic pathway, closely regulates the overall rate of carbon flux through the pathway.[1-3] Oysters are excellent anaerobes, capable of surviving several weeks of anoxia with anaerobic glycogenolysis supplying the energy needs of the animal. In oyster adductor muscle, PFK is closely regulated by the adenylates, arginine phosphate, and pH and is a key site controlling carbon flux during the aerobic–anaerobic transition. PFK from oyster muscle has been purified and characterized[4]; the purification presented here is a relatively simple, high-yield method that combines the ATP–agarose chromatography described by Ramadoss *et al.*[5] with polyethylene glycol fractionation.[4]

Assay Method

Principle. Fructose 1,6-bisphosphate formation is coupled to the oxidation of NADH using either aldolase, triosephosphate isomerase, and glycerol-3-phosphate dehydrogenase or pyruvate kinase and lactate dehydrogenase with the decrease in absorbance at 340 nm monitored spectrophotometrically.

Reagents. All assay reagents are made up in the Tris-HCl buffer with pH readjusted, if necessary, by the addition of solid Tris base.

Tris-HCl, 0.05 *M,* pH 8.0
MgCl$_2$, 0.5 *M*
KCl, 1 *M*
ATP, 0.1 *M*
Fructose 6-phosphate, 0.1 *M*
NADH, 0.01 *M*
and either
  (a) Phosphoenolpyruvate, 0.05 *M*; pyruvate kinase, 75 units/ml; and
      lactate dehydrogenase, 40 units/ml

[1] E. Negelein, *Biochem. Z.* **287,** 329 (1936).
[2] K. Uyeda, *Adv. Enzymol.* **48,** 193 (1979).
[3] A. Ramaiah, *Curr. Top. Cell. Regul.* **8,** 297 (1974).
[4] K. B. Storey, *Eur. J. Biochem.* **70,** 331 (1976).
[5] C. Ramadoss, L. Luby, and K. Uyeda, *Arch. Biochem. Biophys.* **175,** 487 (1976).

or

(b) Aldolase, 4 units/ml; triosephosphate isomerase, 240 units/ml; and glycerol-3-phosphate dehydrogenase, 32 units/ml

Procedure. Coupling enzymes are dialyzed against 1000 volumes of Tris-HCl, $0.05 M$, pH 8.0, for 12 hr at 4° before use. Assays are performed at 24° using a spectrophotometer with a thermostatted cell holder. For all standard assays, the aldolase–triosephosphate isomerase–glycerol-3-phosphate dehydrogenase complex is used; the pyruvate kinase–lactate dehydrogenase couple is useful when examining the effects of fructose 1,6-bisphosphate on the enzyme. To a cuvette add 0.02 ml of $MgCl_2$ (10 mM), 0.1 ml of KCl (100 mM), 0.01 ml of ATP (1 mM), 0.06 ml of fructose 6-phosphate (6 mM), 0.01 ml of NADH (0.1 mM), and either 0.01 ml of aldolase–triosephosphate isomerase–glycerol-3-phosphate dehydrogenase (0.04 : 2.4 : 0.32 units) mixture or 0.02 ml of phosphoenolpyruvate (1 mM) plus 0.01 ml of pyruvate kinase–lactate dehydrogenase (0.75 : 0.40 unit). Total volume is made up to 1 ml with buffer. For unpurified PFK preparations, control assays, omitting fructose 6-phosphate, should be used to test for nonspecific oxidation of NADH. The reaction is started by the addition of the phosphofructokinase preparation. After an initial delay, the reaction rate becomes linear.

Definition of Unit and Specific Activity. One unit of PFK is defined as the amount of enzyme producing 1 μmol of fructose-1,6-bisphosphate per minute at 24°. Specific activity is given in units per milligram of protein. Protein is measured by the method of Bradford[6] using a commercial reagent (Bio-Rad Laboratories, Richmond, California) with bovine γ-globulin as the protein standard.

Purification Procedure

Oyster Adductor Muscle. Choose oysters, *Crassostrea virginica*, that are tightly closed. Gaping oysters, which do not immediately close when touched, are dead. Pry open the oyster and retest for freshness by touching the ciliated mantle edge, which should respond by contracting. Cut out the large adductor muscle; both the phasic (clear) portion and the catch (white) portion can be used. Rinse the muscle twice in homogenization buffer, being especially careful to wash away any flakes of shell. Blot dry. A medium-sized oyster should provide 2–4 g of muscle. The preparation described here utilizes 4 g of muscle with an initial PFK activity of 4.85 units per gram wet weight.

[6] M. Bradford, *Anal. Biochem.* **72**, 248 (1976).

Homogenization. Muscle is weighed, cut into small pieces, and homogenized in 20 volumes (80 ml) of ice-cold Tris-HCl buffer, 0.02 M, pH 7.8, containing 30 mM 2-mercaptoethanol and 1 mM EDTA, using an Ultra-Turrax tissuemizer (Tekmar Co.).

Adductor muscle is easily homogenized; two or three 30-sec bursts at maximum speed, while keeping the sample cold in ice, is sufficient. The homogenate is centrifuged at 39,000 g for 45 min at 4° in a superspeed refrigerated centrifuge; the supernatant is decanted.

Polyethylene Glycol (PEG) Fractionation. All purification steps are now performed at 24°; the supernatant should be allowed to warm up to room temperature. Polyethylene glycol (M_r 6000, Fisher Chemical Co.) is added to the supernatant to a final concentration of 8% w/v (0.0133 M). Polyethylene glycol is added slowly with constant stirring. The solution is further stirred for 45 min followed by centrifugation as described above. The pellet is discarded, and to the supernatant (containing >95% PFK activity) is added a further 4% w/v PEG to bring the total PEG concentration to 12% w/v (0.02 M). The solution is again stirred for 45 min and centrifuged. The supernatant is discarded, and the pellet, which usually measures >100% of initial PFK activity, is gently rinsed once with homogenization buffer, drained well, and then redissolved in 20 ml of Tris-HCl buffer (0.02 M, pH 7.8) containing 30 mM 2-mercaptoethanol.

ATP–Agarose Chromatography. The redissolved pellet is layered onto a column of ATP–agarose [agarose-N^6-(aminohexyl)carbamoylmethyl-adenosine 5'-triphosphate, 2.2 μM 5'-ATP/ml of gel; Sigma Chemical Co., No. A-9624] (3 ml bed volume in a 5 × 10 cm column) previously washed and equilibrated in Tris-HCl buffer (0.02 M, pH 7.8) containing 30 mM 2-mercaptoethanol. The PFK-containing solution is run onto the column at 15 ml/hr. Essentially all PFK is bound to the gel. The column is then washed with equilibration buffer containing 0.05 mM fructose 6-phosphate and 0.05 mM ADP until the eluent protein concentration drops below 1 μg/ml (the limit of detection of the Bio-Rad microassay procedure). Typically this requires at least 200 ml of wash at a flow rate of 40–60 ml/hr. A further 20 ml of buffer (minus fructose 6-phosphate + ADP) is washed through the column to lower metabolite concentrations in preparation for starting the gradient.

Phosphofructokinase is finally eluted from the column by increasing the concentrations of fructose 6-phosphate and ADP above those of the initial wash. The column is eluted with a linear gradient of 0 to 300 μM each of fructose 6-phosphate + ADP in 100 ml of equilibration buffer. The column is eluted at 15 ml/hr; 1-ml fractions are collected. PFK is eluted over 6–20 fractions once fructose 6-phosphate–ADP concentrations reach 65–100 μM; at this point the gradient is somewhat deformed, presumably

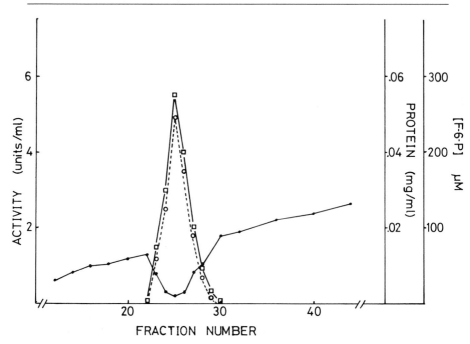

FIG. 1. Elution profile of oyster adductor muscle phosphofructokinase (PFK) from a column of ATP–agarose using a linear gradient, 0 to 300 μM, of fructose 6-phosphate plus ADP. Fractions are 1 ml volume. □——□, PFK activity; ○---○, protein concentration; ●——●, fructose 6-phosphate (F-6-P) concentration.

owing to binding of fructose 6-phosphate and ADP to PFK (Fig. 1). Elution is presumably due to the formation of a ternary complex between these two metabolites and the enzyme.[5] Actual fructose 6-phosphate or ADP levels can be measured using the metabolite assays of Lowry and Passonneau.[7] Peak tubes containing PFK activity are pooled.

Sephacryl S-300 Gel Filtration. Prior to gel filtration, the PFK preparation is concentrated by dialysis (cellulose tubing, M_r cutoff 12,000; Fisher Chemical Co.) against dry, solid polyethylene glycol (M_r 20,000; Fisher Chemical Co.) until the volume of the PFK solution is reduced to 1 ml or less.

The PFK solution is then layered onto a Sephacryl S-300 column (1 × 120 cm) equilibrated in Tris-HCl buffer (0.02 M, pH 7.8) containing 30 mM 2-mercaptoethanol. The column is eluted at 15 ml/hr with the same buffer, and 1-ml fractions are collected. Peak tubes containing PFK activity are

[7] O. H. Lowry and J. V. Passonneau, "A Flexible System of Enzymatic Analysis." Academic Press, New York, 1972.

PURIFICATION OF PHOSPHOFRUCTOKINASE FROM OYSTER ADDUCTOR MUSCLE

Step	Volume (ml)	Total protein (mg)	Total activity (units)	Yield (%)	Specific activity (units/mg protein)
Crude homogenate	80.0	248.4	19.4	—	0.08
Polyethylene glycol	20.0	41.0	22.4	118	0.55
ATP–agarose	5.0	0.14	15.9	82	113.6
Sephacryl S-300	8.0	0.09	13.2	68	146.7

pooled and used for kinetic studies either directly or after a further concentration step using polyethylene glycol.

The purification methods used produce a purified enzyme with a final specific activity of 146 units per milligram of protein and a yield of 68%. The enzyme was judged to be homogeneous using SDS gel electrophoresis.[8] A summary of the purification steps is given in the table.

The enzyme, stored in the concentrated state at 4°, is relatively unstable, with a half-life of 1.5 to 2 days. The addition of 1 mM ATP and/or 1 mM fructose 1,6-bisphosphate increases stability to allow the enzyme to be stored longer. However, subsequent dialysis or Sephadex G-25 gel filtration to remove these effectors prior to kinetic studies can result in large losses of activity.

Comments

Investigations of the effects of polyethylene glycol on rat PFK have shown that PEG dramatically lowers the K_m for fructose 6-phosphate (measured at pH 7.0) and also increases the stability of the enzyme.[9] These effects may be responsible for the apparent >100% recovery of PFK activity after the PEG fractionation.

The purification scheme does not alter any of the key kinetic parameters of oyster muscle PFK compared to the crude preparation, suggesting that the enzyme is unaltered by purification itself.

This purification is relatively simple to perform and has a high yield. Sufficient purified PFK for several thousand assays can be obtained from the muscle of a single oyster using this purification procedure. Other species of the genus *Crassostrea* have similar activities of PFK in the adductor muscle, and these, along with other oyster genera, may be suitable for use in the purification procedure described here.

[8] K. Weber and M. Osborn, *J. Biol. Chem.* **244**, 4406 (1969).
[9] G. Reinhart, *J. Biol. Chem.* **255**, 10576 (1980).

The ATP–agarose column is cleared of protein between uses by washing with 4 M KCl followed by reequilibration in buffer. After several uses of the column, significant numbers of the ATP residues become hydrolyzed. This can be prevented by regenerating the column using a mixture of phosphoenolpyruvate and pyruvate kinase as described by Ramadoss et al.[5]

Properties[4]

Oyster adductor muscle PFK has a pH optimum of 7.9 and a molecular weight of 340,000 ± 20,000. The enzyme displays hyperbolic saturation kinetics with respect to all substrates at both pH 7.9 and pH 6.8 (pH 6.8 is physiologically important as the pH of intracellular fluids during periods of anoxia, and of glycolytic energy production, accompanying shell closure). K_m values for fructose 6-phosphate, ATP, and Mg^{2+} are 0.50, 0.05, and 0.01 mM, respectively, at pH 7.9. AMP activates oyster muscle PFK, lowers the K_m for fructose 6-phosphate, and reverses the effects of inhibitors. ATP and arginine phosphate are inhibitory; other metabolites, such as citrate, Ca^{2+}, fructose 1,6-bisphosphate, and ADP, which inhibit PFK from other sources, do not inhibit oyster PFK.

[9] Phosphofructokinase from *Ascaris suum* Muscle

By BEN G. HARRIS, JANE A. STARLING, and H. WERNER HOFER

Fructose 6-phosphate + ATP → fructose 1,6-bisphosphate + ADP + H^+

Assay Procedure

Principle. The assay for phosphofructokinase[1] is based on the method of Racker,[2] which couples the production of fructose 1,6-bisphosphate (Fru-1,6-P_2) to aldolase, triosephosphate isomerase, and α-glycerolphosphate dehydrogenase.

Reagents

Assay mixture: 100 mM Tris-HCl, pH 8.0, 6 mM MgCl$_2$, 0.5 mM ATP, 1 mM AMP, 30 mM Fru-6-P, 40 mM $(NH_4)_2SO_4$, and 0.25 mM NADH, made up as a single solution and stored at $-20°$

[1] ATP: D-fructose-6-phosphate 1-phosphofructokinase, EC 2.7.1.11.
[2] E. Racker, *J. Biol. Chem.* **167**, 843 (1947).

Auxiliary enzyme solution containing aldolase (6 mg/ml), α-glycerolphosphate dehydrogenase (3 mg/ml), and triosephosphate isomerase (1 mg/ml) in 25 mM triethanolamine-HCl, pH 7.6, and 1 mM dithiothreitol

Enzyme Dilution. The kinase is routinely diluted into a buffer containing 50 mM Tris-potassium phosphate, pH 8.0, 0.2 M KCl, 10 mM KF, 1 mM EDTA, 1 mM ATP, and 5 mM dithiothreitol.

Procedure. One milliliter of assay mixture is placed in a 1-cm light path cuvette in addition to 0.01 ml of coupling enzyme mixture. The cuvette is placed in the thermostatted chamber of a recording spectrophotometer, and the temperature is equilibrated to 30°. The reaction is initiated by the addition of diluted phosphofructokinase, and the oxidation of NADH is monitored at 340 nm. Two moles of NADH are oxidized per mole of Fru-6-P converted to Fru-1,6-P_2.

Definition of Unit and Specific Activity. One unit of phosphofructokinase is that amount of enzyme required to convert 1 μmol of Fru-6-P to Fru-1,6-P_2 per min at 30°. Specific activity is defined as units per milligram of protein.

Protein Determination. Determination of protein concentrations are routinely carried out using the dye-binding method of Bradford[3] with bovine serum albumin as standard. Once the enzyme is purified, the extinction coefficient,[4] $E^{1\%}_{280\,nm} = 6.5$, can be used.

Purification Procedure

Source of Ascaris suum. This helminth resides in the upper small intestine of swine, and the female grows up to 30 cm in length with a weight of 10–12 g. The parasite can be obtained from most slaughterhouses that process swine. The initial contact with the slaughterhouse can be made through the local office of the United States Department of Agriculture (USDA), which can aid the investigator in contacting the USDA Inspector in the packing house, and through him the necessary permission and collecting permits can be obtained. In most cases, the USDA Inspectors are very familiar with the parasitic infestation and can provide valuable information on collection procedures. The number of worms obtained varies with the size of the slaughterhouse and number and type of animals processed, but several kilograms can usually be collected in a few hours. The parasites are transported to the laboratory in a saline solution[5] main-

---

[3] M. Bradford, *Anal. Biochem.* **72**, 248 (1976).

[4] J. A. Starling, B. L. Allen, M. R. Kaeini, D. M. Payne, H. J. Blytt, H. W. Hofer, and B. G. Harris, *J. Biol. Chem.* **257**, 3795 (1982).

[5] M. J. Donahue, N. J. Yacoub, M. R. Kaeini, R. A. Masaracchia, and B. G. Harris, *J. Parasitol.* **67**, 505 (1981).

tained at 37–38°, which consists of the following components: 111 mM NaCl, 10 mM NaHCO$_3$, 24 mM KCl, 1 mM CaCl$_2$, 5 mM MgSO$_4$, 14 mM NH$_4$Cl, 0.5 mM KH$_2$PO$_4$ (pH 7.0).

Although the muscle comprises 50–60% of the weight of the worm, over 90% of the phosphofructokinase in the parasite is in the muscle.[6] Therefore, whole worms are utilized for the purification procedure. In addition, the worms can be stored frozen at $-20°$ for at least a year without significant change in the yield of phosphofructokinase.

Purification. All purification procedures are carried out at 4°. Two hundred grams of frozen, whole, female *Ascaris suum* are homogenized in a Waring blender in 800 ml of a buffer containing 10 mM Tris-potassium phosphate,[7] pH 8.0, 1 mM EDTA, 10 mM NaF, and 40 mM 2-mercaptoethanol (buffer 1). Homogenization is carried out for three 1-min periods separated by 5-min cooling periods. The homogenate is centrifuged for 3 hr at 27,000 g, and the resultant supernatant solution is filtered through glass wool to remove lipid. The pH of the filtrate is carefully adjusted to 8.0 by the slow addition of cold 1 M Tris. This crude filtrate is designated fraction I.

Fraction I ($\sim$750 ml) is mixed with 200 g of damp-cake DEAE-cellulose (DE-52, equilibrated with buffer 1) and allowed to stir for 10 min. This mixture is filtered under reduced vacuum on a Büchner funnel 20 cm in diameter. A volume of 7–8 liters of cold buffer 1 is used to wash the ion exchanger over a period of 30–40 min. Washing is complete when the filtrate has an absorbance of less than 0.05 at 280 nm. Care is taken not to allow the cellulose to come to dryness during the process. The cellulose is washed in the same manner with 4 liters of buffer 1 containing 50 mM NH$_4$Cl (pH 8.0). When the absorbance of the filtrate is less than 0.05 at 280 nm, the DEAE-cellulose is packed into a column 5 cm in diameter under a hydrostatic pressure of 10–15 cm. After the cellulose has settled (height of $\sim$20 cm), the PFK is eluted from the column by the addition of buffer 1 plus 150 mM NH$_4$Cl (pH 8.0). The column is eluted at a rate of 5 ml/min maintained by a peristaltic pump, and the enzyme begins to appear in the first tubes containing protein. All the fractions containing phosphofructokinase activity ($\sim$500 ml) are collected and designated fraction II.

Fraction II is mixed with 30–40 g of damp-cake cellulose phosphate that has been equilibrated with buffer 1 plus 150 mM NH$_4$Cl (pH 8.0). The mixture is allowed to stir for 10 min before filtration on a Büchner funnel

---

[6] B. L. Allen and B. G. Harris, unpublished observations.
[7] The buffer is made up by combining all components (i.e., 10 mM Tris base, 10 mM KH$_2$PO$_4$, 1 mM EDTA, and 10 mM NaF, and then titrating to pH 8.0 by the addition of KOH. Mercaptoethanol is added to the buffer immediately before use.

TABLE I
PURIFICATION OF PHOSPHOFRUCTOKINASE FROM *Ascaris suum*

Step	Fraction number	Volume (ml)	Total units	Total protein (mg)	Specific activity (units/mg protein)	Yield (%)	Purification, overall (fold)
Crude supernatant	I	755	603	5119.0	0.12	100	—
DEAE-cellulose	II	600	588	708.0	0.8	97.5	7.0
Phosphocellulose	III	46.4	350	7.4	47.1	59.5	399.4
$(NH_4)_2SO_4$	IV	4.1	348	7.0	49.6	57.7	421.0

10 cm in diameter. The cellulose is washed slowly on the filter with 3 liters of buffer 1 plus 150 mM NH$_4$Cl. The phosphocellulose is washed with 2 liters of buffer 2, consisting of 50 mM Tris-potassium phosphate, pH 8.0, 1 mM EDTA, and 40 mM 2-mercaptoethanol. When washing is complete, the phosphocellulose is packed into a column 2.5 cm in diameter and allowed to settle under 10–15 cm hydrostatic pressure (~30 cm bed height). The enzyme is eluted from the phosphocellulose by application of buffer 2 plus 5 mM ATP (pH 8.0). Fractions of 10 ml are collected, and those containing enzyme activity are pooled (30–40 ml) and dialyzed overnight against a solution containing saturated ammonium sulfate, 10 mM Tris-potassium phosphate, pH 8.0, and 1 mM EDTA. The precipitated protein is sedimented by centrifugation at 20,000 g for 30 min, resuspended, and stored in a solution containing 50 mM Tris-potassium phosphate, pH 8.0, 1 mM EDTA, 1.4 M $(NH_4)_2SO_4$, 1 mM ATP, 0.2 M KCl, and 10 mM dithiothreitol (buffer 3).

At this stage the PFK is homogeneous, and Table I depicts a typical purification of the enzyme. From 200 g of worms, 7 mg of enzyme are obtained with a specific activity of 50 units/mg. The procedure results in a 420-fold purification with a yield of 58%. The enzyme is stable for at least 6 months when stored at 4° under N$_2$ in buffer 3. The enzyme can also be stored in saturated ammonium sulfate, but the specific activity decreases over a period of time. This is presumably due to sulfhydryl oxidation, since the addition of dithiothreitol (20 mM final concentration) results in a recovery of most of the original activity.

Properties of *A. suum* Phosphofructokinase

The purified enzyme exhibits a single band when subjected to two-dimensional electrophoresis,[4] i.e., gel isoelectric focusing in the first dimension and SDS gel electrophoresis in the second dimension. In coelec-

TABLE II
AMINO ACID ANALYSIS OF *Ascaris suum*
PHOSPHOFRUCTOKINASE

Residue	*Ascaris* PFK	Residue	*Ascaris* PFK
Asx	91.2[a]	Ile	32.1
Thr	45.7	Leu	52.4
Ser	65.5	Tyr	13.4
Glx	137.5	Phe	19.4
Gly	119.4	His	35.5
Ala	127.7	Trp	8.40
Val	53.5	Lys	45.6
Met	19.1	Arg	37.1

[a] Values are based on subunit molecular weight of 95,000.

trophoresis in SDS with rabbit muscle PFK, the ascarid enzyme migrates more slowly, indicating a subunit molecular weight of 95,000. Gel filtration studies of the native enzyme under various conditions using Sepharose 6B and high-pressure liquid chromatography (HPLC) reveals multiple states of aggregation of the enzyme with the slowest migrating peak corresponding to a molecular weight of 400,000. Thus, it is assumed that, as in mammalian systems, the enzyme exists as a tetramer.

The ascarid phosphofructokinase has an apparent isoelectric point of 7.3. The amino acid analysis of the ascarid enzyme is presented in Table II. In comparison with the rabbit muscle enzyme, the parasite phosphofructokinase has larger numbers of Glx, Gly, and Ala residues, but lower numbers of Phe, Tyr, and Trp residues. The enzyme, as purified here, contains 7–9 covalently bound phosphate groups per tetramer.[4] If fluoride is omitted from the purification buffers, a significantly higher yield ($\sim 75\%$ of the enzyme can be obtained, and the enzyme isolated has only 2 or 3 covalently bound phosphate residues per tetramer.[8] ATP inhibits catalytic activity at levels greater than 0.05 mM. This inhibition is more pronounced at pH 6.6, but occurs also at pH 8.0. The enzyme exhibits a sigmoid saturation curve with Fru-6-P as substrate both at pH 6.6 and pH 8.0 in the presence of 1 mM ATP. The ATP inhibition cannot be reversed by increasing the Fru-6-P concentration to 40 mM. However, addition of amounts of AMP equivalent to ATP concentration does overcome the inhibition. The enzyme activity is stimulated by both cations (NH$_4^+$ or K$^+$) and anions (sulfate or phosphate). In the presence of the stimulatory ions

[8] H. W. Hofer, B. L. Allen, M. R. Kaeini, and B. G. Harris, *J. Biol. Chem.* **257**, 3807 (1982).

and AMP at pH 8.0, the apparent values for ATP and Fru-6-P are 5.1 ± 1 μM and $66.0 \pm 8 \mu M$, respectively. Citrate (1 mM) does not significantly inhibit the enzyme; nor does 0.5 mM PEP. Glucose 1,6-bisphosphate stimulates the activity, but only in the presence of AMP.

[10] Phosphofructokinase from Yeast

By EBERHARD HOFMANN and GERHARD KOPPERSCHLÄGER

Fructose 6-phosphate + ATP → fructose 1,6-bisphosphate + ADP + H⁺

Phosphofructokinase (ATP: D-fructose-6-phosphate 1-phosphotransferase; EC 2.7.1.11) is an allosteric enzyme that catalyzes the transfer of the γ-phosphoryl group of MgATP to carbon atom one of $D(-)$-fructofuranose 6-phosphate and produces $D(-)$-fructofuranose 1,6-bisphosphate and MgADP. NH_4^+ or K^+, respectively, are essential for activity.[1] At pH 7.0 (25°) the thermodynamic equilibrium constant K of the overall reaction is 8.0×10^2, ΔG amounts to -3.96 kcal/mol,[2] and the activation energy to -10.2 kcal/mol.[3]

Assay Method

Principle. Phosphofructokinase is assayed in the coupled optical test using aldolase (EC 4.1.2.13), triosephosphate isomerase (EC 5.3.1.1), and glycerol-3-phosphate dehydrogenase (NAD⁺) (EC 1.1.1.8) as auxiliary enzymes. Fructose 1,6-bisphosphate formation is indicated by the NADH-dependent reduction of dihydroxyacetone phosphate to glycerol 3-phosphate. Each mole of fructose 1,6-bisphosphate causes the oxidation of 2 mol of NADH. To produce optimum conditions, the strong allosteric activator AMP is added to the assay mixture.

Reagents

   1. Assay mixture
      Imidazole/HCl, 100 mM, pH 7.2
      Fructose 6-phosphate, 3.0 mM
      ATP, 0.6 mM, pH 7.2
      MgSO₄, 5.0 mM

[1] For reviews, see E. Hofmann, *Rev. Physiol., Biochem. Pharmacol.* **75**, 1 (1976); K. Uyeda, *Adv. Enzymol.* **48**, 193 (1979).

[2] H. J. Böhme, W. Schellenberger, and E. Hofmann, *Acta Biol. Med. Ger.* **34**, 15 (1975).

[3] R. Freyer, M. Kubel, and E. Hofmann, *Eur. J. Biochem.* **17**, 378 (1970).

METHODS IN ENZYMOLOGY, VOL. 90

$(NH_4)_2SO_4$, 5.0 mM
AMP, 1.0 mM
NADH, 0.2 mM
2. Auxiliary enzymes
   Imidazole/HCl, 100 mM, pH 7.2
   Aldolase, 14 units/ml
   Triosephosphate isomerase, 136 units/ml
   Glycerol-3-phosphate dehydrogenase, 12 units/ml
These auxiliary enzymes, usually stored in 3.2 M $(NH_4)_2SO_4$, do not need to be dialyzed before use if phosphofructokinase is assayed in the course of purification. For the investigation of its kinetic properties, however, the auxiliary enzymes have to be dialyzed, and the concentrations of NH_4^+ and sulfate must be checked carefully because high concentrations of sulfate abolish, or at least weaken, the actions of allosteric effectors.
3. Buffer mixture for enzyme dilution
   Imidazole-HCl, 100 mM, pH 7.2
   Fructose 6-phosphate, 1 mM
   2-Mercaptoethanol, 5 mM
   Phenylmethylsulfonyl fluoride (PMSF), 0.5 mM

Procedure. The assay system is composd of 2.00 ml of solution 1 and of 0.05 ml of solution 2. The reaction is started by addition of 0.05 ml of the sample of phosphofructokinase. If necessary the enzyme is appropriately diluted immediately before assay with the ice-cold buffer mixture 3. The rate of NADH oxidation is measured at 340 nm and 25° against a water blank. After a brief lag period (less than 1 min) the change of absorbance proceeds linearly and is recorded for the following 2–3 min. Before use, the auxiliary enzymes have to be checked to determine that they contain no NADH-oxidizing contaminants.

Calculation of Activity. One unit of phosphofructokinase is defined as the amount of catalytic activity producing 1 μmol of fructose 1,6-bisphosphate per minute under the assay conditions. The specific activity is the number of enzyme units per milligram of protein. Protein is determined by applying a modified Lowry procedure[4] with bovine serum albumin as standard or, at later stages of purification, at 279 nm, respectively (see section "Absorption Coefficient"). The molar absorption coefficient (liter mol^{-1} cm^{-1}) for β-NADH at 340 nm is 6.317×10^3.

Purification Procedure

In the 1970s, a number of procedures for the purification of phosphofructokinase from yeast published.[5]

---

[4] H. H. Hess, M. J. Lees, and J. E. Derr, *Anal. Biochem.* **85**, 295 (1978).
[5] E. Stellwagen and H. Wilgus, this series, Vol. 42, p. 78.

Depending on the method applied, two different forms of the enzyme have been isolated, sedimenting either at 18 S or 21 S, respectively.[6] The 18 S enzyme was identified as a rather stable degradation product of the 21 S form originating in limited proteolysis taking place during the purification procedure.[7,8] For preparation of the 21 S form of phosphofructokinase, proteolysis has to be prevented by addition of PMSF as inhibitor of endogenous serine proteases and by acceleration of the isolation procedure.

The application of immobilized Cibacron Blue F3G-A as affinity adsorbent was found to be of great advantage for getting a proteolytically undegraded or only slightly modified enzyme (see Amino Acid Composition and End-Group Analysis under Properties).[6]

Starting from 5 kg of bakers' yeast, the time spent on the purification procedure described does not exceed 2–3 days of continuous work. The procedure was elaborated for bakers' yeast (*Saccharomyces cerevisiae*) as the source of the enzyme available from a yeast factory. The yeast should be in the penultimate state of cultivation before the final marketable product is manufactured. The yeast in this state of cultivation is rich in protein and technologically designated as pitching yeast (Stellhefe I). Commercially available pressed bakers' yeast can also be used as source of phosphofructokinase—however, with smaller yield.

A modification of the purification procedure of yeast phosphofructokinase of Diezel *et al.*[6] is described, which results in a homogeneous and sufficiently stable enzyme with a sedimentation constant of 20.8 S. A similar procedure has been published by Tamaki and Hess.[9]

Materials

Buffer A: 50 mM potassium phosphate, pH 7.2, containing 1 mM EDTA, 5 mM 2-mercaptoethanol, 0.5 mM PMSF
Buffer B: 10 mM potassium phosphate, pH 7.2, containing 1 mM EDTA and 5 mM 2-mercaptoethanol
Buffer C: buffer A, containing 1 mM fructose 6-phosphate

Coupling of Cibacron Blue F3G-A to Sephadex G-100. Sephadex G-100 (100 g) is suspended in 3500 ml of distilled water, and the temperature is adjusted to 60°. Under continuous stirring, 6.0 g of commercially available Cibacron Blue F3G-A (Ciba-Geigy, Basel, Switzerland), dissolved in 400 ml of distilled water, are added dropwise. After addition of 450 g of solid sodium chloride, the suspension is allowed to stir for 1 hr at 60°. Then the

[6] W. Diezel, H.-J. Böhme, K. Nissler, R. Freyer, W. Heilmann, G. Kopperschläger, and E. Hofmann, *Eur. J. Biochem.* **38**, 479 (1973).
[7] M. Taucher, G. Kopperschläger, and E. Hofmann, *Eur. J. Biochem.* **59**, 319 (1975).
[8] G. Kopperschläger, J. Bär, K. Nissler, and E. Hofmann, *Eur. J. Biochem.* **81**, 317 (1977).
[9] N. Tamaki and B. Hess, *Hoppe-Seyler's Z. Physiol. Chem.* **356**, 399 (1975).

temperature is adjusted to 80°, and 40 g of solid sodium carbonate are added; the suspension is allowed to stir for a further 2 hr at the same temperature. Then the gel is cooled, and the free Cibacron Blue F3G-A is removed by washing the gel several times with distilled water on a Büchner funnel. The substituted Sephadex G-100 has a substitution degree of 10 μg of dye per milligram of dried gel and is stable in the frozen state and in the presence of 2 M ammonium sulfate. The gel can be regenerated several times by washing with 0.01 M sodium hydroxide.

Cibacron Blue F3G-A-substituted Sepharose or agarose, as available commercially, can also be applied with success.

Procedure

Step 1. Disruption of the Yeast Cells and Extraction of Phosphofructokinase. Five kilograms of pitching yeast are washed twice with distilled water and suspended in buffer A to give a cell concentration of 60% (v/v). This and all subsequent steps are carried out at 4°. Cell disruption is achieved by ultrasonication. For this the yeast suspension is pumped once through the jacketed flowthrough vessel of an ultrasonic disintegrator at a frequency of 20 kHz, a sonication density of 50 W/cm², and a flow rate of about 2000 ml/hr.

The type 250 ultrasonic disintegrator of Schöller and Co. (Frankfurt/Main, West Germany) was found to fulfill the requirements. The ultrasonic vessel is permanently cooled by circulating ethylene glycol at −8°. This prevents a temperature increase of the yeast suspension in the ultrasonic vessel to more than 4°. An ammonium sulfate solution (3.5 M) is pumped into the plastic tube through which the disrupted cells leave the flowthrough ultrasonication vessel to give a final concentration of 1.17 M, determined by conductivity measurement. After ultrasonication, the suspension is centrifuged at 2000 g for 60 min to remove the insoluble constituents.

Step 2. First Ammonium Sulfate Precipitation. From the supernatant obtained in step 1 (8–10 liters) the phosphofructokinase is quantitatively precipitated by the addition of solid ammonium sulfate to give a concentration of 2.14 M. After 30 min the suspension is centrifuged at 5000 g for 60 min. The precipitate is suspended in buffer A to give a final volume of about 1000 ml. An opaque solution is obtained.

Step 3. Chromatography on Immobilized Cibacron Blue F3G-A. The phosphofructokinase is bound at low ionic strength to Cibacron Blue F3G-A substituted Sephadex G-100. For this, the enzyme solution obtained in the end of step 2 is diluted with buffer B until a conductivity of 6–7 mS is obtained (final volume about 20–30 liters). Then about 2 liters of the packed, substituted gel previously equilibrated with 10 liters of buffer

B, are added to the enzyme solution. After slow and continuous stirring of the suspension for 30 min, the suspension is allowed to stand unstirred for an additional 30 min. The gel is then separated from the medium by suction through a Büchner funnel and washed with buffer B until the absorbance of the washings at 280 nm is less than 0.05. Approximately 50 liters of buffer B are needed.

For elution of the enzyme, the gel is resuspended in 2.0 liters of buffer A containing 2 M ammonium sulfate in addition to the other constituents. The resulting suspension is stirred gently for 30 min. Then the gel is removed by filtration on a Büchner funnel and washed once with 0.5 liter of buffer A containing 1 M ammonium sulfate.

Step 4. Second Ammonium Sulfate Precipitation. From the filtrate obtained in step 3 (2.5 liters), the phosphofructokinase is quantitatively precipitated by addition of solid ammonium sulfate until a concentration of 2.53 M is reached. After standing for about 30 min, the suspension is centrifuged at 25,000 g for 30 min. The supernatant is discarded, and the precipitate is dissolved in a small volume of buffer C (final volume 30–50 ml). Remaining insoluble material is removed by centrifugation at 1000 g for 5 min and discarded.

An additional ammonium sulfate precipitation at 2.53 M is advantageous to reduce the volume of the enzyme solution to about 20 ml.

Step 5. Gel Filtration. The enzyme solution (20 ml) obtained in step 4 is applied to a Sepharose 6B column (5.0 × 100 cm) equilibrated with buffer C having an ammonium sulfate concentration of 1.17 M. Enzyme elution from the gel is performed with the same buffer mixture at a flow rate of 70 ml/hr. The eluate is collected in 15-ml fractions. After a small peak of blue polymeric material, the enzyme begins to be eluted. It is distributed in about 20 fractions. The active fractions are pooled and give a final volume of about 250–300 ml.

Step 6. Third Ammonium Sulfate Precipitation. From the pooled fractions of step 5, the phosphofructokinase is quantitatively precipitated by the addition of solid ammonium sulfate to yield a concentration of 3.11 M. After standing for about 1 hr, the suspension is centrifuged at 50,000 g for 30 min. The supernatant is discarded, and the precipitate is dissolved in a small volume of buffer C.

Step 7. Ion-Exchange Chromatography. The enzyme solution (25 ml) resulting from step 6 is three times dialyzed against 1 liter of buffer C and applied to a DEAE-cellulose column (2.0 × 20 cm) equilibrated with the same buffer. The elution of the enzyme is accomplished by making use of a linear potassium chloride gradient (0.05 to 0.3 M) prepared in 400 ml of buffer C. Phosphofructokinase begins to be eluted at about 150 mM potassium chloride. The phosphofructokinase-containing fractions are pooled (110 ml).

PURIFICATION OF YEAST PHOSPHOFRUCTOKINASE

Step	Volume (ml)	Total activity (units)	Total protein (mg)	Specific activity (units/mg protein)	Purification (fold)	Yield (%)
1. Enzyme extraction	11,000	40,000	133,000	0.3	1	100
2. First ammonium sulfate precipitation	1,500	35,000	77,800	0.45	1.5	88
3. Cibacron Blue F3G-A chromatography	2,500	17,500	2,030	8.6	29	44
4. Second ammonium sulfate precipitation	20	14,000	1,550	9.0	30	35
5. Gel filtration	250	10,000	260	38.4	128	25
6. Third ammonium sulfate precipitation	25	9,000	234	38.4	128	23
7. Ion-exchange chromatography	110	7,000	117	60	200	18
8. Fourth ammonium sulfate precipitation	10	6,200	103	60	200	16

Step 8. Fourth Ammonium Sulfate Precipitation. From the pooled fractions of step 7, phosphofructokinase is quantitatively precipitated by the addition of solid ammonium sulfate until a concentration of 3.11 M is reached. After standing for about 1 hr, the suspension is centrifuged at 50,000 g for 30 min. The supernatant is discarded, and the precipitated phosphofructokinase is suspended in the smallest possible volume of buffer C (pH 7.0) having an ammonium sulfate concentration of 3.11 M. The enzyme suspension is stored at 4°. The data obtained from a typical procedure are shown in the table.

Properties

Purity. The purification procedure yields a phosphofructokinase preparation with a specific activity of 60 units per milligram of protein. The purified enzyme sediments in the analytical ultracentrifuge as a single symmetric boundary and elutes from a Sepharose 4B-column as a single peak. In sodium dodecyl sulfate–polyacrylamide gel electrophoresis, the enzyme shows two bands originating from the two different types of subunits (see below); no further bands are detectable.[8] Other enzyme activities, including glycolytic enzymes, ATP-cleaving enzymes, NADH-oxidizing activities, and proteolytic contaminations, are negligibly low.

Stability. The specific activity, the electrophoretic mobility of the native enzyme, the subunit pattern, and the behavior in the analytical ultracentrifuge remain unchanged for several months after preparation when the enzyme is stored at 4° in the buffer mixture described above. At room temperature about 20% of activity is lost within 1 month. After dilution of the concentrated enzyme to 5 μg/ml in a 50 mM imidazole–HCl buffer, pH 7.0, containing 1 mM fructose 6-phosphate, 5 mM 2-mercaptoethanol, and 0.5 mM PMSF, the enzyme loses within 6 hr at 0° maximally 20% of its initial activity.

Sedimentation Constant and Relative Molecular Mass. At infinite dilution the sedimentation constant $s^0_{20,w}$ of yeast phosphofructokinase is 20.81 S. In the range from 0.1 to 10 mg of protein per milliliter, the concentration dependence of the sedimentation coefficient follows the linear equation $1/s = 1/s^0 (1 + 0.0152 c)$.[8] At very low phosphofructokinase concentrations (less than 10 μg of enzyme per milliliter) the s value decreases sharply as shown by active enzyme centrifugation. Dissociation of the enzyme in half-molecules under these conditions is most probable.[10]

The average partial specific volume calculated from the amino acid composition and determined by density measurements and sedimentation equilibrium analysis in H_2O and $H_2^{18}C$ amounts to 0.742 ml/g (corrected to 20° and water as solvent).[8]

Sedimentation equilibrium analysis yields a relative molecular mass M_r of native yeast phosphofructokinase of 835,000 ± 32,000.[8]

Absorption Coefficient. Yeast phosphofructokinase exhibits an absorption maximum at 279 nm and has a 280 : 260 ratio of absorbance of 1.79. The extinction coefficient $A^{1 mg/ml}_{279}$ is 0.881 ± 0.020; the molar extinction coefficient is 7.36×10^5 liter mol^{-1} cm^{-1}.[8]

Subunit Composition. When yeast phosphofructokinase is subjected to electrophoresis in the denatured state, two protein bands migrating closely together can be distinguished.[6-8] Two-dimensional immunoelectrophoresis demonstrated that these are not cross-reacting.[11] Hence, they represent apparently two nonidentical subunits. Partial proteolysis as a reason for the two-band pattern could be excluded.[7,8,11] The two subunits (designated α and β) present in 1 : 1 ratio[11] could be separated by ion-exchange chromatography on DEAE-cellulose.[12]

By means of pore gradient polyacrylamide gel electrophoresis in sodium dodecyl sulfate, an M_r value of 118,000 ± 3000 has been found for the α subunit, and of 112,000 ± 3000 for the β subunit.[8] Sedimentation

[10] T. Kriegel, Diploma-Thesis, Section of Biological Sciences, Karl-Marx University, Leipzig, 1979.

[11] K. Herrmann, W. Diezel, G. Kopperschläger, and E. Hofmann, *FEBS Lett.* **36**, 190 (1973).

[12] M. N. Tijane, A. F. Chaffotte, F. J. Seydoux, C. Roucous, and M. Laurent, *J. Biol. Chem.* **255**, 10188 (1980).

equilibrium analysis of the sodium dodecyl sulfate complexes of yeast phosphofructokinase yielded an average relative molecular mass of 104,000 referred to the detergent-free subunit.[8] A comparison of the relative molecular mass of the native enzyme with that of the subunits[8,9] and the results of cross-linking experiments[13] make an octameric structure of native yeast phosphofructokinase most likely.

Amino Acid Composition and End-Group Analysis. The amino acid analysis of the octamer[8] shows that it contains 83 ± 2 cysteinyl residues.[8,12] No disulfide bridges could be identified in the native molecule.

The N-terminal amino acids in both subunits appear to be covalently blocked. The C termini of the α and β subunits were found to be identical. In both types of subunit the C-terminal sequence is —Val—Ala—Lys—Tyr—Glu—Thr—Leu—Arg—Ile. Occasionally amino acid differences at the C terminus can be observed. Because they are restricted to amino acids within the C-terminal sequence, they result apparently from the action of a carboxypeptidase in the course of the purification procedure.[14]

Subunit Arrangement. The results of cross-linking and small-angle X-ray scattering studies yield a good fit with a computed model according to which the enzyme consists of two tetramers possessing three twofold axes of symmetry. Evidently, the enzyme structure follows dihedral point group symmetry D_2. Either the four α subunits are located in the center of the octameric molecule and the four β subunits lie peripherally, or vice versa. In consequence the enzyme has an $\alpha_2\beta_4\alpha_2$ or a $\beta_2\alpha_4\beta_2$ subunit arrangement.[13,15]

Dependence on pH and Influences of Inorganic Ions. The kinetic properties of yeast phosphofructokinase depend on the pH level in a rather complicated manner. At decreasing pH the inhibitory action of ATP is weakened while the apparent affinity of the enzyme to fructose 6-phosphate is increased. The sigmoidal characteristics of the fructose 6-phosphate velocity curve as expressed by the Hill coefficient is not changed upon variation of the pH value. The pH optimum is around pH 7.0; it is dependent on the ATP and fructose 6-phosphate concentration.[16]

Ammonium (or potassium) and magnesium ions are essential for activity. NH_4^+ ions increase the maximum activity of the enzyme and its apparent affinity to fructose 6-phosphate without exhibiting a significant effect on the sigmoidal nature of the fructose 6-phosphate velocity curve.

[13] G. Kopperschläger, E. Usbeck, and E. Hofmann, *Biochem. Biophys. Res. Commun.* **71**, 371 (1976).
[14] K. Huse, Ph.D. Thesis, Science Faculty, Karl-Marx University, Leipzig (1981).
[15] P. Plietz, G. Damaschun, G. Kopperschläger, and J. J. Müller, *FEBS Lett.* **91**, 230 (1978).
[16] R. Reuter, K. Eschrich, W. Schellenberger, and E. Hofmann, *Acta Biol. Med. Ger.* **38**, 1067 (1979).

The inhibitory action of ATP is not markedly influenced by ammonium ions. MgATP is the substrate and can also act as an inhibitor of yeast phosphofructokinase.[17] Unchelated ATP has the same inhibitory power as MgATP.[16] The enzyme requires free Mg^{2+} for its activity.[17]

One subunit of yeast phosphofructokinase possesses three independent binding sites for divalent cations as established by means of equilibrium dialysis experiments with Mn^{2+}. Their dissociation constant for binding of this cation is 2.26 mM.[18] Two of the sites can apparently be coordinated to the MgATP binding sites, whereas the third represents possibly the binding site for free Mg^{2+}.

Kinetic Properties. The enzyme exhibits a complex kinetic behavior. Fructose 6-phosphate exerts cooperative kinetics, whereas the action of ATP is biphasic. At low concentrations the enzyme responds to MgATP according to the Michaelis–Menten equation; at high concentrations, however, the enzyme is inhibited either by chelated or unchelated ATP. AMP is a strong allosteric activator, whereas ADP is a weak one. Both nucleotides counteract the inhibition by ATP. Also fructose 6-phosphate is capable of relieving the ATP inhibition. Unlike the muscle phosphofructokinase, the enzyme from yeast is insensitive to fructose 1,6-bisphosphate and citrate as allosteric effectors.[16]

Below pH 7 inorganic phosphate, synergistically with AMP, was found to exhibit a strong activation. Apparently this synergism is correlated with the diminution of the ATP inhibition by AMP and the increase of the enzyme affinity to fructose 6-phosphate by phosphate.[19] Most recently, a strong positive allosteric action of fructose 2,6-bisphosphate on the yeast phosphofructokinase was discovered.[20]

In contrast to other phosphofructokinases,[1] the sigmoidal nature of the activity curve for fructose 6-phosphate as expressed by the Hill coefficient is changed neither by the inhibitor ATP nor by the activators AMP or ADP, although the curve is shifted to the right or to the left, respectively, by both types of effectors. Hence, the strength of cooperative interactions between the fructose 6-phosphate binding sites appears to be independent on ATP and AMP or ADP.

A systematic investigation of the kinetic properties of the yeast enzyme led to the development of a structure-oriented allosteric four-state model capable of describing satisfactorily the substrate and effector ac-

[17] R. D. Mavis and E. Stellwagen, *J. Biol. Chem.* **245**, 674 (1970).
[18] W.-H. Peters, K. Nissler, W. Schellenberger, and E. Hofmann, *Biochem. Biophys. Res. Commun.* **90**, 561 (1979).
[19] M. Banuelos, C. Gancedo, and J. M. Gancedo, *J. Biol. Chem.* **252**, 6394 (1977).
[20] H. G. Hers, personal information of hitherto unpublished results with homogeneous phosphofructokinase obtained by this procedure (1981).

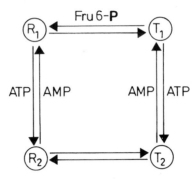

FIG. 1. The four-state model of yeast phosphofructokinase.

tions on the enzyme[16] (Fig. 1). According to this model, the enzyme follows a rapid equilibrium random-order mechanism and has the following rate equation:

$$v_{PFK} = V_{PFK} \; \frac{[ATP]}{([ATP] + K_{ATP})} \; \frac{[F6P]}{([F6P] + K_{F6P})} \; \frac{1}{1 + L}$$

$$K_{F6P} = K_{F6P,S} \; \frac{(1 + [ATP]/K_{ATP,I} + [ADP]/K_{ADP,2} + [AMP]/K_{AMP,2})}{(1 + [ADP]/K_{ADP,1} + [AMP]/K_{AMP,1})}$$

$$K_{ATP} = K_{ATP,S}(1 + [ADP]/K_{ADP,C})$$

$$L = L_0/(1 + [F6P]/K_{F6P})^8$$

At pH 6.8 and in the presence of 20 mM inorganic phosphate, the numerical values of the dissociation constants are as follows.

$K_{F6P,S} = 2 \times 10^{-4} \; M$; $K_{ATP,S} = 3.1 \times 10^{-5} \; M$
$K_{AMP,1} = 1.5 \times 10^{-5} \; M$; $K_{ATP,I} = 2.5 \times 10^{-4} \; M$; $K_{AMP,2} = 2.5 \times 10^{-5} \; M$
$K_{ADP,1} = 2 \times 10^{-4} \; M$; $K_{ADP,2} = 2.5 \times 10^{-4} \; M$
$K_{ADP,C} = 2.5 \times 10^{-4} \; M$; $L_0 = 14.75$

Two principal conformations, R and T, are assumed, the interconversion of which proceeds in a concerted manner. The conformations R and T split into the subconformations R_1/R_2 and T_1/T_2. Fructose 6-phosphate is assumed to bind only to R_1 and T_1, the two states exhibiting different affinities to this substrate but identical maximal activities.

$K_{F6P,S}$ represents the dissociation constant of the R state for fructose 6-phosphate at a distribution of the conformations of R_1 and T_1 existing in the absence of any ligand. ATP is assumed to bind as substrate to all four states with equal affinities by a Michaelis–Menten function; the dissociation constant for this binding is $K_{ATP,S}$. As a negative effector, ATP is bound only to the R_2 and T_2 states ($K_{ATP,I}$). Reciprocally, AMP and ADP possess higher affinities to R_1 and T_1 ($K_{AMP,1}$ and $K_{ADP,1}$) than to R_2 and T_2 ($K_{AMP,2}$ and $K_{ADP,2}$). Hence, their activating actions arise from influence on the R_1/R_2 and T_1/T_2 equilibria in a direction opposite to that caused by ATP as inhibitor. These effectors do not influence the allosteric R/T-equilibrium. In addition, ADP competes with ATP for its substrate binding site ($K_{ADP,C}$). L_0 is the allosteric constant characterizing the ratio of the main conformations R and T in the absence of ligands.

According to the model, the allosteric equilibrium characterized by L is affected only by fructose 6-phosphate, whereas the allosteric actions of ATP, ADP, and AMP are confined to influencing the equilibria between the two pairs of subconformations.

Equilibrium binding studies with substrates and effectors provide strong support for the adequacy of this model. In the absence of ATP, fructose 6-phosphate is cooperatively bound to the octameric molecule.[21] Each subunit was found to have one binding site for fructose 6-phosphate (the dissociation constant for R_1 is equal to $8.4 \times 10^{-5} M$, for T_1, it is equal to $4.25 \times 10^{-4} M$) and two binding sites for ATP (their dissociation constants are $5.2 \times 10^{-6} M$ and $5.1 \times 10^{-5} M$). One of the ATP binding sites can apparently be correlated to the active (substrate) site, the other one to the inhibitory (allosteric) site.[22] AMP is capable of displacing ATP from its high-affinity binding site.[23] The respective dissociation constant for AMP ($1.6 \times 10^{-5} M$) is in good agreement with that resulting from the kinetic model. A difference in the number of binding sites estimated for ATP and fructose 6-phosphate by Laurent et al.[23,24] and the number estimated by Nissler et al.[21,22] remains to be clarified.

General Aspects. Owing to its allosteric properties, phosphofructokinase is the dominant rate-controlling enzyme of glucose degradation via the Embden–Meyerhof pathway.[1] In cooperation with other glycolytic enzymes, phosphofructokinase is involved in the generation of glycolytic oscillations as observable in intact yeast cells and yeast cell extracts.[25] A reconstituted *in vitro* enzyme system containing yeast phosphofructokinase, pyruvate kinase, adenylate kinase, and glucose-6-phosphate isomerase, which was designed to operate under open conditions far from the thermodynamic equilibrium, has been shown to exhibit nonlinear

[21] K. Nissler, R. Kessler, W. Schellenberger, and E. Hofmann, *Biochem. Biophys. Res. Commun.* **79**, 973 (1977).

[22] K. Nissler, W. Schellenberger, and E. Hofmann, *Acta Biol. Med. Ger.* **36**, 1027 (1977).

[23] M. Laurent, A. F. Chaffotte, J.-P. Tenu, C. Roucous, and F. J. Seydoux, *Biochem. Biophys. Res. Commun.* **80**, 646 (1978).

[24] M. Laurent, F. J. Seydoux, and P. Dessen, *J. Biol. Chem.* **254**, 7515 (1979).

[25] A. Boiteux and B. Hess, *in* "Frontiers of Biological Energetics" (P. L. Dutton, J. S. Leigh, and A. Scarpa, eds.), Vol. 1, p. 789. Academic Press, New York, 1978.

kinetic responses toward variations of the substrate influx rates. The system is capable of functional self-organization and of exerting multiple steady states, jump and trigger behavior, self-stabilization of ATP and of the energy charge, as well as the property of hysteresis. It may serve as a simple model system for the investigation of the complexity of metabolic regulation.[26]

[26] W. Schellenberger, K. Eschrich, and E. Hofmann, *Adv. Enzyme Regul.* **19**, 257 (1981).

[11] Phosphofructokinases from *Escherichia coli*

By DENISE KOTLARZ and HENRI BUC

$$\text{ATP} + \text{fructose 6-phosphate} \xrightarrow{\text{Mg}^{2+}} \text{ADP} + \text{fructose 1,6-bisphosphate} + \text{H}^+$$

Two phosphofructokinases (EC 2.7.1.11) have been described in *E. coli* K12. The most studied one,[1-4] PFK1, is coded by the *pfk*A gene located at 87 min on the *E. coli* map.[5] The other enzyme, PFK2, is specified by the *pfk*B gene mapping at 38 min on the *E. coli* genome.[6] The two enzymes differ in their biochemical and catalytic properties.[7-9]

Enzymic Assay

Principle. Phosphofructokinase activity is routinely assayed by following the formation of fructose 1,6-bisphosphate. This product is converted to α-glycerophosphate by aldolase (EC 4.1.2.13), triosephosphate isomerase (EC 5.3.1.1), and glycerol-3-phosphate dehydrogenase (EC 1.1.1.8) in the presence of β-NADH.[10] It is not convenient to follow the formation of ADP by a coupled reaction involving phosphoenolpyruvate, since the latter is an inhibitor of PFK1. A pH-stat assay has also been described.[11]

[1] D. E. Atkinson and G. M. Walton, *J. Biol. Chem.* **240**, 757 (1965).
[2] D. Blangy, H. Buc, and J. Monod, *J. Mol. Biol.* **31**, 13 (1968).
[3] D. Blangy, *FEBS Lett.* **2**, 109 (1968).
[4] D. Blangy, *Biochimie* **53**, 135 (1971).
[5] A. T. E. Morrissey and D. G. Fraenkel, *J. Bacteriol.* **100**, 1108 (1969).
[6] A. T. E. Morrissey and D. G. Fraenkel, *J. Bacteriol.* **112**, 183 (1972).
[7] D. G. Fraenkel, D. Kotlarz, and H. Buc, *J. Biol. Chem.* **248**, 4865 (1973).
[8] D. Kotlarz and H. Buc, *Biochim. Biophys. Acta* **484**, 35 (1977).
[9] J. Babul, *J. Biol. Chem.* **253**, 4350 (1978).
[10] E. Racker, *J. Biol. Chem.* **167**, 843 (1947).
[11] A. Ku and C. C. Griffin, *Arch. Biochem. Biophys.* **149**, 361 (1968).

METHODS IN ENZYMOLOGY, VOL. 90

TABLE I
ASSAY MEDIA FOR THE MEASUREMENT OF THE ENZYMIC ACTIVITIES
ASSOCIATED WITH E. coli PFK1 AND PFK2

Stock solutions	Assay medium		
	1	2	3
Tris-HCl, pH 8.2, 100 mM	9.2 ml	9.18 ml	9.18 ml
MgCl$_2$, 1 M	100 μl	100 μl	100 μl
NADH, 20 mM	100 μl	100 μl	100 μl
NH$_4$Cl, 1 M	100 μl	100 μl	100 μl
Triton X-100, 1%	100 μl	100 μl	100 μl
Fru-6-P, 100 mM	100 μl	20 μl	20 μl
Phosphoenolpyruvate, 200 mM	0	100 μl	0
GDP, 100 mM	0	0	100 μl

Reagents and Stock Solutions. The following stock solutions are prepared in Tris-HCl, 100 mM, pH 8.2 (the same buffer is also used as a diluent in all the assays):

MgCl$_2$, 1 M

NADH, 20 mM

NH$_4$Cl, 1 M (may be omitted for PFK2 assay)

Triton X-100, 1% (may be omitted for PFK1 assay)

Fructose 6-phosphate (Fru-6-P), 100 mM

Phosphoenolpyruvate (P-enolpyruvate), 200 mM

ATP, 100 mM

GDP, 100 mM

Phosphocreatine, 100 mM

Auxiliary enzymes: Aldolase (10 mg/ml) and a mixture of triosephosphate isomerase (1 mg/ml) and glycerol-3-phosphate dehydrogenase (10 mg/ml) are available as ammonium sulfate suspensions. They are mixed in the ratio 0.7 : 0.3 ml and extensively dialyzed against Tris-HCl, 100 mM, pH 8.2. This solution, adjusted to 0.01% Triton X-100, can be kept at 4° for at least 2 weeks.

Creatine kinase, 1 mg/ml; solution made up daily in Tris-HCl, 100 mM, pH 8.2

In Table I, the composition of the three basic assay media is given. These solutions can be kept frozen at −20° for 2 weeks.

Total Activity. Prewarmed assay medium 1 (970 μl) is pipetted into a 1-cm pathlength cuvette; 10 μl of the auxiliary enzymes mixture and 10 μl of PFK solution (20–200 ng/ml, final concentration) are added. The decrease in absorbance at 340 nm is followed in a spectrophotometer thermostatted at 28°. Any oxidation of NADH, due to contamination by mannitol-1-phosphate dehydrogenase or NADH oxidases, is first recorded for 2–3 min. The reaction catalyzed by PFK is then initiated by adding 10 μl of 100 mM ATP. The difference in the initial NADH oxidation rates observed in the presence and in the absence of ATP represents PFK activity.

Discrimination between PFK1 and PFK2 Activities. This can be performed at lower Fru-6-P concentrations (0.2 mM) by adjusting the concentration of PFK1 effectors.[12] In the absence of GDP and at high phosphoenolpyruvate concentrations (2 mM) PFK1 is inactive. These conditions, fulfilled in assay medium 2, give PFK2 activity. By replacing phosphoenolpyruvate by 1 mM GDP (assay medium 3), PFK1 and PFK2 are fully active.

To 970 μl of assay medium 2 or 3 are added 10 μl of auxiliary enzyme mixture, 10 μl of PFK solutions, and 10 μl of 100 mM ATP.

The difference between the activity measured in assay medium 3 and the activity measured in assay medium 2 represents PFK1 activity. In the case of crude preparations of PFK2, it is necessary to run controls by omitting ATP.

Regenerating System. At suboptimal substrate concentrations, accurate initial velocity measurements require that the ATP concentration be maintained constant. Furthermore, as one of the products of the reaction, ADP, is an activator of PFK1, its amount has to be maintained as low as possible during the reaction. These conditions are realized by supplementing the assay media with 1 mM phosphocreatine and 10 μg of creatine kinase per milliliter (final concentrations), so that the recycling reaction

$$\text{Phosphocreatine} + \text{ADP} \rightarrow \text{creatine} + \text{ATP}$$

can take place efficiently.

Units. One unit of phosphofructokinase activity is defined as the amount that catalyzes the formation of 1 μmol of fructose 1,6-bisphosphate per minute. As 1 mol of fructose 1,6-bisphosphate produced corresponds to 2 mol of oxidized NADH in the assay system, phosphofructokinase units are obtained by dividing the initial decrease in absorbance per minute by 12.4.

Strains and Growth Conditions

Any *E. coli* K10 or *E. coli* K12 strain carrying the wild-type genotype *pfk*A$^+$, *pfk*B$^+$ may be used for the preparation of PFK1; for example, strains Hfr3000 or DF1651.[6] In these strains, the expression of PFK2 is severely repressed. Thus, for the purification of this latter enzyme, it is

---

[12] D. Kotlarz, H. Garreau, and H. Buc, *Biochim. Biophys. Acta* **381**, 257 (1975).

more convenient to use strains carrying the mutation *pfk* B1, which dere-presses PFK2 activity: DF1651B1 (*pfk* A⁺, *pfk* B1) or AM1R20 (*pfk* A⁻, *pfk* B1).[6] PFK2 has also been purified from strains carrying a deletion of the *pfk* A locus: DF500 (Δ*pfk* A, *pfk* B⁺) and DF443 (Δ*pfk* A, *pfk* B1).[9] Strains labeled AM or DF can be obtained from D. Fraenkel.

Cells are grown aerobically at 37° until the end of the exponential phase in minimal medium 63[13] supplemented with Difco Bactotryptone (10 g/liter) yeast extract (4 g/liter), and glucose (4 g/liter). This broth is sup-plemented with 50 μg of uracil per milliliter in the case of strains DF1651, DF1651B1, DF500, and DF443, which are *pyr* D⁻; and with 0.2 g of glycerol per liter for strains DF500 or DF443, which are *tpi*⁻.

Cells are spun down by centrifugation at 6000 *g* for 60 min. The bacte-rial pellet can be kept frozen at −20° for months.

Purification of PFK1

Step 1. Crude Extract. Frozen bacteria (100 g) are ground in a mortar in the cold room with 200 g of alumina (type 305 from Sigma Chemical Corp. St. Louis, Missouri) until the paste is homogeneous. The paste is sus-pended in 400 ml of buffer A (Tris acetate, 50 m*M*, pH 7.5; MgCl₂, 5 m*M*; EDTA, 0.1 m*M*; 2-mercaptoethanol, 7 m*M*) and centrifuged for 60 min at 10,000 *g*. The pellet is discarded, and the supernatant is treated with 1 mg of calf thymus deoxyribonuclease I for 10 min at 37°.

Step 2. Affinity Chromatography. Blue Dextran–Sepharose (30 ml)[14] is prepared according to Viktorova *et al.*[15] and poured into a 2.8 × 5.0 cm column. This column is equilibrated at room temperature with buffer A. The supernatant of the first step is applied to the column, which is then rinsed with buffer A until the absorbance of the eluate at 280 nm is lower than 0.020 (flow rate 50 ml/hr). PFK1 enzyme is eluted stepwise from the dye by washing the column with a 0.5 m*M* ATP solution in buffer A (same flow rate); 3-ml fractions are collected.

Step 3. Concentration. The fractions containing PFK1 are absorbed at room temperature on a 3-ml DEAE-cellulose column, previously equili-brated with buffer A. PFK1 is eluted from this column by applying a step of 0.3 *M* NaCl in buffer A. (ATP as well as a small amount of dye coming from the previous chromatography are retained on DEAE-cellulose.) Fractions containing more than 2 mg of protein per milliliter are pooled.

Step 4. Heat Denaturation of Protein Contaminants. The preparation is made 2 m*M* in Fru-6-P and is then heated for 10 min in a water bath at 65°. The denatured contaminants are spun down at 12,000 *g* for 30 min. The

[13] G. N. Cohen and H. V. Rickenberg, *Ann. Inst. Pasteur, Paris* **91**, 693 (1956).
[14] L. D. Ryan and C. S. Vestling, *Arch. Biochem. Biophys.* **160**, 279 (1974).
[15] L. N. Viktorova, B. A. Klyashchitskii, and E. V. Ramensky, *FEBS Lett.* **91**, 194 (1978).

TABLE II
PURIFICATION OF PHOSPHOFRUCTOKINASE 1 FROM 100 g OF *E. coli* DF1651

Step	Total protein (mg)	Total activity (IU)	Specific activity (IU/mg protein)	Yield (%)
Crude extract	8520	2300	0.27	100
Blue Dextran chromatography	10.1	1700	168	74
Heat denaturation	8.5	1610	190	70

supernatant contains pure PFK1, as judged by bidimensional gel electrophoresis.[16]

A typical purification of PFK1 is summarized in Table II. This pure preparation can be kept at 4° as an ammonium sulfate suspension (65% saturation), which is stable for at least 2 months.

Properties of PFK1

Escherichia coli phosphofructokinase 1 is a tetrameric enzyme[3] made up of four identical polypeptide chains as judged by genetic[5] and chemical evidence.[17] The shape of the enzyme in solution has recently been investigated by small-angle X-ray scattering and electron microscopy.[18,19] A tetrahedral arrangement of the subunits has been proposed. Contrary to muscle phosphofructokinases, no association–dissociation behavior has been detected.

The enzymic properties have been described according to the Monod–Wyman–Changeux model implying a concerted transition between a T_4 and a R_4 state differing in their affinity for the substrate Fru-6-P (K system), for nucleoside diphosphate activators and for the inhibitor phosphoenolpyruvate.[2] Equilibrium binding studies indicate the presence of four sites per tetramer for ATP, ADP, and GDP.[8,17] From the observation that neither fructose 6-phosphate nor fructose 1,6-bisphosphate bind to the enzyme in the absence of the other substrate or product, it has been inferred that binding is ordered and sequential.[4]

Other relevant data on the physical, biochemical, and catalytic properties of PFK1 have been adequately reviewed by Kemerer *et al.*,[20] and some are summarized in Table III, which allows a comparison between *E. coli* PFK1 and PFK2.

[16] P. H. O'Farrell, *J. Biol. Chem.* **250**, 4007 (1975).
[17] B. N. Thornburgh, L. L. Wu, and C. C. Griffin, *Can. J. Biochem.* **56**, 836 (1978).
[18] H. H. Paradies, W. Vettermann, and G. Werz, *Protoplasma* **92**, 43 (1977).
[19] A. R. Goldhammer and H. H. Paradies, *Curr. Top. Cell. Regul.* **15**, 109 (1979).
[20] V. F. Kemerer, C. C. Griffin, and L. Brand, this series, Vol. 42, p. 91.

TABLE III

COMPARISON BETWEEN SOME PHYSICOCHEMICAL AND BIOCHEMICAL PROPERTIES OF *Escherichia coli* PFK1 AND PFK2

	PFK1[a]	Reference[b]	PFK2[a]	Reference[b]
Molecular weight	$35,000 \times 4$	3	$37,000 \times 2^c$; $37,000 \times 4^d$	8, 9
$s_{20,w}$	7.5 S	3	4.4 S^c; 7.4 S^d	22
Extinction coefficient, ϵ (278 nm)	0.60 ± 0.03 liter g^{-1} cm^{-1}	8	0.4 liter g^{-1} cm^{-1}	22
Fru-6-P	$K(R) = 12.5\ \mu M$	2	$K_M = 11\ \mu M^f$	22
	$K(T) = 25\ mM^e$			
	$K_D > 100\ \mu M$	2	$K_D = 4\ \mu M$; $\nu = 2$	22
Fru-1,6-P_2	$K_I > 10\ mM$	9	$K_I = 6\ mM^g$	9
			$K_D = 4\ \mu M$; $\nu = 1$	22
ATP	$K(R) = K(T) = 60\ \mu M$	2	$K_M = 50\ \mu M^h$	8, 9
	$K_D = 6\ \mu M$; $\nu = 4$	8	$K_D > 0.4\ mM$	
ADP	$K(R) = 25\ \mu M$	2	Poor inhibitor	22
	$K(T) = 1.3\ mM$			
	$K_D = 18\ \mu M$; $\nu = 4$	17		
GDP	$K(R) = 40\ \mu M$	2	No significant effect	7
	$K(T) > 40\ mM$			
	$K_D = 8\ \mu M$; $\nu = 4$	8		
PEP	$K(R) > 750\ mM$	2	No significant effect	7
	$K(T) = 750\ \mu M$			
NH_4^+	$K(1/2) \sim 0.4\ mM$	24	No significant effect	7

[a] Equilibrium dialysis measurements: ν refers to the number of sites per tetramer (for PFK1) or per dimer (for PFK2); K_D is the dissociation constant measured at half-saturation. Kinetic measurements: For noncooperative responses, K_m is the Michaelis constant (measured at a fixed concentration of the alternative substrate given in the footnote). Cooperative responses are analyzed according to Blangy *et al.*[2] in terms of the binding of the activator or of the inhibitor on the R and T states, $K(R)$ and $K(T)$ being the corresponding dissociation (or Michaelis) constants.
[b] Numbers refer to text footnotes.
[c] In the absence of added effector.
[d] In the presence of 1 mM ATP, 5 mM Mg^{2+}.
[e] Measured in the presence of 0.1 mM ATP.
[f] In the presence of 1 mM ATP.
[g] Measured in the presence of 0.3, 0.4, and 1 mM Fru-6-P.
[h] Measured in the presence of 0.2 mM Fru-6-P.

Purification of PFK2

All the operations are performed in the cold room except when otherwise stated.

Step 1. Crude Extract. The first step is performed as described for PFK1 except that the slurry from 100 g of wet bacteria is suspended in 400 ml of buffer B (Tris acetate, 50 mM, pH 7.5; MgCl$_2$, 5 mM; EDTA, 1 mM; 2-mercaptoethanol, 7 mM; glycerol, 10% w/v).

Step 2. First Blue Dextran–Sepharose Chromatography. The crude extract is passed at a flow rate of 50–80 ml/hr on a 50-ml column of Blue Dextran–Sepharose equilibrated with buffer B. The fractions, containing phosphofructokinase 2 activity, are pooled.

Step 3. Hydroxyapatite Chromatography. BioGel HTP (150 ml) from Bio-Rad are poured into a 2.8 × 25 cm column and equilibrated with buffer B. The preparation is adsorbed on it and washed with buffer C (same as buffer B except that Tris acetate is replaced by 20 mM sodium orthophosphate pH 7.5). A linear gradient from 20 mM to 150 mM sodium orthophosphate is then applied to the column (total volume, 1 liter; flow rate, 30–50 ml/hr). A single peak of PFK2 activity is eluted between 100 and 120 mM orthophosphate. The corresponding 5-ml fractions are pooled.

Step 4. Ammonium Sulfate Fractionation. Ammonium sulfate crystals are gradually stirred into the enzyme solution until the salt concentration reaches 300 g per liter of original solution. After stirring for an additional 15 min, the precipitate is removed by centrifugation at 10,000 g for 30 min and discarded. The volume of the supernatant is measured, more ammonium sulfate crystals are gradually added (130 g per liter of supernatant), and the new suspension is centrifuged as above. The pellet is dissolved in 5–10 ml of buffer B and dialyzed for 4 hr against 2 liters of this buffer.

Step 5. Second Blue Dextran–Sepharose Chromatography. The dialyzed solution is absorbed on a second Blue Dextran–Sepharose column (10 ml) as described in step 2. The column is then washed with buffer B (flow rate 50–80 ml/hr) until the absorbance of the eluate at 280 nm is less than 0.02. PFK2 enzyme is then eluted stepwise from the dye by washing the column with 0.3 mM ATP in buffer B (same flow rate). The 2-ml fractions containing PFK2 enzymic activity are pooled.

Step 6. Concentration and Heat Treatment. The final steps for PFK2 purification are the same as for PFK1 (cf. steps 3 and 4 of the PFK1 preparation); i.e., the protein concentration is adjusted to at least 2 mg/ml and the solution is maintained at 65° for 10 min in the presence of 2 mM Fru-6-P.

PFK2 preparations are kept at 4° in the presence of 10% glycerol in order to reduce limited proteolysis and nonspecific adsorption to glass or plastic vessels.

Comments

1. Unlike PFK1, PFK2 in crude extracts does not bind to Blue Dextran–Sepharose (step 2). After this step, no PFK1 activity should

TABLE IV
PURIFICATION OF PFK2 FROM 100 g OF E. coli AM1R20[a]

Step	Total protein (mg)	Total PFK2 activity (IU)	Specific activity (IU/mg protein)	Yield (%)
Crude extract	7270	2600	0.36	100
Exclusion from Sepharose–Blue Dextran I	7070	2470	0.35	95
Hydroxyapatite	323	2170	6.7	84
$(NH_4)_2SO_4$ fractionation	131	2100	16	81
Blue Dextran II	9.3	1800	192	69
Heat denaturation	8.5	1735	205	67

[a] From a wild-type strain, the preparation yields an enzyme of similar specific activity, but the initial PFK2 activity is reduced by a factor of 20.

remain in the PFK2 preparation. When strains expressing mutated PFK1 are used, the mutated enzyme, characterized as cross-reacting material, as well as active PFK1, remains bound on the first Blue Dextran–Sepharose column.

2. When strains expressing active PFK1 are used for PFK2 preparation (e.g., DF1651B1), PFK2 activity is assayed using media 2 and 3 (see Table I).

3. The heat treatment does not affect the biochemical properties of PFK1 or PFK2.

A typical purification of PFK2 is summarized in Table IV.

Properties of PFK2

Substrate Specificity and Catalytic Mechanism. ATP is the best phosphoryl donor for both PFK1 and PFK2, and furthermore the same specificity is found for various nucleotide triphosphates, as judged from the order of their Michaelis constants: ATP = dATP < ITP < GTP < UTP < CTP.

Fructose 6-phosphate is the best phosphoryl acceptor. Fructose 1-phosphate and glucose 6-phosphate are not substrates for either PFK1 or PFK2. Tagatose 6-phosphate is a poor but real substrate for PFK2 but not for PFK1. However, the corresponding Michaelis constant is so high that it excludes the involvement of PFK2 in the metabolism of tagatose.[9]

Mg^{2+} or Mn^{2+} are the most efficient divalent cations acting as cofactors for the catalytic reaction. As for many other ATP-dependent phosphotransferases, the divalent cation–ATP complex is the true enzyme substrate.[21]

The turnover number of the enzymic reaction has an unusual dependence on pH.[8] Two maxima are observed—at pH 6.5 and 8.5, respectively.[8,21]

Equilibrium dialysis measurements show that PFK2 binds fructose 6-phosphate or fructose 1,6-bisphosphate in the absence of ATP. The converse is not true. While the observed binding stoichiometry for fructose 6-phosphate corresponds to one molecule of substrate per subunit, it is 0.5 for fructose 1,6-bisphosphate. From these observations, it is inferred that fructose 6-phosphate binds first and fructose 1,6-bisphosphate is released last from the enzyme during a given cycle of the catalytic reaction.[22]

Oligomeric Structure. In the absence of substrates, PFK2 is a dimer composed of two subunits having the same molecular weight and the same isoelectric point, as judged from two-dimensional electrophoresis.[16] This dimer aggregates reversibly in the presence of high concentrations of ATP or ATP analogs (ADP or XTP) into a tetramer. This process is reversed by addition of fructose 6-phosphate.[22]

Metabolic Effectors. The specificity of the control of PFK2 differs from that of PFK1. At low fructose 6-phosphate concentration, an inhibition of PFK2 by excess of ATP is observed. This inhibition is cooperatively relieved by fructose 6-phosphate. The same behavior has been observed *in situ* by Domenech and Sols[23] with bacteria permeabilized by toluene. *In vitro* and probably *in situ,* this control is thought to operate through the reversible aggregation of the PFK2 dimer into a less active tetrameric species.[22]

Neither phosphoenolpyruvate nor citrate inhibits the enzyme. PFK2 is not activated by ammonium or potassium ions or by nucleoside mono- or diphosphates.

Immunological Relationship to Other Prokaryotic Phosphofructokinases and Biosynthetic Control

Antibodies have been raised against both PFK1 and PFK2. The absence of cross-reactivity between the two proteins excludes any strong homology between them as well as the occurrence of PFK1–PFK2 hy-

---

[21] K. N. Ewings and H. W. Doelle, *Biochim. Biophys. Acta* **615,** 103 (1980).

[22] D. Kotlarz and H. Buc, *Eur. J. Biochem.* **117,** 569 (1981).

[23] A. Sols, J. G. Castaño, J. J. Aragón, C. Domenech, P. A. Lazg and A. Nieto, *in* "Metabolic Interconversion of Enzymes 1980" (H. Holzer, ed.), p. 111. Springer-Verlag, Berlin and New York, 1981.

TABLE V
EXISTENCE OF ANTIGENS REACTING WITH *E. coli* PHOSPHOFRUCTOKINASE SERA
IN CRUDE EXTRACTS OF VARIOUS BACTERIA[24][a]

Organisms	Cross-reactions with serum prepared against		Footnote
	PFK1	PFK2	
Escherichia coli K10 wt *pfk* A$^+$, *pfk* B$^+$	+++	+	b
E. coli 27-77	+++	+++	b
E. coli 23-77	+++	+++	b
E. coli 20-18	+++	+++	b
E. coli AM1R20 *pfk* A1, *pfk* B1	+++	+++	b
Shigella boydii	+++	+	b
Shigella sonnei	+++	+	c
Salmonella typhimurium	+++	+	b
Klebsiella pneumoniae	+++	+	b
Klebsiella oxytocans	+++	+++	b
Enterobacter agglomerans	++	+	c
Enterobacter aerogenes	++	+	c
Serratia rubicans	++	+	c
Proteus vulgaris	++	−	b
Lactobacillus lactis	−	−	c
Clostridium pasteurianum	−	+	d
Bacillus stearothermophilus	+++	+	d

[a] Antisera have been raised against pure PFK1 and pure PFK2[8] and tested by the Ouchterlony immunodiffusion technique with crude extracts of various bacteria: +++, a strong precipitin line; +, a light precipitin line; −, no precipitin line.

[b] Data qualitatively and quantitatively supported by enzymic assays. The corresponding phosphofructokinase activity has been found in the extracts.

[c] Enzymic assays have not been performed.

[d] The crude extracts contain only a phosphofructokinase activity inhibited by phosphoenolpyruvate.

brids. Table V summarizes the immunological relationships between phosphofructokinases from *E. coli* K12 and other bacteria.[24] It can be concluded that PFK1 has strong structural (and functional) similarities with phosphofructokinase from *B. stearothermophilus*. Additionally, the relative abundance of PFK1 and PFK2 in other strains is subject to large variations.

Growth conditions have no influence on the biosynthesis of PFK2, but the relative amount of PFK1 (expressed in IU per milligram of protein in

[24] D. Kotlarz, thesis, Université de Paris VI (1980).

the crude extract) depends on the nature of the carbon source and on the oxygen tension. Bacteria grown on glucose contain double the amount of PFK1 as do bacteria grown on glycerol. Additionally, following a shift from aerobiosis to anaerobiosis, the level of PFK1 doubles.[7] A similar conclusion was reached earlier by Doelle's group,[25,26] who independently proposed on other bases that two different enzymes were present in *E. coli*.[27,28]

It has to be stressed, however, that strains expressing either PFK1 alone or PFK2 alone have comparable growth rates under all conditions tested[29] and that an identical "Pasteur effect" has been observed with both types of strains.[30]

These two observations, as well as the existence among Enterobacteriaceae of various genera expressing widely different amounts of the two enzymes, lead to the assumption that each type of PFK can ensure, with a comparable efficiency, the regulation of fructose 6-phosphate transphosphorylation.

[25] A. D. Thomas, H. W. Doelle, A. W. Westwood, and G. L. Gordon, *J. Bacteriol.* **112**, 1099 (1972).

[26] H. W. Doelle and N. W. Hollywood, *Eur. J. Biochem.* **83**, 479 (1978).

[27] H. W. Doelle, *FEBS Lett.* **49**, 220 (1974).

[28] K. N. Ewings and H. W. Doelle, *Eur. J. Biochem.* **69**, 563 (1976).

[29] J. P. Robinson and D. G. Fraenkel, *Biochem. Biophys. Res. Commun.* **81**, 858 (1978).

[30] H. W. Doelle and S. McIvor, *FEMS Microbiol. Lett.* **7**, 337 (1980).

[12] Phosphofructokinase from *Bacillus licheniformis*

By Charles K. Marschke and Robert W. Bernlohr

Fructose 6-phosphate + ATP → fructose 1,6-bisphosphate + ADP

Phosphofructokinase   (ATP : D-fructose-6-phosphate   1-phosphotransferase, EC 2.7.1.11) from the mesophile *Bacillus licheniformis* is subject to temperature- and ligand-induced alterations of specific activity. Also, it is unstable at the low protein concentrations required for assay.[1] Therefore, the enzyme's environment must be controlled during purification and kinetic analysis to maintain the structural integrity of the molecule. This report details procedures for the purification and stabilization of *B. licheniformis* phosphofructokinase that allow the determination of valid molecular and kinetic properties of the enzyme.

[1] C. K. Marschke and R. W. Bernlohr, *Arch. Biochem. Biophys.* **156**, 1 (1973).

Culture and Harvest of Cells. A genetically stable rough colonial form of *B. licheniformis* is used for all studies. This is maintained as a spore stock as previously described.[2] Vegetative cells of *B. licheniformis* are grown from this spore stock by the addition of spores to a minimal medium (5×10^6 spores/ml) containing essential inorganic constituents, 30 mM glucose, 10 mM NH_4Cl, and 0.5 mM L-alanine (the L-alanine is present to promote germination). The inorganic constituents of this minimal medium are 0.61 mM $MgSO_4$, 0.61 mM $MgCl_2$, 0.005 mM $MnCl_2$, 0.34 mM $CaCl_2$, and 65 mM phosphate (as the potassium salt). The glucose, NH_4Cl, and inorganic constituents are sterilized separately and added to the sterile phosphate solution.

Cells are cultured in 1 liter of this minimal medium in a 2.8-liter Fernbach flask on a Gyrotory shaker (New Brunswick Co.) at 350 excursions per minute. Growth is continued at 37° until a turbidity of 150–170 Klett units (540 nm) is reached. Then this seed culture is added to 11 liters of sterile minimal medium (without L-alanine) in a fermentation vessel (New Brunswick Scientific Co., Model FS-307). The growth of the vegetative cells is conducted at 37° with the impeller rotation set at 450 rpm and the air flow through the sparger set at the maximum rate.

The bacterial cells are harvested before oxygen or nutrient limitation occurs. The method of Tuominen and Bernlohr[2] is used except that the sedimented cells are washed in a 0.2 M potassium phosphate buffer (pH 7.6).

Preparation of Cell Extracts. The sedimented and washed cells are suspended in 0.2 M potassium phosphate buffer containing 5 mM 2-mercaptoethanol (approximately 300 mg of wet cells per milliliter). These cells are disrupted by sonic treatment at 5° (Measuring & Scientific Equipment, Ltd.) or by passage through a French pressure cell operated at 12,000 psi (2°). Cellular debris is sedimented by a 45-min centrifugation at 105,000 g in a Spinco Model L centrifuge using a fixed-angle rotor. The supernatant solution is removed and dialyzed against 0.1 M potassium phosphate solution containing 1 mM 2-mercaptoethanol (pH 8.0).

Assay System

Phosphofructokinase is assayed at 30° using the procedure described by Racker.[3] The auxiliary enzyme system, which couples the phosphorylation of fructose 6-phosphate (Fru-6-P) to NADH oxidation, is prepared by dialyzing a mixture of 1.5 mg of bovine serum albumin (Cohn's fraction V), 4 mg of aldolase (EC 4.1.2.7), 32 IU of triosephosphate isomerase (EC

---

[2] F. W. Tuominen and R. W. Bernlohr, *J. Biol. Chem.* **246**, 1733 (1971).
[3] E. Racker, *J. Biol. Chem.* **167**, 843 (1947).

5.3.1.1), and 8 IU of glycerol-3-phosphate dehydrogenase (EC 1.1.1.8) against 1.0 mM glycylglycine-NaOH (pH 8.5) containing 1 mM 2-mercaptoethanol (2°). After extensive dialysis the auxiliary enzyme solution is diluted to 2.5 ml with distilled water; 0.1 ml of this solution is used per 1.0 ml of assay solution. The NADH oxidation is monitored at 340 nm with a Zeiss PMQ II spectrophotometer equipped with a continuous recorder. A constant 30° assay temperature is maintained by the use of a refrigerated circulating bath and a flow cell. The assay solution contains 20 mM imidazole-HCl (pH 6.82 at 29°), 0.10 M KCl, 3.0 mM MgCl$_2$, 1.0 mM 2-mercaptoethanol, and the two substrates at various concentrations.

The rate of NADH oxidation is proportional to phosphofructokinase concentration, and the rate is constant with time unless either enzyme inactivation or activation conditions exist.[1] A progressive decay of activity will result from the addition of Mn^{2+} or sodium pyrophosphate to the assay and also with a high ATP concentration at low Fru-6-P concentrations. A time-dependent increase of activity results from the combination of inactivated enzyme with activating ligands and temperature. This occurs, for example, at 30° by combining Mg^{2+}, ATP, and phosphoenolpyruvate (PEP) with a previously inactivated phosphofructokinase.

Purification of Phosphofructokinase

The dialyzed 105,000 g supernatant solution is diluted with a 0.2 M potassium phosphate solution containing 5 mM 2-mercaptoethanol (pH 8.0) to a protein concentration of 10–20 mg/ml. Solid ammonium sulfate is then added (24.3 g/100 ml of solution), and the precipitated protein is collected by centrifugation at 20,000 g for 15 min (2°). This procedure is repeated for subsequent ammonium sulfate fractionations at 3.3 g/100 ml increments. The individual protein samples are suspended in 0.1 M potassium phosphate buffer (pH 7.6) before dialysis against the same phosphate buffer containing 1 mM 2-mercaptoethanol (2°). The greatest specific activity of phosphofructokinase results from the 34.2–37.5 g/100 ml ammonium sulfate fraction (see the table). This fraction is dialyzed against 95 mM potassium phosphate buffer containing 1 mM 2-mercaptoethanol (pH 7.2) before adding it to a 4 × 13 cm hydroxyapatite (BioGel HT) column (4°). The hydroxyapatite column has been previously equilibrated with the same phosphate buffer. After a 170-ml wash of the column with the 59 mM potassium phosphate solution, the phosphate concentration is increased to 115 mM, which elutes the phosphofructokinase. The protein in the region of maximum specific activity is precipitated by ammonium sulfate, collected by centrifugation, and dialyzed against 0.1 M potassium phosphate

PURIFICATION OF PHOSPHOFRUCTOKINASE (PFK) FROM *B. licheniformis* [a,b]

Fraction	Protein (mg/ml)	Volume (ml)	PFK (IU/mg)	Total IU	PFK recovered (%)
Undialyzed cell extract	17.5	342	0.14	840	100
Ammonium sulfate fraction	6.4	23.5	0.71	110	13
Hydroxyapatite eluent (center of PFK fraction)	0.12	30.0	9.6	35	4

[a] PFK was isolated from 50 liters of middle exponential-phase cells grown on glucose and NH$_4$Cl.

[b] PFK was assayed at pH 6.8 with 1.0 mM ATP, 1.0 mM Fru-6-P, 100 mM KCl, 3.0 mM MgCl$_2$, 20 mM imidazole, and 1.0 mM 2-mercaptoethanol.

buffer (pH 7.6) containing 1 mM 2-mercaptoethanol. This solution is further clarified by centrifugation at 20,000 g for 30 min. The 20,000 g supernatant solution is free of contaminating fructose-1,6-bisphosphate 1-hydrolase, adenylate kinase, and adenosine triphosphate phosphatase activities.[1] Glucose-6-phosphate dehydrogenase is present as a low-level contaminant while phosphoglucoisomerase is an easily detected contaminant of this phosphofructokinase preparation. This preparation is judged sufficiently pure for the subsequent molecular and kinetic characterizations.

Results of a representative purification are summarized in the table. The protein assay procedure of Lowry *et al.*[4] is used throughout the purification for determining protein concentrations. Four percent of the activity is recovered in a fraction exhibiting a specific activity of 9.6 IU/mg. This purity is about one-tenth of that observed with homogeneous skeletal muscle phosphofructokinase.[3]

The purified phosphofructokinase can be stored at $-20°$ in phosphate buffer. However, greater stability results from adding glycerol to the phosphate buffer. A 50% solution of glycerol reduces inactivation from 3% per day to 0.3% per day at $-20°$.

Properties

Time-Dependent Alterations in the Activity of Phosphofructokinase. Initial attempts to stabilize the phosphofructokinase in cell extracts by the addition of substrates resulted, surprisingly, in the recovery of negligible enzyme activity. However, normal activities are restored to these "inac-

[4] O. H. Lowry, N. J. Rosebrough, A. L. Farr, and R. J. Randall, *J. Biol. Chem.* **193**, 265 (1951).

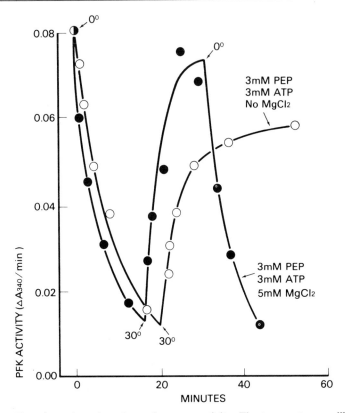

FIG. 1. Time-dependent alterations of enzyme activity. The temperature oscillations of phosphofructokinase activity can be observed by incubating the purified enzyme (5.4 IU/mg) at 0° in the presence of buffer (20 mM potassium phosphate, pH 8) and the indicated effector molecules. The experiment is initiated by mixing the 0° enzyme solution with a 0° solution containing ATP, PEP, and MgCl$_2$ to attain the designated levels of effectors. Two microliters of the enzyme solution (protein = 0.56 mg/ml) are withdrawn at the indicated times and assayed for phosphofructokinase activity. The temperatures for incubation of the phosphofructokinase are as indicated by the arrows in the figure.

tive" extracts by dialyzing them against phosphate buffer at 4° or by raising the temperature of the extract to 30° for 15–30 min before assaying. This reversible inactivation is mediated by either ATP or phosphoenolpyruvate. However, ATP differs from phosphoenolpyruvate in that the ATP also causes an irreversible inactivation of phosphofructokinase.[1]

The reversible inactivation of phosphofructokinase is most apparent upon combining ATP and phosphoenolpyruvate with Mg^{2+} (Fig. 1). In the presence of Mg^{2+}, PEP, and ATP, a rapid inactivation of the enzyme occurs at 0°. This loss is reversed by incubation at 30°, and another cycle of inactivation occurs upon returning the phosphofructokinase to 0°. The omission of Mg^{2+} reduces the extent of activity regain. There may also be a slight decrease in the rate of inactivation with the omission of Mg^{2+} (Fig. 1).

The kinetic order of the activation with respect to enzyme concentration is amenable to analysis. Figure 2A demonstrates that the initial rate of enzyme activation is a nonlinear function of the protein concentration. However, the log(activity gain) is a linear function of log[protein] with a slope near 2 (Fig. 2B). This suggests that the reversible cycles of enzyme activity are due to a dissociation of the molecule into dimers with the subsequent association of these subunits to restore an active molecule. This interpretation is supported by molecular weight estimates of 68,000 and 135,000 for the inactive and active molecules, respectively. These estimates are from sucrose density-gradient centrifugations of the inactive and active phosphofructokinases.[1] If the association of dimers requires Mg^{2+}, this would explain the retarded activation that occurs in the absence of added Mg^{2+} (Fig. 1).

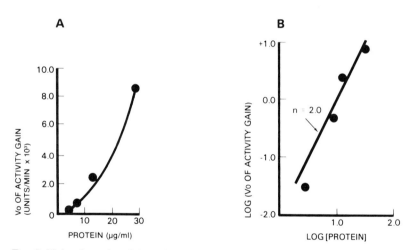

FIG. 2. Molecular order of the activation of phosphofructokinase. The enzyme (1.4 mg of protein per milliliter) is previously inactivated by incubation with 1 mM ATP, 1 mM phosphoenolpyruvate, 5 mM $MgCl_2$, and 20 mM potassium phosphate (pH 7.6) at 0° for 3 hr. Then the indicated amount of the inactivated enzyme is transferred to a 1.0-ml cuvette containing 1.2 mM ATP, 5.0 mM fructose 6-phosphate, 3.0 mM $MgCl_2$, 100 mM KCl, 1 mM 2-mercaptoethanol, and 20 mM buffer (pH 6.8) at 30°. The initial rate of activity gain (A) is determined, and the logarithm of activity gain is plotted against the logarithm of the protein concentration (B). A line with a slope of 2.0 is provided for reference.

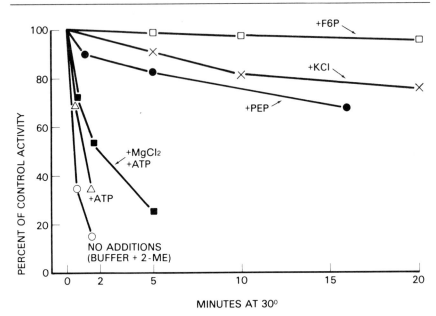

MINUTES AT 30°

FIG. 3. Phosphofructokinase stability under assay conditions. The stability of phospho-fructokinase at assay concentrations can be determined by adding purified enzyme (7.0 IU/mg) to the base solution (20 mM imidazole-HCl with 1 mM 2-mercaptoethanol; pH 6.82) containing the indicated reactant(s). After incubation of this solution at 30° for various times, the enzyme reaction is started by adding the remaining (omitted) reactants. Fructose 6-phosphate (F6P), phosphoenolpyruvate, and ATP are present at 1 mM, and MgCl$_2$ and KCl are present at 3 and 100 mM, respectively. In all cases, the complete assay reaction contains 20 mM buffer, 1 mM 2-mercaptoethanol (2-ME), 1 mM ATP, 100 mM KCl, and 3 mM MgCl$_2$. The control NADH oxidation rate is 0.062 A_{340} per minute.

The temperature-dependent inactivation seen with *B. licheniformis* phosphofructokinase is also found with *B. cereus, B. megaterium, B. mycoides,* and *B. subtilis* 168 M, but not with *Escherichia coli* B.[1] This phenomenon of temperature inactivation appears to correlate with stability of the phosphofructokinases from these bacteria, since 15% of the *E. coli* B activity and 30–35% of the *Bacillus* sp. activity is lost during these experiments. It is also important to note that the molecular weights of these enzymes from *E. coli* and *Bacillus* sp. are 130,000 to 140,000.[1]

Stability of Phosphofructokinase under Assay Conditions. This enzyme is fragile enough to be rapidly inactivated during incubation under assay conditions (Fig. 3). Inactivation is arrested, essentially, by the addition of Fru-6-P. Although both KCl and PEP are stabilizing ligands, they are much less effective than Fru-6-P (Fig. 3). The addition of ATP or ATP with MgCl$_2$ only slightly improves the stability. This general instability dictates that the reaction be initiated by the addition of phosphofructokinase. Even with this precaution, a time-dependent decrease of the reaction rate occurs when a high ATP concentration is combined with a low Fru-6-P concentration. Therefore, the instability of this enzyme under assay conditions can lead to invalid kinetic characterizations.

Kinetics of Phosphofructokinase. Saturation functions for ATP and Fru-6-P are hyperbolic with apparent K_m values of 0.07 and 0.05 mM, respectively.[1] Several divalent cations support catalysis by this enzyme but magnesium is the preferred ion. The pH optimum is in the range of 8.2–8.7. Either NH_4^+ or K^+ activates the *B. licheniformis* phosphofructokinase, whereas Li^+ and Na^+ ions have no effect on the enzyme activity. The apparent dissociation constants for NH_4^+ and K^+ are 0.5 and 10 mM, respectively. The most effective inhibitor of this enzyme is phosphoenolpyruvate. This compound increases the K_m of the enzyme for Fru-6-P. Phosphoenolpyruvate, which is an effective inhibitor at 5–20 μM, has an apparent Hill coefficient of 3. Other inhibitors of this enzyme include ATP, citrate, calcium, and pyrophosphate. Sensitivity to inhibition by ATP and citrate is conditioned by the concentration of Fru-6-P and Mg^{2+}, whereas the inhibition by phosphoenolpyruvate is specifically relieved by Fru-6-P. Calcium is a competitive inhibitor of Mg^{2+}, and its apparent K_i is 0.2 mM. Pyrophosphate has the effect of inactivating the phosphofructokinase under assay conditions.

[13] Phosphofructokinase from *Streptococcus lactis*

By Alison M. Fordyce, C. H. Moore, and G. G. Pritchard

Fructose 6-phosphate + ATP → fructose 1,6-bisphosphate + ADP + H$^+$

Assay Method

Principle. Phosphofructokinase (ATP: D-fructose 6-phosphate 1-phosphotransferase, EC 2.7.1.11) activity can be coupled via the enzymes fructose 1,6-bisphosphate aldolase, triosephosphate isomerase, and glycerol 3-phosphate dehydrogenase to NADH oxidation.[1] Activity can then be determined spectrophotometrically by measuring the decrease in absorbance at 340 nm. Phosphorylation of 1 μmol of fructose 6-phosphate results in oxidation of 2 μmol of NADH in this assay.

[1] E. Racker, *J. Biol. Chem.* **167**, 843 (1947).

METHODS IN ENZYMOLOGY, VOL. 90

An alternative assay is based on measurement of the ADP produced by coupling via pyruvate kinase and lactate dehydrogenase to NADH oxidation.[2] Another assay involves measurement of H^+ production in a pH stat.[3]

Reagents

Stock assay mixture: Tris-HCl buffer, 50 mM, pH 7.5; bovine serum albumin, 0.2 mg/ml; 2-mercaptoethanol, 10 mM; NADH, 0.3 mM; aldolase, 50 μg (0.55 unit)/ml; triosephosphate isomerase–glycerol 3-phosphate dehydrogenase mixture, 5 μg (8 units TPI, 0.8 units GPDH)/ml. This solution is made on the day of use and used within 12 hr. The auxiliary enzymes are dialyzed for 30 min before use to remove ammonium sulfate.

Substrate stock solutions

Fructose 6-phosphate (Fru-6-P) solution, 20 mM

MgATP solution: 20 mM ATP, 25 mM MgCl$_2$ in 50 mM Tris HCl buffer, pH 7.5

Enzyme dilution buffer: Tris-HCl buffer, 50 mM, pH 7.5, containing glycerol, 20% (v/v) and 2-mercaptoethanol, 10 mM

Procedure. The following components are mixed in a 1-ml capacity microcuvette: 0.35 ml of assay mixture, 50 μl of MgATP solution, and 50 μl of suitably diluted enzyme. The reaction is started by addition of 50 μl of Fru-6-P.

In crude extracts some NADH oxidation may occur owing to soluble NADH oxidases or, in the case of the pyruvate kinase–lactate dehydrogenase-coupled assay, to nonspecific phosphatases, ATPase, or adenylate kinase. This "background activity" is corrected for by measuring the rate of NADH oxidation in the absence of Fru-6-P.

Definition of Unit. One unit is defined as the amount of enzyme required to catalyze the conversion of 1 μmol of Fru-6-P to fructose bisphosphate per minute under the above conditions. Specific activity is defined as units of enzyme activity per milligram of protein. Protein concentration is determined by the Coomassie Blue binding assay of Bradford[4] using crystalline bovine serum albumin as standard.

Purification Procedure

Growth of Bacteria. Streptococcus lactis C10 is obtained from the Dairy Research Institute, Palmerston North, New Zealand. It is also available from the CSIRO Division of Food Research, Dairy Research Labora-

[2] W. A. Simon and H. W. Hofer, *Biochim. Biophys. Acta* **481**, 450 (1977).
[3] J. E. Dyson and E. A. Noltmann, *Anal. Biochem.* **11**, 362 (1965).
[4] M. M. Bradford, *Anal. Biochem.* **72**, 248 (1976).

tory, Melbourne, Australia and from the National Institute for Research in Dairying, Shinfield, Reading, England (strain NCDO 509). Cultures are grown in a Fermacell fermentor (New Brunswick Scientific Co., Inc.) in 30-liter batches of medium containing the following constituents per liter: lactose, 20 g; peptone, 10 g; yeast extract, 10 g; beef extract, 2 g; KH_2PO_4, 5 g; $MgSO_4 \cdot 7 H_2O$, 0.2 g; $MnCl_2 \cdot 4 H_2O$, 0.05 g. The medium is inoculated with 500 ml of a starter culture grown overnight. The culture is flushed with 5% CO_2–95% N_2; the pH level is maintained at between 6.2 and 6.8 by addition of 2.5 M NaOH and the temperature at 30°.

The cells are harvested in the late logarithmic phase (after about 6 hr of growth) by centrifugation at 5000 g for 10 min. The cells are washed twice by resuspension in cold 20 mM Tris-HCl buffer (pH 7.5) and stored frozen at $-12°$. The yield of cells averages 15–20 g (wet weight) per liter of medium.

Step 1. Preparation of Crude Extract. All purification steps are carried out at a temperature not exceeding 4°. Frozen cells are thawed and resuspended in two volumes of 50 mM Tris-HCl buffer (pH 7.5) containing 10 mM 2-mercaptoethanol, 20% (v/v) glycerol, 5 mM $MgSO_4$, and 5 mM EDTA (referred to as Tris-glycerol buffer). The cells are broken by two passages through an Aminco French pressure cell at 5500 psi. Cell debris is removed by centrifugation at 13,000 g for 15 min. Nucleic acids are precipitated by addition of a freshly prepared solution of protamine sulfate using 0.1 ml of protamine sulfate solution (10 mg/ml) per 15 mg of protein. The resulting suspension is allowed to stand on ice for 15 min before removing the precipitate by centrifugation at 13,000 g.

Step 2. Ammonium Sulfate Fractionation. Solid ammonium sulfate is added slowly to the supernatant to bring it to a concentration of 24 g/100 ml. The precipitated protein is removed by centrifugation, and the ammonium sulfate concentration of the supernatant is increased by addition of a further 20.5 g of ammonium sulfate per 100 ml. The precipitate is collected by centrifugation and redissolved in a minimum volume of Tris-glycerol buffer. It is dialyzed against the same buffer for 24 hr.

Step 3. Chromatography on Blue Dextran–Sepharose. Blue Dextran (Sigma Chemical Co., St. Louis, Missouri) is covalently bound to cyanogen bromide-activated Sepharose 4B (Pharmacia) following the procedure of Cuatrecasas.[5] The cyanogen bromide-activated Sepharose is stirred slowly for 48 hr at 4° in 0.4 M Na_2CO_3, pH 10, containing 0.25 g of Blue Dextran; excess dye is removed by washing with 3 M KCl and then with distilled water until the wash is no longer blue. The derivatized Sepharose is then suspended in 1 M ethanolamine for 2 hr at 25° to block any unreacted groups. After washing with water, the Blue Dextran–Sepharose

[5] P. Cuatrecasas, *J. Biol. Chem.* **245**, 3059 (1970).

PURIFICATION OF PHOSPHOFRUCTOKINASE FROM 40 g FROZEN CELLS OF
Streptococcus lactis C10

Step	Total activity (units)	Total protein (mg)	Specific activity (units/mg protein)	Purification (fold)	Yield (%)
Crude extract	900	2045	0.44	1.0	100
Ammonium sulfate precipitation	620	900	0.69	1.6	69
Blue Dextran–Sepharose, 0–2 M KCl	514	4.8	107	243	57
Blue Dextran–Sepharose, 0–20 mM ATP	485	3.0	161	364	54

can be stored at 4° in a 0.02% sodium azide solution. After storage the Sepharose should be thoroughly washed with equilibration buffer.

Dialyzed supernatant (100 ml) from step 2 (containing approximately 250 mg of protein) is applied to a 10-cm by 1-cm column of Blue Dextran–Sepharose equilibrated with Tris-glycerol buffer, and the column is washed with buffer until no further protein is eluted as indicated by the absorbance at 280 nm. Phosphofructokinase is then eluted with a linear gradient of KCl from 0 to 2 M in 100 ml of Tris-glycerol buffer. Active fractions are combined, dialyzed against Tris-glycerol buffer and applied to the same column (reequilibrated). Phosphofructokinase is again eluted, this time with a linear gradient of ATP from 0 to 20 mM in 100 ml of Tris-glycerol buffer at pH 7.5. The active fractions are pooled, dialyzed for 48 hr against three changes of Tris-glycerol buffer to remove ATP, then concentrated by ultrafiltration using a PM-30 Diaflo membrane. The purified enzyme is stored at $-12°$ in Tris-glycerol buffer (pH 7.5) with the glycerol concentration increased to 50% (v/v).

A typical purification is shown in the table.

Crystallization. Purified phosphofructokinase (approximately 4 mg/ml in 25 mM sodium phosphate buffer, pH 7.0) is dialyzed at 4° against 25 mM phosphate buffer (pH 7.0) containing 10 mM 2-mercaptoethanol, 1 mM MgSO$_4$, 1 mM EDTA, 2 mM ATP, 2 mM FDP, and ammonium sulfate (228 g/liter). Small crystals form after 8–10 weeks. Crystal formation also occurs in the absence of ATP and FDP.

Properties

Homogeneity. Electrophoresis of 10–50 μg of purified enzyme on polyacrylamide gels at pH 8.9 yields a single protein band that corresponds to the phosphofructokinase activity revealed by an activity stain.

Molecular Weight and Subunit Structure. *Streptococcus lactis* phospho-fructokinase appears to be a tetramer of identical subunits. The native enzyme has a molecular weight (MW) of 144,500 as determined by gel filtration on Sephacryl S-200. Sodium dodecyl sulfate–polyacrylamide gel electrophoresis by the method of Weber and Osborn[6] reveals a single protein band corresponding to a subunit size of MW 33,500 ± 3500.

Stability. The purified enzyme can be stored at $-12°$ in Tris-glycerol buffer for up to 6 months with only 10–20% loss in activity. At 4° the enzyme is less stable at pH 6.0 and pH 8.0 than at neutral pH, and dilution of the enzyme enhances inactivation. Stability is not significantly increased by addition of Fru-6-P, ATP, ADP, citrate, KCl, or ammonium sulfate. However, the enzyme is more stable at 4° if the Tris in the Tris-glycerol buffer is replaced by 50 mM phosphate (pH 7.5). Removal of glycerol results in complete inactivation in 14 days.

pH Optimum. The enzyme has a pH optimum of 7.2–7.8 in a number of different buffers, but in phosphate buffer both the pH optimum and the maximum velocity are lowered.

Michaelis Constants. The K_m values for MgATP and Fru-6-P are 0.18 mM and 0.25 mM, respectively. The enzyme shows a sigmoidal response to varying Fru-6-P concentration with a Hill coefficient of 2.6–3.2 at saturating concentrations of MgATP.

Cation Requirements. Mg^{2+} is required for activity, maximum activity being obtained when Mg^{2+} is present at a 2–5 mM excess over ATP. Mg^{2+} can be replaced by other divalent cations; at optimum cation concentrations Mn^{2+}, Co^{2+}, Zn^{2+}, and Ni^{2+} give 75%, 50%, 35%, and 12%, respectively, of the activity with Mg^{2+}. The enzyme is activated by NH_4^+ at both saturating and nonsaturating substrate concentrations. K^+ increases activity slightly whereas Na^+ and Li^+ are weak inhibitors.

Effectors. ADP and NH_4^+ are activators of *S. lactis* phosphofruc-tokinase shifting the sigmoidal Fru-6-P saturation curve to a more hyperbolic form. The activation constant ($K_{0.5}$) for NH_4^+ at saturating Fru-6-P concentration is 0.3 mM; the $K_{0.5}$ for ADP at nonsaturating (0.25 mM) Fru-6-P is also 0.3 mM. The enzyme is inhibited by AMP, PEP, and Fru-1,6-P_2, 5 mM concentrations of these compounds giving 80, 64, and 25% inhibition, respectively, at nonsaturating concentrations of Fru-6-P. Orthophosphate (10 mM) gives 50% inhibition of activity at nonsaturating Fru-6-P concentration. The enzyme is also inhibited by free ATP or MgATP but only at low Fru-6-P concentrations (less than 0.6 mM) and when the ratio of ATP to Fru-6-P is greater than 1.

Application of Blue Dextran–Sepharose to Purification of Phosphofruc-tokinase from Other Sources. Use of the Blue Dextran–Sepharose "affin-

---

[6] K. Weber and M. Osborn, *J. Biol. Chem.* **244**, 4406 (1969).

ity'' column is a very effective purification step in the isolation of *S. lactis* phosphofructokinase giving high yields of homogeneous enzyme. Similar columns have been used in the purification of phosphofructokinase from yeast[7] and from *E. coli*.[8] Binding of muscle and human erythrocyte phosphofructokinases to Blue Dextrans immobilized on polyacrylamide gels has been demonstrated.[9] However, other attempts to bind mammalian phosphofructokinase to Blue Dextran have been unsuccessful.[10]

[7] H. J. Böhme, G. Kopperschläger, J. Schulz, and E. Hofmann, *J. Chromatogr.* **69**, 209 (1972).
[8] K. N. Ewings and H. W. Doelle, *Eur. J. Biochem.* **69**, 563 (1976).
[9] G. Kopperschläger, W. Diezel, R. Freyer, and S. Liebe, *Eur. J. Biochem.* **22**, 40 (1971).
[10] C. S. Ramadoss, L. J. Luby, and K. Uyeda, *Arch. Biochem. Biophys.* **175**, 487 (1976).

[14] Phosphofructokinase from *Clostridium pasteurianum*

By Gerhard Gottschalk and Hasso v. Hugo

$$\text{D-Fructose 1-phosphate} + \text{ATP} \xrightarrow{\text{Mg}^{2+}} \text{D-fructose 1,6-bisphosphate} + \text{ADP}$$

Assay Method

Principle. D-Fructose-1-phosphate kinase (EC 2.7.1.56, ATP: D-fructose-1-phosphate 6-phosphotransferase) catalyzes the phosphorylation of D-fructose 1-phosphate to yield D-fructose 1,6-bisphosphate.[1] The continuous spectrophotometric assay is based on the conversion of fructose 1,6-bisphosphate via dihydroxyacetone phosphate to α-glycerol phosphate.[1] The rate of NADH oxidation in the glycerol-3-phosphate dehydrogenase reaction is twice the rate of the phosphorylation of D-fructose 1-phosphate.

Because of the presence of high NADH-oxidizing enzyme activities in crude extracts, the continuous spectrophotometric assay I can be applied only to purified enzyme preparations. D-Fructose-1-phosphate kinase activity in crude extracts is determined with assay II. In this assay the ADP formed in the kinase reaction and phosphoenolpyruvate are converted to ATP and pyruvate by pyruvate kinase, and the amount of pyruvate formed is subsequently determined.[2]

[1] R. L. Anderson, T. E. Hanson, and V. L. Sapico, this series, Vol. 42 [11].
[2] H. v. Hugo and G. Gottschalk, *Eur. J. Biochem.* **48**, 455 (1974).

Reagents

Tris(hydroxymethyl)aminomethane(Tris)-HCl buffer, 50 mM, pH 8.0
Potassium phosphate buffer, 100 mM, pH 7.0
MgCl$_2$, 100 mM
NADH, 15 mM
ATP, 15 mM
Phosphoenolpyruvate, sodium salt, 50 mM
D-Fructose 1-phosphate, sodium salt, 40 mM
D-Fructose-1,6-bisphosphate aldolase, 10 mg/ml
Glycerol-3-phosphate dehydrogenase–triosephosphate isomerase, 2 mg/ml
Pyruvate kinase, 2 mg/ml
L-Lactate dehydrogenase, 5 mg/ml

Procedure

ASSAY I. The following solutions are added to a cuvette with a 1.0-cm light path: 0.70 ml of 50 mM Tris-HCl buffer, pH 8.0; 0.03 ml of 100 mM MgCl$_2$; 0.01 ml of 15 mM NADH; 0.05 ml of 15 mM ATP; 0.10 ml of 40 mM D-fructose 1-phosphate; 0.01 ml of D-fructose-1,6-bisphosphate al-dolase; 0.01 ml of glycerol-3-phosphate dehydrogenase–triosephosphate isomerase; a rate-limiting amount of D-fructose-1-phosphate kinase; and water to a final volume of 1 ml. The cuvette compartment is kept at 25°, and the reaction is started by the addition of D-fructose-1-phos-phate kinase. The decrease in absorption is read at 365 nm ($\epsilon = 3.39 \times 10^3$ M^{-1} cm^{-1}).

ASSAY II. The assay mixture contains in a final volume of 1 ml: 0.50 ml of 50 mM Tris-HCl buffer, pH 8.0; 0.03 ml of 100 mM MgCl$_2$; 0.10 ml of 15 mM ATP; 0.10 ml of 40 mM D-fructose 1-phosphate; 0.10 ml of 50 mM phosphoenolpyruvate, sodium salt; 0.01 ml of pyruvate kinase; and cell extract or partially purified D-fructose-1-phosphate kinase and water to a final volume of 1 ml. After the incubation of the complete assay and of controls without D-fructose 1-phosphate in Durham tubes for 10 min at 25°, the reaction is stopped by heating the tubes for 2 min in a boiling water bath. After centrifugation, the amount of pyruvate formed is deter-mined with L-lactate dehydrogenase and NADH. The following solutions are added to a cuvette with a 1.0-cm light path: 0.93 ml of 100 mM potassium phosphate buffer, pH 7.0; 0.01 ml of 15 mM NADH; 0.01 ml of L-lactate dehydrogenase; and 0.05 ml of the solution of assay II. The decrease in absorption at 365 nm is determined.

Definition of Unit and Specific Activity. One unit is defined as the amount of enzyme that catalyzes the phosphorylation of 1 μmol of

D-fructose 1-phosphate per minute under the above conditions. Specific activity is expressed as units per milligram of protein.

Purification Procedure

Growth of Organisms. Clostridium pasteurianum (ATCC 6013; DSM 525) is grown anaerobically in 20-liter carboys at 37°. The medium contains the following components per liter: 20 g of D-fructose; 6% yeast extract; 2 g of peptone; 0.95 g of KH_2PO_4; 5.75 g of K_2HPO_4; 1 g of $(NH_4)_2SO_4$; 0.25 g of $MgSO_4 \cdot 7 H_2O$; 2 mg of $MnCl_2 \cdot 4 H_2O$; 0.10 g of $CaCl_2$; 12 mg of $Na_2MoO_4 \cdot 2 H_2O$; 5.6 mg of $FeSO_4 \cdot 7 H_2O$. Fructose and the phosphates are autoclaved separately. The final pH is 7.8. Prior to inoculation the medium is made anaerobic by passing nitrogen gas through a sterile cotton–wool filter and then through the medium for 15 min. A 10% (v/v) inoculum is used. After 24 hr of growth, the cells are harvested by centrifugation and stored at −20°.

Step 1. Crude Extract. Frozen cells of *C. pasteurianum* (150 g) are thawed in 220 ml of 50 mM potassium phosphate buffer, pH 8.0. The cell suspension is homogenized using an Ultra-Turrax homogenizer (Janke and Kunkel KG, Staufen) and sonicated with a 600-W ultrasonic disintegrator (Schoeller and Co., Frankfurt) for 3 hr in a continuous-flow cell. The temperature is kept below 8° by cooling the cell with ethanol of −5°. Cell debris is removed by centrifugation at 10,000 g for 30 min at 4°. The following steps were performed at 0–4°.

Step 2. Cetyltrimethylammonium Bromide (CTAB) Treatment. A 0.3 volume of a 4% (w/v) solution of CTAB is added slowly to the crude extract with stirring. The precipitate is removed by centrifugation at 10,000 g for 20 min.

Step 3. CM-Sephadex Chromatography. After concentration of the supernatant of the CTAB treatment to 120 ml by ultrafiltration (membrane UM-20 E, Amicon Corp., Lexington, Massachusetts), the extract is freed from precipitated protein by centrifugation at 10,000 g for 20 min. The clear supernatant (pH 6.0) is allowed to flow into a column (2.5 diameter × 20 cm) with CM-Sephadex A-50 equilibrated against 50 mM potassium phosphate buffer, pH 6.0. 1-Phosphofructokinase is not adsorbed on the column. The column is washed with 100 ml of the same buffer. Fractions with 1-phosphofructokinase activity are combined and concentrated by ultrafiltration to 80 ml.

Step 4. Adsorption on Calcium Phosphate Gel.[3] Eighty milliliters of gel (pH 6.5) are added to the solution of step 3, and the mixture is stirred for

[3] D. Keilin and E. F. Hartree, *Proc. R. Soc. London, Ser. B* **124**, 397 (1938).

20 min. After brief centrifugation at 10,000 g the kinase is eluted from the gel by washing it six times with 30 ml each of 0.1 M K_2HPO_4, pH 8.5.

Step 5. Sephadex G-100 Chromatography. The concentrated solution from step 4 (10 ml) is centrifuged at 10,000 g for 20 min and added to a column (2.6 diameter × 90 cm) of Sephadex G-100 that has been equilibrated against 50 mM potassium phosphate buffer, pH 8.0. The column is developed with the same buffer. Fractions of 3 ml are collected at an average flow rate of 4 ml. Fractions 80–93 are combined and concentrated to 20 ml by ultrafiltration. The pH is adjusted to 6.0.

Step 6. DEAE-Sephadex Chromatography. The protein solution is then allowed to flow into a column (1.5 × 20 cm) of DEAE-Sephadex A-50, equilibrated against 50 mM potassium phosphate buffer, pH 6.0. The column is washed with 100 ml of the same buffer and developed with a 200-ml linear gradient from the above buffer to buffer with 0.5 M KCl. Sixty fractions of 3 ml are collected. Fractions 32–42 contain the bulk of kinase activity; they are combined and concentrated by ultrafiltration to 2 ml.

Step 7. Second Sephadex G-100 Chromatography. The protein solution is again subjected to chromatography on Sephadex G-100 as described in step 5. Fractions 75–85 are combined and adjusted to 5 mM potassium phosphate buffer, pH 7.0, by repeated concentration and dilution in an ultrafiltration cell.

Step 8. Chromatography on Hydroxyapatite. The protein solution (5 ml) is then added to a column (1.5 × 11 cm) of hydroxyapatite that has been equilibrated against 5 mM potassium phosphate buffer, pH 7.0. The column is washed with 50 ml of the same buffer and developed with a 200-ml linear gradient from 5 to 300 mM potassium phosphate buffer, pH 7.0; 3-ml fractions are collected, and fractions 43–50 are combined and concentrated to 2 ml by ultrafiltration.

A typical purification of D-fructose-1-phosphate kinase is summarized in the table. The final enzyme preparation is purified 712-fold. The comparatively high yield of 40% is at least partly due to an as yet unexplained increase of total activity during the first steps of purification.

Properties

Homogeneity and Stability. Analytical polyacrylamide gel electrophoresis of the purified enzyme shows that it is homogeneous. Enzyme solutions in 50 mM potassium phosphate buffer, pH 7.5, can be stored at −20° for 1 year without detectable loss of activity.

Molecular Weight. The molecular weight, as determined by Sephadex G-100 chromatography, has been found to be 63,000.[2] D-Fruc-

PURIFICATION OF D-FRUCTOSE-1-PHOSPHATE KINASE FROM *Clostridium pasteurianum*

Step	Volumes (ml)	Total activity (units)	Total protein (mg)	Specific activity (units/mg protein)	Yield (%)
1. Crude extract	220	4318	10,208	0.42	100
2. CTAB supernatant	265	4610	6,148	0.75	106
3. CM-Sephadex	215	4837	3,225	1.5	112
4. Calcium phosphate gel	172	4386	1,290	3.4	101
5. First Sephadex G-100	36	3721	187	19.9	86
6. DEAE-Sephadex	32.5	3507	71	49.4	81
7. Second Sephadex G-100	34.5	3071	37	83	71
8. Hydroxyapatite	2.0	1794	6	299	41

tose-1-phosphate kinase from *C. pasteurianum* consists of two non-identical subunits.

Cofactor Requirements and Inhibitors. Maximum activity of D-fructose-1-phosphate kinase is achieved between pH 7.8 and 9.2. Enzyme activity depends on the presence of divalent metal ions. Mn^{2+}, Mg^{2+}, Co^{2+}, Fe^{2+}, and Zn^{2+} serve as cofactors, the latter two are considerably less effective. Hg^{2+}, Cu^{2+}, and Ni^{2+} are inhibitory, and so is sulfate. Therefore, ammonium sulfate fractionations cannot be done. D-Fructose-1-phosphate kinase is irreversibly inactivated in Tris-HCl buffer. The enzyme is subject to inhibition by its products ADP and D-fructose 1,6-bisphosphate.

Kinetic Properties. D-Fructose-1-phosphate kinase from *C. pasteurianum* follows Michaelis–Menten kinetics. K_m values are 0.14 mM for ATP and 0.62 mM for D-fructose 1-phosphate.

Distribution. D-Fructose-1-phosphate kinase is present in organisms that use the phosphoenolpyruvate phosphotransferase system[4] for D-fructose uptake. This system leads to D-fructose 1-phosphate,[5,6] which is phosphorylated by the kinase to yield the corresponding 1,6-bisphosphate. The enzyme was discovered by Hanson and Anderson[7] in *Enterobacter aerogenes* and by Reeves *et al.*[8] in *Bacteroides symbiosis*. After growth on D-fructose it is also present in *Escherichia coli*,[5] in several *Pseudomonas*[9] and *Clostridium*[10] species, in phototrophic bacteria,[11] and probably in many other bacteria.

[4] W. Kundig, S. Ghosh, and S. Roseman, *Proc. Natl. Acad. Sci. U.S.A.* **52**, 1067 (1964).
[5] D. G. Fraenkel, *J. Biol. Chem.* **243**, 6458 (1968).
[6] T. E. Hanson and R. L. Anderson, *Proc. Natl. Acad. Sci. U.S.A.* **61**, 269 (1968).
[7] T. E. Hanson and R. L. Anderson, *J. Biol. Chem.* **241**, 1644 (1966).
[8] R. E. Reeves, L. G. Warren, and D. S. Hsu, *J. Biol. Chem.* **241**, 1257 (1966).
[9] M. H. Sawyer, P. Baumann, L. Baumann, S. M. Berman, J. L. Canovas, and R. H. Berman, *Arch. Microbiol.* **112**, 49 (1977).

D-Fructose-1-phosphate kinase of *C. pasteurianum* was first detected by Kotzé.[12] Its partial purification and some of its properties have been described by Du Toit *et al.*[13]

[10] H. v. Hugo and G. Gottschalk, *FEBS Lett.* **46**, 106 (1974).
[11] R. Conrad and H. G. Schlegel, *Biochim. Biophys. Acta* **358**, 221 (1974).
[12] J. P. Kotzé, *S. Afr. J. Agric. Sci.* **11**, 439 (1968).
[13] P. J. Du Toit, D. J. J. Potgieter, and V. De Villiers, *Enzymologia* **43**, 285 (1972).

[15] D-Tagatose-6-phosphate Kinase from *Staphylococcus aureus*

By RICHARD L. ANDERSON and DONALD L. BISSETT

$$\text{D-Tagatose 6-phosphate} + \text{MgATP} \xrightarrow{\text{K}^+} \text{D-tagatose 1,6-bisphosphate} + \text{MgADP}$$

This is an inducible enzyme that is instrumental in the catabolism of lactose and D-galactose in *Staphylococcus aureus*.[1-3]

Assay Method

Principle. The continuous spectrophotometric assay is based on coupling ADP formation to pyruvate kinase and lactate dehydrogenase.[3] With the coupling enzymes in excess, the rate of ADP formation is equivalent to the rate of NADH oxidation, which is measured by the absorbance decrease at 340 nm.

Reagents

Glycylglycine-hydroxide buffer, pH 8.5, 0.5 M
KCl, 0.45 M
MgCl$_2$, 0.2 M
ATP, 0.1 M
NADH, 10 mM
D-Tagatose 6-phosphate, 10 mM (see this series, Vol. 89 [15])
Phosphoenolpyruvate, 0.1 M
Pyruvate kinase, crystalline
Lactate dehydrogenase, crystalline

[1] D. L. Bissett and R. L. Anderson, *Biochem. Biophys. Res. Commun.* **52**, 641 (1973).
[2] D. L. Bissett and R. L. Anderson, *J. Bacteriol.* **119**, 698 (1974).
[3] D. L. Bissett and R. L. Anderson, *J. Biol. Chem.* **255**, 8745 (1980).

Procedure. The following constituents are added to a microcuvette with a 1.0-cm light path: 20 μl of glycylglycine buffer, 10 μl of KCl, 5 μl of MgCl$_2$, 5 μl of ATP, 5 μl of NADH, 5 μl of D-tagatose 6-phosphate, 5 μl of phosphoenolpyruvate, non-rate-limiting amounts of pyruvate kinase (500 mU) and lactate dehydrogenase (750 mU), a rate-limiting amount of D-tagatose-6-phosphate kinase, and water to a volume of 0.15 ml. The reaction is initiated by the addition of D-tagatose-6-phosphate kinase. A control cuvette minus D-tagatose 6-phosphate measures ATPase and NADH oxidase rates, which must be subtracted from the total rate. The rates are conveniently measured with a Gilford multiple-sample absorbance recording spectrophotometer. The cuvette compartment should be thermostatted at 30°. Care should be taken to confirm that the rates are constant with time and proportional to the D-tagatose-6-phosphate kinase concentration.

Definition of Unit and Specific Activity. One unit is defined as the amount of enzyme that catalyzes the phosphorylation of 1 μmol of D-tagatose 6-phosphate per minute. Specific activity (units per milligram of protein) is based on protein determinations by the Lowry procedure.

Alternative Assay Procedure. For some determinations, such as phosphoryl donor specificity, an assay based on coupling D-tagatose 1,6-bisphosphate formation to aldolase, triosephosphate isomerase, and glycerol-3-phosphate dehydrogenase may be used.[3]

Purification Procedure[3]

Organism and Growth Conditions. *Staphylococcus aureus* NCTC 8511 is grown at 37° in Fernbach flasks (1500 ml per flask) on a rotary shaker. The medium is the induction medium of McClatchy and Rosenblum[4] supplemented with 1% D-galactose (autoclaved separately). The inoculum is 7 ml of an overnight culture in the same medium except that no carbohydrate is added and the peptone concentration is increased to 2%. The cells are harvested by centrifugation 9 hr after inoculation, washed once by suspension in 0.85% (w/v) NaCl, and collected by centrifugation. The yield is about 4 g (wet weight) of cells per liter of medium.

Preparation of Cell Extracts. Cells (80 g) are suspended in 260 ml of buffer A (20 mM potassium phosphate, pH 7.5; 20%, v/v, glycerol; and 0.2%, v/v, 2-thioethanol). The cells are broken by treating the suspension at 0–2° for 10 min (in 1-min bursts) with the 1.27-cm (diameter) horn of a Heat Systems Ultrasonics W-185C sonifier at 100 W, in the presence of twice the packed-cell volume of glass beads (88–125 μm in diameter).

---

[4] J. K. McClatchy and E. D. Rosenblum, *J. Bacteriol.* **86**, 1211 (1963).

The cell extract is the supernatant fluid resulting from a 10-min centrifugation of the broken-cell suspension at 40,000 g. The pH of the extract is 6.5.

General. The following procedures are performed at 0–4°.

Bentonite Treatment. To the cell extract is added 9.6 g of bentonite (Fisher Scientific Co.) with stirring. After 10 min, the suspension is centrifuged at 14,000 g for 10 min, and the sediment is discarded.

DEAE-Cellulose Chromatography I. A DEAE-cellulose column (2.6 × 10 cm) is equilibrated with buffer A. The supernatant solution from the bentonite treatment is adjusted to pH 7.5 with 0.4 M K$_2$HPO$_4$ and applied to the column, which is then washed with 150 ml of buffer A. The protein is eluted with a linear gradient (530 ml; 75 ml/hr) of 0 to 0.5 M KCl in the same buffer. Fifty-three 10-ml fractions are collected, and those that contain most of the kinase activity (fractions 29 through 37) are combined.

DEAE-Cellulose Chromatography II. A DEAE-cellulose column (1.2 × 7.0 cm) is equilibrated with buffer B (20 mM potassium phosphate, pH 6.0; 20%, v/v, glycerol; and 0.2%, v/v, 2-thioethanol). The combined fractions from the preceding step are adjusted to pH 6.0 with 0.4 M KH$_2$PO$_4$, diluted to 300 ml with buffer B, and applied to the DEAE-cellulose column, which is then washed with 30 ml of the same buffer. The protein is eluted with a linear gradient (80 ml; 30 ml/hr) of 0 to 0.5 M KCl in the same buffer. Forty-eight 1.7-ml fractions are collected, and those that contain most of the kinase activity (fractions 25 through 34) are combined.

Hydroxyapatite Chromatography. A hydroxyapatite (Bio-Rad Laboratories, Richmond, California) column (1.5 × 2.0 cm) is equilibrated with buffer C (5 mM potassium phosphate, pH 6.0; 20%, v/v, glycerol; and 0.2%, v/v, 2-thioethanol). The combined fractions from the preceding step are desalted on a Sephadex G-15 column (4.0 × 12 cm) that has been equilibrated with the same buffer. The Sephadex G-15 fractions containing kinase activity are applied to the hydroxyapatite column, which is then washed with 15 ml of buffer C. The protein is eluted with a linear gradient (40 ml; 8 ml/hr) of 0 to 0.2 M potassium phosphate (pH 6.0) in buffer C. Thirty-five 1.3-ml fractions are collected, and those that contain most of the kinase activity (fractions 20 through 26) are combined.

Sephadex G-100 Chromatography. A Sephadex G-100 column (1.4 × 80 cm) is equilibrated with buffer A. The combined fractions from the preceding step are adjusted to pH 7.5 with 0.4 M K$_2$HPO$_4$, applied to the column, and chromatographed with the same buffer. Fractions (2.6 ml each) are collected, and those that contain the highest specific activity of kinase (fractions 23 and 24) are combined.

The enzyme was 300-fold purified with a recovery of 24%. Electrophoresis on polyacrylamide gels under native and denaturing condi-

PURIFICATION OF D-TAGATOSE-6-PHOSPHATE KINASE FROM *Staphylococcus aureus*

Fraction	Volume (ml)	Total protein (mg)	Total activity[a] (units)	Specific activity (units/mg protein)	Recovery (%)
Cell extract	240	1013	58.6	0.058	(100)
Bentonite	215	314	57.6	0.183	98
DEAE-cellulose I	90	75.4	57.6	0.764	98
DEAE-cellulose II	17	35.7	42.2	1.18	72
Hydroxyapatite	9.1	7.79	32.2	4.13	55
Sephadex G-100	10.4[b]	0.79[b]	13.8[b]	17.5	24[b]

[a] KCl was not added to the assay mixture in these determinations, but an activating amount of NH_4^+ was added with the coupling enzymes.

[b] The actual value was multiplied by 2, since only one-half of the sample from the preceding step was applied to the Sephadex G-100 column.

tions indicated the enzyme was 90% pure. It was free from the constitutive D-fructose-6-phosphate kinase (which has a low K_m for D-fructose 6-phosphate) that was present in the cell extract. A typical purification is summarized in the table.

Properties[3]

Phosphoryl Acceptor Specificity. D-Tagatose 6-phosphate ($K_m = 16\ \mu M$) is apparently the natural substrate. D-Fructose 6-phosphate can also be phosphorylated, but its K_m value ($0.15\ M$) is 10,000 times greater than that for D-tagatose 6-phosphate. Compounds that are not phosphorylated (<1% of the rate with D-tagatose 6-phosphate) at 10 mM concentrations include D-galactose 6-phosphate, D-glucose 6-phosphate, D-mannose 6-phosphate, D-fructose 1-phosphate, and L-sorbose 1-phosphate.

Phosphoryl Donor Specificity. Several nucleoside triphosphates can serve as phosphoryl donors, some giving greater V_{max} values than that obtained with ATP. The four phosphoryl donors that exhibit the greatest activity, in order of decreasing V_{max} values, are GTP ($K_m = 0.4$ mM), UTP ($K_m = 0.9$ mM), ITP ($K_m = 0.17$ mM), and ATP ($K_m = 0.16$ mM).

Activators and Inhibitors. The enzyme does not exhibit allosteric kinetics, and 2 mM citrate, ADP, or AMP did not affect the velocity when added to the standard assay. Also, MgATP up to 100 mM did not inhibit the enzyme. However, monovalent cations can influence the velocity: 33 mM ammonium, rubidium, and potassium activated up to fourfold, whereas sodium and lithium inhibited 31 to 65%.

pH Optimum. Activity as a function of pH is maximal at about pH 8.5 in glycylglycine buffer.

Apparent Molecular Weight. The subunit molecular weight was estimated to be 52,000, and the native enzyme is apparently a dimer.

Stability. The enzyme in the Sephadex G-100 fractions was stable to storage at $-20°$ for several months.

[16] Inorganic Pyrophosphate: D-Fructose-6-phosphate 1-Phosphotransferase from Mung Bean

By RICHARD L. ANDERSON and DARIO C. SABULARSE

$$\text{D-Fructose 6-phosphate} + \text{PP}_i \xrightleftharpoons{\text{Mg}^{2+}} \text{D-Fructose 1,6-bisphosphate} + \text{P}_i$$

Inorganic pyrophosphate: D-fructose-6-phosphate 1-phosphotransferase (EC 2.7.1.90). (PP$_i$: Fru-6-P 1-phosphotransferase) has been found in microbes,[1-4] in the leaves of a crassulacean plant,[5] and in mung bean sprouts.[6-8] The mung bean enzyme is activated by D-glucose 1,6-bisphosphate, D-fructose 1,6-bisphosphate, and D-fructose 2,6-bisphosphate.[6-8]

Assay Methods[6,8]

Method A: Forward Reaction (Standard Assay)

Principle. The continuous spectrophotometric assay is based on coupling D-fructose 1,6-bisphosphate formation to non-rate-limiting amounts of aldolase, triosephosphate isomerase, and glycerol-3-phosphate dehydrogenase. With the three coupling enzymes in excess, the rate of D-fructose 1,6-bisphosphate formation is equivalent to one-half the rate of NADH oxidation, which is measured by the absorbance decrease at 340 nm.

[1] R. E. Reeves, D. J. South, H. J. Blytt, and L. G. Warren, *J. Biol. Chem.* **249**, 7737 (1974).
[2] W. E. O'Brien, S. Bowien, and H. G. Wood, *J. Biol. Chem.* **250**, 8690 (1975).
[3] M. H. Sawyer, P. Baumann, and L. Baumann, *Arch. Microbiol.* **112**, 169 (1977).
[4] J. M. Macy, L. G. Ljungdahl, and G. Gottschalk, *J. Bacteriol.* **134**, 84 (1978).
[5] N. W. Carnal and C. C. Black, *Biochem. Biophys. Res. Commun.* **86**, 20 (1979).
[6] D. C. Sabularse and R. L. Anderson, *Biochem. Biophys. Res. Commun.* **100**, 1423 (1981).
[7] D. C. Sabularse and R. L. Anderson, *Biochem. Biophys. Res. Commun.* **103**, 848 (1981).
[8] D. C. Sabularse and R. L. Anderson, in preparation.

METHODS IN ENZYMOLOGY, VOL. 90

Reagents

HEPES[9]–NaOH buffer (pH 7.8), 170 mM, containing 2.1 mM EDTA
D-Fructose 6-phosphate, 340 mM
Sodium pyrophosphate (PP$_i$), pH 7.8, 17 mM
MgCl$_2$, 102 mM
D-Fructose-1,6-bisphosphate aldolase, 40 mU/μl, previously dialyzed against 1 mM EDTA, pH 7.8
Triosephosphate isomerase, 200 mU/μl, previously dialyzed against 1 mM EDTA, pH 7.8
Glycerol-3-phosphate dehydrogenase, 40 mU/μl, previously dialyzed against 1 mM EDTA, pH 7.8
NADH, 4.1 mM
Activator[10] (17 μM D-fructose 2,6-bisphosphate[11])

Procedure. The following are added to a microcuvette with a 1.0-cm light path: 80 μl of HEPES buffer, 5 μl of D-fructose 6-phosphate, 10 μl of PP$_i$, 10 μl of MgCl$_2$, 5 μl of D-fructose-1,6-bisphosphate aldolase, 5 μl of triosephosphate isomerase, 5 μl of glycerol-3-phosphate dehydrogenase, 10 μl of NADH, 10 μl of D-fructose 2,6-bisphosphate, a rate-limiting amount of PP$_i$: Fru-6-P 1-phosphotransferase, and water to a volume of 170 μl. The reaction is initiated by the addition of PP$_i$: Fru-6-P 1-phosphotransferase. A control cuvette minus PP$_i$ measures NADH oxidase and apparent D-fructose-6-phosphate reductase activities, which must be subtracted from the total rate. The rates are conveniently measured with a Gilford multiple-sample absorbance-recording spectrophotometer. The cuvette compartment should be thermostatted at 30°. Care should be taken to confirm that the rates are constant with time and proportional to the amount of PP$_i$: Fru-6-P 1-phosphotransferase.

Definition of Unit and Specific Activity. One unit is defined as the amount of enzyme that catalyzes the phosphorylation of 1 μmol of D-fructose 6-phosphate per minute in the standard assay. Specific activity (units per milligram of protein) is based on protein determinations by the method of Whitaker and Granum.[12]

Method B: Forward Reaction (Alternative Assay)

Principle. For some determinations, such as measuring the activity with D-fructose 1,6-bisphosphate as the activator, a continuous spec-

[9] *N*-2-Hydroxyethylpiperazine-*N'*-2-ethanesulfonic acid.
[10] D-Glucose 1,6-bisphosphate (10 μl of a 25.5 mM solution) may be used as the activator in place of D-fructose 2,6-bisphosphate in assay method A, but the specific activities will be about halved.
[11] D-Fructose 2,6-bisphosphate may be prepared as described by E. Van Schaftingen and H.-G. Hers, *Eur. J. Biochem.* **117**, 319 (1981).
[12] J. R. Whitaker and P. E. Granum, *Anal. Biochem.* **109**, 156 (1980).

trophotometric assay based on coupling P$_i$ formation to non-rate-limiting amounts of D-glyceraldehyde-3-phosphate dehydrogenase and 3-phosphoglycerate kinase may be used. With the coupling enzymes in excess, the rate of P$_i$ formation is equivalent to the rate of NAD$^+$ reduction, which is measured by the absorbance increase at 340 nm.

Reagents

HEPES–NaOH buffer (pH 7.8), 170 mM, containing 2.1 mM EDTA
D-Fructose 6-phosphate (freed from P$_i$ by chromatography on a column of Sephadex G-10), 170 mM
Sodium pyrophosphate (PP$_i$), pH 7.8, 17 mM
MgCl$_2$, 102 mM
DL-Glyceraldehyde-3-phosphate (prepared from DL-glyceraldehyde 3-phosphate diethylacetal; Sigma Chemical Co.), 17 mM of the D-enantiomorph
ADP (vanadium free, Sigma Chemical Co.), 17 mM
D-Glyceraldehyde-3-phosphate dehydrogenase, 40 mU/μl, previously dialyzed against 1 mM EDTA, pH 7.8
3-Phosphoglycerate kinase, 80 mU/μl, previously dialyzed against 1 mM EDTA, pH 7.8
NAD$^+$, 8.2 mM
Activator (3.4 mM D-fructose 1,6-bisphosphate)

Procedure. The following are added to a microcuvette with a 1.0-cm light path: 80 μl of HEPES buffer, 10 μl of D-fructose 6-phosphate, 10 μl of PP$_i$, 10 μl of MgCl$_2$, 5 μl of DL-glyceraldehyde 3-phosphate, 10 μl of ADP, 5 μl of D-glyceraldehyde-3-phosphate dehydrogenase, 5 μl of 3-phosphoglycerate kinase, 10 μl of NAD$^+$, 5 μl of D-fructose 1,6-bisphosphate, a rate-limiting amount of PP$_i$: Fru-6-P 1-phosphotransferase, and water to a volume of 170 μl. The reaction is initiated by the addition of PP$_i$: Fru-6-P 1-phosphotransferase. A control cuvette minus the enzyme preparation measures P$_i$ release due to nonenzymic hydrolysis of glyceraldehyde 3-phosphate, which must be subtracted from the total rate. This assay is valid only when used with phosphatase-free preparations. In addition, the D-fructose 6-phosphate reagent must be essentially free of P$_i$ so that the initial absorbance values are kept low. The rates are conveniently measured with a Gilford multiple-sample absorbance-recording spectrophotometer. The cuvette compartment should be thermostatted at 30°.

Method C: Reverse Reaction

Principle. This continuous spectrophotometric assay is based on coupling D-fructose 6-phosphate formation to non-rate-limiting amounts of phosphoglucose isomerase and D-glucose-6-phosphate dehydrogenase.

With the coupling enzymes in excess, the rate of D-fructose 6-phosphate formation is equivalent to the rate of NADP$^+$ reduction, which is measured by the absorbance increase at 340 nm.

Reagents

HEPES–NaOH buffer, 170 mM, pH 7.8, containing 2.1 mM EDTA
D-Fructose 1,6-bisphosphate, 17 mM
Sodium phosphate (P$_i$), 17 mM, pH 7.8
MgCl$_2$, 102 mM
Phosphoglucose isomerase, 80 mU/μl, previously dialyzed against 1 mM EDTA, pH 7.8
D-Glucose-6-phosphate dehydrogenase, 40 mU/μl, previously dialyzed against 1 mM EDTA, pH 7.8
NADP$^+$, 8.2 mM

Procedure. The following constituents are added to a microcuvette with a 1.0-cm light path: 80 μl of HEPES buffer, 10 μl of D-fructose 1,6-bisphosphate, 10 μl of P$_i$, 10 μl of MgCl$_2$, 5 μl of phosphoglucose isomerase, 5 μl of D-glucose-6-phosphate dehydrogenase, 10 μl of NADP$^+$, a rate-limiting amount of PP$_i$: Fru-6-P 1-phosphotransferase, and water to a volume of 170 μl. D-Fructose 1,6-bisphosphate serves both as the substrate and activator. The reaction is initiated by the addition of PP$_i$: Fru-6-P 1-phosphotransferase. A control cuvette minus P$_i$ measures D-fructose-1,6-bisphosphatase activity, which must be subtracted from the total rate. The rates are conveniently measured with a Gilford multiple-sample absorbance-recording spectrophotometer. The cuvette compartment should be thermostatted at 30°.

Purification Procedure

Preparation of Germinated Seeds. Commercially available mung bean seeds (475 g) are soaked in distilled water for 12 hr in the dark at 30°. The soaked seeds are spread between 8 layers of moist cheesecloth and allowed to germinate for an additional 12 hr in the dark at 30°.

General. All purification steps are conducted at 0–4°. A typical purification is summarized in the table.

Preparation of Crude Extract. The germinated seeds are separated from the seed coat by repeated washing with distilled water and handpicking. The germinated beans, substantially free of their seed coats, are suspended about 1 : 1 (v/v) in buffer A (50 mM Tris-acetate buffer, pH 7.0; 5 mM dithiothreitol; 1 mM EDTA; and 180 mM sodium acetate) and ground with a mortar and pestle. The homogenate is filtered (hand-squeezed) through eight layers of cheesecloth and clarified by centrifugation at

PURIFICATION OF INORGANIC PYROPHOSPHATE : D-FRUCTOSE-6-PHOSPHATE
1-PHOSPHOTRANSFERASE FROM MUNG BEAN

Fraction	Volume (ml)	Total protein (mg)	Total activity[a] (units)	Specific activity[a] (units/mg protein)	Recovery (%)
Crude extract	2200	18,260	330	0.018	(100)
$(NH_4)_2SO_4$	98	4,480	275	0.061	83
DEAE-cellulose concentrate	36	1,355	244	0.180	74
BioGel A-1.5m	41	451	189	0.419	57
Phosphocellulose I	41	39.2	81.2	2.07	25
Phosphocellulose II	21	2.01	24.5	12.2	7.4

[a] The values reported were obtained with 1.5 mM D-glucose 1,6-bisphosphate replacing 1 μM D-fructose 2,6-bisphosphate in the standard assay. With D-fructose 2,6-bisphosphate as the activator, the specific activities would have been doubled.

13,000 g for 20 min. The supernatant portion is designated the crude extract.

Ammonium Sulfate Precipitation. To the crude extract is added $(NH_4)_2SO_4$ (162 g/liter), and the resulting precipitate is removed by centrifugation at 13,000 g for 20 min. To the supernatant is added another portion of $(NH_4)_2SO_4$ (87 g/liter). The resulting precipitate is collected by centrifugation, dissolved in sufficient buffer A, and dialyzed against buffer B (10 mM Tris-acetate buffer, pH 7.3; 0.1 mM EDTA; and 0.5 mM dithiothreitol).

DEAE-Cellulose Chromatography. A DEAE-cellulose column (4.6 × 25 cm) is equilibrated with buffer B. The dialyzed ammonium sulfate fraction is applied to the column, which is then washed with 2.0 liters of buffer B. The protein is eluted with a linear gradient (3300 ml; 150 ml/hr) of 0 to 0.4 M KCl in the same buffer. Three hundred thirty 10-ml fractions are collected, and those that contained most of the PP_i : Fru-6-P 1-phosphotransferase activity (fractions 80 through 130) are combined.

BioGel A-1.5m Chromatography. A BioGel A-1.5 m column (2.5 × 90 cm) is equilibrated with buffer B containing 100 mM KCl. The combined fractions from the DEAE-cellulose chromatography step are concentrated to 36 ml by pressure dialysis using a PM-30 ultrafiltration membrane (Amicon Corp., Lexington, Massachusetts). One-third of the concentrate is applied to the column and chromatographed by elution with the same buffer. Fractions (3.6 ml each) are collected, and those that contain high specific activity of PP_i : Fru-6-P 1-phosphotransferase (fractions 75 through 90) are combined. The remainder of the DEAE-cellulose concentrate is divided into two equal portions, which are separately chromatographed on the BioGel A-1.5m column as above. The PP_i: Fru-6-P 1-phosphotransferase-containing fractions from the three BioGel chromatography runs are combined.

Phosphocellulose Chromatography I. A phosphocellulose column (4.0 × 16 cm) is equilibrated with buffer C (5 mM PIPES[13] buffer, pH 6.6; 0.1 mM EDTA; and 0.5 mM dithiothreitol). Buffer B in the combined fractions from the BioGel chromatography step is replaced with buffer C by repeated dilution and concentration by pressure dialysis through a PM-30 membrane. The resulting preparation (about 150 ml) is then applied to the phosphocellulose column at a flow rate of 30 ml/hr. The column is washed with 800 ml of buffer C, and then the protein is eluted with 17 mM $Na_4P_2O_7$ in the same buffer (1600 ml; 160 ml/hr). The pyrophosphate in the resulting preparation is then removed by repeated dilution with buffer C and concentration by pressure dialysis.

Phosphocellulose Chromatography II. The enzyme from the above step was applied to a phosphocellulose column (4.0 × 8 cm) previously equilibrated with buffer C. The column was washed with 500 ml of buffer C, and then the protein is eluted with a linear gradient (1440 ml, 150 ml/hr) of 0 to 400 mM KCl in the same buffer. Two hundred and forty 6-ml fractions are collected, and those that contain most of the PP_i: Fru-6-P 1-phosphotransferase activity (fractions 170 through 192) are combined. To enhance stability of the enzyme, the buffer in the combined fractions is replaced with buffer B containing 17 mM KCl, by repeated dilution and concentration by pressure dialysis. The enzyme is 678-fold purified with an overall recovery of 7.4%. It is free of enzymes that would interfere with any of the three assays described.

Properties[6-8]

Phosphoryl Donor Specificity. Inorganic pyrophosphate ($K_m = 0.1$ mM) is the only known phosphoryl donor in the forward reaction. Compounds that will not substitute for PP_i at 1.0 mM concentrations include ATP, UTP, ADP, UDP, and phosphoenolpyruvate.

Activators and Kinetic Constants. Hexose bisphosphates activate the enzyme both by decreasing the K_m for D-fructose 6-phosphate and by increasing the V_{max}. K_a values for various hexose bisphosphates are as follows: D-glucose 1,6-bisphosphate, 0.4 mM; D-fructose 1,6-phosphate, 17 μM; and D-fructose 2,6-bisphosphate, 50 nM. K_m values for D-fructose 6-phosphate, and the relative V_{max} values, respectively, in the absence and in the presence of various activators, are as follows: no hexose bisphosphate, 20 mM, 1.0; D-glucose 1,6-bisphosphate, 5 mM,

[13] Piperazine-N,N'-bis(2-ethanesulfonic acid).

9.2; D-fructose 1,6-bisphosphate, 0.56 m*M*, 9.2; and D-fructose 2,6-bisphosphate, 0.1 m*M*, 15.2.

Inhibitors. The enzyme is inhibited by various salts, possibly by the increased ionic strength. Inhibition of 40–78% is achieved by additional 30 m*M* KCl, 30 m*M* NH$_4$Cl, 10 m*M* Na$_2$SO$_4$, 10 m*M* (NH$_4$)$_2$SO$_4$, and 5 m*M* sodium phosphate.

Divalent Metal Ion Requirement. The enzyme has an absolute requirement for a divalent metal ion. Mg^{2+} (6 m*M*) gives the highest V_{max} of the metal ions tested. Co^{2+} and Mn^{2+} are inferior substitutes for Mg^{2+}.

pH Optimum. Activity as a function of pH is maximal at about pH 7.8 in HEPES–NaOH buffer.

Stability. The enzyme (in 10 m*M* Tris-acetate buffer, pH 7.3; 0.1 m*M* EDTA; 0.5 m*M* dithiothreitol; and 17 m*M* KCl) is stable (100% in 6 months) to storage at −20° with added 20% (v/v) glycerol, and at −80° without added glycerol. It is unstable to storage at −20° in the absence of glycerol (94% decrease in 1 week). At 4° in the absence of glycerol, it retains its activity (100%) for about a week and then the activity decreases slowly over a period of weeks.

[17] 6-Phosphofructokinase (Pyrophosphate) from *Entamoeba histolytica*

By RICHARD E. REEVES, PATRICIA LOBELLE-RICH, and WILLIAM B. EUBANK

Fructose 6-phosphate^{2-} + MgPP$_i$$^{2-}$ $\rightleftharpoons$ fructose 1,6-bisphosphate^{4-} + Mg^{2+} + P$_i$$^{2-}$

A pyrophosphate-utilizing phosphofructokinase[1] functions in the glycolytic pathway of *Entamoeba histolytica*.[2] It has also been found in *Propionibacterium shermanii*,[3] in various species of *Alcaligenes* and in *Pseudomonas marina*,[4] in *Bacteroides fragilis*,[5] and in pineapple leaves.[6] Its function is less certain in some of the latter organisms. The low K_m (PP$_i$) of the amoeban enzyme suggests its use as an assay enzyme and as a linking

[1] EC 2.7.1.90.

[2] R. E. Reeves, D. J. South, H. J. Blyth, and L. G. Warren, *J. Biol. Chem.* **249**, 7737 (1974).

[3] W. E. O'Brien, S. Bowien, and H. G. Wood, *J. Biol. Chem.* **250**, 8690 (1975).

[4] M. H. Sawyer, P. Baumann, and L. Baumann, *Arch. Microbiol.* **112**, 169 (1977).

[5] J. M. Macy, L. G. Lungdahl, and G. Gottschalk, *J. Bacteriol.* **134**, 84 (1978).

[6] N. W. Carnal and C. C. Black, *Biochem. Biophys. Res. Commun.* **86**, 20 (1979).

enzyme to monitor the progress of reactions in which pyrophosphate is produced. An example of such use is given elsewhere.[7]

Assay Method

Principle. The enzyme has been assayed spectrophotometrically, in either direction, by the use of suitable linking enzymes and a pyridine nucleotide substrate.[2,3] The method given below assays reaction left to right,[2] and, with purified or partially purified enzyme, it is suitable for the assay of pyrophosphate or to monitor the rate of its formation by other systems if pyrophosphate is omitted from the assay. The assay is linear with enzyme up to 10 milliunits of enzyme per milliliter in the cuvette.

Reagents

Imidazole-HCl buffer, 0.2 M, pH 7
$MgCl_2$, 0.1 M
Inorganic pyrophosphate (PP_i), tetrasodium salt, 50 mM
Fructose 6-phosphate, disodium salt, 60 mM
NADH, 10 mM
A solution of sulfate-free rabbit muscle fructose-bisphosphate aldolase in 1 mM NaEDTA, pH 7, 100 units/ml
A solution containing sulfate-free rabbit muscle glycerol-3-phosphate dehydrogenase, 300 units/ml and triosephosphate isomerase, 1000 units/ml, in 1 mM NaEDTA, pH 7

Procedure. To a quartz cuvette of 1-cm light path is added 100 μl of imidazole buffer, 10 μl of magnesium chloride, 10 μl of PP_i, 5 μl of NADH, 5 μl of each of the solutions of assay enzymes, up to 4 milliunits of enzyme, and water to a volume of 0.39 ml. After equilibration of the cuvette at 30°, the reaction is started by the addition of 10 μl of fructose 6-phosphate. The reaction is monitored at 30° in a spectrophotometer. A control cuvette lacking pyrophosphate serves to correct for NADH oxidase activity in crude preparations. This correction is negligible following the Sephacryl 200-S column fractionation (see below).

Definition of Unit and Specific Activity. One unit is defined as that amount of enzyme necessary to catalyze the oxidation of 2 μmol of NADH per minute under the above conditions. Specific activity is defined as units of enzyme per milligram of protein. Protein concentration is determined by the method of Lowry *et al.*[8]

---

[7] See this volume [88].
[8] O. H. Lowry, N. J. Rosebrough, A. L. Farr, and R. J. Randall, *J. Biol. Chem.* **193**, 265 (1951).

Purification Procedure

Growth. The K-9 strain of *Entamoeba histolytica* was obtained from the American Type Culture Collection, ATCC 30015. It is grown aseptically at 36° in Diamond's TPS-1 broth base (North American Biologicals, Miami, Florida, Stock No. 73-9504) supplemented by Diamond's TPS-1, 40× vitamin solution (NAB, Stock No. 72-2315); 10% bovine serum, inactivated for 30 min in a 56° water bath; and 800 units of penicillin G, potassium salt, per milliliter. Stock cultures are maintained by twice-weekly transfers in well-filled 16 × 125 mm screw-cap culture tubes. Large amounts of cells are grown in well-filled, screw-cap Erlenmeyer flasks containing freshly prepared medium. Cell counts are made in a Coulter counter equipped with a 100-μm aperture tube or in a Sperry-Levy counting chamber. Inoculation, from growing cultures, is made at the level of 2 × 10^4 cells ml for 3-day growth periods, one-half that amount for 4-day growth periods. Optimal inoculation cell density varies over a severalfold range with different lots of bovine serum.

Harvesting. Harvesting is conducted in the spent growth medium. Cells are dislodged from the vessel walls by vigorous agitation and centrifuged for 4000 g × minutes in calibrated tubes to determine fresh cell volume. They are then resuspended, washed in a salt buffer,[2] and pelleted in ampoules. The pellets are frozen in a Dry Ice–acetone mixture, lyophilized, sealed under vacuum, and stored at $-20°$. (See comment 1 below.)

Step 1. Crude Enzyme. Lyophilized cells representing 6 ml of fresh cells are suspended in 30 ml of cold 20 mM potassium phosphate buffer (pH 6.5) containing 0.1 mM EDTA and centrifuged for 1 × 10^6 g × minutes. The supernatant solution is stirred with 351 mg/ml solid (NH$_4$)$_2$SO$_4$, chilled for one-half hour, and then centrifuged for 1.2 × 10^5 g × minutes. This pellet is reserved for the preparation of galactose-1-phosphate uridylyltransferase.[7] The supernatant solution is placed in a cellophane bag and dialyzed overnight against four volumes of 4.1 M ammonium sulfate. The precipitated protein is centrifuged, and the pellet is dissolved in 3 ml of the buffer. This constitutes the crude extract. Subsequent steps are conducted at 0 to 4° unless otherwise noted.

Step 2. Sephacryl 200-S Column Fractionations. The crude extract is applied to a 51-cm × 5.73-cm^2 column of Sephacryl 200-S, previously equilibrated with 20 mM imidazole-HCl buffer, pH 7, containing 0.1 mM EDTA. The column is fitted with a flow adaptor. Enzyme is eluted with the imidazole buffer in 10-ml fractions. The two fractions comprising the enzyme peak are combined.

Step 3. DEAE-Cellulose Fractionation. A column, 20 cm × 3.8 cm², is prepared from DEAE-cellulose that has been pretreated according to a published method[9] and then equilibrated with 20 mM imidazole-HCl, pH 7. Enzyme from step 2 is applied to the column, which is then washed with 120 ml of the imidazole buffer. Enzyme is eluted with 0.1 M NaCl in the buffer. Its specific activity decreases across the enzyme peak (see comment 2). Enzyme from the peak fractions was pooled and subjected to the further purification steps described below. Enzyme from this step is stable and useful for many assay purposes (see comment 2, and see Stability under Properties).

Step 4. Desalting on BioGel. A column, 28 cm × 7 cm², is prepared from BioGel P-10. It is equipped with a flow adaptor and equilibrated with deionized water. Enzyme from step 3 is applied, and elution is with water. All of the applied enzyme activity is recovered before the breakthrough of chloride ions occurs (see comment 3).

Step 5. Fractionation on Blue Sepharose. A column, 15 cm × 4.2 cm², is prepared from Blue Sepharose and equilibrated with 20 mM potassium phosphate buffer, pH 6.5. Enzyme from step 4 is applied to the column, is then washed with 50 ml of the buffer. Enzyme is eluted with 1 mM $MgPP_i$ in the buffer (see comment 4).

Step 6. Concentration on Hydroxyapatite. A column of 8 ml bed volume is prepared from BioGel HT hydroxyapatite and equilibrated with 20 mM potassium phosphate buffer, pH 7. The pooled fractions from step 5 are adjusted to pH 7 with 1 N KOH and applied to this column. The column was washed with 20 ml of the buffer, and enzyme is then eluted with 0.3 M potassium phosphate, pH 7, collecting 2-ml fractions. All of the applied activity was eluted in two adjacent fractions.

Step 7. Desalting on BioGel. A column, 16 cm × 2.54 cm², of BioGel P-10 is prepared and equipped with a flow adaptor. It is equilibrated with water. Enzyme from step 6 is applied to this column, and enzyme is eluted with water; 2-ml samples are collected. Most of the enzyme activity is collected before the breakthrough of phosphate ion occurs. Samples of this phosphate-free enzyme are lyophilized for acrylamide gel chromatography and for protein assay.

Comments on the Procedure

1. One milliliter of fresh cells packed by the standard 4000 g times minutes of centrifugation weighs 1.05 g and contains about 8 × 10⁷ cells and 92 ± 5 mg of amoeban protein. One liter of medium will support cell harvests of 2 ml of fresh cells. Cells should pack firmly on centrifugation. Opalescence in the supernatant medium or a layer of colorless cells on the

---

[9] D. J. South and R. E. Reeves, this series, Vol. 42 [31].

PURIFICATION OF PHOSPHOFRUCTOKINASE (PP$_i$) FROM LYOPHILIZED CELLS
REPRESENTING 6.2 ml OF FRESH CELLS[a]

Step and treatment	Enzyme (units)	Volume (ml)	Protein (mg)	Specific activity (units/mg protein)	Recovery (%)
1. Crude extract	159	3	83.7	1.9	(100)
2. Sephacryl 200-S	129	20	20.9	6.6	81
3. DEAE-cellulose	126	23	3.4	36.8	79
4. First desalting	114	45	2.36	48.4	72
5. Blue Sepharose	54	64	NA[b]	—	34
6. Concentration on hydroxyapatite	54	4	NA	—	34
7. Second desalting	33	8.6	0.14	240	21

[a] Experimental data are corrected for samples removed from the preparation.
[b] NA, Not assayed.

pellet indicates that the culture has passed its prime. A normal yield of enzyme may be obtained from such cells, but serial transfers made from such a culture will grow poorly.

2. Enzyme from the first one-half of the enzyme peak of step 3 is free from most interfering enzyme activities except glucose phosphate isomerase and glucose-1-phosphate uridylyltransferase. Such an enzyme may be used for assays of pyrophosphate and as a linking enzyme to monitor the rate of its formation by many enzyme systems.

3. Step 4 removes most of the contaminating glucose-1-phosphate uridylyltransferase activity and 40% of the protein.

4. Ten percent of the applied enzyme appears in the run-through and wash effluents from the Blue Sepharose column, and these fractions are discarded. Enzyme is slowly eluted by the buffer alone, more rapidly by the incorporation of MgPP$_i$, and still more rapidly by incorporation of 0.1 M NaCl in the buffer.

A typical purification is summarized in the table.

Properties

Homogeneity. Lyophilized enzyme from step 7, 25 μg of protein, gave a single protein band upon electrophoresis in 7.5% polyacrylamide gel at pH 7.[10] The protein band coincided with enzyme activity localized in a second gel. Localization of enzyme on the second gel followed treatment with a solution containing 2.5 mM MnCl$_2$, 1 mM fructose 1,6-bisphos-

[10] See this series, Vol. 22 [39].

phate, and 2 mM phosphate in 20 mM imidazole-HCl buffer, pH 7. Enzyme activity is identified by a white precipitate of manganese pyrophosphate that forms in the gel at the site of the enzyme.

Stability. Enzyme does not withstand freezing in 20 mM imidazole buffer, but salt-containing solutions from step 3 with 25% added glycerol retained 50% of their activity after 1 year in storage at $-20°$. Enzyme samples from step 4 with 10 mg of added sucrose or 2 μmol of added sodium citrate, pH 6.5, retained 50–90% of their enzyme activity upon lyophilization. Less activity was retained in samples lyophilized after step 7.

Molecular Weight. The molecular weight of the native enzyme has been reported as 8.3×10^4, as determined by chromatography on Sephadex G-200.[11]

Specificity. It has been reported that no substance other than pyrophosphate acts as a phosphoryl donor with this enzyme.[11] In the reverse reaction, arsenate may substitute for orthophosphate. Carbohydrate substrates are fructose 6-phosphate, sedoheptulose 7-phosphate, and 2,5-anhydro-D-mannitol 6-phosphate. In the reverse reaction, substrates are the same carbohydrates phosphorylated additionally at position 1. Mn^{2+}, Mg^{2+}, and Co^{2+} stimulate the reaction in the glycolytic direction, whereas calcium, nickel, and zinc ions inhibit the stimulation by magnesium ions.

Kinetic Values for Substrates and Activators. Published kinetic values[2,11] for K_m in the direction of fructose 1,6-bisphosphate formation are PP_i, 14 μM; Mg^{2+}, 8 μM; fructose 6-phosphate, 38 μM. In the direction of fructose 6-phosphate formation they are Mg^{2+}, 500 μM; fructose 1,6-bisphosphate, 18 μM; P_i, 800 μM; and arsenate, 3.6 mM. In this work the K_m for sedoheptulose 7-phosphate was found to be 60 μM, and its V_{max} was 0.26 that of fructose 6-phosphate.

[11] R. E. Reeves, R. Serrano, and D. J. South, *J. Biol. Chem.* **251**, 2958 (1976).

[18] Phosphoglycerate Kinase from Animal Tissue

By GÜNTER W. K. KUNTZ and WOLFGANG K. G. KRIETSCH

$$1,3\text{-Diphosphoglycerate} + \text{MgADP} \rightleftharpoons 3\text{-phosphoglycerate} + \text{MgATP}$$

Assay

Phosphoglycerate kinase (EC 2.7.2.3) activity is measured in the backward reaction combined with the glyceraldehyde-3-phosphate dehydrogenase (EC 1.2.1.12) reaction.[1] To avoid inactivation, the enzyme is diluted in 100 mM triethanolamine-HCl buffer (pH 7.6) containing 1 g of bovine serum albumin per liter and 1 mM dithioerythritol.

Reagents. The reaction mixture contains:
Triethanolamine-HCl buffer, 100 mM, pH 7.6
NADH, 0.3 mM
ATP, 3.8 mM
MgSO$_4$, 10 mM
Glyceraldehyde-3-phosphate dehydrogenase, 8 units/ml
Procedure. The reaction is monitored at 365 nm and 25°. The unit used is 1 μmol of NADH consumed per minute with a molar absorption coefficient for NADH of 3.4 × 10^3 M^{-1} cm^{-1}.[2]

Protein Determination

The protein concentration of the purified enzyme from vertebrates can be determined spectrophotometrically at 280 nm with an $E_{280\,\text{nm}}^{1\%} = 6.9$ (1-cm light path).[3] In contaminated phosphoglycerate kinase solutions, the protein concentration is determined by the biuret method according to Beisenherz *et al.*[4] with one-tenth of the given volumes and a color protein converting factor of 37 (vertebrate phosphoglycerate kinase).

Purification

The procedure consists of two steps: affinity chromatography on Sepharose-bound ATP and molecular sieving chromatography on

[1] T. Bücher, this series, Vol. 1, p. 415.
[2] J. Ziegenhorn, M. Senn, and T. Bücher, *Clin. Chem.* (*Winston-Salem, N.C.*) **22**, 151 (1976).
[3] W. K. G. Krietsch and T. Bücher, *Eur. J. Biochem.* **17**, 568 (1970).
[4] G. Beisenherz, H. Boltze, T. Bücher, R. Czok, K.-H. Garbade, E. Meyer-Arendt, and G. Pfleiderer, *Z. Naturforsch. B: Anorg. Chem., Org. Chem., Biochem. Biophys., Biol.* **8B**, 555 (1953).

Sephadex G-75 superfine. All stages of the isolation (including affinity gel preparation) are carried out at 4° unless otherwise specified. All pH adjustments are done at the temperature at which the buffer is used.

Preparation of the Affinity Gel[5]

1. Hexamethylenediamino-Sepharose. Washed Sepharose 4B (Pharmacia) is suspended in an equal volume of water in a glass beaker. Activation is performed under a well-ventilated hood through the addition of ground cyanogen bromide (250 mg per milliliter of agarose suspension). During the reaction, the pH is kept at 11 by titration with 8 M NaOH. A temperature of 20–25° should be maintained by adding crushed ice. The end of the reaction is evident when the consumption of base decreases. The slurry is poured into a chilled Büchner funnel, filtered by suction, and then immediately washed with a 15-fold volume of 0.2 M sodium carbonate solution (pH 10). After washing, the gel (which should be about the consistency of a moist cake) is quickly transferred into a plastic beaker. The same volume of ice-cold adipic acid dihydrazide solution (85 mg/ml, pH 9.5) as the beginning gel volume is added and is stirred for at least 4 hr. Solid NaCl is added to reach a concentration of 1 M, and the suspension is stirred for about 30 min at room temperature. Finally, the gel is washed with 20 volumes each of 0.1 M NaCl and water.

2. Coupling of Periodate-Oxidized Adenosine Triphosphate to Adipoyldihydrazo-Sepharose. Adenosine triphosphate (64 mM) and sodium metaperiodate (62 mM) are dissolved separately in water in filter flasks. Both solutions are adjusted to pH 8 and deaerated. After cooling to 0°, the two solutions are combined in equal volumes and are stirred for 1 hr in the dark. The periodate-oxidized nucleotide solution (31 mM) is added to an equal volume of ice-cold adipoyldihydrazo-Sepharose previously equilibrated with 0.5 M sodium acetate (adjusted with concentrated HCl to pH 4.8), and deaerated. After at least 4 hr of stirring, the gel is washed as described above and is ready for use.

Preparation of Muscle Extracts[5]

Soluble protein is extracted from minced (using a household blender) muscle tissue from man, mouse, trout, and rabbit for 30 min with a threefold volume of 50 mM Tris-HCl, 5 mM $MgSO_4$, pH 8.0 (standard buffer). The crude extract is centrifuged at 13,000 g for 20 min, and the supernatant is filtered through quartz wool to remove fat. A second centrifugation is performed at 100,000 g for 1 hr. Finally, the pH of the supernatant is adjusted to 8.0 by adding small portions of solid Tris.

[5] G. W. K. Kuntz, S. W. Eber, W. Kessler, H. Krietsch, and W. K. G. Krietsch, *Eur. J. Biochem.* **85**, 493 (1978).

Preparation of Hemolysate[5]

Packed erythrocytes are washed twice with 2 volumes of 0.9% NaCl. The supernatant and buffy coat are removed by suction after each centrifugation (20 min at 13,000 g). The washed erythrocytes are suspended in 3 volumes of standard buffer and disrupted under cooling (5–10°) in 0.5-liter batches by sonication at maximum power for 60 sec (Branson Model S75 sonifier). The hemolysate is centrifuged at 13,000 g for 30 min to remove cell debris, and the pH is adjusted to 8.0 with Tris.

Preparation of Mouse Testis Extract

Frozen testes, including epididymides, are finely minced using a household blender. A twofold volume of standard buffer is added, and the mixture is stirred for 30 min. This is followed by centrifugation at 100,000 g for 1 hr. The supernatant is filtered through quartz wool, and the pH is adjusted to 8.0 by adding solid Tris.

Affinity Chromatography[5]

The affinity gel is equilibrated with standard buffer and poured into a column. The bed volume chosen depends on the amount of phosphoglycerate kinase to be bound. One milliliter of the affinity gel binds optimally 5 mg of phosphoglycerate kinase. It is recommended to use approximately double the amount of gel actually needed in order to minimize the loss of unbound enzyme. The ratio of column diameter to column length is not a critical factor and may vary between four and eight. The extract is pumped onto the column with a flow rate of 1.5 ml per milliliter of bed volume per hour.

After application of the extract, the column is rinsed with twice the bed volume of standard buffer. This is followed by a wash (two column volumes) with 0.7 M NaCl dissolved in standard buffer to remove unspecifically bound protein. Another buffer rinse (one column volume) is performed to eliminate the NaCl from the affinity column. The elution of phosphoglycerate kinase can now begin by applying a solution of 40 mM ATP, 40 mM MgSO$_4$, and 50 mM Tris-HCl (pH 8.0, flow rate a half to a third of the application flow rate). The procedure ends with another standard buffer wash until no more ATP is present in the eluate. All fractions containing phosphoglycerate kinase activity are pooled.

Gel Filtration[5]

Usually a Sephadex G-75 superfine column (12.5 cm^2 × 100 cm) equilibrated with standard buffer is used. In this case, a maximal phosphoglycerate kinase pool volume of 50 ml can be applied. However, if the

TABLE I
PURIFICATION OF DIFFERENT PHOSPHOGLYCERATE KINASES[a]

Source	Extract specific activity (units/mg protein)	Affinity chromatography		Chromatography on Sephadex G-75 superfine		Purification (fold)
		Specific activity (units/mg protein)	Recovery of activity (%)	Specific activity (units/mg protein)	Recovery of activity (%)	
Trout muscle	5.2	640	96	700	90	135
Mouse muscle	7.2	635	87	710	80	99
Rabbit muscle	10.0	660	95	695	89	70
Human muscle	7.0	620	100	710	91	101
Human erythrocytes	0.2	620	88	680	85	3780
Mouse testes	1.8	110	87	600	74	333

[a] The best recovery is made with a once-used affinity gel. New gels yield a poorer recovery. The reason for this is under investigation.

volume is larger, it is advantageous to precipitate the proteins by dialysis against 3.6 M ammonium sulfate, dissolved in 50 mM Tris-HCl buffer (pH 8.0). This precipitate is then centrifuged, dissolved in a minimal volume of buffer, poured onto the column, and chromatographed with a flow rate of 30 ml/hr. With especially small affinity chromatography pool volumes, a column with a smaller diameter, but of the same length, should be used, with a correspondingly adjusted flow rate. The pooled phosphoglycerate kinase-containing fractions are again dialyzed against 3.6 M ammonium sulfate (in 50 mM Tris-HCl buffer, pH 8.0) and are thus salted out.

The specific activities of all purification steps are summarized in Table I.

The Isolation of Multiple Forms

From Human Skeletal Muscle Phosphoglycerate Kinase. CM-Sepharose CL-6B is washed, equilibrated with 10 mM sodium phosphate buffer (pH 7.0), and filled into a column (3.2 cm² × 40 cm). Purified phosphoglycerate kinase is dialyzed against the same buffer and applied to the column. Elution is performed by a linear gradient from 10 to 40 mM of sodium phosphate buffer, pH 7.0 (350 ml in each vessel).

From Mouse Testis Phosphoglycerate Kinase. A column (1 cm² × 26 cm) is filled with DEAE-Sephadex CL-6B previously equilibrated with 20 mM Tris-HCl, pH 7.8 (starting buffer). The phosphoglycerate kinase mixture is dialyzed against the same buffer and applied to the column. The separation of the phosphoglycerate kinase isozymes is performed via a hyperbolic gradient, as follows. An Erlenmeyer flask with a base diameter of 6.4 cm is filled with 70 ml of starting buffer. The shape of the flask must be such that the resulting meniscus displays a diameter of 3.2 cm. This flask is then connected with tubing to a beaker having a base diameter of 5 cm. The beaker is filled with 85 ml of starting buffer containing 100 mM NaCl. If the geometry of the vessel is correct, the menisci should be at the same level. A connecting tube is placed between the Erlenmeyer flask and the column, and the chromatography can now be performed at a flow rate of 5.6 ml/hr. With this experimental design, first phosphoglycerate kinase-1, then phosphoglycerate kinase-2B, and last phosphoglycerate kinase-2A will appear completely separated.

Properties

Purity. It is advantageous to test purity at various stages of isolation by polyacrylamide gel electrophoresis in the presence of sodium dodecyl sulfate (SDS).[5] (In the native state, polyacrylamide gel electrophoresis of phosphoglycerate kinase offers only unsatisfactory results.) When phosphoglycerate kinase is isolated according to the described method, it is homogeneous after Sephadex G-75 chromatography in SDS–gel electrophoresis.

Because they have the same molecular weight, different phosphoglycerate kinase isozymes cannot be detected using the above-mentioned electrophoresis. Therefore, electrophoresis should be carried out on Cellogel sheets and be made visible by an NADPH-generating activity stain.[6,7]

Stability. All phosphoglycerate kinases discussed in this chapter, except for the human enzyme, remain stable for years without loss of specific activity when stored as a precipitate in 3.3 M ammonium sulfate.

Molecular Weight. Based on their amino acid sequences, molecular weights of 44,515 for horse muscle phosphoglycerate kinase[8] and 44,657 for the human erythrocyte enzyme[9] have been calculated. (The sequence of the human enzyme contains one more lysine residue.) Using polyacrylamide gel electrophoresis in the presence of SDS, the molecular weight of various phosphoglycerate kinases was determined to be 46,000 ± 2000. Regardless of the source, the enzyme consists of one polypeptide chain. Through X-ray diffraction studies of yeast and horse muscle enzyme, this chain has been found to display a bilobal structure.[10,11]

[6] F. J. Oelschlegel and G. J. Brewer, *Experientia* **28**, 116 (1972).

[7] T. Bücher, W. Bender, R. Fundele, H. Hofner, and I. Linke, *FEBS Lett.* **115**, 319 (1980).

[8] R. D. Banks, C. C. F. Blake, P. R. Evans, R. Haser, D. W. Rice, G. W. Hardy, M. Merrett, and A. W. Phillips, *Nature (London)* **279**, 773 (1979).

[9] I.-J. Huang, C. D. Welch, and A. Yoshida, *J. Biol. Chem.* **255**, 6412 (1980).

[10] P. L. Wendell, T. N. Bryant, and H. C. Watson, *Nature (London), New Biol.* **240**, 134 (1972).

[11] C. C. F. Blake, P. R. Evans, and R. K. Scopes, *Nature (London), New Biol.* **235**, 195 (1972).

TABLE II
THIOL GROUPS, ISOELECTRIC POINTS, AND pH OPTIMA

| | Thiol groups | | | |
Source	Measured[a]	Nearest integer	Isoelectric points	pH Optima
Trout[13]	8.63	9	8.1	6–9
Mouse[13]	7.03	7	8.0	6.5–8.5
Rabbit[3,13]	7.0	7	8.5	6–9.3
Human[13,14]	6.0	7	8.3	6–8.5

[a] With 5,5'-dithiobis(2-nitrobenzoic acid).

Thiol Groups, Isoelectric Points, and pH Optima. The isoelectric points of vertebrate tissue phosphoglycerate kinases so far studied are slightly alkaline, and, as measured by thin-layer isoelectric focusing (pH gradient 3–9), they are listed in Table II. Electrophoretic mobility studies with a great variety of phosphoglycerate kinases[12] are in agreement with the data of Table II.[3,13,14]

When the most reactive thiol groups are not effectively protected, phosphoglycerate kinase immediately loses its enzymic activity. To determine the specific activity, therefore, a complete reduction of the enzyme, e.g., with dithioerythritol, is necessary.[15]

Substrate Specificity. The only known analog of 3-phosphoglycerate that can replace the substrate is an artificial 3-phosphoglycerate in which the phosphate group $-O-PO_3H_2$ is replaced by the phosphonomethyl group $-CH_2-PO_3H_2$ (2-hydroxy-4-phospho-DL-butyric acid).[16] The nucleotide specificity is summarized in Table III.[3,13,17]

Inhibitors and Michaelis–Menten Constants. The only known natural inhibitors of phosphoglycerate kinase are AMP, IMP, GMP,[3] and 2,3-diphosphoglycerate. The latter inhibits competitively with 3-phosphoglycerate ($K_i = 3.8$ mM) and with MgATP ($K_i = 2.3$ mM).[18] The soluble ligand-spacer–Mg complex (adipoyldihydrazo-ATP) from the affinity gel is a very strong competitive inhibitor with ATP, displaying a K_i

[12] T. Fifis and R. K. Scopes, *Biochem. J.* 175, 311 (1978).
[13] Unpublished results.
[14] W. K. G. Krietsch, S. W. Eber, B. Haas, W. Ruppelt, and G. W. K. Kuntz, *Am. J. Hum. Genet.* 32, 364 (1980).
[15] W. K. G. Krietsch, I. U. Freier, and S. W. Eber, *Arch. Biochem. Biophys.* 193, 415 (1979).
[16] G. A. Ort and J. R. Knowles, *Biochem. J.* 141, 721 (1974).
[17] W. K. G. Krietsch, H. Krietsch, W. Kaiser, M. Dünnwald, G. W. K. Kuntz, J. Duhm, and T. Bücher, *Eur. J. Clin. Invest.* 7, 427 (1977).
[18] J. Ponce, S. Roth, and D. R. Harkness, *Biochim. Biophys. Acta* 250, 63 (1971).

TABLE III
NUCLEOSIDE TRIPHOSPHATE SPECIFICITY[a]

| Nucleoside | Percentage of activity of phosphoglycerate kinase from | | | |
	Mouse[13]	Human[17]	Rabbit[3]	Trout[13]
ATP	100	100	100	100
dATP	16	—	—	—
GTP	55	53	54	55
dGTP	21	—	—	—
ITP	65	64	64	53
dITP	3	—	—	—
CTP	0	0	0	0
UTP	9.6	6.8	8.8	7.1
dTTP	0	0	0	0

[a] The specificity of nucleoside diphosphates is the same as for triphosphates.

TABLE IV
MICHAELIS–MENTEN CONSTANTS (mM)

| Enzyme | MgATP | 3-PG[a] | | MgADP | 1,3-DPG[b] |
		0.08–1.5 mM	1.5–13 mM		
Mouse[13]	0.43	0.65	—	0.109	0.0025
Human[14,17]	0.35	0.4	0.85	0.083	0.0019
Rabbit[3]	0.42	0.42	1.37	0.15	0.0022
Trout[13]	0.46	0.35	—	—	—

[a] 3-PG, 3-Phosphoglycerate; two substrate concentration ranges were measured.
[b] 1,3-DPG, 1,3-Diphosphoglycerate.

of 2 μM.[13,19] Other nonphysiological inhibitors are, e.g., Cibacron Blue,[20] Congo Red,[21] and Cr(III).[22] The Michaelis–Menten constants are summarized in Table IV.

Multiple Forms. Since phosphoglycerate kinase is a monomeric enzyme for which only one cistron is known, multiple forms are the consequence of epigenetic factors involving modifications of the primary chain.[15] The

[19] G. W. K. Kuntz and W. K. G. Krietsch, *Fresenius' Z. Anal. Chem.* **290**, 186 (1978).
[20] S. T. Thompson and E. Stellwagen, *Proc. Natl. Acad. Sci. U.S.A.* **73**, 361 (1976).
[21] R. A. Edwards and R. W. Woody, *Biochemistry* **18**, 5197 (1979).
[22] C. A. Janson and W. W. Cleland, *J. Biol. Chem.* **249**, 2567 (1974).

phosphoglycerate kinase cistron of mammalian somatic and female germ cells are linked to the X chromosome (phosphoglycerate kinase-1).[23,24] However, in sperm cells, phosphoglycerate kinase is expressed from an autosomal cistron (phosphoglycerate kinase-2).[25]

Acknowledgment

The authors wish to extend their special thanks to Ms. M. L. Everett for linguistic assistance and for typing the manuscript.

[23] W. N. Valentine, H. S. Hsieh, D. E. Paglia, H. M. Anderson, M. A. Baughan, E. R. Jaffé, and O. M. Garson, *N. Engl. J. Med.* **280**, 528 (1969).
[24] P. M. Khan, A. Westerveld, K. H. Grzeschik, B. F. Deys, O. M. Garson, and M. Siniscalco, *Am. J. Hum. Genet.* **23**, 614 (1971).
[25] J. L. VandeBerg, D. W. Cooper, and P. J. Close, *Nature (London), New Biol.* **243**, 48 (1973).

[19] Phosphoglycerate Kinase from Spinach, Blue-Green Algae, and Yeast

By GÜNTER W. K. KUNTZ and WOLFGANG K. G. KRIETSCH

$$1,3\text{-Diphosphoglycerate} + \text{MgADP} \rightleftharpoons 3\text{-phosphoglycerate} + \text{MgATP}$$

Assay

The assay principle and procedure are as given in this volume [18].

Protein Determination

Protein determination of purified plant phosphoglycerate kinases is performed spectrophotometrically using an $E_{280\,\text{nm}}^{1\%} = 5.0$ with a 1-cm light path.[1] In contaminated phosphoglycerate kinase solutions, protein determination is performed by the biuret method (see this volume [18]).

Purification

Preparation of Spinach Extract [2]

Fresh spinach leaves are homogenized with a Waring blender. To 1 kg of leaves is added 500 ml of 0.3 M Tris-HCl (pH 8.0), and the suspension is

[1] T. Bücher, this series, Vol. 1, p. 415.
[2] G. W. K. Kuntz, S. W. Eber, W. Kessler, H. Krietsch, and W. K. G. Krietsch, *Eur. J. Biochem.* **85**, 493 (1978).

stirred for 1 hr. The homogenate is passed through a common household fruit press, then centrifuged at 46,000 g for 30 min. The supernatant is dialyzed twice against five volumes of 50 mM Tris-HCl (pH 8.0) and then centrifuged at 100,000 g. (Dialysis is necessary because of the high organic acid content in the crude extract, which prevents adsorption to the affinity gel.) Magnesium sulfate is added until a concentration of 5 mM is reached, and the pH is then adjusted to 8.0 using solid Tris.

Preparation of Yeast Extract[3]

Brewers' yeast (260 g) is washed several times with tap water and allowed to dry at room temperature for a few days. The cells are disrupted in a ball mill and suspended in 490 ml of 1 M ammonia. The suspension is kept at room temperature for 20 hr and then diluted with 1.2 liters of distilled water and 34 ml of 0.5 M EDTA and 84 g of ammonium sulfate. The mixture is incubated at 60° for 10 min and subsequently centrifuged. The supernatant is then dialyzed against a buffer containing 50 mM Tris, 5 mM MgSO$_4$, pH 8.0 (standard buffer).

Preparation of Extract from Blue-Green Algae

From Fresh Algae. After a 14-day cultivation,[4] *Spirulina platensis* is collected by centrifugation at 2800 g for 30 min. The pellet is suspended in a threefold volume of standard buffer. The cells are then disrupted by sonication with three bursts of full power for 1 min (Branson Model S75 sonifier), and cell debris is removed by centrifugation at 100,000 g for 2 hr. The pH must now be adjusted to 8.0 using solid Tris. Before being applied to the affinity column, the extract has to be dialyzed against standard buffer.

From Spray-Dried Algae.[5] Spray-dried *Spirulina geitleri*[6] is poured slowly into a ninefold volume (w/v) of a vigorously stirred 0.8 M ammonium sulfate solution. The pH of the suspension is adjusted to 8.0 by the addition of solid Tris. After stirring for 5 hr, ammonium sulfate is added to achieve a concentration of 2.0 M. The suspension is stirred overnight and then centrifuged at 46,000 g for 30 min. From the slightly green supernatant, phosphoglycerate kinase is precipitated by the addition of another portion of ammonium sulfate until a concentration of 3.2 M is reached. By this method, about 200 units of phosphoglycerate kinase are

---

[3] W. K. G. Krietsch, P. G. Pentchev, H. Klingenburg, T. Hofstötter, and T. Bücher, *Eur. J. Biochem.* **14**, 289 (1970).

[4] W. K. G. Krietsch and G. W. K. Kuntz, *Anal. Biochem.* **90**, 829 (1978).

[5] Unpublished results.

[6] Available from Sosa Texcoco, S. A., Sullivan 51, Mexico 4, D.F. or from Dr. W. Behr, Friedrich-Breuer-Strasse 86, 5300 Bonn 3, West Germany.

TABLE I

PURIFICATION OF DIFFERENT PHOSPHOGLYCERATE KINASES[a]

Source	Extract specific activity (units/mg protein)	Affinity chromatography		Chromatography on Sephadex G-75 superfine		
		Specific activity (units/mg protein)	Recovery (%)	Specific activity (units/mg protein)	Recovery (%)	Purification (fold)
Spirulina geitleri (spray-dried)[5]	20[b]	450	85	720	80	36
Spinach[2]	6.6	480	92	702	102[c]	106
Yeast[5]	6.0	430	87	785	85	131

[a] The best recovery is made with a once used affinity gel. New gels yield a poorer recovery. The reason for this is under investigation.
[b] After ammonium sulfate precipitation.
[c] This recovery of more than 100% is due to the presence of tannin-like substances that act as phosphoglycerate kinase inhibitors. These are not removed until the gel chromatography step, which leads to an increase in the recovered activity.

derived from 1 g of spray-dried algae. After centrifugation, the pellet is dissolved with and dialyzed against standard buffer, and the protein solution is then ready for affinity chromatography.

Further Purification Steps

The preparation of the affinity gel, affinity chromatography, and gel chromatography are performed as described in this volume [18]. Note an exception in the affinity chromatography of yeast phosphoglycerate kinase: the NaCl wash is performed with 0.4 M NaCl instead of 0.7 M NaCl.

The specific activities of all purification steps are summarized in Table I.

Properties

Purity. Purity testing is performed as described in this volume [18] for phosphoglycerate kinase isolated from vertebrates. Phosphoglycerate kinases isolated from plants by the above method are homogeneous in polyacrylamide gel electrophoresis in the presence of sodium dodecyl sulfate.

TABLE II
THIOL GROUPS, ISOELECTRIC POINTS, AND pH OPTIMA

	Thiol groups			
Source	Measured[a]	Nearest integer	Isoelectric points	pH Optima
Yeast[7,8]	0.99	1	5.3[b]	6–9.2
Spinach[5]	2.2	2	4.3	6.5–9.5
Spirulina[5]	4.8	5	4.8	—

[a] Measured with 5,5′-dithiobis(2-nitrobenzoic acid).
[b] Another method yielded a value of 7.0.[8]

Stability. All phosphoglycerate kinases discussed in this chapter remain stable for years without loss of specific activity when stored as precipitate in 3.3 M ammonium sulfate.

Molecular Weight. On the basis of sodium dodecyl sulfate–polyacrylamide gel electrophoresis studies, the molecular weight of plant phosphoglycerate kinases has been found to be the same as for the vertebrate enzyme: 46,000 (see this volume [18]).

TABLE III
NUCLEOSIDE TRIPHOSPHATE SPECIFICITY[a]

	Percentage of activity of phosphoglycerate kinase from		
Nucleosides	*Spirulina*[4]	Yeast[5]	Spinach[5]
ATP	100	100	100
dATP	4	20	—
GTP	0.4	55	4.8
dGTP	0	35	—
ITP	0.1	70	7.7
dITP	0	40	—
CTP	0	0	0
dCTP	0	0	—
UTP	0	1	7.7
dUTP	0	0	—
dTTP	0	0	0

[a] The specificity of nucleoside diphosphates is the same as for monophosphates.

TABLE IV
MICHAELIS–MENTEN CONSTANTS (mM)

Enzyme	MgATP	3-PG[a]		MgADP	1,3-DPG[b]
		0.08–1.5 mM	1.5–13 mM		
Spirulina geitleri[5]	0.48	—	1.5	—	—
Yeast[8]	0.48	0.69	1.28	0.2	0.0018
Spinach[5]	0.3	—	1.1	0.27	0.0002

[a] 3-Phosphoglycerate; two substrate concentration ranges were measured.
[b] 1,3-DPG, 1,3-Diphosphoglycerate.

Thiol Groups, Isoelectric Points, and pH Optima. The isoelectric points of plant phosphoglycerate kinases so far studied are in the acidic range. Using thin-layer isoelectric focusing (pH gradient 3–9), the results given in Table II are obtained. These data are in agreement with electrophoretic mobility studies with a great variety of phosphoglycerate kinases.[7]

The sole thiol group of the yeast enzyme can be modified by 5,5′-dithiobis(2-nitrobenzoic acid) (DTNB) or p-chloromercuribenzoate without loss of activity.[8] In the native *Spirulina* enzyme, one of the five thiol groups reacts with DTNB with no loss of activity. Of the two thiol groups of the native spinach enzyme, one is modified by DTNB with concomitant loss of activity.[5]

Substrate Specificity. The information on substrate specificity of plants is the same as that given for vertebrates in this volume [18]. Worthy of mention is the high nucleotide specificity of the spinach enzyme and especially of the *Spirulina* enzyme (Table III). The latter enzyme allows the specific enzymic determination of adenine nucleotides.[9]

Inhibitors and Michaelis–Menten Constants. For information about inhibitors of phosphoglycerate kinase, see this volume [18]. The Michaelis–Menten constants of plant phosphoglycerate kinases mentioned here are summarized in Table IV.

Acknowledgment

The authors wish to extend their special thanks to Ms. M. L. Everett for linguistic assistance and for typing the manuscript.

[7] T. Fifis and R. K. Scopes, *Biochem. J.* **175**, 311 (1978).
[8] W. K. G. Krietsch and T. Bücher, *Eur. J. Biochem.* **17**, 568 (1970).
[9] T. P. M. Akerboom, W. K. G. Krietsch, G. W. K. Kuntz, and H. Sies, *FEBS Lett.* **105**, 90 (1979).

[20] 3-Phosphoglycerate Kinase from Bovine Liver and Yeast

By Klaus D. Kulbe and Marina Bojanovski

$$1,3\text{-BPG} + \text{MgADP} \rightleftharpoons 3\text{-PG} + \text{MgATP}$$

Assay Method

Principle. 3-Phosphoglycerate kinase (ATP:3-phospho-D-glycerate 1-phosphotransferase, EC 2.7.2.3) activity was assayed spectrophotometrically at 366 nm in a test system coupled with D-glyceraldehyde-3-phosphate dehydrogenase (EC 1.2.1.12) as described by Bücher.[1]

Reagents. Prepare a stock solution of the following composition:
   38.00 ml of triethanolamine buffer, 100 mM
   0.60 ml of MgSO$_4$, 100 mM
   0.75 ml of EDTA, Na$_2$, 34 mM
   75.0 mg of 3-phospho-D-glycerate, tricyclohexylammonium salt
   30.0 mg of ATP, Na$_2$
   7.5 mg of NADH, Na$_2$
This mixture is adjusted to pH 7.5 with NaOH and should be consumed on the day of preparation.

Procedure. To 0.75 ml of the stock solution in an 1-cm cuvette, 5 μl of D-glyceraldehyde-3-phosphate dehydrogenase (10 mg/ml; 4 units) are added; after 1 min the reaction is started by addition of 10 μl of suitably diluted kinase to give an A_{366} of 0.1–0.3 per minute at 25°.

The concentrations of the reaction participants in the final volume of 0.765 ml are as follows: 1.5 mM MgSO$_4$, 0.65 mM EDTA, 0.27 mM NADH, 3.5 mM 3-phospho-D-glycerate, and 1.25 mM ATP.

One unit of enzyme activity is defined as the amount of the enzyme required for the formation of 1 μmol of D-glyceraldehyde 3-phosphate or NAD$^+$ per minute at 25°. Specific activity is expressed as units/mg protein. Protein is determined by the method of Lowry *et al.*[2] Ammonium sulfate concentrations are measured by titration with BaCl$_2$.[3] Conductivity of solutions is determined with a Philips PW 9501 conductivity meter. Centrifugations are carried out at 0° and either at 23,000g for 30 min

---

[1] T. Bücher, *Biochim. Biophys. Acta* **1**, 292 (1947).
[2] O. H. Lowry, N. J. Rosebrough, A. L. Farr, and R. J. Randall, *J. Biol. Chem.* **193**, 265 (1951).
[3] H. U. Bergmeyer, G. Holz, E. M. Kauder, H. Möllering, and O. Wieland, *Biochem. Z.* **333**, 471 (1961).

or at 13,700 g for 40 min. Small volumes are centrifuged for 20 min at 48,000 g.

Purification of Bovine Liver 3-Phosphoglycerate Kinase

This procedure has already been reported in detail,[4] but in this description some improvements have been included. A scheme, similar to the following one, was also applied to the pig[5-7] and human liver enzymes. All steps are carried out at 4–6° unless otherwise indicated.

During the purification procedure the following buffers are used.

Buffer I: 5 mM triethanolamine-HCl, 5 mM EDTA, and 20 mM 2-mercaptoethanol titrated to pH 7.5 with NaOH (conductivity 1.2 mmho)

Buffer II: 20 mM Tris, 2 mM EDTA, and 20 mM 2-mercaptoethanol titrated to pH 7.0 with NaOH (conductivity 0.32 mmho)

Preparation of Crude Extract and Ammonium Sulfate Fractionation. One kilogram of bovine liver is washed with cold water, cut into small pieces, and homogenized in a commercial blender for 3 min with three volumes of a solution containing 10 mM EDTA, 20 mM 2-mercaptoethanol, and 1.8 M ammonium sulfate per liter (pH 7.0). The water content of this homogenate is calculated from the portion of water in liver according to Bergmeyer *et al.*[8] and the volume of buffer added. The solution obtained is slowly stirred for 1 hr. Insoluble materials are separated by centrifugation. To remove particles of fat, the supernatant was filtered through quartz wool. Ammonium sulfate was added to the filtrate in small portions to a concentration of 3.25 M within 30 min. The suspension is further stirred for 3 hr and then centrifuged for 40 min. The sediment is collected and dissolved in a minimum volume of buffer I.

Heat Denaturation in the Presence of Magnesium Sulfate. After addition of 10% (v/v) 0.1 M MgSO$_4$, the protein solution is placed in a water bath maintained at 53–54° and heated with continuous shaking to 50°. At this temperature the mixture is incubated for 10 min. The suspension is then rapidly cooled to 10° in an ice-water bath. The denatured proteins are separated by centrifugation.

DEAE-Cellulose Batch Chromatography. The supernatant is dialyzed exhaustively against several changes of buffer II, and a small insoluble

[4] M. Bojanovski, K. D. Kulbe, and W. Lamprecht, *Eur. J. Biochem.* **45,** 321 (1974).
[5] H. Foellmer, M.-R. Kula, and K. D. Kulbe, *Hoppe-Seyler's Z. Physiol. Chem.* **358,** 232 (1977).
[6] J. Fuchs and K. D. Kulbe, *Hoppe-Seyler's Z. Physiol. Chem.* **360,** 1146 (1979).
[7] K. D. Kulbe, H. Foellmer, and J. Fuchs, this volume [81].
[8] H. U. Bergmeyer, E. Bernt, M. Grassl, and G. Michal, *in* "Methoden der enzymatischen Analyse" (H. U. Bergmeyer, ed.), 2nd Ed., Vol. 1, p. 273. Verlag Chemie, Weinheim, 1970.

residue is removed by brief centrifugation. The solution is diluted with about its own volume of deionized water to give a conductivity of 0.6 to 0.7 mmho. DEAE-cellulose is equilibrated with buffer II adjusted to pH 8.0. For each kilogram of the DEAE-cellulose cake, enzyme solution corresponding to 7 g of protein is added, mixed thoroughly, and allowed to stand for 30 min at 4° before filtration through a large Büchner funnel. The cake is washed three times with one volume of buffer II, pH 7.0. By this procedure about 80% of the original activity is recovered in the filtrate and washes.

Sephadex G-75 Gel Filtration. The concentrated enzyme solution (30–40 mg of protein per milliliter) is centrifuged and the clear supernatant is pumped onto a Sephadex G-75 column (90 × 5 cm) equilibrated with buffer I. Elution is started with the same buffer. Fractions containing more than 10 units/ml are pooled and then concentrated by dialysis against a solution of polyethylene glycol 6000 (30%, w/v) in buffer I and finally against this buffer alone.

DEAE-Sephadex A-50 Ion-Exchange Chromatography. Any precipitate observed after the concentration step is separated by centrifugation, and the clear solution, which should have a conductivity of not more than 1.2 mmho, is placed on a DEAE-Sephadex A-50 column (130 × 5 cm) previously equilibrated with buffer I. Fractions containing the kinase activity are combined, and protein is salted out with solid ammonium sulfate.

Sephadex G-100 Gel Filtration. The enzyme suspension is centrifuged for 20 min, and the precipitate is redissolved in a minimum volume of buffer I (approximately 5 ml). The solution so obtained is again cleared by centrifugation as above and poured onto a Sephadex G-100 column (110 × 3 cm). This column is equilibrated and eluted with buffer I, pH 7.5.

Quantitative details of a typical purification procedure are presented in Table I. A specific activity of 130–150 units/mg was routinely obtained. The overall purification was 250-fold, and 17% of the original enzyme activity was recovered. The maximum specific activity attained is about one-fifth of that gained for the best preparations of muscle and yeast phosphoglycerate kinase under these conditions. Liver phosphoglycerate kinase crystallizes in fine needles from ammonium sulfate. A more detailed description of the crystallization process and optimum storage conditions for bovine liver 3-phosphoglycerate kinase has been published elsewhere.[4]

Properties

Chemical and physicochemical properties of bovine liver 3-phosphoglycerate kinase have been described by our group in great detail.[9] An

---

[9] K. D. Kulbe, M. Bojanovski, and W. Lamprecht, *Eur. J. Biochem.* **52**, 239 (1975).

TABLE I
PURIFICATION OF PHOSPHOGLYCERATE KINASE FROM BOVINE LIVER[a]

Step	Total activity (units)	Total protein (mg)	Specific activity (units/mg protein)
1. Extraction	42,600	75,300	0.57
2. $(NH_4)_2SO_4$ fractionation, 1.8–3.25 M	33,000	40,000	0.83
3. Heat treatment	20,000	15,000	1.3
4. DEAE-cellulose batch	16,600	2,400	7
5. Sephadex G-75	14,900	865	17
6. DEAE-Sephadex A-50	8,000	125	64
7. Sephadex G-100	7,300	50	143
Yield	17%	—	—

[a] A typical procedure starting with 1 kg of tissue is shown.

alternative preparation scheme for bovine liver phosphoglycerate kinase is discussed elsewhere.[7]

Purification of Yeast 3-Phosphoglycerate Kinase

A previous report on yeast phosphoglycerate kinase preparation was given by Scopes.[10] Based on a previous publication[11] our contribution presents an improved purification scheme for phosphoglycerate kinase from bakers' yeast. In the new procedure anion-exchange and hydroxyapatite chromatography are replaced by reversible salting-out chromatography on Sepharose CL-6B.

Autolysis. Autolysis of bakers' yeast is achieved at room temperature by slowly adding 1 kg of the air-dried material[12] over a period of 30 min to 2 liters of a 0.75 M ammonium solution containing 1 g of EDTA per liter. After 6 hr of vigorous stirring, the yeast sludge is left overnight at room temperature. Stirring is recommenced the next morning. Then 1.6 liters of 0.5 M lactic acid are added, and the pH is adjusted to 7 ± 0.5 with either concentrated ammonia or 5 M lactic acid; the pH is measured on a 10-fold diluted sample. The suspension is centrifuged for 40 min, and the precipitate is discarded.

[10] R. K. Scopes, this series, Vol. 42, p. 134.
[11] K. D. Kulbe and R. Schuer, *Anal. Biochem.* **93**, 46 (1979).
[12] Fresh bakers' yeast (*Saccharomyces cerevisiae*) was purchased from Deutsche Hefewerke-Universal Hefe Company (Hamburg/Wandsbek, Germany) and air dried at room temperature for about 1 week. One kilogram of this material corresponds to about 3.5 kg of the original yeast.

Ammonium Sulfate Precipitation. Solid ammonium sulfate is added to the orange-brown supernatant liquid to a concentration of 1.85 M. After centrifugation the supernatant is adjusted to pH 5.7.

Reversible Salting Out on Sepharose CL-6B. The slightly turbid enzyme solution is layered directly onto a 5 × 40 cm column of Sepharose CL-6B equilibrated with 2.35 M ammonium sulfate in 10 mM triethanolamine-HCl at pH 5.7. This column is developed by a linear gradient of decreasing ammonium sulfate concentration by mixing 10 mM triethanolamine-HCl, pH 5.7, into the buffer used for equilibration of the column. Phosphoglycerate kinase activity is recovered almost quantitatively at an ammonium sulfate concentration of 1.7 M.

Heat Treatment. The enzyme solution is adjusted to pH 7.5, heated in a water bath of 53°, and held at 50° for 10 min. The suspension is cooled in ice and centrifuged for 30 min.

Pseudoaffinity Chromatography on Cibacron Blue 3G-A-Substituted Sepharose 4B. Kinase activity is applied directly to a 5 × 25 cm column of Cibacron Blue 3G-A-substituted Sepharose 4B[13] equilibrated with 10 mM triethanolamine-HCl at pH 7.5. After washing with 1 liter of this buffer, phosphoglycerate kinase activity is eluted either stepwise by addition of 0.4 M sodium chloride to the pH 7.5 buffer (5.2 mmho) or by a linear gradient to 1 M sodium chloride in a total volume of 4 liters. Fractions containing more than 250 units/ml are combined. About 90% of the kinase activity is recovered at a conductivity of approximately 4.3 mmho at 6°, which corresponds to about 0.33 M NaCl. The enzyme solution is concentrated by dialysis against a 25–35% solution of polyethylene glycol (type 20,000) in 10 mM sodium phosphate at pH 6.5.

Sephadex G-100 Gel Filtration. The concentrated sample is centrifuged and applied to a column (5 × 130 cm) of Sephadex G-100 and eluted by 10 mM triethanolamine-HCl at pH 7.5. Three well-separated protein peaks are detected; fractions with phosphoglycerate kinase activity (second peak) exceeding 320 units/ml are pooled.

Detailed results of a typical purification of phosphoglycerate kinase from yeast are given in Table II. Only one band of protein was observed on disc gel electrophoresis. The enzyme was stored in either a lyophilized state or as a suspension in 2.3 M ammonium sulfate soltuion at 4–6°.

Comments on the Purification Procedures

Interaction between Cibacron Blue-substituted Sepharose 4B and proteins possessing nucleotide folds was found to be relatively strong. This

---

[13] H. J. Böhme, G. Kopperschläger, J. Schulz, and E. Hofmann, *J. Chromatogr.* **69**, 209 (1972).

TABLE II

PURIFICATION SCHEME FOR PHOSPHOGLYCERATE KINASE[a]

Step	Total activity (k units)	Total protein (g)	Specific activity (units/mg protein)
1. Autolysis by ammonia	2924	113	26
2. Ammonium sulfate precipitation (supernatant, 1.85 M, pH 5.7)	2886	90.7	31
3. Sepharose CL-6B	2643	15.6	170
4. Heat treatment	2185	7.15	306
5. Cibacron Blue pseudoaffinity chromatography	2010	2.780	722
6. Sephadex G-100	1940	2.060	945
Yield	64%	—	—

[a] A typical procedure starting from 1 kg of yeast is shown.

material is characterized by good flow properties and a very high capacity for nucleotide-dependent enzymes; e.g., a column of 500-ml volume can bind about 7 g of protein. Free binding capacity is easily detected by observing the shift in color from dark to light blue during the binding process.

Dye-substituted Sepharose CL-6B and Sephadex G-100 are not well suited for purification of yeast phosphoglycerate kinase. Whereas interactions between the Sepharose derivative and the enzyme were too strong, the Sephadex derivative copurifies other proteins that are only partially separable in the purification steps that follow.

Contrary to the findings of Stellwagen et al.[14,15] it was not possible to bind yeast phosphoglycerate kinase on Blue Dextran–Sepharose in large-scale experiments. However, phosphoglycerate kinase from a liver extract could be purified to homogeneity on Blue Dextran–Sepharose 4B by biospecific elution with 1 mM MgATP in 0.1 M Tris at pH 7.5 after preliminary ammonium sulfate fractionation, heat treatment, and Sephadex G-75 gel filtration. An alternative new way for purification of yeast and bovine liver enzymes by affinity elution chromatography was published by Fifis and Scopes.[16]

[14] S. T. Thompson, K. H. Cass, and E. Stellwagen, Proc. Natl. Acad. Sci. U.S.A. 72, 669 (1975).
[15] E. Stellwagen, Acc. Chem. Res. 10, 92 (1977).
[16] T. Fifis and R. K. Scopes, Biochem. J. 175, 311 (1978).

[21] 3-Phosphoglycerate Kinase Isozymes and Genetic Variants from Mouse

By Chi-Yu Lee

$$\text{3-Phosphoglycerate} + \text{ATP} \overset{\text{Mg}^{2+}}{\rightleftharpoons} \text{1,3-diphosphoglycerate} + \text{ADP}$$

Two isozymes of 3-phosphoglycerate kinase (EC 2.7.2.3; ATP:3-phospho-D-glycerate 1-phosphotransferase) in mice are generally designated as PGK-1 and PGK-2. PGK-1 is found in all somatic tissues[1-3] and is coded by a gene on the X chromosome. On the contrary, PGK-2 is mapped on chromosome 17 in mice and is present only in mature testis and spermatozoa.[4] Three genetic variants of PGK-2 were identified among inbred strains of mice. They were designated as PGK-2B, PGK-2A, and PGK-2C based on increasing anodal electrophoretic mobility by starch gel electrophoresis.[4-6]

Assay Method

3-Phosphoglycerate kinase is routinely assayed by the formation of 1,3-diphosphoglycerate.[7] The reaction is detected as the consumption of NADH in the reduction of 1,3-diphosphoglycerate to glyceraldehyde 3-phosphate using excess glyceraldehyde-3-phosphate dehydrogenase as the coupling enzyme. In a final volume of 1 ml, the reaction mixture contains 100 mM Tris-HCl, pH 8.0, 10 mM MgCl$_2$, 0.15 mM NADH, 2 mM ATP, 6 mM 3-phosphoglycerate, and 10 μl of glyceraldehyde-3-phosphate dehydrogenase (10 mg/ml in 1.2 M ammonium sulfate). A suitable amount of phosphoglycerate kinase is added to cause a change in absorbance at 340 nm of 0.05 to 0.1 A/min at 25°. One unit of enzyme activity is defined as the amount of enzyme that catalyzes the formation of 1 μmol of 1,3-diphosphoglycerate per minute under the described assay conditions.

[1] J. L. VandeBerg, D. W. Cooper, and P. J. Close, *Nature (London), New Biol.* **243**, 48 (1973).
[2] S.-H. Chen, L. A. Malcolm, A. Yoshida, and E. R. Giblett, *Am. J. Hum. Genet.* **23**, 87 (1971).
[3] L. P. Kozak, G. K. Mclean, and E. M. Eicher, *Biochem. Genet.* **11**, 41 (1974).
[4] E. M. Eicher, M. Cherry, and L. Flaherty, *Mol. Gen. Genet.* **158**, 225 (1978).
[5] J. L. VandeBerg, *Genetics* **86**, s66 (1977).
[6] J. L. VandeBerg and S. V. Blohm, *J. Exp. Zool.* **201**, 479 (1977).
[7] W. K. G. Krietsch and T. Bücher, *Eur. J. Biochem.* **17**, 568 (1970).

Affinity Column

3-Phosphoglycerate kinase utilizes ATP as the coenzyme. Hence this enzyme can be substantially enriched by general ligand affinity chromatography using 8-(6-aminohexyl)amino-ATP as the ligand. The preparation, maintenance, and commercial sources of this affinity agent and column have been described elsewhere.[8,9]

Purification Procedure

All the mouse strains employed in the purification of isozymes and genetic variants of 3-phosphoglycerate kinase are from the Jackson Laboratory, Bar Harbor, Maine. Phosphate buffer(s) employed in different purification steps are the potassium form.

3-Phosphoglycerate Kinase-2 from Mouse Testis

In this procedure, we take advantage of the fact that an 8-(6-amino-hexyl)amino-ATP–Sepharose column is able to adsorb 3-phosphoglycerate kinase (kinase) as well as lactate dehydrogenase (NAD^+-dependent dehydrogenase) from mouse testicular extract. A sequential biospecific elution with ATP and reduced NAD^+-pyruvate adduct results in a substantial enrichment of these two sperm-specific enzymes: 3-phosphoglycerate kinase-2 and lactate dehydrogenase-X.

Step 1. Frozen testes (20 g) from C57BL/6J mice are homogenized in 100 ml of 10 mM phosphate, pH 6.5. After centrifugation at 27,000 g for 20 min, the supernatant is filtered through glass wool to remove the suspended lipid and loaded on an 8-(6-aminohexyl)amino-ATP–Sepharose column (2 × 20 cm, 50 ml) at 4°. Both 3-phosphoglycerate kinase and lactate dehydrogenase are adsorbed. After the wash with two column volumes of 10 mM phosphate, pH 6.5, 3-phosphoglycerate kinase is eluted biospecifically with 2 mM ATP in the same buffer. Lactate dehydrogenase is subsequently eluted with reduced NAD^+-pyruvate adduct (0.2 absorbance per milliliter at 340 nm) in 10 mM phosphate, pH 6.5. Reduced NAD^+-pyruvate adduct is a known specific inhibitor of lactate dehydrogenase and is prepared according to the procedure described elsewhere.[8]

Step 2. Fractions containing PGK activity are pooled and concentrated to 5 ml. After adjusting the pH to 8.0 by a slow addition of 1 M NH$_4$OH, the enzyme solution is loaded on a DEAE-Sephadex column (2 × 10 ml, 30 ml) equilibrated with 10 mM phosphate, pH 8.0. About 20% of the total

[8] C.-Y. Lee, J. H. Yuan, and E. Goldberg, this series, Vol. 89 [61].
[9] D. W. Niesel, G. Bewley, C.-Y. Lee, and F. B. Armstrong, this series, Vol. 89 [51].

PGK activity is eluted in broad fractions with the same buffer. The remaining PGK activity is eluted with a 0 to 0.1 M NaCl gradient (200 by 200 ml) in the same buffer. The salt-eluted PGK fractions are pooled and concentrated to 1 ml by ultrafiltration. It is then loaded on a Sephadex G-75 gel filtration column (1.5 × 75 cm) equilibrated with 0.1 M phosphate, pH 7.5, containing 1 mM 2-mercaptoethanol. After elution with the same buffer, fractions containing homogeneous PGK (judged by sodium dodecyl sulfate–acrylamide gel electrophoresis) are pooled and concentrated to 1 ml. 3-Phosphoglycerate kinase-2 purified by this procedure has a specific activity of 450 units/mg under the described assay conditions.

Lactate dehydrogenase eluted from the ATP column is further purified by DEAE-Sephadex ion-exchange chromatography according to the procedure described previously.[8]

3-Phosphoglycerate Kinase-1 from Mouse Muscle

Purification of 3-phosphoglycerate kinase-1 from mouse muscle follows essentially the same procedure as that for 3-phosphoglycerate kinase-2 (above). Frozen abdominal muscle (5 g) is homogenized in 20 ml of 10 mM phosphate, pH 6.5, containing 1 mM 2-mercaptoethanol. After centrifugation, the supernatant is loaded on an 8-(6-aminohexyl)amino-ATP–Sepharose column (47 ml, 2 × 15 cm). The adsorbed 3-phosphoglycerate kinase is eluted biospecifically with 2 mM ATP in the same buffer. Fractions containing PGK activity are pooled and concentrated to 1 ml. After adjusting the pH to 8.0 with 1 M NH$_4$OH, the enzyme preparation is loaded on a DEAE-Sephadex column (30 ml, 2 × 10 cm) equilibrated with 10 mM phosphate, pH 8.0, and 1 mM 2-mercaptoethanol. 3-Phosphoglycerate kinase-1 is eluted with the same buffer. The specific activity of the purified enzyme is 500 units/mg.

A typical preparation of these two PGK isozymes from mouse tissue is summarized in Table I.[11]

Purification of PGK-2B and PGK-2C

3-Phosphoglycerate kinase-2B can be purified from frozen testes of a genetic variant from C3H/HeJ mice according to the procedure described for PGK-2A from the C57BL/6J strain.

PGK-2C from testis of the C57L/J strain is a low activity variant that can be purified by a different procedure.

Frozen testes (50 g) from C57L/J mice is homogenized in 200 ml of 5 mM phosphate, pH 6.0, containing 1 mM each of EDTA and dithio-

[10] P. Böhlen, S. Stein, W. Dairman, and S. Udenfriend, *Arch. Biochem. Biophys.* **155**, 312 (1973).

TABLE I

PURIFICATION OF TWO ISOZYMES OF 3-PHOSPHOGLYCERATE KINASE FROM
TESTIS AND MUSCLE OF C57BL/6J MICE[a,b]

Step	Total activity (units)	Specific activity (units/mg protein)	Conditions of elution
Testis			
Crude extract	371	0.43	—
Affinity column 8-(6-Aminohexyl)amino-ATP–Sepharose	305	35	2 mM ATP
DEAE-Sephadex	133	260–450	10 mM phosphate, pH 8.0, and NaCl gradient
G-75 Sephadex	105	450	0.1 M phosphate, pH 7.5
Muscle			
Crude extract	300	3.0	—
ATP affinity column	240	80	2 mM ATP
DEAE-Sephadex	200	500	Negative adsorption

[a] Twenty grams of frozen testes and 5 g of frozen abdominal muscle are used to purify PGK-2A and PGK-1, respectively.
[b] Protein determination was by fluorescamine assays of Böhlen *et al.*,[10] using bovine serum albumin as the protein standard.

threitol. The homogenate is centrifuged at 27,000 g for 20 min. The resulting supernatant (170 ml) is passed through a CM-Sepharose column (5 × 30 cm) equilibrated with the homogenization buffer. The column is subsequently washed with 1 liter of the same buffer. Less than 10% of the original PGK activity appears in the void volume. This void volume is loaded directly onto an 8-(6-aminohexyl)amino-ATP–Sepharose column (2.5 × 30 cm) equilibrated with 10 mM phosphate, pH 6.5, and 1 mM EDTA and 1 mM dithiothreitol. After washing with 1 liter of the equilibration buffer, PGK-2C is eluted from the column with 2 mM ATP in the same buffer. Active enzyme fractions are pooled and concentrated to 10 ml by an Amicon unit fitted with a PM-10 membrane. The concentrated fraction is then loaded in a preparative isoelectric focusing column (LKB, Model 8100, 110 ml). A mixed Ampholine solution that generates a pH gradient of 4 to 6 (2% ampholytes) is used for isoelectric focusing at 1600 V for 18 hr. Fractions containing PGK-2C activity are recovered at an isoelectric point of 4.9. The purified PGK-2C has a specific activity of 8.5 units/mg. A summary for the purification of PGK-2C from C57L/J mice is presented in Table II.

TABLE II

PURIFICATION OF PGK-2C FROM TESTIS OF C57L/J MICE

Step	Total volume (ml)	Total protein (mg)	Total activity (unit)	Specific activity (unit/mg protein)
1. Crude homogenate[a]	170	3269	850	0.26
2. CM-Sepharose (filtrate)	310	2258	70	0.031
3. ATP affinity column	55	622	56	0.9
4. Isoelectric focusing	5.4	1.76	15[b]	8.5[b,c]

[a] Fifty grams of frozen testes from C57L/J mice are employed for this experiment.

[b] Refers only to the peak fractions (70–75) containing homogeneous PGK-2C.

[c] Protein determination by fluorescamine assays of Böhlen et al.,[10] using bovine serum albumin as the standard.

TABLE III

COMPARISON OF BIOCHEMICAL PROPERTIES OF THREE PGK-2 VARIANTS AND PGK-1 FROM THE MOUSE

Biochemical properties	3-Phosphoglycerate kinase isozymes and variants			
	PGK-2A	PGK-2B	PGK-2C	PGK-1
Specific activity (units/mg protein)	450	430	8.5	500
K_m (mM)				
ATP	0.42	0.37	0.038	0.22
ADP	0.15	0.19	0.039	0.23
3-PG[a]	0.20	0.15	0.028	
	(0.7)	(0.4)	(0.4)	(0.96)
Relative activity				
ATP	100	100	100	100
2'-Deoxy-ATP	7	7	119	3
GTP	93	94	20	75
UTP	3	4	58	1
CTP	3	4	16	1

[a] K_m values for 3-phosphoglycerate (3-PG) are biphasic. Those presented in parentheses were determined under 3-phosphoglycerate concentrations ranging from 1 to 10 mM.

General Comments

For large-scale purification of PGK-2A or PGK-2B from frozen mouse testes, the ATP affinity column step and DEAE-Sephadex column step are reversed to improve the overall enzyme recovery.[11] By using DEAE-Sephadex chromatography as the first step, the effective capacity and the lifetime of ATP affinity column is increased.[11]

Properties

Both PGK isozymes and PGK-2 variants are monomers with a molecular weight of about 47,000. PGK-1 and PGK-2A exhibit quite similar biochemical properties.[12] Together with PGK-2B and PGK-2C, their properties are summarized in Table III. Among three PGK-2 variants, PGK-2C has only 2% of the apparent specific activity of PGK-2A or PGK-2B. Comparative biochemical and immunological studies have revealed that the low activity of PGK-2C is the result of structural gene mutations that affect the active site of this genetic variant.[11]

Immunologically, antiserum to PGK-1 partially cross-reacts with PGK-2, but antiserum to PGK-2 does not cross-react with PGK-1. There is a high degree of homology in amino acid composition between the two PGK isozymes; however, no apparent similarity of tryptic peptide maps was found. Postmeiotic expression of PGK-2 in testis has been postulated to be the consequence of X chromosome inactivation during spermatogenesis.[13]

[11] C.-Y. Lee, D. Niesel, B. Pegoraro, and R. P. Erickson, *J. Biol. Chem.* **255**, 2590 (1980).
[12] B. Pegoraro and C.-Y. Lee, *Biochim. Biophys. Acta* **522**, 423 (1978).
[13] R. P. Erickson, *Dev. Biol.* **53**, 134 (1976).

[22] Phosphoglycerate Kinase from *Bacillus stearothermophilus*

By KOICHI SUZUKI and KAZUTOMO IMAHORI

$$\text{MgATP} + \text{3-phosphoglycerate} = \text{1,3-diphosphoglycerate} + \text{MgADP}$$

Assay Method[1]

Principle. 1,3-Diphosphoglycerate formed from ATP and 3-phosphoglycerate is converted to glyceraldehyde 3-phosphate by a coupled en-

[1] K. Suzuki and K. Imahori, *J. Biochem. (Tokyo)* **76**, 771 (1974).

zyme assay with glyceraldehyde-3-phosphate dehydrogenase in the presence of NADH (1,3-diphosphoglycerate + NADH = glyceraldehyde 3-phosphate + NAD + P_i). The concomitant oxidation of NADH is measured spectrophotometrically as the decrease of absorption at 340 nm.

Reagents

Tris-HCl buffer, 0.1 M, pH 7.6, containing $MgSO_4$, 10 mM
Na_2ATP, 0.12 M
Na_3-3-phosphoglycerate, 0.12 M
NADH, 12.5 mM
Glyceraldehyde-3-phosphate dehydrogenase (rabbit muscle, 10 mg/ml, suspension in ammonium sulfate, Boehringer Mannheim)

Procedure. The assay mixture for 20 assays is prepared by mixing the following: 4.3 ml of Tris-HCl buffer, 0.1 ml of NADH, 0.1 ml of ATP, 0.5 ml of 3-phosphoglycerate, and 25 μl of glyceraldehyde-3-phosphate dehydrogenase. A portion (0.4 ml) of this mixture is placed in a cuvette of 1-cm light path, and the reaction is started by addition of the enzyme, phosphoglycerate kinase (PGK). Progress of the reaction is followed at 340 nm with a Gilford spectrophotometer.

Units. One unit of enzyme is defined as the amount of enzyme that converts 1 μmol of 3-phosphoglycerate, i.e., oxidation of 1 mol of NADH, in 1 min.

Purification [1]

Growth of Bacterial Cells. Bacillus stearothermophilus, strain NCA 1503, is grown in a 40-liter jar fermentor. The growth medium contains per liter: polypeptone, 5 g; yeast extract, 5 g; sucrose, 10 g; $Na_2HPO_4 \cdot 12 H_2O$, 6.44 g; $FeCl_3 \cdot 6 H_2O$, 7 mg; $MgSO_4 \cdot 7 H_2O$, 0.32 g; K_2SO_4, 1.3 g; $MnCl_2 \cdot 4 H_2O$, 15 mg; and citric acid, 0.27 g. The pH of the medium is adjusted to 7.1 with 10 N KOH. The complete medium is sterilized for 30 min at 120° and then allowed to cool to 60°. A 5% (v/v) inoculum (2 liters),[2] prepared by growing cells for 2 hr at 60° on a shaker, is added, and the cells are grown with vigorous aeration to the late log phase (ca. 2–3 hr) at 60°. The cells are harvested, then washed with 0.1 M Tris-HCl, pH 7.6, containing 0.1 M KCl and 10 mM $MgCl_2$. The cell paste is stored at $-20°$ until use. Approximately 150 g of cells (wet weight) are obtained from a 40-liter culture.

All the following steps are performed at 4° and all centrifugations are at 15,000–20,000 g for 20 min.

---

[2] The original 3-hr tube culture (10 ml) is used to inoculate the medium, 100 ml. About 3 hr later, this second culture is used to prepare the final inoculum. The first and the second cultures can be stored overnight at 4°.

Step 1. Extraction. Frozen cells, 280 g, are suspended in 400 ml of 0.1 M Tris-HCl, pH 7.6, containing 10 mM MgCl$_2$ and 0.1 M KCl and disrupted by sonication for 5 min at 10 KHz in 50-ml aliquots. The mixture is centrifuged, and the supernatant (fraction I, 575 ml, crude extract) is used for further purification.

Step 2. Streptomycin Treatment. A 10% solution of streptomycin sulfate, 57.5 ml (0.1 volume of the crude extract), is added slowly with stirring to the crude extract. The precipitate is removed by centrifugation after 20 min, and the supernatant is retained (fraction II, streptomycin supernatant, 605 ml).

Step 3. Ammonium Sulfate Fraction. Solid ammonium sulfate is added to the streptomycin supernatant to 70% at 0° (436 g/liter). After standing overnight at 4°, the precipitate is removed by centrifugation and the supernatant is brought to 100% at 0° (697 g/liter) by addition of solid ammonium sulfate. The resulting precipitate, collected after standing for ca 12 hr, is dissolved in 30 ml of buffer A (10 mM Tris-HCl buffer, pH 7.5, containing 2 mM EDTA and 2 mM 2-mercaptoethanol) and dialyzed against the same buffer (fraction III, ammonium sulfate fraction).

Step 4. DEAE-Cellulose Chromatography. The dialyzed sample is applied to a DEAE-cellulose column (3 × 66 cm, DE-52, Whatman) equilibrated with buffer A. The column is developed with a linear NaCl gradient from 0 to 0.8 M in a total volume of 1.5 liters of buffer A. Fractions of 12 ml are collected at a flow rate of 30–40 ml/hr and analyzed for absorption at 280 nm and for enzyme activity. The enzyme appears at about tube No. 90. Active fractions are pooled and concentrated by ultrafiltration (fraction IV, DEAE-cellulose fraction, 6 ml).

Step 5. Gel Filtration. The concentrated fraction is applied to a Sephadex G-100 column (2.2 × 135 cm) equilibrated with buffer A, which is also used for elution. Fractions of 2.9 ml are collected at a flow rate of 18 ml/hr. The PGK fractions are pooled and concentrated by ultrafiltration (fraction V, Sephadex G-100 fraction, 6 ml).

Further Purification. In most cases, the pooled enzyme fraction after purification step 5 is homogeneous. However, in some cases when the separation of the PGK peak from that eluted immediately before is incomplete, minor band(s) other than PGK are observed on disc gel electrophoresis. In these cases, further purification can be achieved by either or both of the following two methods: rechromatography on Sephadex G-100 or hydroxyapatite[3] column chromatography (1.5 × 35 cm, linear gradient from 5 mM to 500 mM potassium phosphate buffer, pH 6.8, containing 5 mM 2-mercaptoethanol in a total volume of 600 ml). The results of the purification are summarized in the table. The specific activ-

---

[3] For preparation of hydroxyapatite, see this volume [28].

PURIFICATION OF PHOSPHOGLYCERATE KINASE OF *Bacillus stearothermophilus*[a]

Fraction	Volume (ml)	Total protein (mg)	Total activity (units)	Specific activity (units/mg protein)	Yield (%)
I. Crude extract	575	81,800	11,900	0.15	100
II. Streptomycin supernatant	605	19,600	11,100	0.56	93
III. Ammonium sulfate	37	3,400	8,470	2.5	71
IV. DEAE-cellulose	6	200	6,240	32.0	52
V. Sephadex G-100	6	11	6,170	557	51

[a] From 280 g of cells. The systematic name for phosphoglycerate kinase is ATP:3-phospho-D-glycerate 1-phosphotransferase (EC 2.7.2.3).

ity of the enzyme (557 units/mg at 25°) is slightly lower than those of PGKs from other sources (650–1000 units/mg) measured under similar conditions.[4] However, the activity at 50° is 1400 units/mg. Thus at the physiological temperature of *B. stearothermophilus* (55–60°), the enzyme is probably 2–3 times more active than ordinary mesophilic enzymes.

Properties

Purity.[1] The enzyme purified by the above procedure is homogeneous as judged by polyacrylamide gel electrophoresis with or without SDS and by electrophoresis on cellulose acetate strips at pH 3.9–8.6.

Molecular Properties.[1] The molecular weight of the enzyme is 42,000, based on the results of gel filtration, SDS–polyacrylamide gel electrophoresis, SH titration, and amino acid analyses. It is clearly smaller than the yeast enzyme (M_r 47,000). The extinction coefficient ($E_{280 \text{ nm}}^{0.1\%}$) is 0.56. The isoelectric pH is 4.9, which is much more acidic than that for enzymes from other sources. The isoelectric pH values of rabbit and yeast enzymes are 7.0 and 7.2, respectively.[4]

The amino acid composition calculated, based on a molecular weight of 42,000, is as follows: Lys, 29; His, 8; Arg, 18; Asp, 44; Thr, 14; Ser, 13; Glu, 36; Pro, 8; Gly, 35; Ala, 49; Val, 35; Met, 9; Ile, 22; Leu, 39; Tyr, 9; Phe, 16; Trp, 2; and Cys, 1; the total residue number is 387. The composition is similar to that of PGKs from microorganisms but distinct from that of mammalian PGKs. The former have a single SH group that is apparently not essential for the enzyme activity, whereas the SH content of the latter is high and the modification of SH groups leads to inactivation. The N terminus is methionine, and 1 mol each of glycine and threonine and 2

[4] W. K. G. Krietsch and T. Bücher, *Eur. J. Biochem.* **17**, 568 (1970).

mol of serine are liberated on carboxypeptidase A digestion of this PGK. The single SH group is buried and is titrated only in the presence of a denaturing reagent. The SH group is not essential for the activity, because carboxymethylated PGK in urea regains full activity on renaturation. Chemical modification reveals that 2 mol of amino group and 1 mol of imidazole group are required for the activity.

The secondary structure of this PGK is α-helix, 20%; β-structure, 45%; and unordered structure, 35%. This enzyme is stable at pH values above 6, although inactivation is noticed at pH values below 5.5. No loss of activity is observed at temperatures up to 60° after heating for 10 min.

Substrate Specificity. GTP and ITP can replace ATP as a phosphoryl donor in the enzyme reaction. Their activities are 27% and 42%, respectively, of that measured with ATP when compared at nucleotide concentrations of 2.4 mM. Pyrimidine nucleotide triphosphates cannot substitute for ATP. Purine nucleotide mono- and diphosphate inhibit the reaction partially. The specificity for D-3-phosphoglycerate is absolute, and none of the substrates tested as a phosphoryl acceptor are active in the enzyme reaction. Mg^{2+} or Mn^{2+} ions are essential for the activity. Co^{2+} and Ca^{2+} are partially active in the reaction at 8 mM; i.e., 58% and 15% of the activity with Mg^{2+}, respectively. Zn^{2+}, Ba^{2+}, Ni^{2+}, and monovalent cations are inactive.

Kinetic Properties. The enzyme shows a broad pH activity profile with maximum activity at pH 5.5–8.5. The activity decreases rapidly at pH values below 5.5, while about 50% of the maximum activity is observed even at pH 10. The K_m values at 25° for ATP and 3-phosphoglycerate are 2.9 mM and 2.2 mM, respectively. The logarithmic value of maximum velocity is linear with the reciprocal of the absolute temperature between 30° and 50°. Above 50°, accurate determination of the enzyme activity is difficult, presumably owing to the decomposition of the substrate(s). The activation energy and the temperature coefficient are 7.5 kcal/mol and 1.5, respectively.

[23] Pyruvate Kinases from Human Erythrocytes and Liver

By AXEL KAHN and JOËLLE MARIE

$$\text{Phosphoenolpyruvate} + \text{ADP} \rightarrow \text{pyruvate} + \text{ATP}$$

Assay Method

Principle. Pyruvate kinase[1] activity is measured by coupling with lactate dehydrogenase, which transforms pyruvate into lactate and oxidizes NADH into NAD. Oxidation of NADH is followed at 340 nm.[1a]

Procedure.[2-4] The assay mixture contains, in a final volume of 1.0 ml: 100 mM Tris-HCl buffer (pH 8), 0.5 mM EDTA, 100 mM KCl, 10 mM MgCl$_2$, 0.2 mM NADH, 1 IU of lactate dehydrogenase per milliliter,[5] 1.5 mM neutralized ADP, and 5 mM phosphoenolpyruvate. The reaction is started by the addition of this last substrate, after incubation of the reaction mixture mixed with 20 μl of appropriate prediluted enzyme sample for 5 min at the reaction temperature (30 or 37°). For erythrocyte pyruvate kinase, the enzyme sample consists of a 1:20 dilution of hemolysate in water containing 1 mM EDTA and 1 mM 2-mercaptoethanol. For liver pyruvate kinase, the sample is prepared by diluting liver extract in 100 mM Tris-HCl buffer, pH 8, 1 mM EDTA, 1 mM 2-mercaptoethanol, 100 mM KCl, and 10% (v/v) glycerol so that a final rate of change in absorbance at 340 nm of 0.01–0.2 per minute is observed. Since erythrocyte and liver L-type pyruvate kinases are allosteric enzymes[6-10]

---

[1] ATP : pyruvate phosphotransferase, EC 2.7.1.40.

[1a] T. Bücher and G. Pfleiderer, this series, Vol. 1, p. 435.

[2] E. Beutler, *in* "Red Cell Metabolism: A Manual of Biochemical Methods," 2nd Ed. Grune & Stratton, New York, 1975.

[3] International Committee for Standardization in Haematology, *Br. J. Haematol.* **35**, 331 (1977).

[4] International Committee for Standardization in Haematology, *Br. J. Haematol.* **43**, 275 (1979).

[5] E. D. Sprengers, J. Marie, A. Kahn, K. Punt, and G. E. J. Staal, *Hum. Genet.* **41**, 61 (1978).

[6] P. Cartier, H. Temkine, and J. P. Leroux, *C.R. Hebd. Seances Acad. Sci., Ser. D* **266**, 1680 (1968).

[7] P. Cartier, A. Najman, J. P. Leroux, and H. Temkine, *Clin. Chim. Acta* **22**, 165 (1968).

[8] T. Tanaka, Y. Harano, H. Morimura, and R. Mori, *Biochem. Biophys. Res. Commun.* **21**, 65 (1965).

[9] T. Tanaka, Y. Harano, F. Sue, and H. Morimura, *J. Biochem. (Tokyo)* **62**, 71 (1967).

[10] T. Tanaka, F. Sue, and H. Morimura, *Biochem. Biophys. Res. Commun.* **29**, 444 (1967).

activated, in particular, by fructose-1,6-P_2 and glucose-1,6-P_2,[11] it is useful to dialyze the enzyme samples (especially hemolysate) twice against 100 volumes of 100 mM Tris-HCl buffer, pH 8, 1 mM EDTA, 1 mM 2-mercaptoethanol, and 100 mM KCl in order to obtain the activity of the nonactivated enzyme.

Activity of the enzyme is expressed as micromoles of pyruvate formed (i.e., of NADH oxidized into NAD) per minute (IU) per gram of hemoglobin for the hemolysate or per milligram of protein for liver extract. In addition to this activity assayed under "standard conditions," it is possible to measure the "maximum velocity" in the presence of 10 mM phosphoenolpyruvate (instead of 5 mM) and 0.5 mM fructose-1,6-P_2.[12,13] Pyruvate kinase can also be assayed under "low substrate" conditions[2] in the presence of 0.25 mM phosphoenolpyruvate. These last two measurements may be useful for detecting any defective red cell pyruvate kinase variants with apparently unmodified activity under "standard conditions" but that have abnormal affinity for phosphoenolpyruvate.[14]

Purification Procedure

Unless otherwise indicated, all steps of the purification are carried out at +4° and all the buffers contain 10% (v/v) glycerol, 2 mM 2-mercaptoethanol, 10 mM EDTA, 10 mM 4-aminocaproic acid, and 0.1 mM diisopropylphosphofluoridate.

Preparation of the Dextran Blue–Sepharose Column

Cyanogen-activated Sepharose 4B is covalently coupled with Dextran Blue using the method of Ryan and Vestling,[15] except that the Dextran Blue concentration of the solution mixed with the activated Sepharose is reduced to 5 mg/ml (instead of 20 mg/ml). We found, in fact, that excess ligand binding strongly increases the nonspecific interactions between absorbent and enzyme.

Isolation of Human Erythrocyte Pyruvate Kinase

The pyruvate kinase purification procedure described here is slightly modified from our previously described methods.[16-20]

[11] J. F. Koster, R. G. Slee, G. E. J. Staal, and T. J. C. Van Berkel, Biochim. Biophys. Acta 258, 763 (1972).
[12] A. Kahn, J. Marie, C. Galand, and P. Boivin, Hum. Genet. 29, 271 (1975).
[13] J. Marie, J. L. Vives Corrons, and A. Kahn, Clin. Chim. Acta 81, 153 (1977).
[14] K. G. Blume, H. Arnold, G. W. Lohr, and E. Beutler, Clin. Chim. Acta 43, 443 (1973).
[15] L. D. Ryan and C. S. Vestling, Arch. Biochem. Biophys. 160, 279 (1974).
[16] A. Kahn, J. Marie, H. Garreau, and E. D. Sprengers, Biochim. Biophys. Acta 523, 59 (1978).

Hemolysate. Whole blood collected on anticoagulant (heparin or citrate phosphate dextrose) is mixed with 1 : 2 volume of a gelatin solution (Plasmagel sodique, laboratoire Rober Bellon, containing 10 mg of heparin per milliliter) and 1 volume of 145 mM NaCl, then allowed to stand at room temperature until red cells are totally sedimented. The upper layer, containing leukocytes and platelets, is eliminated, and erythrocytes are washed three times in 145 mM NaCl; after each centrifugation (4000 g for 15 min at 4°), residual buffy coat is carefully removed. If pyruvate kinase is to be purified from small amounts of blood, the above procedure for elimination of leukocytes is better replaced by filtration of whole blood on a cellulose column, as reported by Beutler *et al.*[21]

Packed, washed erythrocytes are mixed with 2 volumes of distilled water containing 2 mM 2-mercaptoethanol, 10 mM EDTA, 10 mM 4-aminocaproic acid, and 2 mM diisopropylphosphofluoridate, or phenylmethylsulfonyl fluoride and with 1 : 2 volume of cold toluene. They are then homogenized in a Waring blender for 1 min. After centrifugation for 30 min at 10,000 g, the upper fat layer is sucked off and the hemolysate is decanted.

Ammonium Sulfate Fractionation. Solid $(NH_4)_2SO_4$ (23 g/100 ml) is slowly added to the hemolysate under continuous gentle agitation. After 2 hr of incubation at 4°, the precipitate is collected by centrifugation (10,000 g for 30 min) and dissolved in an $(NH_4)_2SO_4$ solution (11 g/100 ml) adjusted to pH 8 with solid Tris. The residual precipitate collected by centrifugation is discarded, and the supernatant is brought to 22 g of $(NH_4)_2SO_4$ per 100 ml, the pH being adjusted to 6.6 with a 10% (v/v) acetic acid solution. The precipitate obtained under these conditions is collected by centrifugation and washed once in a solution of 21 g of $(NH_4)_2SO_4$ per 100 ml adjusted to pH 6.6 with solid Tris. The mutant erythrocyte pyruvate kinase variants are partially purified by this method before being characterized.[4]

Dextran Blue–Sepharose 4B Chromatography. The enzyme preparation is desalted by chromatography on a Sephadex G-25 column equilibrated with a 10 mM Tris-HCl buffer, pH 7.5, 20 mM KCl, and applied to a Dextran Blue–Sepharose 4B column (containing 1 ml of swollen absorbent per IU of pyruvate kinase) previously equilibrated with the same buffer. The KCl concentration is then raised to 80 mM, and the column is washed until all absorbance at 280 nm has disappeared. Elution of pyru-

[17] J. Marie and A. Kahn, *Enzyme* **22**, 407 (1977).
[18] J. Marie and A. Kahn, *Biochem. Biophys. Res. Commun.* **91**, 123 (1979).
[19] J. Marie, A. Kahn, and P. Boivin, *Biochim. Biophys. Acta* **481**, 96 (1977).
[20] J. Marie, L. Tichonicky, J. C. Dreyfus, and A. Kahn, *Biochem. Biophys. Res. Commun.* **87**, 862 (1979).
[21] E. Beutler, C. West, and K. G. Blume, *J. Lab. Clin. Med.* **88**, 329 (1976).

vate kinase is provoked at $+18°$ by adding 0.1 mM fructose-1,6-P_2 to this buffer. The active fractions are made 70 g/liter in $(NH_4)_2SO_4$ (in order to stabilize the enzyme), concentrated to a protein concentration higher than 0.2 mg/ml by ultrafiltration over a PM-30 Amicon membrane, then precipitated by dialysis against a 50 mM Tris-HCl buffer, pH 7.5, containing 35 g of $(NH_4)_2SO_4$ per 100 ml.

The enzyme can be stored frozen at $-70°$ in the form of a precipitate in the above $(NH_4)_2SO_4$ solution without significant inactivation for several years. The overall yield of this procedure is 45–60%. It should be noted that properties of different batches of Dextran Blue–Sepharose can vary slightly, therefore each new batch must be tested before being used for large-scale purification; it is particularly important to check for the maximum KCl concentration that does not elute pyruvate kinase in the absence of fructose-1,6-P_2 but permits elution in its presence.

Separation of the Different Forms of Erythrocyte Pyruvate Kinase

In erythroblasts, pyruvate kinase is synthesized as a L'_4 tetramer [L' has a molecular weight (M_r) of 63,000], which is partially proteolysed and transformed into active heterotetrameric forms containing two partially proteolysed subunits with red cell aging.[16,22] The most important of these "aged" forms is L'_2Lc_2 (M_r of Lc being 57,000–58,000).[18] L'_4 and L'_2Lc_2 can be separated in the following manner. After fixation of pyruvate kinase on Dextran Blue-linked absorbent, the column is washed with buffer containing 80 mM KCl as described above. After disappearance of all absorbance at 280 nm, the column is equilibrated with a buffer lacking KCl, and pyruvate kinase is eluted by fructose-1,6-P_2 and a linear gradient of ionic strength between 6 volumes of the non-KCl buffer and 6 volumes of this buffer plus 80 mM KCl, both solutions containing 0.02 mM fructose-1,6-P_2. Two peaks of activity are clearly separated, which, on analysis by SDS–polyacrylamide gel electrophoresis, correspond to L'_2Lc_2 for the first peak and to L'_4 for the second.

Isolation of Human Liver Pyruvate Kinase

Extraction and $(NH_4)_2SO_4$ Precipitation. Human liver samples are usually obtained from kidney transplantation donors; they are taken within 30 min of circulatory arrest and immediately frozen in liquid nitrogen and stored at $-70°$ until use. Homogenization is carried out in a Potter–Elvehjem homogenizer in 3 volumes of cold 50 mM Tris-HCl buffer, pH 7.5, containing, in addition to the protectors reported above, 2% (v/v) aprotinin, 0.1 mM pepstatin and 0.01 mM leupeptin as protease inhibitors. The cell debris is removed by centrifugation at 35,000 g for 30 min.

[22] J. Marie, H. Garreau, and A. Kahn, *FEBS Lett.* **78**, 91 (1977).

Then the supernatant is ultracentrifuged at 105,000 g for 1 hr. The upper fat layer and the pellet are eliminated, and the soluble fraction is fractionated by $(NH_4)_2SO_4$ precipitation as described for the erythrocyte enzyme.

CM-Sephadex Chromatography.[17] The final $(NH_4)_2SO_4$ precipitate is dissolved in a minimum volume of 10 mM sodium phosphate buffer, pH 6.1, and desalted by passage through a G-25 column equilibrated with the same buffer. CM-Sephadex C-50 in 10 mM phosphate buffer, pH 6.1, is added to the eluate and gently stirred for 30 min until all pyruvate kinase activity is bound to the exchanger (about 1 ml of resin per 15 IU of pyruvate kinase activity). The absorbent is then gathered in a Büchner funnel and washed with 50 mM phosphate buffer, pH 6.1, until disappearance of absorbance at 280 nm. The enzyme is eluted by the same buffer containing 4 mM ATP and 0.1 mM fructose-1,6-P_2. The eluate is precipitated by adding 30 g of solid $(NH_4)_2SO_4$ per 100 ml.

This CM-Sephadex chromatography step can be omitted when the amount of pyruvate kinase to be purified is relatively low (less than 10 mg).

Dextran Blue–Sepharose 4B Chromatography. Liver L-type pyruvate kinase is purified by Dextran Blue–Sepharose 4B chromatography as described above for erythrocyte enzyme except that elution is achieved by addition of fructose-1,6-P_2 at a KCl concentration of 40 mM instead of 80 mM.

Tables I and II show typical results of pyruvate kinase purification from human erythrocyte and from liver. The enzymes isolated by these methods have been found to be homogeneous by SDS–polyacrylamide electrophoresis and immunological analysis.[17–19,22,23]

Properties of Human Erythrocyte and Liver L-Type Pyruvate Kinases

Subunit Structure and Molecular Weight. Erythrocytes contain two active forms of pyruvate kinase that can be electrophoretically distinguished; these were designated PKR_1 (the only form in erythroblasts and predominant in young red cells[16,24,25]) and PKR_2 (the predominant form in older red cells[16,24,25]) by Nakashima *et al.*[24,25]

PKR_1 corresponds to a homotetramer L'_4 (M_r of L' = 63,000[16]), and PKR_2 is a heterotetramer of formula L'_2Lc_2 (M_r of Lc = 57,000–58,000[18]). In fact, SDS–polyacrylamide gel electrophoresis of PKR_2 shows that a third type of subunit (called Lb) can be detected, representing less than 10% of the total; its molecular weight is about 60,000.[18] Liver L-type

[23] J. Marie, A. Kahn, and P. Boivin, *Biochim. Biophys. Acta* **438**, 393 (1976).
[24] K. Nakashima, *Clin. Chim. Acta* **55**, 245 (1974).
[25] K. Nakashima, S. Miwa, S. Oda, T. Tanaka, K. Imamura, and T. Nishina, *Blood* **43**, 537 (1974).

TABLE I
PURIFICATION OF PYRUVATE KINASE FROM HUMAN ERYTHROCYTES

Fraction	Activity (IU)	Proteins (mg)	Specific activity (IU/mg protein)	Purification (fold)	Recovery (%)
Hemolysate	7500	825,000	0.0091	1	100
Ammonium sulfate fractionation	6000	3000	2	220	80
Dextran Blue–Sepharose 4B column, elective elution by fructose-1,6-P_2	3600	11	330	36,264	48

TABLE II
PURIFICATION OF L-TYPE PYRUVATE KINASE FROM HUMAN LIVER

Fraction	Activity (IU)	Proteins (mg)	Specific activity (IU/mg protein)	Purification (fold)	Recovery (% yield)
Liver 105,000 g soluble fraction		45,000		—	—
Total activity	11,000		0.24		
L-type activity[a]	7,280		0.16		
Ammonium sulfate fractionation	5,824	1941	3	19	80
CM-Sephadex batchwise, elective elution by ATP	4,732	237	20	125	65
Dextran Blue–Sepharose 4B column, elective elution by fructose-1,6-P_2	2,912	6.9	420	2625	40

[a] L-type pyruvate kinase activity of the liver extract was determined by measuring the ra of the total activity inhibited by excess anti-human L-type pyruvate kinase serum.

pyruvate kinase is composed of four identical subunits of about M_r 60,000[17] (Fig. 1).

On isoelectric focusing, nondissociated enzymes exhibit multiple active forms with isoelectric pH values between 5.85 and 6.69[13,26]; liver enzyme forms are slightly more acidic than the erythrocyte forms.[26]

Sensitivity of L-Type Pyruvate Kinases to Proteolysis. When erythrocyte pyruvate kinase, stabilized under its active conformation,[16,22] is subjected to mild tryptic attack, the L' subunits are sequentially transformed into

[26] A. Kahn, J. Marie, and P. Boivin, *Hum. Genet.* 33, 35 (1976).

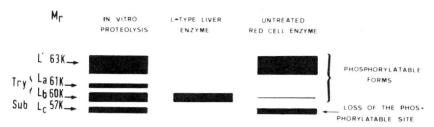

FIG. 1. Schematic representation of the different types of L-type pyruvate kinase subunits found in the liver and red cells, spontaneously and after mild proteolytic attack. M_r, molecular weight, in thousands; Try, mild tryptic attack of L'_4; Sub, mild subtilisin attack of L'_4.

subunits of M_r 61,000–62,000 (La), then M_r 60,000 (Lb)[18] (Fig. 1). Liver L-type kinase subunits are slightly affected by such treatment, which seems to split out an M_r 500–1000 fragment. Subtilisin, in the same conditions, transforms L' and liver L pyruvate kinase subunits into forms similar to Lc (i.e., of M_r 57,000–58,000[18]). Liver and erythrocyte enzymes partially proteolysed with either trypsin or subtilisin exhibit similar peptide maps as analyzed in SDS–polyacrylamide gel according to Cleveland.[27] Therefore, the following conclusions may be drawn.

1. In red cells, pyruvate kinase can undergo a partial proteolytic cleavage occurring with red cell aging, which can be mimicked *in vitro* by a partial subtilisin treatment. When erythrocyte pyruvate kinase is purified without eliminating leukocytes and without addition of anti-proteolytic agents to the buffers, L'_4 can be artifactually transformed into L'_2Lc_2 and Lc_4 forms during purification.[16,19,22]

2. The structure of liver and erythrocyte L-type enzymes seems to be identical, with an additional fragment present in erythrocyte pyruvate kinase.

Immunological Properties. There is complete immunological cross-reactivity between erythrocyte and liver L-type pyruvate kinase[19,23,26] and a total absence of cross-reactivity between both of them and K or M pyruvate kinase isozymes.[26,28] The only tissues synthesizing pyruvate kinase recognized by an anti L-type pyruvate kinase antiserum are erythrocytes, liver, and, to a very small extent, kidney (for review, see Ibsen[29]). Enzymes from erythrocytes and liver are both quantitatively and qualitatively affected in patients with congenital erythrocyte pyruvate

[27] D. W. Cleveland, S. G. Fischer, M. W. Kirschner, and U. K. Laemmli, *J. Biol. Chem.* **252**, 1102 (1977).

[28] J. Marie, A. Kahn, and P. Boivin, *Hum. Genet.* **31**, 35 (1976).

[29] K. H. Ibsen, *Cancer Res.* **37**, 341 (1977).

kinase deficiency[30-34]; this finding indicates that both are probably encoded by the same structural gene.[35]

Messenger RNAs Encoding L' and L Subunits. Liver messenger RNAs direct, in a cell-free system, the synthesis of 60,000-M_r pyruvate kinase subunits whereas reticulocyte messenger RNAs direct synthesis of 63,000-M_r subunits[35]; a mixture of liver and reticulocyte RNAs directs synthesis of both 60,000- and 63,000-M_r subunits. It appears therefore that one gene, but two messenger RNAs, encodes liver L and erythrocyte L' pyruvate kinase subunits.[35] The sedimentation coefficient of these RNA species is about 22 S.

It is not yet known whether this finding results from gene remodeling or from a differential splicing of nuclear RNAs in erythroid and liver cells.

Phosphorylation, Dephosphorylation, and Kinetics. Erythrocyte and liver L-type pyruvate kinases are allosteric enzymes with fructose-1,6-P_2 as the major allosteric activator and ATP and alanine as the major allosteric inhibitors[6-10,36] (Table III). As compared to liver enzyme, L_4' is characterized by a decreased affinity for phosphoenolpyruvate and less pronounced positive homotropic interactions.[16,37] A mild proteolysis by trypsin makes the kinetics of erythrocyte pyruvate kinase very similar to those of liver enzyme. Both enzymes are phosphorylatable in a single identical site by soluble 3'5'-cyclic-AMP-dependent protein kinases present in both erythrocytes and liver.[18,20,38-40] The maximum velocity of the phosphorylation reaction is about twice as high with liver L_4 as with red cell L_4' pyruvate kinase as substrate, whereas the affinity of protein kinase for either form is identical.[41] La, Lb, and liver L subunits possess the phosphorylatable site, but Lc does not. Subtilisin, indeed, cleaves this site[18] (Fig. 1). Pyruvate kinase phosphorylation results in a shift toward a

[30] R. H. Bigley and R. D. Koler, *Ann. Hum. Genet.* **31**, 383 (1965).
[31] A. Kahn, J. Marie, C. Galand, and P. Boivin, *Scand. J. Haematol.* **16**, 250 (1976).
[32] K. Nakashima, S. Miwa, H. Fujii, K. Shinohara, K. Yamauchi, Y. Tsuji, and M. Yanai, *J. Lab. Clin. Med.* **90**, 1012 (1977).
[33] K. Shinohara, S. Miwa, K. Nakashima, E. Oda, T. Kegeoka, and G. Tsujino, *Am. J. Hum. Genet.* **28**, 474 (1976).
[34] G. E. J. Staal, G. Rijksen, A. M. C. Vlug, B. Vromen Van den Bos, J. W. N. Akkerman, G. Gorter, J. Dierick, and M. Petermans, *Clin. Chim. Acta* **118**, 241 (1982).
[35] J. Marie, M. P. Simon, J. C. Dreyfus, and A. Kahn, *Nature (London)* **292**, 70 (1981).
[36] W. Seubert and W. Schonor, *Curr. Top. Cell. Regul.* **3**, 237 (1971).
[37] E. D. Sprengers and G. E. J. Staal, *Biochim. Biophys. Acta* **570**, 259 (1979).
[38] J. Marie, H. Buc, M. P. Simon, and A. Kahn, *Eur. J. Biochem.* **108**, 251 (1980).
[39] G. B. Van den Berg, T. J. C. Van Berkel, and J. F. Koster, *Biochem. Biophys. Res. Commun.* **82**, 859 (1978).
[40] P. A. Kiener, C. V. Massaras, and E. W. Westhead, *Biochem. Biophys. Res. Commun.* **91**, 50 (1979).
[41] J. Marie and A. Kahn, *Biochem. Biophys. Res. Commun.* **94**, 1387 (1980).

TABLE III

KINETIC CHARACTERISTICS OF PHOSPHORYLATED AND DEPHOSPHORYLATED ERYTHROCYTE
AND LIVER PYRUVATE KINASE[a,b]

Parameter	Erythrocyte pyruvate kinase		Liver L-type pyruvate kinase	
	Dephospho	Phospho	Dephospho	Phospho
$_{0.5}$ Phosphoenolpyruvate (mM)				
Without effector	0.6–0.76	1.1–1.51	0.4	0.85
+3 mM alanine	1.1–1.7	2.2–3.3	1.2	2.6
+1 mM ATP	1.5–1.6	2.2–2.3	1.05	1.5
+0.5 mM Fru-1,6-P$_2$	0.08–0.14	0.09–0.18	0.04	0.06
Ill coefficient				
Without effector	1.0–1.4	1.5–2.0	1.7	2.2
+3 mM alanine	1.5–2.2	1.7–3.3	2.5	3.3
+1 mM ATP	1.4–1.5	2.0–3.3	2.2	2.9
+0.5 mM Fru-1,6-P$_2$	0.8–1.05	0.8–1.1	1.0	1.0
TP inhibition (mM ATP for a 30% inhibition)	1.2–1.3	0.7–0.8	1.3	0.4
lanine inhibition (mM Ala for a 30% inhibition)	2.8–3.2	1.1–1.25	1.5	0.5
u-1,6-P$_2$ activation (μM Fru-1,6-P$_2$ for a 50% activation)	0.11–0.14	0.14–0.16	0.065	0.15

[a] Reproduced from Marie *et al.*,[38] with permission of the publisher.
[b] All the kinetic determinations using red cell enzyme were performed in triplicate, using different preparations. The extreme values of the results found are given. The results concerning liver L-type pyruvate kinase correspond to a typical experiment. The calculated phosphate incorporation in phosphorylated pyruvate kinase used for these kinetic studies varied from 3.1 to 3.5 phosphates per tétramer. Fru-1,6-P$_2$, fructose 1,6-bisphosphate.

T allosteric conformation[38,42−45] characterized by increased Hill coefficient and affinity for the allosteric inhibitors ATP and alanine and decreased affinity for phosphoenolpyruvate and allosteric activator fructose-1,6-P$_2$ (Table III).

This type of regulation probably plays a major role in regulation of the dynamic balance between glycolysis and gluconeogenesis in the liver; in contrast, whether it plays any regulatory function in mature erythrocytes is not certain.[38]

[42] P. Ekman, U. Dahlqvist, E. Humble, and L. Engström, *Biochim. Biophys. Acta* **429**, 374 (1976).
[43] J. E. Feliu, L. Hue, and H. G. Hers, *Proc. Natl. Acad. Sci. U.S.A.* **73**, 2762 (1976).
[44] O. Ljunström, L. Berglund, and L. Engström, *Eur. J. Biochem.* **68**, 497 (1976).
[45] T. J. C. Van Berkel, J. K. Kruijt, and J. F. Koster, *Eur. J. Biochem.* **81**, 423 (1977).

Genetic Mutations. As indicated above, erythrocyte pyruvate kinase is frequently affected by mutations resulting in qualitatively, and usually quantitatively, abnormal activity. Through qualitative characterization of the residual enzyme activity, several mutant pyruvate kinase variants can be distinguished (for review, see Kahn *et al.*[46] and Miwa[47]). The clinical consequences are usually congenital nonspherocytic hemolytic anemias of variable severity according to the type of mutation. The anomalies of the erythrocyte enzyme are found again at the level of liver L-type pyruvate kinase, but do not seem to be associated with a clinically detectable liver disease.

[46] A. Kahn, J. C. Kaplan, and J. C. Dreyfus, *Hum. Genet.* **50**, 1 (1979).
[47] S. Miwa, *Clin. Haematol.* **10**, 57 (1981).

[24] Pyruvate Kinase from Bovine Muscle and Liver

By JANET M. CARDENAS

Phosphoenolpyruvate + ADP → pyruvate + ATP

Bovine tissues, like those from other mammals, contain at least three distinct and electrophoretically separable isozymes of pyruvate kinase (ATP: pyruvate phosphotransferase, EC 2.7.1.40).[1] One of the three, type K, predominates in all fetal tissues early in development and persists in many adult tissues. During development, two additional forms of pyruvate kinase appear in certain tissues. One of these is type L, which is the predominant form in adult liver but also occurs in kidney; another is type M, which is the main isozyme of adult brain, skeletal and cardiac muscle.

Like pyruvate kinases from other organisms, bovine types L and M each contain four subunits.[2,3] Each of the three main forms of the enzyme are thought to contain four identical subunits, but hybrid isozymes result *in vivo* if a cell synthesizes two or more subunit types simultaneously or *in vitro* if one denatures and then renatures mixtures of two pyruvate kinase isozymes.[4]

[1] J. J. Strandholm, R. D. Dyson, and J. M. Cardenas, *Arch. Biochem. Biophys.* **173**, 125 (1976).
[2] J. M. Cardenas and R. D. Dyson, *J. Biol. Chem.* **248**, 6938 (1973).
[3] J. M. Cardenas, R. D. Dyson, and J. J. Strandholm, *in* "The Isozymes" (C. L. Markert, ed.), Vol. 1, p. 523. Academic Press, New York, 1975.
[4] R. D. Dyson and J. M. Cardenas, *J. Biol. Chem.* **248**, 8482 (1973).

METHODS IN ENZYMOLOGY, VOL. 90

All three isozymes require both a monovalent cation, normally K^+, and a divalent cation, optimally Mg^{2+}, for catalytic activity, and they catalyze the transfer of a phosphate group from phosphoenolpyruvate to ADP to form pyruvate and ATP, respectively. However, the isozymes are electrophoretically separable from each other and differ somewhat in their chemical, physical, and kinetic properties. Type K has a high isoelectric point (near pH 9) and has slightly cooperative kinetics with the substrate phosphoenolpyruvate.[5] Type L pyruvate kinase exhibits considerably greater cooperativity with this substrate but is activated to produce hyperbolic kinetic profiles by low concentrations of fructose 1,6-bisphosphate. Type M, on the other hand, has hyperbolic kinetics with phosphoenolpyruvate and, like type K, its kinetics are normally unaffected by the presence of fructose bisphosphate.

Large amounts of fresh bovine tissues are generally readily available at low cost from local slaughterhouses. Thus, such tissues are convenient and inexpensive starting materials for the isolation of pyruvate kinase isozymes.

Assay of Enzymic Activity

This procedure is based on that of Bücher and Pfleiderer[6] and is most conveniently and economically performed in cuvettes of 4-mm sample width and 10-mm light path. Standard assays are performed at 25° and pH 7.0 in 1-ml volumes containing $0.05\,M$ imidazole-HCl, $0.10\,M$ KCl, 10 mM $MgCl_2$, 2 mM ADP, 1 mM phosphoenolpyruvate, 0.16 mM NADH, and approximately 20 units (μmol/min) of lactate dehydrogenase; the decrease in absorbance is monitored at 340 nm. For optimal activity of type L pyruvate kinase, 1.0 mM fructose 1,6-bisphosphate should also be included in the assay medium described above.

For samples that have not yet been submitted to a chromatography step, protein concentrations may be estimated by the method of Folin and Ciocalteau as described by Clark and Switzer.[7] For samples of type L pyruvate kinase from chromatography on DE-52, protein concentrations are estimated roughly from the absorbance at 280 nm, assuming an average extinction coefficient of 1.0 absorbance unit per milligram of protein per milliliter. For all chromatographed samples of type M and for samples of type L that have been chromatographed on phosphocellulose, protein concentrations are most accurately and conveniently determined from the absorbance at 280 nm and using the previously determined extinction

[5] J. M. Cardenas, J. J. Strandholm, and J. M. Miller, *Biochemistry* **14**, 4041 (1975).
[6] T. Bücher and G. Pfleiderer, this series, Vol. 1 [66].
[7] J. M. Clark, Jr. and R. L. Switzer, "Experimental Biochemistry," 2nd Ed., p. 12. Freeman, San Francisco, California, 1977.

coefficients for the pure proteins—i.e., 0.55 absorbance unit per milligram per milliliter for type M,[8] and 0.57 absorbance unit for type L.[2]

Materials. Distilled, deionized water should be used for making all solutions. Enzyme-grade ammonium sulfate and sucrose can be obtained from Schwarz-Mann. Substrates, lactate dehydrogenase, agarose (type used for immunoelectrophoresis), and chromatography resins can be obtained from Sigma Chemical Co. (St. Louis, Missouri). Cellulose acetate electrophoresis supplies and equipment are obtainable from the Gelman Instrument Co.

Electrophoresis. Prior to electrophoresis, all enzyme samples should be dialyzed at 4° in 0.25-inch-diameter dialysis casing for at least 3 hr against an electrophoresis buffer containing 0.5 M sucrose, 0.02 M Tris-HCl, 8.0 mM 2-mercaptoethanol, 1.0 mM EDTA, and 1.0 mM fructose 1,6-bisphosphate, pH 7.5, at room temperature. After dialysis, samples should be diluted to 4–9 activity units per milliliter if more concentrated than that.

Electrophoresis is performed on Sepraphore III strips, 1 × 6 inches, at 4°. Application of 250 V for 3 hr is sufficient to separate types M and L (or K and L) pyruvate kinases and their hybrids, but 200 V for approximately 24 hr are normally required to resolve types M and K.

Preparation of Electrophoresis Assay Plates. Plates for the localization of pyruvate kinase activity after electrophoresis can be prepared by a modification of methods described by Susor and Rutter.[9] Photographic glass plates salvaged from analytical ultracentrifugal use are cut in half to produce pieces 2 × 5 × 0.04 inches, and the photographic emulsion is scraped off. The plates are acid-washed until, when rinsed, water forms a sheet, *not* beads, on the glass. Half of the cleaned glass plates are prepared for use as the lower plates for enzymic activity detection by the following procedure. Two layers of Time tape[10] cut to a width of 0.25 inch are wrapped around each end of the lower plates to serve as spacers. The upper surface of these plates is then wiped with 0.5% melted agarose and allowed to air dry thoroughly. The dried agarose coating serves to enhance binding to the lower plate of the agarose assay film that will be applied later. The treated lower plates should be prepared at least 4 hr before use but can be stored in a clean, dry place for several months.

An electrophoretic assay medium is prepared that contains 0.1 M imidazole-HCl (pH 7.0 at room temperature), 0.2 M KCl, 0.02 M MgCl$_2$, 4 mM ADP, 4 mM phosphoenolpyruvate, and 2 mM NADH. This solution

---

[8] J. M. Cardenas, R. D. Dyson, and J. J. Strandholm, *J. Biol. Chem.* **248**, 6931 (1973).

[9] W. A. Susor and W. J. Rutter, *Anal. Biochem.* **43**, 147 (1971).

[10] Available from most laboratory supply companies. Most kinds of adhesive-backed tape are satisfactory.

can be stored frozen for up to 3 months but loses its photographic sensitivity on very long storage or after repeated freezing and thawing.

To prepare an electrophoresis assay plate, a 2% agarose solution is melted in a boiling water bath and cooled to 50°. For each electrophoresis assay plate, 15 units of lactate dehydrogenase are added to 1.5 ml of electrophoresis assay medium immediately before use, and this solution is warmed to 50° by immersion for a few seconds in a water bath at that temperature. Then 1.5 ml of 2% agarose solution, also at 50°, is added using a pipet whose tip has been cut off to produce a wider opening and hence faster flow. After thorough mixing of assay and agarose solutions, the mixture is pipetted onto a prewarmed lower glass plate and quickly spread to a thin, even layer by placing an acid-cleaned glass plate on top, taking care to exclude bubbles. The assay "sandwich" is then wrapped in plastic wrap and allowed to solidify by cooling, first to room temperature and then to 4°. Such plates can be stored for up to a few days at 4° before use.

Detection of Enzymic Activity and Recording of Electrophoretic Patterns. This is accomplished after electrophoresis by removing the top glass plate from the assay plate, placing an electrophoresed cellulose acetate strip onto the agarose assay film, then covering the sandwich with the top glass plate to retard drying. Visualization of the isozymic patterns is accomplished in a dark room by placing the sandwich over a longwave ultraviolet lamp (one should wear goggles to protect the eyes). Bands of pyruvate kinase activity appear as dark bands on a light background. Patterns can be recorded by contact photography with Kodabromide F-5 paper and the ultraviolet lamp as a light source as presented schematically in Fig. 1. When a Kodak wratten filter 18A is used, appropriate exposure times are generally 1–5 sec.

Isolation of Type M Pyruvate Kinase

This procedure is similar to that described by Cardenas *et al.*[8] Bovine neck muscle from a recently killed animal is trimmed to remove most of the fat and connective tissue. All steps are performed at 4° unless otherwise indicated. The preparation may conveniently be carried out with 200–900 g of trimmed muscle, and excess muscle may be stored for several months at −80°.

Extraction. The muscle is passed through a chilled meat grinder and then blended for 30 sec with two volumes (2 ml/g muscle) of 0.02 M Tris-HCl, 10 mM 2-mercaptoethanol, pH 7.0. The mixture is stirred for 1 hr, then centrifuged at 8000 g for 20 min. The precipitate is discarded. The supernatant is filtered through a loose plug of glass wool and adjusted to pH 7.0 with 2 N HCl or KOH if necessary.

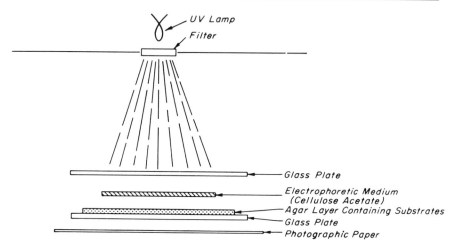

UV Lamp
Filter

Glass Plate
Electrophoretic Medium
(Cellulose Acetate)
Agar Layer Containing Substrates
Glass Plate
Photographic Paper

Fig. 1. Schematic diagram of the system used for detecting and photographing the electrophoretic patterns of pyruvate kinase. Procedures for preparing this assay system are described in the text.

Heat Treatment. The filtered supernatant from the preceding step is rapidly brought to 60° by swirling in an 80° water bath, maintained at 60° for 5 min with constant swirling, then rapidly cooled to at least 10° by swirling in an ice-water bath. The precipitate is removed by centrifugation for 20 min at 10,000 g and is discarded.

Ammonium Sulfate Fractionation. The supernatant from the heat step is placed in a container in an ice-water bath, and 32 g of solid ammonium sulfate per 100 ml of supernatant are added gradually with constant stirring. After all the ammonium sulfate has been added, stirring is continued for 30 min before centrifuging the suspension for 20 min at 10,000 g. The precipitate is discarded, and to the supernatant are added 8.35 g of solid ammonium sulfate per 100 ml of supernatant, stirring and centrifuging as before, but this time the supernatant is discarded and the precipitate is retained. The precipitate can be stored for several days as a suspension at a protein concentration of 20–40 mg/ml in a solution containing 3.2 M ammonium sulfate, 0.02 M imidazole-HCl, and 10 mM 2-mercaptoethanol, pH 7.0 at room temperature.

Carboxymethyl Sephadex Chromatography. Carboxymethyl Sephadex (type C-50) is allowed to swell for 48 hr in deionized water and then washed successively with several volumes of 0.1 M KOH, then deionized water, then 0.1 M HCl, then deionized water, and finally with chromatography buffer consisting of 0.03 M potassium phosphate, 10 mM 2-mercaptoethanol, pH 6.0. The gel is allowed to equilibrate for several hours at 4° before readjusting the pH to 6.0, if necessary, with 2 N HCl or KOH and preparing a 5.0 × 40-cm column.

The final precipitate from ammonium sulfate fractionation is collected by centrifugation and dissolved in a minimal volume (50 ml or less) of chromatography buffer, then dialyzed overnight against at least three changes, 20 volumes each, of the same buffer. The sample is washed into the column with approximately 100 ml of chromatography buffer before applying a linear gradient formed with 2 liters of the buffer and 2 liters of 0.3 M KCl in the same buffer. Pyruvate kinase typically elutes toward the end of the gradient following a large protein peak containing red material. Fractions having the highest specific activities, generally greater than 200 μmol/mg/min at 25°, are pooled and dialyzed against a solution containing 3.2 M ammonium sulfate, 10 mM 2-mercaptoethanol, and 5 mM MgCl$_2$, pH 7.0, at room temperature, and stored at 0–4°.

Gel Filtration. Trace amounts of contaminating proteins can be removed from the pyruvate kinase preparation by filtration on BioGel A-1.5m. Small quantities (i.e., 15 mg) of protein from the preceding step are collected by centrifugation (10,000 *g* for 20 min) of the ammonium sulfate suspension and are dissolved in 1–2 ml of gel filtration buffer consisting of 0.05 M potassium phosphate, 0.1 M KCl, 10 mM 2-mercaptoethanol, pH 7.0. The sample is applied to a 2.5 × 90-cm column of BioGel A-1.5m and filtered at a rate of 17 ml/hr, collecting 2.5-ml fractions. Samples having the highest specific activity (about 225 units per milligram of protein) are dialyzed against a 3.2 M ammonium sulfate solution containing 10 mM 2-mercaptoethanol and 5 mM MgCl$_2$ and stored as described above.

The results of a typical purification procedure of bovine type M pyruvate kinase are presented in Table I.

TABLE I
PURIFICATION OF BOVINE SKELETAL MUSCLE PYRUVATE KINASE[a]

Fraction	Volume (ml)	Total pyruvate kinase activity (μmol/min)	Total protein (mg)	Specific activity (μmol/ min/mg)	Yield (%)
Extract from 232 g of neck muscle	450	18,000	5360	3.4	100
Heat step supernatant	440	17,200	2680	6.4	95
Ammonium sulfate precipitate	27	14,500	383	37.7	80
CM-Sephadex chromatography	—	6,090	26.1	233	34

[a] Reprinted, with permission, from Cardenas *et al.*[8]

Purification of Type L Pyruvate Kinase

This isozyme, like its counterparts from other mammals, is somewhat less stable than type M. However, sucrose and dithiothreitol stabilize it somewhat, as do higher enzyme concentrations. Because of its relative instability, solutions of the enzyme should not be allowed to stand for long periods of time, and dialyses should be as rapid as feasible to accomplish dialysis equilibrium. Preparations of bovine type L pyruvate kinase can be performed reasonably conveniently with up to 4 lb of liver.

Extraction. Liver stored at $-80°$ or obtained from a freshly killed animal is passed through a prechilled meat grinder and then blended with 3 volumes (3 ml per gram of liver) of solution containing $0.15 M$ KCl, $0.02 M$ Tris-HCl (pH 7.5 at room temperature), 5 mM MgCl$_2$, 2 mM dithiothreitol, and 1 mM EDTA. After stirring for 30 min at 0–4°, the suspension is centrifuged at $14,000 g$ for 30 min, the precipitate is discarded, and the supernatant is filtered through a loose plug of glass wool.

Acetone Fractionation. The supernatant from step 1 is adjusted if necessary to pH 7.5 at 0° with 2 N HCl or 2 N KOH. Success of the acetone step depends on both minimizing exposure time of the enzyme to acetone and keeping solutions very cold. Therefore, for this step the extract is divided into several portions that are small enough to permit centrifugation of the whole portion at once.

Acetone (precooled to $-20°$) is added rapidly with vigorous stirring to give a final concentration (v/v) of 33%. The suspension is immediately centrifuged at 7000 g for 8 min at $-10°$, and the precipitate is discarded. Using the same procedure, the supernatant is then brought to a final acetone concentration of 45% (v/v) and centrifuged as before. The supernatant is discarded, and the precipitate is dissolved in a volume of $0.5 M$ sucrose, 10 mM potassium phosphate, 10 mM mercaptoethanol, pH 7.1, sufficient to produce 200 ml per pound of liver used for the preparation. The solution is stirred gently overnight in a relatively flat, open container at 4° to remove remaining acetone.

Ammonium Sulfate Fractionation. The product of the preceding step is fractionated by adding 19.9 g of solid ammonium sulfate per 100 ml of enzyme solution with constant stirring in an ice-water bath at 0°. The suspension is stirred for at least 30 min after adding the ammonium sulfate and is centrifuged at $10,000 g$ for 20 min. The precipitate is discarded, and to the supernatant 7.5 g of solid ammonium sulfate are added per 100 ml of supernatant, using the same procedures as described above. After centrifuging as before, the supernatant is discarded and the precipitate is dissolved in a minimum volume of DE-52 buffer ($0.5 M$ sucrose, 10 mM potassium phosphate, 10 mM 2-mercaptoethanol, pH 7.1) and dialyzed for 8 hr against three changes, 2 liters each, of this buffer.

Chromatography on DE-52. DE-52 cellulose is washed as described above for carboxymethyl Sephadex except that 0.5 N HCl and 0.5 N KOH are substituted for 0.1 N KOH and 0.1 N HCl, respectively, and a 5 × 60-cm column is prepared. After application of the sample, the column is washed with approximately 50 ml of DE-52 buffer before applying a gradient formed with 2 liters each of 0.09 M KCl in DE-52 buffer and 0.3 M KCl in the same buffer. Fractions of about 10 ml each are collected, and those having the highest specific activities, generally greater than 4 units/ml, are pooled. The protein is precipitated by the addition, with stirring in an ice-water bath, of 44.8 g of solid ammonium sulfate per 100 ml of enzyme solution. Precipitated protein is collected by centrifugation at 14,000 g for 30 min. This precipitate can be dissolved in a small amount of DE-52 buffer and stored for a day or two at 0–4° or frozen and stored for several weeks at −80°.

Chromatography on Phosphocellulose. Phosphocellulose is washed with several volumes of 2 N KOH and then with distilled, deionized water before equilibrating it in 0.02 M sodium acetate buffer, pH 5.2, containing 1 mM dithiothreitol and 0.5 M sucrose. A 2.5 × 40-cm column is prepared. The enzyme sample is dialyzed for *only* 3–4 hours in 0.25-inch diameter dialysis casing against three changes, at least 20 volumes each, of the acetate buffer described above. The sample is applied to the column, then a gradient consisting of 400 ml *each* of the acetate buffer and of 0.05 M potassium phosphate buffer, pH 7.5, containing 0.5 M sucrose and 1 mM dithiothreitol, is applied. The column is eluted at a flow rate of 30–40 ml/hr, and 3-ml fractions are collected. Samples exhibiting the highest specific activity, typically greater than 40 units per milligram of protein, are pooled, and the protein is precipitated by ammonium sulfate, collected, and stored by the same procedures as those used subsequent to the DE-52 chromatography. Additional purified enzyme can be obtained by rechromatography of the less pure side fractions from both DEAE and phosphocellulose chromatography.

Small amounts of contaminating proteins can be removed and the specific activity increased to 63–75 units per milligram of protein by chromatography on Sephadex G-200 or BioGel A-1.5m as described for the type M isozyme with the exception that 0.5 M sucrose should be included in buffers for the chromatography of type L pyruvate kinase.

The results of a typical purification procedure of type L pyruvate kinase are presented in Table II.

Hybridization of Bovine Types L and M Pyruvate Kinases

Denaturation–renaturation procedures such as those required for the hybridization of bovine pyruvate kinases should be performed in clean

TABLE II
PURIFICATION OF BOVINE LIVER PYRUVATE KINASE[a]

Fraction	Volume (ml)	Total pyruvate kinase activity (μmol/min)	Total protein (mg)	Specific activity (μmol/ min/mg)	Yield (%)
Extract from 1700 g of liver	4200	9310	333,000	0.029	100
Acetone fractionation	580	4980	57,400	0.087	59
Ammonium sulfate fractionation	100	2160	9,100	0.237	23
Chromatography on DE-52	230	660	150	4.5	7.1
Chromatography on phosphocellulose	61	390	9.8	39.8	4.2
Filtration on BioGel A-1.5m	12	253	4.0	63.4	2.7

[a] Reprinted, with permission, from Cardenas and Dyson.[2]

polycarbonate or polyethylene containers to minimize loss of enzyme via adsorption to the container. Approximately equal quantities of enzyme activity (50–100 units/ml of each) are dissolved in, or dialyzed into, 0.05 M Tris-HCl (pH 7.5 at room temperature), 0.5 M sucrose, and 5 mM dithiothreitol at 0°. To the enzyme solution is added an equal volume of precooled 7 M guanidine-HCl (ultrapure). The solution is gently mixed and allowed to stand 5–10 min at 0°. The solution is then diluted with

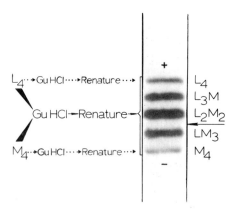

FIG. 2. Electrophoretic pattern obtained from the *in vitro* hybridization of types M and L pyruvate kinases. Procedures for hybridization, electrophoresis, and detection of pyruvate kinase activity are described in the text. The arrow on the right side of the pattern denotes the electrophoretic origin. Reprinted, with permission, from Cardenas et al.[3]

TABLE III
SUMMARY OF THE PROPERTIES OF BOVINE TYPES M AND L PYRUVATE KINASES[a]

Property	Type M	Type L
Kinetic		
K_m (ADP)	0.35	0.18
Maximum activity (μmol/min/mg)	225	75
Hill coefficient	1.0	2.5 (1.0 with FBP[b])
K_m or $S_{0.5}$ for PEP	0.03 mM	0.5 (0.09 with FBP)
Physical		
Electrophoretic mobility at pH 7.5	Toward ($-$)	Toward ($+$)
Molecular weight	230,000	215,000
$s_{20,w}$ (S)	9.95	9.3
Subunits	4(57,000)	4(52,000)
Isoelectric point	8.9	5.1
Phosphoenolpyruvate binding sites	4	4

[a] Reprinted, with permission, from Cardenas *et al.*[3]
[b] FBP, fructose 1,6-bisphosphate.

gentle swirling into a sufficient volume of renaturation buffer (0.5 M sucrose, 0.05 M Tris-HCl, 0.1 M KCl, 5 mM MgCl$_2$, 5 mM dithiothreitol, pH 7.5) at 0° to give a final protein concentration of about 0.05 mg/ml. The solution is incubated at 12–15° for 4 hr, dialyzed against 20 volumes of renaturation buffer to remove the guanidine HCl, and finally dialyzed against saturated ammonium sulfate solution, pH 7.0, containing 1 mM dithiothreitol to precipitate the protein. Figure 2 shows the electrophoretic pattern obtained after bovine types L and M pyruvate kinases have been hybridized *in vitro*.

Comments

Table III lists some of the properties of the two isozymes, including the striking differences in kinetic properties and isoelectric points. The facts that (*a*) type M pyruvate kinase typically exhibits hyperbolic kinetics with its substrates while type L may exhibit sigmoidal kinetics with phosphoenolpyruvate, and (*b*) the two isozymes can be easily hybridized *in vitro* to form what appear to be random combinations of two subunit types[2] suggest some very interesting experiments that can be performed with these isozymes and their hybrids. A beginning has been made toward using this system as a model to study mechanisms of allostery.[3,4] However, much is yet to be learned.

[25] Pyruvate Kinase Isozymes from Rat

By Kiichi Imamura and Takehiko Tanaka

Phosphoenolpyruvate + ADP → pyruvate + ATP

Since multiple forms of pyruvate kinase (PK)[1] (EC 2.7.1.40) were first found in different tissues of the rat by Tanaka *et al.*,[1a,2] there have been many studies[3-8] on these multiple forms, or isozymes, of PK in various systems to obtain information on their nature and on the mechanisms of control of carbohydrate metabolism. These studies have thrown much light on differentiation, deviation of gene expression during neoplasia or aging, genetic disorders, evolution, etc. Judging from the chemical, physical, electrophoretic, kinetic, and immunological parameters for the isozymes, there first appeared to be three unique forms of PK in mammalian systems, including humans,[9-11] named types L, M_1, and M_2 by Imamura and Tanaka.[9] In addition, another type in human erythrocytes was later separated electrophoretically from human L type by Imamura *et al.*[11] and named type R by Nakashima *et al.*[12] On the basis of the various properties of the four types of PK (see Table VII) and results on their amino acid compositions[13] and on peptide mapping of limited proteolysis products,[14] the question arises whether the four types of PK are products of four distinct genes, since M_2 type is very similar to M_1, and R type to L.

[1] Abbreviations used: PK, pyruvate kinase; LDH, lactate dehydrogenase; PEP, phosphoenolpyruvate; Fru-1,6-P_2, fructose 1,6-bisphosphate; NADH, reduced nicotinamide adenine dinucleotide; cAMP, adenosine 3′,5′-cyclic monophosphate; Tris, tris(hydroxymethyl)aminomethane; EDTA, ethylenediaminetetraacetic acid.

[1a] T. Tanaka, Y. Harano, H. Morimura, and R. Mori, *Biochem. Biophys. Res. Commun.* **21**, 55 (1965).

[2] T. Tanaka, Y. Harano, F. Sue, and H. Morimura, *J. Biochem.* (*Tokyo*) **62**, 71 (1967).

[3] K. H. Ibsen, *Cancer Res.* **37**, 341 (1977).

[4] E. R. Hall and G. L. Cottam, *Int. J. Biochem.* **9**, 785 (1978).

[5] J.-C. Dreyfus, A. Kahn, and F. Schapira, *Curr. Top. Cell. Regul.* **14**, 243 (1978).

[6] L. Engström, *Curr. Top. Cell. Regul.* **13**, 29 (1978).

[7] K. A. Munday, I. G. Giles, and P. C. Poat, *Comp. Biochem. Physiol. B* **67**, 403 (1980).

[8] T. H. Claus and S. J. Pilkis, *in* "Biochemical Actions of Hormones" (G. Litwack, ed.), Vol. 8, p. 209. Academic Press, New York, 1981.

[9] K. Imamura and T. Tanaka, *J. Biochem.* (*Tokyo*) **71**, 1043 (1972).

[10] K. Imamura, T. Taniuchi, and T. Tanaka, *J. Biochem.* (*Tokyo*) **72**, 1001 (1972).

[11] K. Imamura, T. Tanaka, T. Nishina, K. Nakashima, and S. Miwa, *J. Biochem.* (*Tokyo*) **74**, 1165 (1973).

[12] K. Nakashima, S. Miwa, S. Oda, T. Tanaka, K. Imamura, and T. Nishina, *Blood* **43**, 537 (1974).

[13] K. Harada, S. Saheki, K. Wada, and T. Tanaka, *Biochim. Biophys. Acta* **524**, 327 (1978).

[14] S. Saheki, K. Saheki, and T. Tanaka, *FEBS Lett.* **93**, 25 (1978).

TABLE I
NOMENCLATURE OF PYRUVATE KINASE (PK) ISOZYMES[a]

Nomenclature first used by	Adult liver (major peak)	Muscle	Adult liver (minor peak)	Erythrocytes
Imamura and Tanaka[9]	L	M_1	M_2	RBC[b]
Tanaka et al.[1a]	L	M	M	—
Faron et al.[15]	L	M	K	—
Susor and Rutter[16]	B	A	C	—
Bigley et al.[17]	I	III	II	—
Farina et al.[18]	II	—	I	—
Schloen et al.[19]	V	Muscle	I	—
Weinhouse et al.[20]	II	—	I	—
Walker and Potter[21]	I	II	III_A, III_B	—
Nakamura et al.[22]	L	M	S	—
Osterman et al.[23]	1	3	4	—
Carbonell et al.[24]	L	M	A	—
Nakashima et al.[12]	—	—	—	$R(R_1, R_2)^{b,c}$
Marie et al.[25]	—	—	—	L', $L^{b,d}$
Marie and Kahn[26]	—	—	—	L', L_a, L_b, $L_c^{b,d}$
Harada et al.[13]	—	—	—	$R(R_1, R_2, R_3)^{c,e}$

[a] Adapted from S. Yanagi et al., Cancer Res. **34**, 2283 (1974).
[b] Human erythrocytes PK.
[c] R_1, R_2, and R_3, respectively, in order from the cathodic side.
[d] L', L, L_a, L_b, and L_c: Normal and partially proteolysed human erythrocytes PK.
[e] Rat erythrocytes PK.

This question remains to be resolved. However, it was demonstrated by Marie et al.[14a] that the L- and R-type isozymes were translated from different mRNAs, and by Noguchi and Tanaka[14b] that the M_1 and M_2 types are also from different mRNAs.

This chapter describes the tissue distribution, methods of assay, electrophoretic behavior, purification, and characterization of PK isozymes. The isozymes are called, L, M_1, M_2, the nomenclature of Imamura and Tanaka, and R, used by Nakashima et al. for erythrocyte PK. However, to prevent confusion, the various PK designations used by other investigators are summarized in Table I.[1a,9,12,13,15-26]

[14a] J. Marie, M.-P. Simon, J.-C. Dreyfus, and A. Kahn, Nature (London) **292**, 70 (1981).
[14b] T. Noguchi and T. Tanaka, J. Biol. Chem. **257**, 1110 (1982).
[15] F. Faron, H. H. T. Hsu, and W. E. Knox, Cancer Res. **32**, 302 (1972).
[16] W. E. Susor and W. J. Rutter, Biochem. Biophys. Res. Commun. **30**, 14 (1968).
[17] R. H. Bigley, R. Stenzel, R. T. Jones, J. O. Compos, and R. D. Koler, Enzymol. Biol. Clin. **9**, 10 (1968).
[18] F. A. Farina, R. C. Adelman, C. H. Lo, H. P. Morris, and S. Weinhouse, Cancer Res. **28**, 1897 (1968).

Assay Method

Principle. Two different assay methods are described here. One is the simple, colorimetric 2,4-dinitrophenylhydrazone method. Pyruvic acid formed from PEP by PK is determined with the reagent for carbonyl residues, 2,4-dinitrophenylhydrazine, by a slight modification of the method of Kimberg and Yielding.[27] The other method is the NADH–LDH coupled method in which PK activity is measured by following the oxidation of NADH at 340 nm using a recording spectrophotometer with a coupled assay system similar to that described by Bücher and Pfleiderer.[28]

The 2,4-Dinitrophenylhydrazone Method

Dilution medium: homogenization buffer (described below) containing 0.5 mM Fru-1,6-P_2 and 1% bovine serum albumin, which stabilize PK activity

Reagent A: 2,4-dinitrophenylhydrazine (0.0125% in 2 N HCl)

Reagent B: 2.4 N NaOH containing 1 mM EDTA

Assay mixture for routine assays: 50 mM Tris-HCl buffer, pH 7.4, 0.1 M KCl, 5 mM MgSO$_4$, 2 mM PEP, 2 mM ADP, and 0.5 mM Fru-1,6-P_2

The reaction is started by adding 0.5 ml of assay mixture (preincubated at 37°) to 0.05 ml of the enzyme solution suitably diluted with dilution medium (preincubated at 37° for 5 min). After incubation for 3 min at 37°, 1.5 ml of reagent A is added, and the mixture is allowed to stand for 10 min at 37°. Then 4 ml of reagent B is added, and after full color development (in about 10 min) the absorbance is measured at 510 nm. This method is very convenient for routine assay of PK activity when there are many samples, but the absorbance should be measured within 10 min after full color development because the color is not stable. If the absorbance cannot be measured in this time, the mixture should be kept at the stage after adding reagent A. This method is not suitable for kinetic studies because it is less sensitive than the coupled method described below.

---

[19] L. H. Schloen, J. R. Bamburg, and H. J. Sallach, *Biochem. Biophys. Res. Commun.* **36**, 823 (1969).

[20] S. Weinhouse, J. B. Shatton, W. E. Criss, and H. P. Morris, *Biochimie* **54**, 685 (1972).

[21] P. R. Walker and V. R. Potter, *Adv. Enzyme Regul.* **10**, 339 (1972).

[22] N. Nakamura, K. Hosoi, N. Nishikawa, and T. Horio, *Gann* **63**, 239 (1972).

[23] J. Osterman, P. J. Fritz, and T. Wunch, *J. Biol. Chem.* **248**, 1011 (1973).

[24] J. Carbonell, J. E. Feliu, R. Marco, and A. Sols, *Eur. J. Biochem.* **37**, 148 (1973).

[25] J. Marie, H. Garreau, and A. Kahn, *FEBS Lett.* **78**, 91 (1977).

[26] J. Marie and A. Kahn, *Biochem. Biophys. Res. Commun.* **91**, 123 (1979).

[27] D. V. Kimberg and K. L. Yielding, *J. Biol. Chem.* **237**, 3233 (1962).

[28] T. Bücher and G. Pfleiderer, this series, Vol. 1 [66].

NADH–LDH Coupled Method. The assay mixture consists of 50 mM Tris-HCl buffer, pH 7.5, 0.1 M KCl, 5 mM MgSO$_4$, 2 mM ADP, 2 mM PEP, 0.5 mM Fru-1,6-P$_2$, 0.18 mM NADH, and 8 units of LDH.

One milliliter of assay mixture (preincubated at 25 or 37°) is pipetted into a microcuvette (1-cm light path and 0.5-cm width), which is placed in the recording spectrophotometer for equilibration at 25° or 37°. The reaction is started by adding 10 μl of enzyme solution suitably diluted with dilution medium, and the oxidation of NADH is followed at 340 nm in a spectrophotometer equipped with a recorder and multisample programmer. This method is good for kinetic studies and for detecting enzyme activity of the eluate during column chromatography. The presence of Fru-1,6-P$_2$ in the assay mixture in both methods gives the maximum activity of PK, especially of L and M$_2$ with the concentration of PEP used.

Definition of Unit. One unit of enzyme is defined as the amount catalyzing the formation of 1 μmol of pyruvate (or oxidizing 1 μmol of NADH in the coupled system) in 1 min under the assay conditions. Specific activity is defined as units of enzyme per milligram of protein.

Tissue Distributions of the Four Types of Pyruvate Kinase

M$_1$ is the main type in specifically differentiated tissues, such as adult skeletal muscle, heart, and brain. L is predominant in gluconeogenic tissues, especially liver, but is a minor type in kidney. R is present in erythrocytes and hematopoietic tissues. Imamura *et al.*[11] found that human erythrocyte PK differed in mobility from human L PK on thin-layer polyacrylamide gel electrophoresis, though at that time it was believed that erythrocyte PK was identical to L PK. Later, Nakashima *et al.*[12] found that human erythrocyte PK gives two bands, R$_1$ PK and R$_2$ PK, by the same method. M$_2$ is predominant in the fetus, in neoplasias, and in undifferentiated or proliferating tissues, but it is also widely distributed in adult tissues. It is a minor isozyme in liver, but the major one in kidney. Table II summarizes the tissue distributions of the four types of PK, which were classified and identified mainly by electrophoretic (zymograms of tissues), kinetic, and immunological methods.

The question of whether an isolated enzyme is a hybrid of different subunits is determined by *in vitro* hybridization and by electrophoresis in the absence and the presence of antibodies of the suspected components as described below. The electrophoretic pattern of the reassociated components of the isolated enzyme dissociated by a denaturant is compared with that of dissociated mixtures of pairs of homologous PK preparations, such as L$_4$ and (M$_1$)$_4$, or L$_4$ and (M$_2$)$_4$. If the isolated enzyme is a hybrid form, the electrophoretic pattern will show five spots of PK, and in the

TABLE II

TISSUE DISTRIBUTION OF RAT PYRUVATE KINASE ISOZYMES[a,b]

Tissue	Type of isozyme					
	L	R	M_1	M_2	$L–M_2$ hybrid	$M_1–M_2$ hybrid
Skeletal muscle			+++	±		Yes (fetu
Heart			+++	+		Yes (fetu
Brain			+++	+		Yes (fetu
Liver	+++			+	★	
Kidney	+			+++	Yes	
Intestine	+			+++	Yes	
Stomach				+++		
Spleen				+++		
Lung				+++		
Adipose tissue				+++		
Testis				+++		
Uterus				+++		
Placenta				+++		
Leukocytes				+++		
Fetal tissues				+++		
During differentiation	Increased in liver, etc.		Increased in skeletal muscle, heart and brain	Decreased in liver, skeletal muscle, heart and brain		
Regenerating liver	Decreased			Increased		
Tumor				+++		
Host liver of tumor-bearing animal	Decreased			Increased		
Erythrocytes	?	+++				
Hematopoietic tissues, intrahepatic erythropoietic cells of fetus or newborn rat	?	+++		?		

[a] Based on data in the references cited in footnotes 2, 9, 11, 13, 23, and 29.

[b] +++, Large amount; +, small amount; ±, occasionally detected; ★, often observed in hum hepatomas and normal tissue around hepatomas.

presence of antibody to one of the expected subunits it should show only one spot, which is the activity of the homologous oligomer of the partner subunit of the expected subunit. If the enzyme is not heterologous, but homologous, the electrophoretic pattern will show only one spot of PK. The electrophoretic pattern of PK can be detected as both protein and activity in the purified enzyme system, but only as activity in the un-purified enzyme system. As listed in Table II, the presence of L–M_2 hybrids in kidney and intestine, and of M_1–M_2 hybrids in fetal tissues was confirmed, but no evidence has been obtained for the existence of L–M_1 hybrid forms *in vivo*.

Thin-Layer Polyacrylamide Gel Electrophoresis

The method of thin-layer polyacrylamide gel electrophoresis originally employed by Imamura and Tanaka[9] to separate multiple forms of PK has been modified by Harada et al.[13] and Saheki et al.[29] The improved method is described here.

Reagents

Reagent A: 40% (w/v) acrylamide monomer solution containing 1.5% (w/v) *N,N'*-methylene bisacrylamide.

Reagent B: 0.5% (w/v) *N,N,N',N'*-tetramethylethylenediamine solution

Reagent C: 1% (w/v) ammonium persulfate solution (prepared just before use)

Running buffer: 10 mM Tris-HCl, pH 7.5, containing 20 mM KCl, 2 mM $MgSO_4$, 0.4 mM Fru-1,6-P_2, and 20 mM 2-mercaptoethanol

Staining mixture of PK: PK assay mixture containing 3 mM NADH

Agarose solution, 3% (w/v)

Procedure. Reagents A (4.2 ml), B (2.5 ml), and C (5 ml) and distilled water (38.3 ml) are gently mixed, and air in the mixture is removed by aspiration. A thin layer (1 mm thick) of polyacrylamide gel (3.4%, w/v, acrylamide with 0.13%, w/v, bisacrylamide) is prepared on a glass plate (on which a rectangular acryl resin frame, 210 × 120 mm and 1 mm thick, is fixed) with an acryl resin cover that holds a slot-former. The prepared slab gel is dialyzed against deionized water for a few hours and then equilibrated with the running buffer for about 6 hr. Then excess buffer on the gel surface is drained off by standing the gel slab vertically until its surface is half dry. Excess buffer in the sample slots is carefully removed with filter paper. Then the gel plate is placed horizontally, and 3-μl samples of suitably diluted enzyme (1–4 mU) are applied to sample slots with capillary pipettes or Eppendorff pipets. The gel plate is placed horizontally on the flat glass surface of the cell cooled with crushed ice and water. Electrical connections between the gel plate and running buffer are made with glass fiber filter paper, which greatly reduces the electro-osmotic effect of the bridges. The electrode vessels are each divided into two compartments. One is the chamber for the running buffer and another is the

[29] S. Saheki, K. Saheki, Y. Sanno, and T. Tanaka, *Biochim. Biophys. Acta* **526**, 116 (1978).

electrode chamber, which is filled with 5% K_2SO_4 solution. Electrical contact between two chambers is made with 1% agar large block containing 5% K_2SO_4. A plastic plate is placed 2 mm above the gel surface to prevent the gel from drying during electrophoresis.

Horizontal electrophoresis is performed for 5–6 hr at a constant current of 4–5 mA per centimeter of width of the gel layer in a cold room. After electrophoresis, PK activity is detected by a modification of the method of Susor and Rutter.[30] Then 5 ml of 3% agarose (solubilized in a boiling water bath) is rapidly added to 19 ml of the staining mixture, and 50 μl of LDH suspension (5500 U/ml) is added to the mixture. The gel plate is fixed to another glass plate with a frame (1 mm thick) of the same size as the gel plate, but with one side open. Thus, a space about 1 mm thick is made over the gel. A film of staining mixture of uniform thickness (about 1 mm thick) on the gel surface is obtained by introducing the mixture into the space from the open side with a Pasteur pipet. This mixture is allowed to solidify at room temperature, and then the gel plate with the film of staining mixture is incubated at 37° to allow the enzyme reaction. After 10–20 min, observation under ultraviolet light (to detect fluorescence) shows areas in which fluorescence is lost owing to oxidation of NADH by the coupled PK–LDH reaction.

The gel plate is photographed from 30 cm above the gel using the same ultraviolet light source, with the photographic paper placed under the gel plate (contact printing). When the applied sample is pure, the electrophoretic pattern can be detected by measuring both activity and stained protein. Under these conditions, L PK has the highest mobility to the anode, following in order by R, M_1, and M_2. R PK migrates rather more slowly than L, but much faster than M_1, and M_1 migrates rather faster than M_2. FDP in the running buffer is necessary to separate M_2 clearly from M_1, as first reported by Susor and Rutter[16] for cellulose acetate membrane electrophoresis and later confirmed by Imamura and Tanaka[9] for the original thin-layer polyacrylamide gel electrophoresis method. With this method, Harada et al.[13] and Saheki et al.[29] separated rat R PK into three bands, R_1, R_2, and R_3, in increasing order toward the anode.

Purification of PK Isozymes

Methods for purification of PK isozymes have been reported by Tanaka et al.,[2] Imamura et al.,[10] and others.[31-34] However, the yields of

---

[30] W. E. Susor and W. J. Rutter, *Anal. Biochem.* **43**, 147 (1971).
[31] O. Ljunström, G. H. Hjelmquist, and L. Engström, *Biochim. Biophys. Acta* **358**, 289 (1974).

the purified enzymes by these methods were not always good. Improved methods for purification of PK isozymes were developed by Harada *et al.*[13] in our laboratory. Methods based on the latter methods are described here.

Phosphate buffer, pH 6, containing 2 mM MgSO$_4$ and 10 mM 2-mercaptoethanol is used throughout, and 0.2 mM Fru-1,6-P$_2$ is also used as a stabilizing agent for all isozymes except M$_1$. This method has several advantages; for example, a high yield can be obtained and column chromatography can be carried out at room temperature without loss of activity under these conditions, because the phosphate buffer (pH 6.0), MgSO$_4$, 2-mercaptoethanol, and Fru-1,6-P$_2$ described above act as stabilizers. Application of affinity elution chromatography (with PEP) also greatly simplifies the purification procedure.

Reagents

Buffer A: homogenizing buffer containing 20 mM Tris-HCl, pH 7.5, 100 mM KCl, 5 mM MgSO$_4$, and 1 mM EDTA

Buffer B: 10 mM potassium phosphate buffer, pH 6.0, containing stabilizer consisting of 2 mM MgSO$_4$, 10 mM 2-mercaptoethanol, and 0.2 mM Fru-1,6-P$_2$. Fru-1,6-P$_2$ is omitted from the stabilizer during purification of M$_1$ because the stability of M$_1$ is not affected by this reagent.

Buffer C: 5 mM potassium phosphate buffer, pH 6.0, containing the stabilizer

Buffer D: 33 mM potassium phosphate buffer, pH 6.0, containing the stabilizer

Buffer E: 40 mM potassium phosphate buffer, pH 6.0, containing the stabilizer

Buffer F: 70 mM potassium phosphate buffer, pH 6.0, containing the stabilizer

Buffer G: 100 mM potassium phosphate buffer, pH 6.0, containing the stabilizer

Cold acetone: Acetone for chromatography is cooled at $-15°$ in Dry Ice–acetone

General Procedure. Detailed procedures for each isozyme are described later. In each procedure the following general conditions are used unless otherwise stated. All procedures except column chromatography are carried out at 0–4°. Activated phosphocellulose is equilibrated with

[32] Y. Nagao, T. Toda, K. Miyazaki, and T. Horio, *J. Biochem.* (*Tokyo*) **82**, 1331 (1977).
[33] J. P. Riou, T. H. Claus, and S. J. Pilkis, *J. Biol. Chem.* **253**, 656 (1978).
[34] E. R. Hall, V. McCully, and G. L. Cottam, *Arch. Biochem. Biophys.* **195**, 315 (1979).

buffer B with or without Fru-1,6-P_2 and packed into a column. Fraction-ation with ammonium sulfate is done by slow addition of the calculated amount of the solid salt with stirring. The resulting suspensions are equilibrated with stirring for 20 min and then centrifuged. Fractions obtained with ammonium sulfate or acetone are dissolved in, and dialyzed against, buffer B or applied to a Sephadex G-25 column equilibrated with buffer B according to the purification step. The dialyzed enzyme solution (about 10 mg/ml column bed volume) is applied to a phosphocellulose column. All four types of PK (L, R, M_1, and M_2) are absorbed similarly to the column at the low ionic strength of buffer B in the presence of stabiliz-ing agents. The column is washed with a linear concentration gradient between buffer B and buffer D, and then with buffer D until the absor-bance of the eluate at 280 nm becomes almost zero. For batchwise chromatography on phosphocellulose, phosphocellulose previously equilibrated with buffer B is added to the enzyme solution with stirring for 10 min. Then it is packed into a column and washed with buffer B with stepwise increasing concentrations of phosphate buffer to 33 mM (2 mS, corresponds to buffer D) until the absorbance of the eluate at 280 nm becomes almost zero. Then the enzyme is eluted with a suitable concen-tration of phosphate buffer.

Enzyme Assay. PK activity during the purification is measured with the coupled system described in the preceding section.

Purification of M_1 PK from Skeletal Muscle

Step 1. Crude Extract. About 3.5 kg of rat skeletal muscle, which has been stored at $-20°$, is minced with a meat chopper and homogenized with three volumes of buffer A containing 10 mM 2-mercaptoethanol in a War-ing blender for 3 min. The homogenate is centrifuged, and the crude extract is used as starting material for purification of PK.

Step 2. First Ammonium Sulfate Fractionation. Ammonium sulfate (32.5 g/100 ml, 55% saturation) is added to the crude extract; insoluble material is removed by centrifugation. Further ammonium sulfate (9.3 g/100 ml, 70% ammonium sulfate saturation) is added to the resulting supernatant solution; the resulting precipitate is recovered by centrifuga-tion and dissolved in, and dialyzed against, buffer E.

Step 3. Acetone Fractionation. Cold acetone ($-15°$) containing 10 mM 2-mercaptoethanol is added to the dialyzed enzyme solution to 30% (v/v). Insoluble material is removed by centrifugation at $-15°$, and the resulting supernatant is adjusted to 38% (v/v) of the cold acetone ($-15°$). The 30–38% acetone fraction is recovered by centrifugation at $-15°$.

Step 4. Second Ammonium Fractionation. The dialyzed acetone frac-tion is refractionated with ammonium sulfate (55–70% saturation).

TABLE III
PURIFICATION OF TYPE M_1 PYRUVATE KINASE FROM RAT MUSCLE[a]

Step	Total protein (mg)	Total activity (units)	Specific activity (units/mg protein)	Recovery (%)
Crude extract	138×10^3	690×10^3	5	100
55–70% $(NH_4)_2SO_4$ fraction	607×10^2	621×10^3	10	90
30–38% Acetone fraction	907×10	483×10^3	53	70
55–70% $(NH_4)_2SO_4$ fraction	650×10	455×10^3	70	66
Phosphocellulose chromatography (KCl elution)	146×10	366×10^3	251	53
Phosphocellulose chromatography (phosphoenolpyruvate elution)	763	290×10^3	380	42

[a] Reprinted, with permission, from Harada *et al.*[13]

Step 5. First Phosphocellulose Chromatography (KCl Elution). The dialyzed ammonium sulfate fraction is applied to a phosphocellulose column (280 ml). The column is washed with buffer of stepwise increasing electric conductance to 9 mS obtained by addition of KCl to buffer B, keeping the pH at 6. Then, the enzyme is eluted with buffer B containing 170 mM KCl (12 mS).

Step 6. Second Phosphocellulose Chromatography (Affinity Elution with PEP). The eluate is condensed and dialyzed and again applied to a phosphocellulose column. The column is washed with a linear gradient to ionic strength of 0–80 mM (6 mS) KCl in the same buffer, and PK is eluted with the latter buffer containing 0.5 mM PEP.

The procedures are summarized in Table III.

Purification of M_2 PK from AH-130 Yoshida Ascites Hepatoma Cells

M_2 enzyme is prepared from AH-130 Yoshida ascites hepatoma cells maintained in female Sprague–Dawley albino rats. The cells are washed with saline.

Step 1. Crude Extract. About 500 g of frozen AH-130 Yoshida ascites hepatoma cells are thawed and lysed by adding two volumes of buffer A containing 10 mM 2-mercaptoethanol and 0.2 mM Fru-1,6-P$_2$. Crude extract is obtained as the supernatant on centrifugation at 0°.

Step 2. Ammonium Sulfate Fractionation. The crude extract is fractionated with ammonium sulfate (45–75% saturation).

Step 3. Phosphocellulose Chromatography (Affinity Elution with PEP). The dialyzed ammonium sulfate fraction is purified by batchwise

TABLE IV
PURIFICATION OF TYPE M₂ PYRUVATE KINASE FROM AH-130 CELLS[a]

Step	Total protein (mg)	Total activity (units)	Specific activity (units/mg protein)	Recovery (%)
Crude extract	120×10^2	677×10^2	5.6	100
45–75% $(NH_4)_2SO_4$ fraction	338×10	643×10^2	19	95
Phosphocellulose chromatography (phosphoenolpyruvate elution)	78	406×10^2	520	60

[a] Reprinted, with permission, from Harada et al.[13]

chromatography on phosphocellulose (400 ml, equilibrated with buffer B containing 0.2 mM Fru-1,6-P₂) by elution with buffer F containing 0.5 mM PEP. The procedure is summarized in Table IV.

Purification of L PK from Rat Liver

Sprague–Dawley albino rats of both sex are used. A high level of L type is induced in the liver by feeding the animals on synthetic high-carbohydrate diet consisting (per 100 g) of 68 g of dextrin, 15 g of sucrose, 10 g of casein, 4 g of inorganic salt mixture, 1 g of vitamin mixture, 2 g of cellulose powder, and 0.01 g of choline chloride. The animals are fed the diet for 3 days before sacrifice.

Step 1. Crude Extract. About 3 kg of chopped rat liver are homogenized with three volumes of buffer A containing 0.2 mM Fru-1,6-P₂ and 10 mM 2-mercaptoethanol in a Waring blender for 1 min, and the homogenate is centrifuged (2×10^4 g) to obtain the crude extract.

Step 2. Ammonium Sulfate Fractionation. The crude extract is fractionated with ammonium sulfate (33–45% saturation).

Step 3. Acetone Fractionation. The fraction from step 2 is dissolved in, and dialyzed against, buffer E. The dialyzed enzyme solution is fractionated with 20% (v/v) to 40% (v/v) saturation of acetone ($-15°$) containing 1 mM dithiothreitol.

Step 4. Phosphocellulose Chromatography (Affinity Elution with PEP). The dialyzed acetone fraction is purified by batchwise chromatography on phosphocellulose (1000 ml). The enzyme is eluted with buffer D containing 0.5 mM PEP.

Step 5. Crystallization. Fractions with PK activity are combined and brought to 50% saturation of ammonium sulfate. The resulting precipitate is dissolved in a small volume of 50 mM phosphate buffer, pH 6, containing stabilizing agents (0.4 mM Fru-1,6-P_2, 2 mM MgSO$_4$, 10 mM 2-mercaptoethanol) and 1 mM PEP, which results in a protein concentration of about 10 mg/ml. Then the enzyme solution is slowly mixed with solid ammonium sulfate until the solution becomes slightly turbid, and it is stored at 4°. After a few weeks, crystals appear as hexagonal plates. In the absence of PEP, very thin, needle-shaped crystals are formed.

The purification procedure is summarized in Table V.

Purification of R PK from Rat Erythrocytes

R enzyme is prepared from erythrocytes, collected by decapitation of animals, and washed with saline to remove leukocytes, etc.

Step 1. Hemolysate. About 4 liters of frozen erythrocytes are thawed and lysed by adding three volumes of buffer C containing 0.2 mM Fru-1,6-P_2. The hemolysate is adjusted to pH 5 with 2 M acetic acid and centrifuged at 2 × 10^4 g for 30 min at 0°.

Step 2. First Ammonium Sulfate Fractionation. The resulting supernatant is rapidly adjusted to pH 6 with 2 M NaOH and brought to 50% saturation of ammonium sulfate. The precipitate is collected by centrifugation and dissolved in 3 liters of buffer E.

Step 3. Second Ammonium Sulfate Fractionation. The resulting enzyme solution is refractionated with ammonium sulfate (25–45% saturation). The fraction is dissolved in, and dialyzed against, buffer B at 4°.

Step 4. First Phosphocellulose Chromatography. The dialyzed enzyme solution is purified by batchwise chromatography on phosphocellulose (600 ml) in buffer G. The precipitate from the eluate with 50% saturation of ammonium sulfate is dissolved in, and dialyzed against, buffer B.

TABLE V

PURIFICATION OF TYPE L PYRUVATE KINASE FROM RAT LIVER[a]

Step	Total protein (mg)	Total activity (units)	Specific activity (units/mg protein)	Recovery (%)
Crude extract	227 × 10³	340 × 10³	1.5	100
33–45% (NH$_4$)$_2$SO$_4$ fraction	126 × 10³	272 × 10³	2.0	80
20–40% Acetone fraction	163 × 10²	163 × 10³	10	48
Phosphocellulose chromatography (phosphoenolpyruvate elution)	504	126 × 10³	250	37
Crystallization	157	816 × 10²	520	24

[a] Reprinted, with permission, from Harada *et al.*[13]

TABLE VI
PURIFICATION OF TYPE R PYRUVATE KINASE FROM RAT ERYTHROCYTES[a]

Step	Total protein (mg)	Total activity (units)	Specific activity (units/mg protein)	Recovery (%)
Hemolysate	382×10^4	34,380	0.009	100
50% $(NH_4)_2SO_4$ fraction	79×10^3	34,020	0.43	99
25–45% $(NH_4)_2SO_4$ fraction	34×10^3	31,680	0.93	92
Phosphocellulose chromatography	2×10^3	26,880	13.4	78
Phosphocellulose chromatography (phosphoenolpyruvate elution)	62	19,870	320	58
Crystallization	55	16,850	307	49

[a] Reprinted, with permission, from Harada et al.[13]

Step 5. Second Phosphocellulose Chromatography (Affinity Elution with PEP). The dialyzed preparation is applied to a phosphocellulose column. The enzyme is specifically eluted with buffer D containing 0.5 mM PEP. Table VI summarizes the purification of R PK.

Properties of the Isozymes

Various properties of the four types of PK are summarized in Table VII.[1a,2,4,6,8,32,35,36] As described above, the four types of PK differ clearly in electrophoretic mobilities. However, as shown in Table VII, L and R, and M_1 and M_2, respectively, are closely similar in certain properties, including their immunological and kinetic properties, phosphorylations by cAMP-dependent protein kinase (L and R are phosphorylated, but M_1 and M_2 are not), amino acid compositions, and peptide maps of limited proteolysis products (not shown in Table VII). However, M_2 shows partially similar kinetic properties to L and R, and partially different properties from M_1. M_2, like L and R but unlike M_1, shows sigmoidal kinetics with respect to PEP and is allosterically activated by Fru-1,6-P_2; moreover, unlike M_1, M_2 is strongly inhibited by low concentrations of amino acids, such as Ala and Phe. Therefore, the question arises whether the four types of PK are products of four distinct genes.

Studies of genetic disorders or hereditary PK deficiency should give a clue to this question. In 1968, Bigley et al.[17] found that there was little or

[35] U. Dahlqvist-Edberg, *FEBS Lett.* **88**, 139 (1978).
[36] C. Cladaras and G. L. Cottam, *Arch. Biochem. Biophys.* **200**, 426 (1980).

TABLE VII

CHARACTERISTICS OF PYRUVATE KINASE ISOZYMES OF RAT[a]

Characteristics	Type of isozyme			
	L	R	M_1	M_2 (AH-130 cells)
Molecular weight (M_r)	208,000– 220,000	211,800– 220,700[b]	250,000	216,000– 240,000[d]
$s_{20,w}$	9.4 S	7.8 S	9.6 S	10.8 S
Subunit M_r	57,000	62,000 (major) 57,000 (minor)	57,000– 59,000	60,000[d] 61,000
Crystallization	Yes	Yes	Yes	Yes (also d)
Specific activity (units)	520 (25°)	307 (25°)	780 (37°) 380 (25°)	770 (37°) 520 (25°)
Kinetics with regard to PEP	Sigmoidal	Sigmoidal	Hyperbolic	Sigmoidal
K_m for PEP (mM)	0.3–0.96	1.0,[b] 1.4	0.08	0.4
Hill coefficient for PEP	1.6–2	1.2–1.6[c]	1.0	1.4–1.5
K_m for ADP (mM)	0.1–0.4	0.44–0.6[c]	0.3	
Activation by Fru-1,6-P_2	Yes, hyperbolic	Yes, hyperbolic	No	Yes, hyperbolic
K_a(app) for Fru-1,6-P_2 (μM)	0.06–0.1	0.4	—	0.1–0.4
K_i(app) for ATP (mM)	0.1–0.15	0.04–0.06[c]	3–3.5	2.5
Phosphorylation by cAMP- dependent protein kinase	Yes	Yes[c]	No	No
Mole P_i/mole subunit	1	1[c]	—	—
Activity of phosphoenzyme in low concentration of PEP	Decreased	Decreased[c]	—	—
V_{max} of phosphoenzyme	Unchanged	Unchanged[c]	—	—
Inhibition of phosphoenzyme by ATP	Stimulated	Stimulated[c]	—	—
Effect of Fru-1,6-P_2 on phosphorylation	Inhibition	Inhibition[c]	—	—
Inhibition by amino acid				
K_i(app) for L-Ala (mM)	1.0	—	—	0.6
K_i(app) for L-Phe (mM)	5	—	—	0.5
Half-life (hr)	59	—	—	—
Neutralization of antibody				
Anti-L	Yes	Yes	No	No
Anti-R	Yes	Yes	No	No
Anti-M_1	No	No	Yes	Yes
Anti-M_2	No	No	Yes	Yes
Acute hormonal regulation				
Glucagon	Inhibition	—	—	—
Catecholamines	Inhibition	—	—	—
Insulin	Stimulation	—	—	—

(*continued*)

TABLE VII (*continued*)

| | | | | M_2 |
| | | | | (AH-130 |
Characteristics	L	R	M_1	cells)
Type of isozyme				
Chronic adaptation				
Starvation	Decreased	—	Unchanged	Unchanged
Diabetes	Decreased	—	Unchanged	Unchanged
High-carbohydrate diet	Increased	—	Unchanged	Unchanged

a Based on data in the references cited in footnotes 1a, 2, 4, 6, 8, 10, 32, 35, and 36.
b S. Saheki *et al.*, unpublished data.
c Data for human R PK.
d Data for rhodamine sarcoma of rats.

no PK activity in the erythrocytes and liver of a patient with hereditary hemolytic anemia, a so-called classical type quantitative PK deficiency. In 1973, Imamura *et al.*[11] found that PK zymograms of erythrocytes and liver of a patient with PK deficiency (PK "TOKYO"), who was found by Miwa *et al.*, showed abnormal patterns: both PKs from the patient showed lower mobilities than normal, and the kinetic properties of erythrocyte PK of the patient were also abnormal. These results suggest that erythrocyte PK and liver L PK are not identical, but that both have a subunit that is a product of a gene coding for either erythrocyte PK or L PK.

Investigations on the L–R subunit relationship were developed and extended by Nakashima *et al.*[37] and Marie *et al.*[25] Since 1977, Marie *et al.*[25] and Kahn *et al.*[38] have reported that the L' subunit, which is a major form of erythrocyte PK, might be a precursor of the L subunit and that the L subunit might be a product of proteolytic processing of precursor L'. However, Marie *et al.*[14a] examined *in vitro* protein synthesis experiments using RNA extracted from rat reticulocytes and liver which express the L- and R-type isozymes, respectively, and demonstrated that the difference was reflected in tissue-specific mRNAs. Thus, they withdrew their hypothesis that the difference was due to posttranslational processing. On the other hand, little is known about the relationship between the M_1 and M_2 isozymes. Marie *et al.*[39] reported that M-type isozymes in human tissues seemed to be molecular forms of the same gene product. Ibsen *et*

[37] K. Nakashima, *Clin. Chim. Acta* **55**, 245 (1974).
[38] A. Kahn, J. Marie, H. Garreau, and E. D. Sprenger, *Biochim. Biophys. Acta* **523**, 59 (1978).
[39] J. Marie, A. Kahn, and P. Boivin, *Hum. Genet.* **31**, 35 (1976).

al.[40] also reported data suggesting that the M_2 subunit may be a precursor of the M_1 subunit in the mouse PK isozyme system. However, Noguchi and Tanaka[14b] examined *in vitro* protein synthesis experiments isolating total RNA from rat skeletal muscle and AH-130 Yoshida ascites hepatoma cells, which express the M_1- and M_2-type isozymes, respectively, and demonstrated that the M_1- and M_2-type isozymes were translated from different mRNAs. Thus, the detailed relationships between L and R, and M_1 and M_2 still remain to be studied.

[40] K. H. Ibsen, R. H. Chiu, H. R. Park, D. A. Sanders, S. Roy, K. N. Garratt, and M. K. Mueller, *Biochemistry* **20**, 1497 (1981).

[26] Pyruvate Kinase from *Streptococcus lactis*

By V. L. CROW and G. G. PRITCHARD

Phosphoenolpyruvate + ADP → pyruvate + ATP

The lactic streptococci, *Streptococcus lactis* and *S. cremoris*, rely almost exclusively on glycolysis to fulfill their energy requirements. Pyruvate kinase (EC 2.7.1.40) is a key regulatory enzyme in these bacteria, being subject to feed-forward activation by a number of phosphorylated sugars.[1-4] The following procedure is based on a previously published account of the purification and properties of pyruvate kinase from *S. lactis* C10.[3]

Assay Method

Pyruvate kinase activity is measured in a coupled assay system with excess lactate dehydrogenase according to the principle described by Bücher and Pfleiderer.[5] Oxidation of NADH is followed at 340 nm.

Reagents

Triethanolamine buffer, 0.1 M, pH 7.5, containing 10 mM $MgCl_2$ and 100 mM KCl

[1] L. B. Collins and T. D. Thomas, *J. Bacteriol.* **120**, 52 (1974).
[2] T. D. Thomas, *J. Bacteriol.* **125**, 1240 (1976).
[3] V. L. Crow and G. G. Pritchard, *Biochim. Biophys. Acta* **438**, 90 (1976).
[4] J. Thompson and T. D. Thomas, *J. Bacteriol.* **130**, 583 (1977).
[5] T. Bücher and G. Pfleiderer, this series, Vol. 1, p. 435.

METHODS IN ENZYMOLOGY, VOL. 90

Phosphoenolpyruvate (PEP), trisodium salt, 30 mM
ADP, sodium salt, 75 mM
Fructose 1,6-bisphosphate (Fru-1,6-P$_2$), tetrasodium salt, 30 mM
NADH, 5 mM
Lactate dehydrogenase, 2000 units/ml, dialyzed overnight against triethanolamine buffer, 0.1 M, pH 7.5, containing 20% (v/v) glycerol

Cyclohexylammonium salts of PEP and Fru-1,6-P$_2$ and the sulfates of Mg^{2+} and K$^+$ should be avoided, as these ions are inhibitory. Tris, maleate, and phosphate also inhibit pyruvate kinase activity, and so should not be used in buffers. Stock solutions of PEP, ADP, and NADH are used within 12 hr and of Fru-1,6-P$_2$ within 2 days of preparation. All stock solutions are stored on ice.

Procedure. The activity is determined at 25° in a final volume of 3 ml containing 80 mM triethanolamine, 80 mM KCl, 8 mM MgCl$_2$ (2.4 ml of stock buffer solution), 5 mM ADP (0.2 ml stock solution), 1 mM Fru-1,6-P$_2$ (0.1 ml stock solution), 0.167 mM NADH (0.1 ml stock), 1 mM PEP (0.1 ml stock), 20 units of dialyzed lactate dehydrogenase (10 μl stock), and 0.1 ml of appropriately diluted enzyme solution. The enzyme is diluted in 20% (v/v) glycerol in deionized, distilled water at 4°. Fru-1,6-P$_2$ (or other activator) is essential for activity.

NADH oxidase activity is present in crude extracts and early purification steps. This can be corrected for by measurement of the rate of NADH oxidation prior to addition of ADP that is used to start the pyruvate kinase reaction.

Definition of Unit. One unit is defined as the amount of enzyme required to catalyze the oxidation of 1 μmol of NADH per minute under the above conditions. The specific activity is defined as units of activity per milligram of protein. Protein concentration is determined by the method of Lowry et al.[6] using crystalline bovine serum albumin as standard.

Purification

Growth of Bacteria. Streptococcus lactis C10 is obtained from the Dairy Research Institute, Palmerston North, New Zealand. It is also available from the CSIRO Division of Food Research, Dairy Research Laboratory, Melbourne, Australia and from the National Institute for Research in Dairying, Shinfield, Reading, England (strain NCDO 509). Cultures are grown at 30° in the medium of Jago et al.[7] containing the following constit-

[6] O. H. Lowry, N. J. Rosebrough, A. L. Farr, and R. J. Randall, *J. Biol. Chem.* **193**, 265 (1951).
[7] G. R. Jago, L. W. Nichol, K. O'Dea, and W. H. Sawyer, *Biochim. Biophys. Acta* **250**, 271 (1971).

uents per liter of distilled water: lactose, 30 g; tryptone, 30 g; yeast extract, 10 g; beef extract, 2 g; and KH_2PO_4, 5 g. Cultures are grown without aeration in 5-liter Erlenmeyer flasks containing 3 liters of medium. The pH is maintained between 6.0 and 6.5 by periodic addition of 2.5 M NaOH during growth. The cells are harvested near the end of the logarithmic phase of growth (after about 6 hr of growth) by centrifugation at 6000 g for 15 min at 0° and then washed three times in 0.005 M phosphate buffer, (pH 7.0) containing 1% (w/v) NaCl. The washed cells are stored frozen for no longer than 16 hr before disruption. Storage of frozen cells for even a few days appears to render the partially purified enzyme unstable, and this trend becomes more noticeable with longer periods of storage.

Step 1. Preparation of Crude Extract. All purification steps are carried out at a temperature not exceeding 4°. The frozen cells are thawed and suspended in 0.01 M phosphate buffer (pH 7.0) containing 0.05% (v/v) 2-mercaptoethanol and disrupted by two passages through an Aminco French pressure cell at 5500 psi. Unbroken cells and cell debris are removed by centrifugation at 13,000 g at 4°. Nucleic acids are precipitated from the cell-free extract by dropwise addition of streptomycin sulfate using 3 ml of a 10% (w/v) solution for every 100 mg of protein. The resulting suspension is allowed to stand for 2 hr before removing the precipitate by centrifugation at 13,000 g. The supernatant is then dialyzed against 0.01 M phosphate buffer (pH 7.0) containing 0.1% (w/v) 2-mercaptoethanol for 15 hr.

Step 2. Ammonium Sulfate Fractionation. Powdered ammonium sulfate is added slowly to bring the supernatant to a concentration of 29.1 g/100 ml. The resulting precipitate is removed immediately by centrifugation. A further 15.9 g of powdered ammonium sulfate per 100 ml of supernatant are slowly added. The precipitate is collected by centrifugation at 13,000 g for 15 min, redissolved in 0.01 M phosphate buffer (pH 6.7) containing 0.1 M KCl plus 0.1% (w/v) 2-mercaptoethanol, and dialyzed against the same buffer for 24 hr.

Step 3. Chromatography on DEAE-Cellulose. The dialyzed sample is applied to a DEAE-cellulose column (12 × 4.5 cm) preequilibrated with 0.01 M phosphate buffer (pH 6.7) containing 0.1 M KCl and 0.1% (w/v) 2-mercaptoethanol. The column is washed with the same buffer until the absorbance at 280 nm has fallen to zero. This wash removes a considerable amount of protein, but the pyruvate kinase activity remains bound. The pyruvate kinase activity is then eluted with 0.02 M phosphate buffer (pH 6.6) containing 0.15 M KCl and 0.1% (w/v) 2-mercaptoethanol. All fractions containing pyruvate kinase at an activity greater than 15 units per milligram of protein are pooled and concentrated by ultrafiltration using an XM-50 (Diaflo) membrane.

PURIFICATION OF PYRUVATE KINASE FROM *Streptococcus lactis* C10 HARVESTED FROM
6 LITERS OF MEDIUM (80–90 g OF FROZEN CELLS)

Step	Total activity (units)	Total protein (mg)	Specific activity (units/mg protein)	Purification (fold)	Yield (%)
Cell-free extract	6400	5280	1.21	1.00	100
Streptomycin sulfate	6500	5000	1.30	1.09	101
Ammonium sulfate precipitation	4350	1640	2.62	2.18	68
DEAE-cellulose	2720	124	22.0	18.05	42
BioGel filtration	1850	24.6	75.0	62.0	29

Step 4. Gel Filtration on BioGel. The concentrated fractions from the
DEAE-cellulose column are dialyzed against 0.025 M phosphate buffer
(pH 7.3) containing 0.1% (w/v) 2-mercaptoethanol for 15 hr. An aliquot of
the dialyzed sample (containing not more than 80 mg total protein in a
volume of 7 ml) is loaded onto a 70 × 2.5-cm column of BioGel A-0.5m
(100–200 mesh) preequilibrated with 0.025 M phosphate buffer (pH 7.3)
containing 0.1% 2-mercaptoethanol and eluted with the same buffer. The
fractions containing pyruvate kinase with a specific activity greater than
70 units per milligram of protein are bulked and concentrated to 2 mg of
protein per milliliter by ultrafiltration using an XM-50 (Diaflo) membrane.
The concentrated solution is diluted to a concentration of 1 mg/ml by
addition of an equal volume of glycerol and then dialyzed for 15 hr against
a solution containing 0.005 M phosphate buffer (pH 7.0), 0.1%
2-mercaptoethanol, 0.005 M $MgCl_2$, and 50% (v/v) glycerol. After
dialysis, the purified enzyme solution is stored in 1-ml aliquots either at 4°
or frozen at −20°. The samples stored at 4° can be used for up to 2 weeks
without loss of activity. Samples stored frozen are used within 2 months of
storage. On further storage there is a slight change in certain cooperativity
properties of the enzyme. For kinetic studies the purified enzyme is
diluted with 20% (v/v) glycerol in deionized water at 0°.

Typical results of the purification procedure are summarized in the
table.

Properties

Homogeneity. The purified enzyme runs as a single major band after
electrophoresis of 40 μg of the enzyme on 7% (w/v) polyacrylamide gels
with only two very faint additional bands. The major band corresponds to
pyruvate kinase revealed by activity staining. Electrophoresis of 50 μg of

purified enzyme on SDS–polyacrylamide gels also yields a single protein band.

Molecular Weight. Equilibrium sedimentation in an analytical ultracentrifuge gives a molecular weight for the purified enzyme of 235,000. The subunit molecular weight determined from SDS–polyacrylamide gel electrophoresis by the method of Weber and Osborn[8] is 60,750. The native enzyme is therefore a tetramer.

Cation Requirements. The *S. lactis* pyruvate kinase has an obligatory requirement for both monovalent and divalent cations.[9] K^+ or NH_4^+ are equally effective in fulfilling the monovalent cation requirement ($K_{0.5}$ for $K^+ = 9.4$ mM and for $NH_4^+ = 7.0$ mM). Mg^{2+}, Mn^{2+}, or Co^{2+} will function as divalent cations, the $K_{0.5}$ values being 0.9 mM, 0.46 mM, and 0.9 mM, respectively. The V_{max} with Co^{2+} is lower than that with Mg^{2+} or Mn^{2+}.

Nucleotide Specificity. Guanosine diphosphate (GDP) can be substituted for ADP as phosphate group acceptor, the K_m for GDP (0.1 mM) being at least 10 times lower than that for ADP while the V_{max} values for the two acceptors are similar.

Activators. Streptococcus lactis pyruvate kinase is activated by a wide range of sugar mono- and diphosphates including glucose 6-phosphate, fructose 6-phosphate, tagatose 6-phosphate, ribose 5-phosphate, erythrose 4-phosphate, glyceraldehyde 3-phosphate, dihydroxyacetone phosphate, fructose 1,6-bisphosphate, and tagatose 1,6-bisphosphate.[2] Of 20 different activators tested, 9 are more effective than Fru-1,6-P_2 in that they have lower $K_{0.5}$ values and give similar V_{max} values. With Mg^{2+} as the divalent cation the enzyme has no activity in the absence of activator. If Mg^{2+} is replaced by Mn^{2+}, the enzyme has 50% of maximum activity in the absence of any activator and the $K_{0.5}$ value for the activator is much lower than with Mg^{2+} as divalent cation.[9]

Kinetic Properties. Pyruvate kinase shows a sigmoidal response to varying PEP concentration at low activator concentrations ($n_H = 1.92$ and an $S_{0.5}$ for PEP $= 0.38$ mM at 0.15 mM Fru-1,6-P_2), but as the activator concentration is increased the response becomes more hyperbolic and the affinity for PEP increases ($n_H = 1.37$ and $S_{0.5} = 0.11$ mM at 1 mM Fru-1,6-P_2). The response to varying ADP is also influenced by activator concentration. At very low activator concentration the response to varying ADP is sigmoidal ($n_H = 1.5$ and $S_{0.5}$ for ADP $= 2.1$ mM at 0.1 mM Fru-1,6-P_2), but at higher activator concentrations the response becomes hyperbolic and affinity for ADP increases ($n_H = 1.0$ and $S_{0.5} = 1.3$ mM at 1 mM Fru-1,6-P_2).

[8] K. Weber and M. Osborn, *J. Biol. Chem.* **244,** 4406 (1969).
[9] V. L. Crow and G. G. Pritchard, *Biochim. Biophys. Acta* **481,** 105 (1977).

Inhibitors. The enzyme is strongly inhibited by phosphate and to a lesser extent by sulfate. Phosphate at 1 mM increases the $S_{0.5}$ value for PEP 25-fold and the $S_{0.5}$ value for Fru-1,6-P$_2$ 5-fold. Both ATP and AMP inhibit enzyme activity in a cooperative manner. Under standard assay conditions, 50% inhibition of activity occurs with 7.0 mM AMP or 2.3 mM ATP. Inhibition by these compounds is not due to chelation of the divalent cation to any significant extent.

pH Optimum. The pH optimum under standard assay conditions in triethanolamine buffer is 7.5.

[27] AMP- and Fructose 1,6-Bisphosphate-Activated Pyruvate Kinases from *Escherichia coli*[1]

By MASSIMO MALCOVATI and GIOVANNA VALENTINI

$$\text{Phosphoenolpyruvate}^{2-} + \text{ADP}^{3-} + 5\,\text{H}^+ \xrightarrow{\text{Mg}^{2+},\,\text{K}^+} \text{pyruvate} + \text{ATP}^{4-} + 4\,\text{H}^+$$

Two noninterconvertible forms of pyruvate kinase (ATP:pyruvate 2-*O*-phosphotransferase, EC 2.7.1.40) have been detected in *Escherichia coli*.[1a,2] They differ in physical and chemical properties[3] as well as in their kinetic behavior: both show positive cooperative effects with respect to the substrate phosphoenolpyruvate, but one of them is activated by fructose 1,6-bisphosphate[1a,2,4,5] and inhibited by ATP[4,6,7] and succinyl-CoA,[4,6] whereas the second is activated by AMP[1a,2,8] and by several intermediates of the hexose phosphate pathway.[8] Moreover, although the two forms are under independent genetic control[9] they do coexist in a wide range of nutritional and metabolic states.[1a,2,5,10,11]

[1] This work was supported by grants from the Italian National Research Council (C.N.R.), Rome.

[1a] M. Malcovati and H. L. Kornberg, *Biochim. Biophys. Acta* **178**, 420 (1969).

[2] M. Malcovati, G. Valentini, and H. L. Kornberg, *Acta Vitaminol. Enzymol.* **27**, 96 (1973).

[3] G. Valentini, P. Iadarola, B. L. Somani, and M. Malcovati, *Biochim. Biophys. Acta* **570**, 248 (1979).

[4] E. B. Waygood and B. D. Sanwal, *J. Biol. Chem.* **249**, 265 (1974).

[5] A. Y. Gibriel and H. W. Doelle, *Microbios* **12**, 179 (1975).

[6] E. B. Waygood and B. D. Sanwal, *Biochem. Biophys. Res. Commun.* **48**, 402 (1972).

[7] M. Markus, T. Plesser, A. Boiteux, B. Hess, and M. Malcovati, *Biochem. J.* **189**, 421 (1980).

[8] E. B. Waygood, M. K. Rayman, and B. D. Sanwal, *Can. J. Biochem.* **53**, 444 (1975).

[9] A. Garrido Pertierra and R. A. Cooper, *J. Bacteriol.* **129**, 1208 (1977).

[10] H. L. Kornberg and M. Malcovati, *FEBS Lett.* **32**, 257 (1973).

[11] D. Kotlarz, H. Garreau, and H. Buc, *Biochim. Biophys. Acta* **381**, 257 (1975).

METHODS IN ENZYMOLOGY, VOL. 90

Different criteria have been suggested for their nomenclature.[1a,4,8,10] The fructose 1,6-bisphosphate-activated form (Pk F), on the basis of its chromatographic behavior on DEAE-cellulose, has also been indicated as type I pyruvate kinase (Pk I). For the same reason, the form activated by AMP (Pk A), ribose 5-phosphate, glucose 6-phosphate, etc., has also been named type II pyruvate kinase (Pk II).

The presence of the two forms within the cell can be demonstrated by *in situ* studies.[10]

Assay Method

Principle. The approaches successfully used for the continuous assay of *E. coli* pyruvate kinase include both "direct" and "coupled" methods. Among the former are proton uptake measured with a pH-stat[7,12] and the spectrophotometric recording of the decrease in absorbance at 230 nm due to the loss of phosphoenolpyruvate[13]; the latter include the "classical" coupled reaction with lactate dehydrogenase, in which the oxidation of NADH is followed spectrophotometrically[14]; and the increase of absorbance at 315 nm due to pyruvate phenylhydrazone formation in the presence of phenylhydrazine at pH 6.8.[13]

Although the pH-stat method is undoubtedly the most suitable for kinetic studies,[7] we will describe here the lactate dehydrogenase coupled assay because it is the most widely used and requires minimal amount of enzyme.

The conditions given below[15] are suitable for routine assays of both forms of the enzyme.

Procedure. The standard assay mixture contains in a final volume of 1 ml: 10 mM Tris or HEPES (N-2-hydroxyethylpiperazine-N'-2-ethane-sulfonic acid), pH 7.5; 10 mM MgCl$_2$; 50 mM KCl; 2 mM ADP; 10 mM phosphoenolpyruvate, tricyclohexylammonium salt; 22 units of crystalline lactate dehydrogenase; 0.12 mM NADH, and 0.03–0.1 unit of *E. coli* pyruvate kinase. The reaction is started by the addition of the enzyme and carried out at 25° in 1-cm light path cuvettes. The decrease in absorbance at 340 nm is followed in a recording spectrophotometer. A molar extinction coefficient for NADH of 6.22 × 10^3 M^{-1} cm^{-1} is used for converting changes in absorbance into moles of pyruvate formed per minute.

Definition of Unit and Specific Activity. One unit of pyruvate kinase is the amount of enzyme that catalyzes the formation of 1 μmol of pyruvate

[12] A. Boiteux, B. Hess, M. Malcovati, M. Markus, and T. Plesser, *Abstr. 10th Int. Congr. Biochem., Hamburg, July 25–31, 1976*, Abstr. 07-4-101 (1976).
[13] H. L. Kornberg and M. Malcovati, unpublished results (1969).
[14] T. Büchner and G. Pfleiderer, this series, Vol. 1 [66].
[15] B. L. Somani, G. Valentini, and M. Malcovati, *Biochim. Biophys. Acta* **482**, 52 (1977).

TABLE I
ASSAY OF THE TWO FORMS OF PYRUVATE KINASE FROM *Escherichia coli*
IN CRUDE EXTRACTS[a,b]

Cuvette	Phosphoenolpyruvate (mM)	5'-AMP (mM)	Fructose 1,6-bisphosphate (mM)
a	0	0	0
b	1	0	0
c	1	1	0
d	1	0	1

[a] Adapted from Kotlarz *et al.*[11]
[b] Each cuvette contains: 50 mM Tris-acetate, pH 7.0; 0.2 mM NADH; 2.75 units of lactate dehydrogenase per milliliter; 50 mM KCl; 5 mM MgCl$_2$; 2 mM ADP. Further additions are made according to the scheme in this table.

per minute under the above conditions. Throughout the initial steps of the purification procedure, protein is estimated by the method of Lowry *et al.*[16] In the case of the fructose 1,6-bisphosphate-activated form, a good agreement was found, in the final steps of purification, between values obtained by this method and the extinction coefficient at 280 nm reported by Waygood and Sanwal[4] ($E_{1cm}^{1\%} = 1.8$).

Differential Assay of the Two Forms in Crude Extracts. The different physical and kinetic properties of the two forms of pyruvate kinase from *E. coli* allow differential assay in crude extracts containing both. Two strategies have been adopted.

1. One method[13] takes advantage of the different stability of the two forms at 55° in 5 mM phosphate buffer, pH 7.5. Under these conditions, both activities decay according to first-order kinetics,[2] with $t_{1/2} > 1$ hr for Pk I and $t_{1/2} = 2$ min for Pk II. Crude extracts are first assayed for total pyruvate kinase activity, then heated for 20 min at 55° in tubes closed with glass marbles, cooled, and assayed for residual activity, which is assumed to be wholly contributed by type I. The amount of type II can be calculated by the difference between the activity before and after heat treatment.

2. The second method[2,11] is based on the different kinetic and regulatory properties of the two forms. A modified lactate dehydrogenase-coupled assay is used for this purpose (Table I). Under these conditions, at 1 mM phosphoenolpyruvate and in the absence of fructose 1,6-bisphosphate, type I exhibits less than 1% of its maximal velocity. Upon

[16] O. H. Lowry, N. J. Rosebrough, A. L. Farr, and R. J. Randall, *J. Biol. Chem.* **193**, 265 (1951).

addition of 1 mM fructose 1,6-bisphosphate, the rate of the reaction catalyzed by this type of enzyme is brought to 80% of its maximal value. On the other hand, at the same phosphoenolpyruvate concentration, the addition of 1 mM AMP brings Pk II to its maximal velocity without affecting the activity of Pk I.

Four cuvettes (a, b, c, d) are set up according to Table I, and the rates of the reaction (v_a, v_b, v_c, v_d) are measured. The activities of the two forms can then be calculated using the following formulas[11]: pyruvate kinase I = $1.2(v_d - v_b)$; pyruvate kinase II = $v_c - v_a$.

Purification Procedure

The method described here allows the purification of both forms of enzyme from the same starting material.

Buffers and General Conditions. All buffers contain 1 mM EDTA and 2 mM 2-mercaptoethanol.

Buffer A: 5 mM phosphate, pH 7.5
Buffer B: 10 mM Tris, pH 7.5
Buffer C: 10 mM Tris, 100 mM KCl, pH 8.5
Buffer D: 100 mM phosphate, 150 mM KCl, pH 8.0

Unless otherwise stated, all steps are performed at 0–4°.

Diluted enzyme solutions are concentrated by ultrafiltration under N_2 pressure in an Amicon ultrafiltration apparatus equipped with PM-30 membranes (or PM-10 in late stages of purification).

Strains and Culture Conditions. The two forms of pyruvate kinase have been mainly studied on derivatives of *E. coli* K12,[1a,3,7,8,15] but have been demonstrated also in other strains, such as *E. coli* B.[4,13] Bacterial cells can be grown under the conditions described below or can be purchased as frozen cell paste.[17]

Since the relative levels of the two forms of pyruvate kinase vary according to the type of carbon source used for growth, culture conditions should be chosen according to the purpose of the work. Aerobic growth on glucose or glycerol yields cells containing equivalent amounts of the two forms and is therefore recommended for the purification procedures described here.

Strains are routinely maintained on nutrient agar. Cultures are grown aerobically in a fermentor equipped with a 10-liter vessel, on the synthetic medium of Ashworth and Kornberg,[18] supplemented with the required

---

[17] The British Public Health Laboratory Service, Center for Applied Microbiology and Research, Porton, Salisbury SP4 OJG, Wiltshire, U.K., supplies several suitable strains both as washed or unwashed cell paste.

[18] J. M. Ashworth and H. L. Kornberg, *Proc. R. Soc. London, Ser. B* **165**, 179 (1966).

amino acids, if any (40 μg/ml each), and nutrient broth (Oxoid CM1, 0.5% v/v); 25 mM glucose is used as carbon source. Sterile medium (10 liters) is inoculated with 200 ml of a liquid culture (approximately 1 mg dry weight of cells/ml)[19] and the culture is grown at 37° under constant mixing (500 rpm) with an air supply of 10 liters/min.

Growth is interrupted in the late logarithmic phase at a cell density of 1.0–1.2 mg dry weight per milliliter. Bacterial cells are harvested by centrifugation at 9000 g for 8 min; the precipitate is washed twice with buffer A and finally resuspended in the same buffer at a concentration of approximately 100 mg dry weight per milliliter.

When a frozen unwashed cell paste is used, this is thawed, resuspended in the same buffer, and washed as above.

Preparation of the Crude Extract. Cells are broken by exposure in small aliquots (approximately 15 ml) to the output of a 100 W sonicator equipped with a standard probe at a frequency of 20,000 Hz and an amplitude of 8 μm, in a refrigerated vessel, for 6 min (six 1-min bursts with 1-min intervals). Unbroken cells and cell debris are removed by centrifugation at 49,000 g for 20 min. The supernatant is collected and stored at $-60°$.

DEAE-Cellulose Chromatography. Pooled extracts obtained from 30–50 g (dry weight) of cells are applied to a 6 × 35-cm DEAE-cellulose column equilibrated with buffer A. After washing with 2 liters of the same buffer, a linear gradient of KCl from 0 to 0.5 M in the same buffer, for a total volume of 1800 ml, is applied to the column. The flow rate is 70 ml/hr. Two peaks of pyruvate kinase activity appear in the effluent at approximately 0.12 M and 0.18 M KCl, respectively.[1a]

Further Purification of Type I Pyruvate Kinase

Heat Treatment. Fractions corresponding to the first peak and containing more than 10 units/ml are pooled. The solution is heated at 55° for 1 hr; the precipitate is removed by centrifugation at 46,000 g for 10 min; the supernatant is thoroughly dialyzed against 6 × 2 liters of buffer B[20] and finally concentrated to a small volume (approximately 7 ml) by ultrafiltration.

Because of the stability of the enzyme, the following chromatographic steps can be performed at room temperature.

Affinity Chromatography on Phosphocellulose. The concentrated solution is applied to a 2.5 × 50-cm phosphocellulose column equilibrated in

---

[19] The density of liquid cultures is measured turbidimetrically at 680 nm; absorbance readings are converted into cell density, expressed as milligrams dry weight per milliliter, by multiplying by a factor of 0.68.[1a]

[20] The complete removal of KCl is essential for the binding of type I pyruvate kinase to phosphocellulose used in the following step.

TABLE II

PURIFICATION OF THE FRUCTOSE 1,6-BISPHOSPHATE-ACTIVATED FORM OF PYRUVATE
KINASE FROM *Escherichia coli*

Step	Volume (ml)	Protein (mg/ml)	Activity (units/ml)	Total units	Specific activity (units/mg protein)	Purification (fold)	Yield (%)
Crude extract[a]	378	49.7	25.8	9752	0.52	1	100
DEAE-cellulose	166	10.0	27.9	4631	2.8	5.4	47
Heat treatment	158	4.5	28.6	4519`	6.4	12.3	46
Phosphocellulose	108	—[b]	15.6	1685	—[b]	—[b]	17
Sephacryl S-200	17	0.68	84.5	1437	124.3	239	14.7

[a] Both forms of pyruvate kinase are present in crude extracts.
[b] A_{280} was not measured because of the high dilution of protein and interference by fructose 1,6-bisphosphate.

buffer B. After washing with 1 liter of the same buffer, PK I is eluted with a solution of 2 mM fructose 1,6-bisphosphate in buffer B. The flow rate is 30 ml/hr. The enzyme appears in the effluent as a single sharp peak. Fractions containing more than 1 unit/ml are pooled and concentrated by ultrafiltration to less than 2 ml.

Sephacryl S-200 Chromatography. The solution is applied to a 1.7 × 115-cm Sephacryl S-200 column equilibrated in buffer B containing 100 mM KCl. Elution is performed with the same buffer, at a flow rate of 13 ml/hr. Fractions containing more than 10 units/ml are pooled.

In most preparations, at this stage the enzyme gives a single band in polyacrylamide gel electrophoresis. A typical purification is summarized in Table II. In some cases, an additional final purification step on DEAE-Sephadex is required.

DEAE-Sephadex Chromatography. Aliquots of approximately 5 mg of protein, dissolved in buffer C, are applied to a 1.6 × 22-cm column of DEAE-Sephadex A-50, equilibrated with the same buffer. Elution is performed with 80 ml of buffer C, followed by a linear gradient of KCl from 0.1 to 0.5 M in the same buffer (total volume of the gradient: 200 ml), at a flow rate of 15 ml/hr. Fractions showing constant specific activity (units/ A_{280}) are pooled.

Alternative Step. If only type I pyruvate kinase is needed, a less time-consuming procedure, based on the same principles, can be used.[21] Crude extracts are prepared as described above (with the exception that buffer C is used instead of buffer A) and heated at 55° for 1 hr. The precipitate is

---

[21] G. Valentini, S. Bartolucci, and M. Malcovati, *Ital. J. Biochem.* **28**, 345 (1979).

removed by centrifugation at 49,000 g for 15 min; the supernatant is brought to pH 8.5 by addition of NaOH and left overnight at 0–4°. The precipitate formed during this period is removed by centrifugation in the same conditions as above.

The heat treatment is followed by a DEAE-Sephadex step, which can be performed either on a column (4 × 40 cm, equilibrated in buffer C; eluting buffers: 1800 ml of buffer C followed by a linear gradient (2 liters) of KCl from 0.1 to 0.5 M in the same buffer) or batchwise. In this case, the clear enzyme solution obtained from the preceding step is titrated with a slurry of DEAE-Sephadex A-50 equilibrated in buffer C, until no more pyruvate kinase activity is detectable in the supernatant. The exchanger is first washed with the same buffer until the effluent is colorless and then with a small volume (approximately half of that of the slurry) of the same buffer containing 300 mM KCl. The enzyme is finally eluted with the same volume of buffer C containing 700 mM KCl. The fraction with pyruvate kinase is then exhaustively dialyzed against buffer B.

The subsequent steps (affinity chromatography on phosphocellulose and Sephacryl S-200 chromatography) are performed as described above.

Further Purification of Type II Pyruvate Kinase

Affinity Chromatography on Phosphocellulose. Fractions from the second DEAE-cellulose peak containing more than 7 units/ml are pooled and dialyzed against 3 × 2 liters of buffer B containing 150 mM KCl and then concentrated to a small volume (approximately 7 ml) by ultrafiltration. The precipitate formed during concentration is removed by centrifugation at 46,000 g for 30 min.

The following step is the most tricky of the whole procedure. The conditions we adopted are the result of a compromise between the need for high ionic strength for stability of this form and the need for low ionic strength for its binding to the exchanger.

The supernatant is applied to a 2.5 × 65-cm phosphocellulose column equilibrated with buffer B. The column is washed with 1250 ml of buffer B containing 150 mM KCl and 1 mM fructose 1,6-bisphosphate; elution is completed with 600 ml of buffer D. The flow rate is 30 ml/hr.

As shown in the chromatogram reproduced in Fig. 1, the washing buffer elutes two major peaks of protein. Type II pyruvate kinase is generally eluted as a very sharp peak by phosphate. But in some preparations, a different pattern is observed: pyruvate kinase is eluted together with the second protein peak by the washing buffer, in the absence of phosphates. In both cases, fractions with specific activity (units/$A_{280\,nm}$) higher than 20 are pooled and dialyzed against 2 × 1 liter of buffer C.

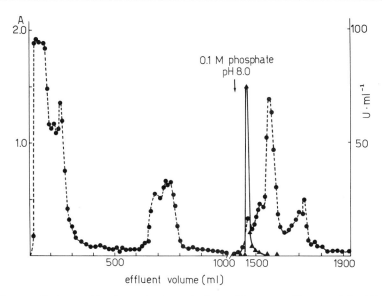

Fig. 1. Elution profile of protein (●) and of type II pyruvate kinase activity (▲) from phosphocellulose. From Somani et al.[15]

DEAE-Sephadex Chromatography. The enzyme preparation is applied to a DEAE-Sephadex A-50 column (1.8 × 90-cm) equilibrated with buffer C. Elution is performed with a convex gradient of KCl (875 ml) from 0.1 to 0.5 *M*, whose shape is reproduced in Fig. 2, together with the elution profiles of pyruvate kinase and protein. The flow rate is 15 ml/hr. Fractions showing, within experimental error, constant specific activity (units/$A_{280\,nm}$) are pooled.

The purification procedure is summarized in Table III.

Properties

Homogeneity. The homogeneity of the enzyme preparations obtained by the procedures described above has been checked by several criteria[3,15,21]: (*a*) amino acid analysis: different enzyme preprations gave consistent results upon amino acid analysis; (*b*) N-terminus analysis: only a single N-terminal amino acid was detectable after dansylation and hydrolysis of enzyme preparations; (*c*) polyacrylamide gel electrophoresis in sodium dodecyl sulfate: both enzymes gave a single band even on overloaded gels; (*d*) polyacrylamide gel electrophoresis at different polyacrylamide concentrations: a single band was detectable at all concentrations tested (within the range of 7 to 10% acrylamide).

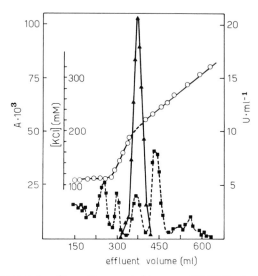

FIG. 2. Chromatography of type II pyruvate kinase on DEAE-Sephadex. ■, absorbance at 280 nm; ▲, pyruvate kinase activity; ○, KCl concentration. From Somani et al.[15]

Stability. Type I pyruvate kinase dissolved in buffer B at concentrations ranging from 1 to 10 mg/ml can be stored at 0–4° for up to 9 months with no loss of activity. Type II pyruvate kinase requires KCl for stability: when dissolved in 10 mM Tris, 1 mM EDTA, 2 mM 2-mercaptoethanol, 150 mM KCl, pH 8.0, at concentrations ranging from 0.5 to 10 mg/ml, this form of enzyme can be stored at 0–4° for up to 6 months with no appre-

TABLE III

PURIFICATION OF THE AMP-ACTIVATED FORM OF PYRUVATE KINASE FROM
Escherichia coli

Step	Volume (ml)	Protein (mg/ml)	Activity (units/ml)	Total units	Specific activity (units/mg protein)	Purification (fold)	Yield (%)
Crude extract[a]	378	49.7	25.8	9752	0.52	1	100
DEAE-cellulose + dialysis	191	9.26	17.9	3413	1.93	3.7	35
Phosphocellulose + dialysis	32	0.56	34.9	1117	62.4	120	11.5
DEAE-Sephadex	62	0.149	16.37	1015	109.8	211	10.4

[a] Both forms of pyruvate kinase are present in crude extracts.

ciable loss of activity. NH_4^+ and Na^+ are less effective than K^+ in stabilizing the enzyme.[15]

Molecular Weight. Both forms of enzyme appear to be tetramers of identical subunits. For type I, molecular weights of 240,000 (subunit M_r = 60,000)[4] and 225,000 ± 12,000 (subunit M_r = 56,000)[3] have been reported. Type II pyruvate kinase has a molecular weight of 190,000 ± 10,000 (subunit M_r = 51,000).[3]

NH_2-Terminal Amino Acid. A free N-terminal amino acid can be detected in both forms of pyruvate kinase: it corresponds to methionine for type I[3] and to serine for type II.[15]

Amino Acid Analysis. Type I pyruvate kinase is made up of 522 amino acids; type II, of 484. Neither form contains tryptophan.[3] Both amino acid composition and analysis of tryptic peptides suggest that the two forms of enzyme are different proteins.[3]

[28] Acetate Kinase from *Bacillus stearothermophilus*

By Koichi Suzuki, Hiroshi Nakajima, and Kazutomo Imahori

$$\text{MgADP} + \text{acetyl phosphate} = \text{MgATP} + \text{acetate}$$

Acetate kinase (AK) ATP : acetate phosphotransferase, EC 2.7.2.1) is an enzyme found only in microorganisms and has been shown to function in the metabolism of pyruvate or synthesis of acetyl-CoA coupling with phosphoacetyltransacetylase.[1] Although some reports have been published on this enzyme from certain sources (e.g., *Escherichia coli*,[2] *Veillonella alcalescens*[3]), *Bacillus stearothermophilus* AK is the only one that has been purified thus far to homogeneity and crystallized.[4]

Assay Method[4]

Principle. The formation of ADP from acetate and ATP is measured spectrophotometrically after NADH oxidation in the presence of pyruvate kinase (PK) and lactate dehydrogenase (LDH). This method is described here in detail. The production of ATP can be determined by a coupled

[1] I. A. Rose, *in* "The Enzymes" (P. D. Boyer, ed.), Vol. 6, p. 115. Academic Press, New York, 1962.

[2] I. A. Rose, this series, Vol. 1 [97].

[3] F. Yoshimura, *Arch. Biochem. Biophys.* **189**, 424 (1978).

[4] H. Nakajima, K. Suzuki, and K. Imahori, *J. Biochem. (Tokyo)* **84**, 193 (1978).

assay with hexokinase and glucose 6-phosphate dehydrogenase.[4] Acetyl phosphate can be measured nonenzymically after conversion into its hydroxamate.[4]

Reagents

Imidazole-HCl buffer, 100 mM, pH 7.2
ATP, 100 mM, neutral
Phosphoenolpyruvate (PEP), 32 mM
NADH, 22.5 mM
MgCl$_2$, 1 M
KCl, 2.5 M
Sodium acetate, 2 M
Pyruvate kinase (PK) from rabbit muscle, 10 mg/ml in 50% glycerol (Boehringer)
Lactate dehydrogenase (LDH) from pig muscle, 10 mg/ml in 50% glycerol (Boehringer)

Procedure. The assay mixture (for 20 assays) contains the following constituents: 5.7 ml of imidazole buffer, 1 ml of ATP, 0.6 ml of PEP, 0.2 ml each of NADH and MgCl$_2$, 0.3 ml of KCl, and 20 μl each of PK and LDH. This assay mixture (0.4 ml) and 0.1 ml of sodium acetate are placed in a cell of 1-cm light path. The assay is started at 30° by the addition of AK. Oxidation of NADH is followed spectrophotometrically at 340 nm with a Gilford spectrophotometer. The decrease of NADH is calculated from the linear portion of the progress curve, although nonlinear lower initial rates are usually observed for the first 2–3 min of the reaction.

Application to Crude Extract ATPase and NADH dehydrogenase are also detected as background activity by this procedure. In order to estimate this activity, an assay mixture to which water is added in place of sodium acetate is used. ATPase and NADH oxidase are separated from AK by DEAE-cellulose column chromatography. Thus, after this step of purification the background activity is negligible.

Definition of Unit. One unit of AK catalyzes the conversion of 1 μmol of substrate per minute under the standard assay conditions ($\epsilon = 6.22 \times 10^3\,M^{-1}\,cm^{-1}$ at 340 nm for NADH). Specific activity is expressed as units per milligram of protein, calculated using $E_{280\ nm}^{0.1\%} = 0.90$.

Purification[4]

Growth of Bacterial Cells. Bacillus stearothermophilus, strain NCA 1503, is cultured as previously described.[5] Cells grown to the late log phase are used, when both the total and specific activities of AK are maximum.

[5] K. Suzuki and K. Imahori, this volume [22].

Preparation of Hydroxyapatite.[6] Five liters of 0.5 M $CaCl_2$ are added dropwise with constant stirring to 6 liters of 0.5 M sodium phosphate, pH 6.8, prepared by mixing 3 liters each of 0.5 M NaH_2PO_4 and 0.5 M Na_2HPO_4. The stirring must be just enough to avoid sedimentation of the precipitate in order to avoid its disintegration. The mixture is stirred for an additional 1 hr after addition of the $CaCl_2$ solution, and then the precipitate is allowed to settle. The supernatant, which has a pH of ca 3.3, is decanted and the precipitate is washed 3 times with 5 liters of water. The washed precipitate is transferred to a 5-liter conical flask with 3 liters of water. After addition of 1 ml of 1% phenolphthalein solution to the mixture, concentrated ammonia (25%) is added to give a pink color. The mixture is then boiled for 30 min with gentle stirring. The pink color is maintained during the boiling by addition of sufficient ammonia. The precipitate is allowed to settle, and the supernatant is removed by decantation while still hot. The precipitate is washed 5 times with 10 liters of 5 mM sodium phosphate, pH 6.8. The final precipitate is stored in the same phosphate buffer at 4°. The yield is 800–1000 ml of packed precipitate.

Hydroxyapatite, commercially available from Seikagakukogyo, Tokyo, Japan, can be used without any significant changes.

The following purification steps are carried out in potassium phosphate buffer at 4°, and all centrifugations are at 15,000 g for 30 min except when otherwise indicated.

Step 1. Extraction of Enzyme. Frozen cells, 500 g, are suspended in 1 liter of 0.1 M phosphate buffer, pH 7.1, disrupted in a Dinomill cell mill (W. A. Bachofen, Basel, Switzerland). The suspension is centrifuged, and the supernatant (fraction I, crude extract, 1100 ml) is used for further purification.

Step 2. Streptomycin Treatment. A 10% solution of streptomycin sulfate (500 ml) is added slowly with stirring to the crude extract. The mixture is stirred for 1 hr, and after centrifugation to remove the precipitate, the supernatant (fraction II, streptomycin supernatant) is retained.

Step 3. Ammonium Sulfate Fractionation. Solid ammonium sulfate is added to the supernatant to 50% at 0° (291 g/liter), and the precipitate is collected by centrifugation. This is dissolved in 400 ml of 0.1 M phosphate buffer, pH 7.5, containing 2 mM EDTA, and dialyzed against 20 mM phosphate buffer, pH 6.9, overnight (fraction III, ammonium sulfate fraction).

Step 4. DEAE-Cellulose Column Chromatography. A column (5 × 43 cm) of DEAE-cellulose (Whatman DE-52) is equilibrated with 20 mM phosphate buffer, pH 6.9, containing 2 mM EDTA and 50 mM KCl, and the dialyzed ammonium sulfate fraction is applied to the column. The column is developed with a linear salt gradient composed of 2 liters each

[6] R. K. Main, M. J. Wilkins, and L. Cole, *J. Am. Chem. Soc.* **81**, 6490 (1959).

of the equilibration buffer and 450 mM KCl in the same buffer. The flow rate is 120 ml/hr, and fractions of 20 ml are collected. The AK activity is eluted at about 0.25 M KCl.[7] Fractions containing the enzyme activity are combined (fraction IV, DEAE-cellulose fraction, 522 ml) and dialyzed overnight against 3 liters of 0.1 M phosphate buffer to which solid ammonium sulfate is added to 70% saturation at 0° (436 g/liter).[8] The resulting precipitate in the dialysis bag collected by centrifugation is dissolved in 50 ml of 0.1 M phosphate buffer, pH 7.5, and dialyzed against 5 liters of 5 mM phosphate buffer with two changes of the outer solution.

Step 5. Hydroxyapatite Column Chromatography. A hydroxyapatite column (3.2 × 42 cm) is equilibrated with 5 mM phosphate buffer, pH 7.1. The dialyzed enzyme solution is applied to the column and eluted with a 5 to 400 mM phosphate linear concentration gradient in a total volume of 4 liters at a flow rate of 40–50 ml/hr. Acetate kinase is eluted at 50–60 mM phosphate. The active fractions are pooled (fraction V, hydroxyapatite fraction) and concentrated to about 2 ml by ultrafiltration.

Step 6. Ultrogel Chromatography. The concentrated solution is loaded onto a column (1.5 × 180 cm) of Ultrogel AcA-34 (LKB Laboratories) equilibrated with 20 mM phosphate buffer containing 100 mM KCl, which is also used for elution. Active fractions are pooled (fraction VI, Ultrogel fraction), and dialyzed against 30 mM phosphate buffer containing 100 mM KCl.

Step 7. DEAE-Sephadex A-50 Chromatography. The dialyzed active fraction is fractionated on a DEAE-Sephadex A-50 column (1.9 × 35 cm) equilibrated with the same buffer used for dialysis. The column is eluted with an increasing gradient of KCl from 100 mM to 600 mM in a total volume of 1 liter of the same buffer. The flow rate is 20 ml/hr, and 5-ml fractions are collected. The enzyme fractions are pooled (fraction VII, DEAE-Sephadex fraction) and concentrated by ultrafiltration. This enzyme solution can be stored as a 50% glycerol solution without loss of activity for at least 1 year at 4°.

Crystallization. If necessary, crystallization can be performed as follows: The concentrated enzyme solution (6 mg in 1 ml) is dialyzed against 500 ml of 50 mM imidazole-HCl buffer, pH 7.3, containing 1.28 M sodium acetate and 5 mM MgCl$_2$ at 4° overnight. The dialysis bag is placed in a 100-ml graduated cylinder together with 100 ml of the outer solution. The

---

[7] Usually three peaks are detected with the ordinary assay procedure at NaCl concentrations of about 0.10 M, 0.25 M, and 0.28 M. The first and the last peaks are active with the assay mixture without acetate, and the second peak is inactive. The first and the last peaks are ATPase and NADH oxidase.

[8] Volume of the sample in the dialysis bag is also taken into account in calculating the amount of ammonium sulfate.

TABLE I
PURIFICATION OF ACETATE KINASE OF *Bacillus stearothermophilus* [a]

Fraction	Volume (ml)	Total protein (mg)	Total activity (units)	Specific activity (units/mg protein)	Yield (%)
I. Crude extract	1100	123,000	44,200	0.36	100
II. Streptomycin supernatant	1500	76,700	50,100	0.65	113
III. Ammonium sulfate fraction	800	7560	26,000	3.44	58.8
IV. DEAE-cellulose fraction	522	1670	22,700	13.6	51.4
V. Hydroxyapatite fraction	135	91.0	17,100	188	38.7
VI. Ultrogel fraction	35.5	25.3	16,700	660	37.8
VII. DEAE-Sephadex fraction	65	8.7	13,300	1530	30.2

[a] Purification was from 500 g of cells.

cylinder covered with Parafilm is left at 4° without stirring. Crystals normally form in a few days. The specific activity of the enzyme does not change on crystallization.

Table I summarizes the purification procedure. This procedure results in a 4270-fold purification with a 30% yield.

Properties

Purity.[4] The purified enzyme is homogeneous as judged by polyacrylamide gel electrophoresis with or without SDS and by cellulose acetate strip electrophoresis at pH 5.5–8.5. The preparation sediments as a single symmetrical peak on ultracentrifugation.

Molecular Weight and Subunit Structure.[9] The molecular weight estimated by gel filtration and sedimentation analyses is 160,000–170,000. Since the subunit molecular weight estimated by SDS–polyacrylamide gel electrophoresis is 43,000 and 4 mol of alanine are detected as the N terminus, *B. stearothermophilus* AK is composed of four identical subunits of a molecular weight of ca. 40,000.

Molecular Properties.[4] The amino acid composition of the AK subunit calculated on the basis of a molecular weight of 43,000 is Lys, 19; His, 9; Arg, 18; Asp, 33; Thr, 19; Ser, 24; Glu, 42; Pro, 16; Gly, 39; Ala, 35; Val, 30; Met, 10; Ile, 31; Leu, 30; Tyr, 11; Phe, 15; Trp, 2; and Cys, 0. The extinction coefficient ($E_{280\,nm}^{0.1\%}$) is 0.90, based on a molecular weight of 160,000. The secondary structure estimated by circular dichroism is

[9] H. Nakajima, K. Suzuki, and K. Imahori, *J. Biochem. (Tokyo)* **84**, 1139 (1978).

TABLE II
KINETIC PARAMETERS OF ACETATE KINASES

| Substrate | Bacillus stearothermophilus AK[a] | | | | Escherichia coli AK + Fru-1,6-P_2 | |
| | − Fru-1,6-P_2 | | + Fru-1,6-P_2 | | | |
	$S_{0.5}$[c]	n_H[d]	$S_{0.5}$	n_H	$S_{0.5}$	n_H
ATP	1.2	1.7	0.36	1	0.17	1
Acetate	120	1	60	1	59	1
ADP	0.8	1	0.8	1	1.3	1
Acetyl phosphate	2.3	1	2.1	1	1.3	1

[a] Acetate kinase.
[b] Fructose 1,6-bisphosphate.
[c] Concentration of substrate required to give half-maximum velocity.
[d] Hill constant.

α-helix, 21%; β-structure, 36%; and unordered structure, 43%. Chemical modification shows that 1 mol each (per AK subunit) of amino and imidazole groups are essential for the enzyme activity. Modification of carboxyl group(s) also inactivates the enzyme. The essential role of these three groups is common to other kinases.

Stability.[4] The enzyme is stable at pH 7–8. Rapid inactivation is observed at pH values lower than 6 even at 4°, although it is rather stable at alkaline pH. The enzyme is quite heat stable up to 65°, whereas the *E. coli* enzyme is gradually inactivated even at 4°.

Substrate Specificity.[4] As phosphoryl donors, ATP and GTP (112%, activity relative to ATP) are most effective, whereas UTP (20%) and CTP (7.3%) are active only partially when compared at nucleotide concentrations of 13.3 mM. As a phosphoryl acceptor, the enzyme is highly specific for acetate. Propionate acts only as a poor acceptor (5% relative to acetate at 500 mM). The enzyme absolutely requires Mg^{2+} or Mn^{2+} ions for activity. Co^{2+}, Ca^{2+}, Cd^{2+}, and Zn^{2+} can replace them only partially. The overall feature of the substrate specificity and metal ion requirement are similar to those for the *E. coli* enzyme when examined together.

Optimum pH.[4] Thermophilic AK shows nearly the same activity at pH values between 7 and 8, with a maximum at pH 7.3.

Kinetic Properties.[4] Kinetic constants for *B. stearothermophilus* AK together with those for *E. coli* AK are summarized in Table II. Both AKs show normal Michaelis–Menten kinetics with respect to acetate, acetyl phosphate, and ADP. The thermophilic AK gives a sigmoidal saturation curve with ATP in contrast to the *E. coli* enzyme.

Activators.[10] Fructose 1,6-bisphosphate (Fru-1,6-P_2) activates formation of ADP. The activation, saturated at 2 mM Fru-1,6-P_2, is about 260% when assayed with 1 mM ATP and 400 mM acetate. α-Glycerophosphate and glucose 1,6-bisphosphate activate the reaction only slightly at 10 mM (about 30% activation). These activators have no effect on the *E. coli* enzyme.

Allosteric Properties.[10] Acetate kinase shows a homotropic allosteric effect for ATP. This allosteric property can be explained in terms of the two-state model of Monod–Wyman–Changeux. The activation by Fru-1,6-P_2 is ascribed to changes of the reaction curve for ATP from a sigmoidal type to a hyperbolic type in the presence of Fru-1,6-P_2. In the absence of Fru-1,6-P_2, AK is in the T state. Binding of ATP causes a T $\rightarrow$ R transition. Thus the binding curve for ATP is sigmoidal. Whereas, in the presence of Fru-1,6-P_2, since thermophilic AK is in the R state, hyperbolic binding of ATP to AK occurs. The binding of Fru-1,6-P_2 to AK is much easier in the presence of ATP than in its absence, and the dissociation constants are 40 μM and 640 μM, respectively. Kinetic constants in the presence of Fru-1,6-P_2 are also summarized in Table II.

The functional unit of AK is not a protomer but a dimer, as shown by the results of hybridization of native and succinylated AKs and other experiments.

[10] H. Nakajima, K. Suzuki, and K. Imahori, *J. Biochem.* (*Tokyo*) **86**, 1169 (1979).

[29] Purification of Human and Canine Creatine Kinase Isozymes

By ROBERT ROBERTS

Creatine kinase (CK) (EC 2.7.3.2) is an enzyme that predominates in muscular tissue and is responsible for the reversible transfer of high-energy phosphate as shown below:

$$\text{Creatine P} + \text{ADP} \rightleftharpoons \text{creatine} + \text{ATP}$$

The high concentration of the enzyme in muscle tissue is related to the rapid high energy turnover of these tissues. Creatine kinase is composed of two monomers or subunits of equal molecular weight of approximately 41,000. The subunits were named B and M, since the predominant enzyme in the brain is BB CK and in the muscle MM CK. Three molecular forms (isozymes) exist in the cytoplasm, namely, MM, MB, and BB CK.

A separate CK isozyme exists in the mitochondria, which is also a dimeric molecule but is made up of a subunit distinctly different from that of M or B.[1]

Assay Method

Principle. Creatine kinase activity is assayed by a coupled-enzyme system according to the method of Oliver[2] and Rosalki[3] as shown below[3a]:

$$\text{Creatine phosphate} + \text{ADP} \xrightarrow{\text{CK}} \text{creatine} + \text{ATP}$$
$$\text{ATP} + \text{glucose} \xrightarrow{\text{hexokinase}} \text{G-6-P} + \text{ADP}$$
$$\text{G-6-P} + \text{NADP} \xrightarrow{\text{G-6-PDH}} \text{NADPH} + \text{6-phosphogluconate}$$

The reaction is monitored by measuring the reduction of NADP spectrophotometrically at 340 nm. The assay is best performed so that one can record the change in absorbance and determine the linear portion of the reaction.

Reagents. The necessary ingredients are commercially available in a single reagent solution that contains PIPES buffer, adenosine diphosphate, adenosine monophosphate, diadenosine pentaphosphate, ethylenediaminetetraacetic acid, glucose, glucose-6-phosphate dehydrogenase (microbial), hexokinase (microbial), magnesium, nicotinamide adenine dinucleotide phosphate, creatine phosphate, N-acetyl-L-cysteine, and dithiothreitol.

Procedure. The reaction is initiated by adding the sample to the single reagent. Sample size may vary from 25 to 50 μl. Assays are performed in cuvettes maintained at 30° containing 3.0 ml or 1.5 ml of creatine kinase reagent, depending on whether the sample size is 25 or 50 μl. If dilution of sample is necessary, this may be performed with Tris-HCl buffer (0.5 M, pH 8.5).

Creatine Kinase Isozymes. The widespread interest in CK isozymes in the medical field is primarily related to their use in the assessment of patients with acute myocardial infarction. Elevated plasma MB CK has been shown to be the most sensitive and specific enzymic marker of cardiac injury. While traces of MB CK are present in tissue such as the prostate and gastrointestinal mucosa, the heart is the only organ with

---

[1] R. Roberts and A. M. Grace, *J. Biol. Chem.* **225**, 2870 (1980).
[2] I. Oliver, *Biochem. J.* **61**, 116 (1955).
[3] S. Rosalki, *Proc. Assoc. Clin. Biochem.* **4**, 23 (1966).
[3a] ADP = adenosine diphosphate; ATP = adenosine triphosphate; CK = creatine kinase; G-6-P = glucose 6-phosphate; G-6-PDH = glucose-6-phosphate dehydrogenase; NADP = nicotinamide adenine dinucleotide phosphate; and NADPH = nicotinamide adenine dinucleotide phosphate, reduced.

significant MB CK. In human myocardium about 85% of the CK activity is MM CK and the remainder is MB CK.[4]

MB CK has been conventionally assessed qualitatively by electrophoresis, which has a lower limit of detection of 10 to 15 IU/liter.[5] Normal plasma levels range from 1 to 12 IU/liter, and thus a severalfold increase may be necessary to detect plasma MB CK by electrophoresis. The need for greater sensitivity coupled with the utilization of enzymic estimates of infarct size based on total MB CK released have stimulated several different approaches to the quantitation of plasma MB CK.[6] The most recent approach, one that offers greater sensitivity with similar specificity, is the radioimmunoassay.[7,8] The radioimmunoassay based on detection of the B subunit has a sensitivity for MB CK of 0.01 IU/liter, which is about 1000-fold greater than that of conventional electrophoresis and severalfold greater than the sensitivity of more recently developed assays based on detection of enzyme activity.[6] Furthermore, the radioimmunoassay detects enzyme protein concentration rather than activity and thus is independent of the many factors that may influence enzymic activity, such as temperature, conditions of storage, and various inhibitors that may give rise to spurious results.

The radioimmunoassay stimulated further the need for highly purified human CK isozymes. Procedures for the isolation of MM and BB CK have been available for some time[9,10] but are somewhat cumbersome and yield low specific enzyme activity. Purification of human MB CK remained a problem primarily owing to contamination with albumin, which coprecipitates and cofractionates with MB CK. Over the past 5 years we have modified the overall approach for purification of human MM and BB CK to provide more rapid isolation techniques and purer preparations; most recently we have developed a method to separate albumin from MB CK that provides a highly pure preparation.[11]

Purification of Human CK Isozymes

Preparation of Tissue. Creatine kinase isozymes MM and MB are obtained from human myocardium, and BB CK is obtained from brain.

[4] R. Roberts, K. S. Gowda, P. A. Ludbrook, and B. E. Sobel, *Am. J. Cardiol.* **36**, 433 (1975).
[5] R. Roberts, "Clinical Enzymology." Masson Publ., Inc., California, 1979.
[6] R. Roberts and B. E. Sobel, *Am. Heart J.* **95**, 521 (1978).
[7] R. Roberts, B. E. Sobel, and C. W. Parker, *Science* **194**, 855 (1976).
[8] R. Roberts, C. W. Parker, and B. E. Sobel, *Lancet* **2**, 319 (1977).
[9] H. J. Keutel, K. Okabe, and H. K. Jacobs, *Arch. Biochem. Biophys.* **150**, 658 (1972).
[10] E. Carlson, R. Roberts, and B. E. Sobel, *J. Mol. Cell. Cardiol.* **8**, 159 (1976).
[11] C. A. Herman and R. Roberts, *Anal. Biochem.* **106**, 244 (1980).

Tissues are obtained at necropsy within 4 hr of death, but preferably within 2 hr. Attempts to isolate CK isozymes from tissue that was removed 6–24 hr after death resulted in very low yields and is recommended only if no other source is available. As soon as the tissue is obtained, it is emersed in cold saline to remove the blood. Then the fat and connective tissue are trimmed from the myocardium and the tissue is processed immediately or frozen at −70°. We have obtained good yields with high specific enzyme activity and high degree of purity from heart and brain frozen at −70° for up to 2 years. It is best to store the tissue in 50- to 150-g portions to avoid refreezing of thawed tissue to be used later. The crude homogenates of myocardial tissue if obtained reasonably fresh will contain a total creatine kinase activity ranging from 400 to 1000 IU/g, wet weight.

Homogenization of Tissue. The tissue is homogenized in a buffer consisting of Tris-HCl (0.05 M, pH 7.5) and 2-mercaptoethanol (0.005 M), which will be referred to as the buffer medium. Prior to homogenization the tissue is minced with scissors into fine chunks and then homogenized in a Waring blender or Brinkmann Polytron. The following protocol is based on starting with 150 g of myocardium. Buffer is added at a volume of 2 ml per gram wet weight of tissue. Homogenization is performed on the Warning blender with 4 bursts of 15 sec, each being 1 min apart. The total volume is usually around 450–500 ml; for convenience of centrifugation it is then divided into 50-ml aliquots and centrifuged at 31,000 g for 15 min at 4°. Prior to centrifugation a sample is taken and refrigerated for later analysis of protein and creatine kinase activity. After centrifugation the pellet is discarded and the supernatant is filtered through four layers of cheesecloth to remove any larger particles.

Alcohol Extraction. Ethanol extraction is performed on the supernatant by adding ethanol (95%) dropwise, very slowly to a final concentration of 50%. The procedure should be performed in a cold room or in a refrigerator, or otherwise the sample must be kept in an ice bath. It is absolutely essential that the supernatant be constantly stirred during the adding of ethanol to assure homogeneous mixing. Since the reaction is exothermic, considerable heat is given off, which must be minimized by keeping the supernatant at 4°. A marked increase in temperature or lack of homogeneous mixing inactivates the enzyme and leads to very low yields. To obtain a final concentration of 50%, add 111 ml of alcohol (95%) per 100 ml of supernatant. After the addition of ethanol, the reaction is constantly stirred for another 30 min at 4°. The total procedure should require from 90 to 120 min.

The preparation is again centrifuged (1000 g) for 15 min (4°), and the supernatant containing the creatine kinase activity is recovered; a second alcohol extraction is performed by adding ethanol (95%) dropwise to a final concentration of 70%. To achieve a final concentration of 70% ethanol, add 169 ml of ethanol (95%) to every 100 ml of supernatant based on the volume measured prior to the initial alcohol extraction. The preparation is again centrifuged (1000 g) for 15 min (4°). Prior to discarding the supernatant a sample is obtained for protein and creatine kinase activity. It is advised that the creatine kinase activity be determined before proceeding to the next step, as occasionally one may not obtain the creatine kinase in the pellet. The pellet containing the creatine kinase is then resuspended in the buffer medium. The pellets are now pooled together in a total volume of 50–75 ml. It is important at this point to rinse the flask carefully for complete recovery of the pellets. To be certain that the pellet is completely dissolved, it is helpful to hand homogenize for 3 or 4 strokes with a glass pestle. The preparation is then centrifuged at 31,000 g for 15 min at 4°. In preparation for column chromatography on DEAE-Sephadex A-50, the preparation is dialyzed for 1 hr against the buffer medium to remove any remaining ethanol. This step is not necessary if one uses DEAE-Sephadex A-50.

Ion-Exchange Column Chromatography. The gel for the column and the necessary eluting buffer should be made up several days ahead of time prior to initiating the preparative steps. DEAE-Sephadex A-50 gel should be properly equilibrated and degassed under vacuum at room temperature. Adequate gel for optimum separation is about 2 ml of gel per 5 mg of protein. The gel is equilibrated with buffer medium, and usually about six changes are required to get complete equilibration of pH to 7.5. In the initial equilibration it is advised that the liquid portion above the gel, which often contains fine particles, be aspirated; this will provide a better flow rate, since these particles tend to clog the column. The column should be packed under a flow rate that is to be used during the fractionation procedure. It is also important that the sample, eluting buffer, and gel be at the same temperature of about 4° and that fractionation be performed at this temperature. One should aim for a flow rate of 30–40 ml/hr and collect fractions of 2–5 ml. The fractions are analyzed spectrophotometrically for protein at an absorbance of 280 nm. The fractions are also analyzed spectrophotometrically for total creatine kinase activity and by electrophoresis for isozyme determinations. Elution is initiated with buffer medium, and MM CK appears in the void volume, after which a NaCl gradient is initiated (0.05 to 0.5 M NaCl) and at a concentration of about 0.250 M, a second peak of enzyme activity appears in the eluent, representing MB creatine kinase. The fractions containing MM creatine kinase are pooled, as are the fractions containing MB creatine kinase, and are separately dialyzed for 24 hr against buffer medium to remove the salt. The dialysis usually requires three to four changes.

Further Purification of MM CK on CM-Sephadex. The MM creatine kinase preparation undergoes final purification on CM-Sephadex eluting with 0.5 M Tris-barbital with a pH gradient of 6.0 to 8.0. The CM-Sephadex gel is equilibrated overnight with 0.05 M Tris-barbital, pH 6.0, containing 0.005 M 2-mercaptoethanol. This usually requires two or three changes of buffer to reach a pH of 6.0. One should aim for a flow rate of 30–40 ml/min and collect in fraction of 2–5 ml. The fractions are assayed for protein, total creatine kinase activity, and CK isozymes. The elution profile shows a small protein peak without creatine kinase activity followed by a larger peak that contains the MM creatine kinase. After elution of MM CK there is usually an orange zone essentially devoid of protein or creatine kinase activity. The fractions containing enzyme activity are pooled and dialyzed against Tris-buffer (0.01 M, pH 7.5 mercaptoethanol, 0.005 M). After dialysis the preparation is concentrated with an UM-10 filter (Amicon Corp., Lexington, Massachusetts) and stored in small aliquots at −70°. The MM creatine kinase stored under such conditions retains its activity for at least 2 years.

Further Purification of MB CK on Affigel Blue. After fractionation of DEAE-Sephadex A-50 the fractions containing MB CK are pooled and dialyzed for 24 hr to remove the salt, as in the case of MM CK. After dialysis, the preparation was concentrated by ultrafiltration in an Amicon chamber using a UM-10 filter in preparation for affinity chromatography. After concentration the preparation is applied to an Affigel Blue column (0.9 × 15 cm, 100–200 mesh) with a void volume of 2 ml equilibrated with Tris-HCl (0.05 M, pH 8.0) containing 0.005 M mercaptoethanol previously rinsed with 2–3 liters of the same buffer. The sample, which should not exceed 12 ml of protein or a volume of 2 ml, is eluted at a rate of 60 ml/hr, and fractions are collected in 1-ml aliquots. After the first peak of protein, elution is initiated with 0.250 M NaCl and MB creatine kinase is eluted after about 3 void volumes. The fractions containing MB creatine kinase are pooled, concentrated by an Amicon filter as previously stated, and stored at −70°. It is often necessary to repeat the Affigel Blue step to exclude traces of albumin that have not been removed by the first fractionation.

Purity of MM and MB CK

Results of a typical procedure for purification of MM and MB creatine kinase from human heart are shown in Table I, outlining the changes in

TABLE I
PREPARATION OF MM AND MB CREATINE KINASE FROM HUMAN HEART (150 g)

Fraction	Total activity (IU)	Total protein (mg)	Specific activity (IU/mg protein)	Recovery activity (%)	Increase in specific activity (fold)
Homogenate	123,330	31,443	6	—	—
Supernatant (31,000 g)	96,585	5,922	16	78	3
50% Ethanol supernatant	59,400	2,217	27	48	5
MM fraction					
After DEAE A-50	27,287	131	209	22	36
After CM-Sephadex	12,750	30	425	10	73
MB fraction					
After DEAE A-50	1,758	82	21	1	
After Affigel Blue	897	2	435	1	75

specific enzyme activity during each step. The specific enzyme activity for MM or MB CK usually averages between 400 and 500 IU/mg. By means of a similar procedure, MM creatine kinase can be isolated from skeletal muscle. Electrophoresis on sodium dodecyl sulfate–polyacrylamide gel exhibits a single protein band for MM and MB creatine kinase with a molecular weight of about 41,000.[11] The molecular weight of the intact molecule determined by gel filtration on Sephadex G-150 is 82,000. Electrophoresis of either the MM or MB creatine kinase preparation on cellulose acetate shows a single isozyme band. Immunological analysis by the Ouchterlony technique and by immunoelectrophoresis show a single precipitant line between MM antiserum and MM creatine kinase or MB creatine kinase.[11] Antiserum to BB creatine kinase shows no reaction to MM creatine kinase but exhibits a single precipitant line to MM creatine kinase.

Since albumin has been the major contaminant of the MB creatine kinase preparation in previous approaches to purification, special studies were performed on the MB creatine kinase preparation to determine whether even traces of albumin are present. Electrophoresis on polyacrylamide gel (8%) of the MB creatine kinase preparation after anion exchange on DEAE-Sephadex A-50 showed 50% of the protein to be albumin in contrast to complete absence of albumin in the final preparation after Affigel Blue as shown in Fig. 1. The ^{125}I-labeled MB creatine kinase preparation from the DEAE-Sephadex A-50 exhibited 86% binding to human albumin antiserum and was virtually similar to that observed in ^{125}I-labeled human albumin. However, no binding to human albumin antiserum was observed to ^{125}I-labeled MB creatine kinase after the Affigel

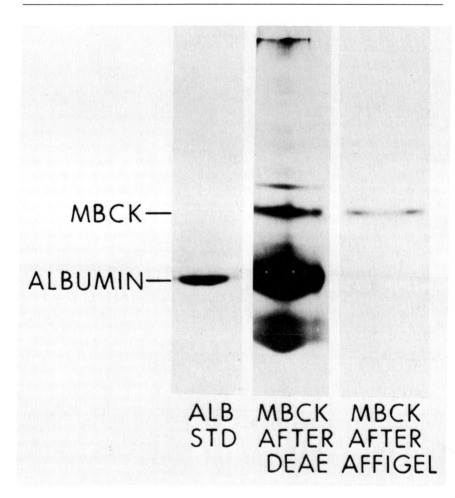

MBCK—

ALBUMIN—

ALB MBCK MBCK
STD AFTER AFTER
     DEAE AFFIGEL

FIG. 1. The polyacrylamide gels (8%) comparing the MB creatine kinase (CK) preparation after anion exchange chromatography on DEAE-Sephadex A-50 (middle gel) and the
final preparation after Affigel Blue are shown on the extreme right. The gel on the left
represents an albumin standard. As shown here, the preparation following DEAE A-50
exhibits a single band in the MB region, but significant staining is evident in the region of
albumin. The preparation after Affigel Blue exhibits a single protein band in the MB position,
which is free of albumin.

Blue step, as shown in Fig. 2. Similarly, while marked inhibition was
observed with unlabeled MB creatine kinase from the DEAE-Sephadex
A-50 gel, no inhibition was observed with the preparation after Affigel
Blue. It is worth emphasizing that our first attempt with Affigel Blue was
unsuccessful when performed according to the instructions supplied by

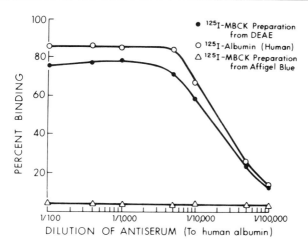

FIG. 2. The binding shown between ^{125}I-labeled albumin and ^{125}I-labeled MB CK following anion exchange on DEAE A-50 indicates that the preparation is markedly contaminated with albumin, so that the binding curves to human albumin antiserum are virtually superimposable. In contrast, ^{125}I-labeled MB CK after purification on Affigel Blue exhibits essentially no specific binding to albumin antiserum.

the manufacturer, which recommended a large column (4 × 2.6 cm) and a NaCl gradient of 0.1 to 1 M NaCl. However, with the procedure described only 50% of the initial MB creatine kinase placed on the column was recovered. Thus, although a high degree of purity is obtained, the yield is very low.

Purification of Human BB Creatine Kinase

The BB CK is extracted from the human brain in the same fashion as for MM and MB CK from myocardium. After chromatography on DEAE-Sephadex A-50, instead of on Affigel Blue as for MB CK, or on CM-Sephadex as for MM CK, a second DEAE-Sephadex A-50 purification procedure is performed utilizing Tris-buffer (0.5 M, pH 8.0) containing mercaptoethanol (0.005 M). The preparation is then concentrated and stored at $-70°$ as for MM and MB CK. The specific enzyme activity is also between 400 and 500 IU/mg, and typical results after each step are shown in Table II. It is even more important to obtain human brain within 4 hr of death, since BB CK is much more labile than MM or MB creatine kinase and the amount of BB present in human brain is much less, on an equivalent basis, than MM or MB in the heart. The usual enzyme activity in brain tissue obtained within 4 hr of death is 50–100 IU/g wet weight. Brains obtained after 4 hr contain very low enzyme activity, and the yield is extremely small.

TABLE II
PREPARATION OF BB CREATINE KINASE FROM HUMAN BRAIN (150 g)

Fraction	Total activity (IU)	Total protein (mg)	Specific activity (IU/mg protein)	Recovery activity (%)	Increase in specific activity (fold)
Homogenate	16,869	9000	2	100	—
Supernatant (31,000 g)	14,765	2015	7	88	4
50% Ethanol supernatant	13,714	455	30	81	16
70% Ethanol pellet	13,176	239	55	78	29
After DEAE A-50 at pH 7.5	13,510	60	225	80	119
After DEAE A-50 at pH 9.0	9,114	19	487	54	257

TABLE III
PREPARATION OF MM CREATINE KINASE FROM CANINE HEART (150 g)

Fraction	Total activity (IU)	Total protein (mg)	Specific activity (IU/mg protein)	Recovery activity (%)	Increase in specific activity (fold)
Homogenate	273,657	13,638	20	100	—
Supernatant (31,000 g)	168,114	3,710	45	61	2
50% Ethanol supernatant	167,011	1,036	161	61	8
70% Ethanol supernatant	145,841	831	176	53	9
After DEAE A-50	98,447	439	224	36	11
After CM-Sephadex	39,476	90	439	14	22

TABLE IV
PREPARATION OF BB CREATINE KINASE FROM CANINE BRAIN (150 g)

Fraction	Total activity (IU)	Total protein (mg)	Specific activity (IU/mg protein)	Yield (%)	Increase in specific activity (fold)
Homogenate	39,584	10,888	4	100	
Supernatant (31,000 g)	36,350	3,400	11	92	3
50% Ethanol supernatant	25,807	590	44	65	12
70% Ethanol supernatant	25,619	240	107	65	30
After DEAE A-50 at pH 7.5	20,194	70	288	51	80
After DEAE A-50 at pH 9.0	20,788	45	462	53	128

Purification of Canine MM and BB Creatine Kinase

Procedures for isolation and purification of MM and BB creatine kinase from canine heart and brain are identical to those for purification of human MM and BB creatine kinase. Canine myocardium contains about 1800–2100 IU/g wet weight, and brain contains 100–200 IU/g wet weight. Results for MM are shown in Table III and for BB in Table IV. Both MM and BB creatine kinase have a specific enzyme activity of 400–500 IU after purification and can be used in the radioimmunoassay for canine isozyme.[12] MB creatine kinase is virtually absent from canine myocardium, and no attempt has been made to purify it from this source. Using the same procedures outlined for the isolation of human and canine isozymes, similar results have been obtained in the purification of MM, MB, and BB CK from the monkey and baboon.[13]

Acknowledgment

This work was supported in part by SCOR in Ischemic Heart Disease, National Institutes of Health Grant HL 17646.

[12] R. Roberts and A. Painter, *Biochim. Biophys. Acta* **480**, 521 (1977).
[13] C. A. Herman, A. M. Grace, B. E. Sobel, and R. Roberts, *Fed. Proc., Fed. Am. Soc. Exp. Biol.* **38**, 889 (1979).

[30] Pyruvate Dehydrogenase Kinase from Bovine Kidney

By Flora H. Pettit, Stephen J. Yeaman, and Lester J. Reed

Pyruvate dehydrogenase (E_1) + ATP → phosphopyruvate dehydrogenase + ADP

A distinctive feature of the pyruvate dehydrogenase complex from mammalian cells, and apparently from other eukaryotic cells as well, is the presence of two regulatory enzymes, a kinase and a phosphatase, that modulate the activity of the pyruvate dehydrogenase component (E_1) by phosphorylation (inactivation) and dephosphorylation (activation), respectively.[1] The kinase (EC 2.7.1.99) is tightly bound to the dihydrolipoyl transacetylase component (E_2) of the complex and copurifies with the complex, whereas the phosphatase is loosely associated with the complex.[2]

[1] T. C. Linn, F. H. Pettit, and L. J. Reed, *Proc. Natl. Acad. Sci. U.S.A.* **62**, 234 (1969).
[2] T. C. Linn, J. W. Pelley, F. H. Pettit, F. Hucho, D. D. Randall, and L. J. Reed, *Arch. Biochem. Biophys.* **148**, 327 (1972).

Assay Method

Principle. Kinase activity in preparations of the pyruvate dehydrogenase complex can be detected by monitoring inactivation of the complex in the presence of ATP and Mg^{2+} or incorporation of ^{32}P-labeled phosphoryl groups from $[\gamma\text{-}^{32}P]ATP$ into the complex, i.e., into its E_1 component. Quantitative assay of kinase activity is based on measurement of the initial rate of incorporation of ^{32}P-labeled phosphoryl groups into crystalline pyruvate dehydrogenase in the presence of dihydrolipoyl transacetylase (E_2). The latter enzyme stimulates the rate of phosphorylation three- to fivefold. The method is modified from that of Linn *et al.*[2]

Reagents

Assay buffer: 0.02 M potassium phosphate, pH 7.0, 1 mM MgCl$_2$, 0.1 mM EDTA, 2 mM dithiothreitol

Dihydrolipoyl transacetylase, prepared from pyruvate dehydrogenase complex as described below

Pyruvate dehydrogenase, crystalline, prepared from pyruvate dehydrogenase complex as described below

$[\gamma\text{-}^{32}P]ATP$ (Amersham), 5 mM, in assay buffer, with a specific radioactivity of about 10,000 cpm/nmol

Trichloroacetic acid, 10%

Aqueous counting scintillant (Amersham, Arlington Heights, Illinois)

Procedure. The reaction mixture contains 0.3 mg of crystalline pyruvate dehydrogenase and 0.2–2 units of kinase in 0.18 ml of the assay buffer.[3] After equilibration of the solution at 30° for 30 sec, 0.02 ml of the radioactive ATP solution is added to start the reaction. Three 0.05-ml aliquots are withdrawn at 20-sec intervals and applied to 2.2-cm (diameter) disks of Whatman 3 MM chromatography paper. The paper disks are placed immediately in cold 10% trichloroacetic acid. The disks are washed four times with cold 10% trichloroacetic acid, twice with 95% ethanol, and once with diethyl ether. The papers are air-dried and placed in vials containing 5 ml of scintillant; radioactivity is determined in a scintillation counter.

Units. Units are expressed as nanomoles of ^{32}P incorporated per minute at 30°; and specific activities as units per milligram of protein. Protein is determined by the biuret method[4] with crystalline bovine serum albumin as standard.

---

[3] When free kinase, i.e., kinase not complexed with dihydrolipoyl transacetylase, is assayed, 50 μg of the latter enzyme are included in the assay mixture. When possible, kinase is added to the assay mixture from concentrated solution to minimize dilution of the enzyme. If dilution is necessary, assay buffer containing 0.3% bovine serum albumin is used.

[4] A. G. Gornall, C. J. Bardawill, and M. M. David, *J. Biol. Chem.* **177**, 751 (1949).

PURIFICATION OF PYRUVATE DEHYDROGENASE KINASE FROM BOVINE KIDNEY

Fraction	Volume (ml)	Protein (mg)	Specific activity[a]	Recovery (%)
Pyruvate dehydrogenase complex	10	347	11.8	100
Transacetylase–kinase subcomplex	4.3	78	28.2	54
Kinase	1.0	4	332.0[b]	32

[a] Nanomoles of ^{32}P incorporated into pyruvate dehydrogenase per minute per milligram of protein.
[b] Dihydrolipoyl transacetylase (50 μg) was added to the assay mixture.

Purification Procedure

Pyruvate dehydrogenase kinase is tightly bound to the dihydrolipoyl transacetylase component of the pyruvate dehydrogenase complex. The kidney complex contains 3–5 times as much kinase activity as the heart complex. Therefore, the highly purified pyruvate dehydrogenase complex from bovine kidney is the preferred source for isolation of the kinase. The complex is separated into pyruvate dehydrogenase (E_1), dihydrolipoyl dehydrogenase (E_3), and a dihydrolipoyl transacetylase (E_2)–kinase subcomplex. The subcomplex is then resolved into E_2 and the kinase. The purification procedure is modified from that of Linn et al.[2] and is summarized in the table.[5] Unless specified otherwise, all operations are performed at 4°.

Step 1. Resolution of Pyruvate Dehydrogenase Complex at pH 9.0. To a concentrated solution of the highly purified pyruvate dehydrogenase complex (about 300 mg in <10 ml) from bovine kidney[6] is added solid NaCl and 1 M glycine (adjusted to pH 9.5 with NaOH) to make the final concentrations 1 M and 0.1 M, respectively. The volume is adjusted to 10 ml; the pH of the solution is 9.0. The solution is kept in an ice bath for at least 30 min and then applied to a column (2.5 × 50 cm) of Sepharose 6B that has been equilibrated with buffer A (0.1 M glycine, pH 9.0; 1 M NaCl; 1 mM MgCl$_2$; 0.1 mM EDTA). The column is developed with this buffer. The flow rate is about 8 ml/hr, and fractions of about 2.6 ml are collected. Two major protein peaks are eluted, as determined by monitoring A_{280} of the fractions. The first peak to emerge from the column contains the dihydrolipoyl transacetylase–kinase subcomplex. The second peak contains pyruvate dehydrogenase and dihydrolipoyl dehydrogenase.

Step 2. Ammonium Sulfate Precipitation. The fractions containing the transacetylase–kinase subcomplex are combined, and solid ammonium

---

[5] The procedure is also applicable to resolution of the pyruvate dehydrogenase complex from bovine heart.
[6] F. H. Pettit and L. J. Reed, this series, Vol. 89 [65].

sulfate (0.12 g/ml) is added with stirring. The solution is stirred for 15 min, then the precipitate is collected by centrifugation at 30,000 g for 15 min. The precipitate is dissolved in about 3 ml of 0.05 M potassium phosphate, pH 7.5, containing 1 mM MgCl$_2$ and 0.1 mM EDTA, and the solution is dialyzed for about 30 hr with stirring against two 500-ml portions of this same buffer.

   Step 3. Treatment with p-Hydroxymercuriphenyl Sulfonate. The dialyzed protein from step 2 is diluted to 5 mg/ml with buffer A containing 2 mM dithiothreitol and dialyzed against this same buffer for 6 hr.[7] The dialyzed solution is placed in a centrifuge tube, and 0.01 volume of 0.25 M monosodium$p$-hydroxymercuriphenyl sulfonate (Sigma Chemical Co., St. Louis, Missouri) is added. After mixing, the preparation is kept on ice for 30 min. If no precipitation occurs, another 0.03 ml of the mercurial is added. After 30 min, 0.01-ml aliquots of the mercurial are added at 10-min intervals until no further local precipitation occurs. The mixture is kept overnight on ice, and the precipitate is collected by centrifugation at 30,000g for 20 min. The precipitate contains dihydrolipoyl transacetylase, and the supernatant fluid contains the kinase. The precipitate is dissolved overnight in a minimal amount of 0.05 M potassium phosphate, pH 7.5, containing 30 mM dithiothreitol, 0.1 mM MgCl$_2$, and 0.01 mM EDTA; the solution is clarified by centrifugation at 30,000 g for 10 min. The yield of transacetylase is about 50 mg.[8] The kinase solution is dialyzed for 16 hr, with stirring, against two 1-liter portions of buffer B (0.01 M imidazole adjusted to pH 7.3 with solid DL-asparagine, 0.1 mM MgCl$_2$, and 0.01 mM EDTA). The solution is centrifuged twice at 144,000g for 2.5-hr periods to remove trace amounts of dihydrolipoyl transacetylase. The supernatant fluid is concentrated by vacuum dialysis against buffer B in a Micro-ProDiCon concentrator (Bio-Molecular Dynamics, Beaverton, Oregon).

   Step 4. DEAE-Cellulose Chromatography. Kinase preparations from step 3 are usually at least 90% pure as judged by sodium dodecyl sulfate (SDS)–polyacrylamide gel electrophoresis. Preparations of lesser purity are purified further by chromatography on Whatman DE-52. The enzyme is applied to a column (0.9 × 1.5 cm) of DE-52 that has been equilibrated with buffer B. The column is washed with 3 ml of this buffer, and then developed with 3-ml portions of this buffer containing 0.05 M, 0.1 M, and 0.2 M NaCl; 0.5-ml fractions are collected. The kinase elutes at 0.2 M NaCl. The active fractions are combined and dialyzed against buffer B. The recovery of kinase activity in this step is about 85%.

   Purification and Crystallization of Pyruvate Dehydrogenase. The pyruvate dehydrogenase-containing fractions from step 1 are made 10 mM

---

[7] Treatment of the transacetylase–kinase subcomplex with dithiothreitol at alkaline pH is essential for subsequent resolution of the subcomplex in the presence of mercurial.

[8] Protein is determined after precipitation with trichloroacetic acid to remove dithiothreitol.

with respect to dithiothreitol and kept on ice for 30 min. Solid ammonium sulfate (0.142 g/ml) is added with stirring. The suspension is stirred for 5 min, then the precipitate is removed by centrifugation at 30,000 g for 15 min; the pellet is discarded. A second portion of solid ammonium sulfate (0.107 g/ml) is added with stirring. The precipitate is collected by centrifugation. It contains pyruvate dehydrogenase; the supernatant fluid contains dihydrolipoyl dehydrogenase.[9] The precipitate is dissolved in about 2.2 ml of buffer C (0.02 M potassium phosphate, pH 7.0; 0.1 mM MgCl$_2$; 0.01 mM EDTA; 2 mM dithiothreitol) and dialyzed overnight against this same buffer. The solution is dialyzed over a 3-day period against buffer C that has been adjusted consecutively to pH 6.8, 6.7, and 6.6. The dialyzed preparation is warmed to 30° for 10 min and then clarified by centrifugation for 15 min at 20°. The solution contains about 100 mg of protein. It is kept on ice for about 2 weeks. The needle-like crystals of pyruvate dehydrogenase are collected by centrifugation and redissolved in 1.2 ml of buffer C by warming at 30° for 15 min. Any insoluble material is removed by centrifugation at 20°. The yield of crystalline pyruvate dehydrogenase is 20–66%. The solution is divided into small portions and stored at −50°.

Properties

Pyruvate dehydrogenase kinase obtained from step 3 or step 4 is at least 90% pure as judged by analytical ultracentrifugation. The sedimentation velocity pattern shows a major peak with a sedimentation coefficient (s_{20}) of about 5.5 S in 0.01 M imidazole buffer, pH 7.5.[2] The kinase consists of two subunits with molecular weights of about 48,000 and 45,000 as estimated by SDS–polyacrylamide gel electrophoresis.[10] The kinase doublet is difficult to detect on SDS–polyacrylamide gels of the kidney pyruvate dehydrogenase complex because of the small amount of kinase present in the complex (about 5% by weight). The kinase exhibits a pronounced tendency to aggregate in buffers other than buffer B.

Catalytic Properties.[11] Pyruvate dehydrogenase kinase requires a divalent cation (Mg^{2+} or Mn^{2+}). Its activity is not affected by cyclic AMP or cyclic GMP. This kinase catalyzes transfer of the γ-phosphoryl group of ATP to three serine residues in the α subunit (M_r 41,000) of pyruvate dehydrogenase (E$_1$).[12,13] Phosphorylation proceeds markedly faster at site

[9] Highly purified dihydrolipoyl dehydrogenase (5–8 mg) is collected between 0.45 and 0.65 ammonium sulfate saturation.
[10] F. H. Pettit, S. J. Yeaman, and L. J. Reed, unpublished results.
[11] F. Hucho, D. D. Randall, T. E. Roche, M. W. Burgett, J. W. Pelley, and L. J. Reed, *Arch. Biochem. Biophys.* **151**, 328 (1972).
[12] C. R. Barrera, G. Namihira, L. Hamilton, P. Munk, M. H. Eley, T. C. Linn, and L. J. Reed, *Arch. Biochem. Biophys.* **148**, 343 (1972).
[13] S. J. Yeaman, E. T. Hutcheson, T. E. Roche, F. H. Pettit, J. R. Brown, D. C. Watson, and G. H. Dixon, *Biochemistry* **17**, 2364 (1978).

1 than at sites 2 and 3, and phosphorylation at site 1 correlates closely with inactivation of E_1. The apparent K_m for $MgATP^{2-}$ is about 20 μM. The apparent K_m for E_1 is about 20 μM. In the presence of E_2 this value is decreased to about 0.6 μM, and the rate of phosphorylation of E_1 is increased three- to fivefold. This acceleration in rate could be due to favorable topographical positioning of E_1 with respect to its kinase or possibly to a change in conformation of either E_1, the kinase, or both enzymes, induced by binding to E_2.

Kinase activity is stimulated by acetyl-CoA and by NADH if K^+ or NH_4^+ ions are present,[14] and kinase activity is inhibited by ADP and pyruvate. ADP is competitive with respect to ATP, and this inhibition apparently requires the presence of a monovalent cation.[15]

Pyruvate dehydrogenase kinase appears to be highly specific for pyruvate dehydrogenase. It exhibits little activity, if any, toward rabbit skeletal muscle phosphorylase b or glycogen synthase a, various histones, or casein.

[14] F. H. Pettit, J. W. Pelley, and L. J. Reed, *Biochem. Biophys. Res. Commun.* **65**, 575 (1975).
[15] T. E. Roche and L. J. Reed, *Biochem. Biophys. Res. Commun.* **59**, 1341 (1974).

[31] cAMP-Dependent Protein Kinase, Soluble and Particulate, from Swine Kidney

By Joseph Mendicino, K. Muniyappa, and Fredrich H. Leibach

The general reaction catalyzed by the catalytic subunit of protein kinase is shown in the following equation.

$$\text{Dephosphoenzyme} + n\text{ATP} \xrightarrow{\text{protein kinase}} \text{phosphoenzyme} + n\text{ADP}$$

The homogeneous preparations of protein kinase isolated from swine kidney catalyze the phosphorylation of purified homologous glycogen synthase, phosphorylase kinase, phosphofructokinase, pyruvate kinase, and fructose-1,6-bisphosphatase.[1] The regulation of these enzymes through the action of the cyclic AMP-dependent forms of protein kinase may mediate the intracellular effects of hormones on the adenylate cyclase system in kidney.[2] Since the discovery of cyclic AMP-dependent forms of

[1] J. Mendicino, F. Leibach, and S. Reddy, *Biochemistry* **17**, 4662 (1978).
[2] F. R. DeRubertis and P. A. Craven, *J. Clin. Invest.* **57**, 1442 (1976).

protein kinase by Walsh *et al.*,[3] in rabbit skeletal muscle, these enzymes have been shown to have a very widespread distribution in mammalian tissues.[4]

Protein kinase catalyzes the transfer of the terminal phosphate of ATP to serine residues of a number of particulate and soluble substrates in kidney.[5] It consists of two dissimilar subunits, and activation involves the dissociation of the regulatory subunit by cAMP to yield a free catalytic subunit with phosphotransferase activity. Our earlier studies showed that protein kinase is present in the cytosol as well as in particulate fraction of swine kidney.[5] The homogeneous catalytic subunit has been purified from the soluble and particulate fractions of swine kidney in high yield.

Assay Methods

Both enzymic and isotopic procedures are used to measure the activity of these enzymes.[5] The enzymic assay is based on the ability of protein kinase to catalyze conversion of glycogen synthase to a form that is dependent on the presence of glucose 6-phosphate for activity. The rate of decrease in the activity of glycogen synthase is measured in the absence of glucose 6-phosphate. Two incubations are required in this assay. In the first reaction mixture the protein kinase is allowed to phosphorylate glycogen synthase. This reaction is terminated by the addition of EDTA, which binds magnesium ion and completely inhibits the protein kinase. Then aliquots are incubated in a second reaction mixture in the absence of magnesium to determine the glycogen synthase activity remaining.

In order to examine the activity with proteins that have no enzymic activity, an assay method based on the rate of phosphorylation of the protein by [γ-^{32}P]ATP is used. The reaction mixture is incubated under the same conditions used in the enzymic assay, so that the results obtained in the two assays would be comparable.

Reagents

Sucrose, 1.0 M
Tris-HCl, 1 M, pH 7.0
2-(N-morpholino)ethanesulfonic acid, 1 M, pH 6.5
2-Mercaptoethanol, 0.1 M
Cysteine, 0.1 M, pH 7.0
MgCl$_2$, 0.5 M
Cyclic AMP, 0.1 mM

[3] D. A. Walsh, J. P. Perkins, and E. G. Krebs, *J. Biol. Chem.* **243**, 3763 (1968).
[4] C. S. Rubin and O. M. Rosen, *Annu. Rev. Biochem.* **44**, 831 (1975).
[5] H. A. Issa, N. Kratowich, and J. Mendicino, *Eur. J. Biochem.* **42**, 461 (1974).

[γ-^{32}P]Adenosine triphosphate, 2 mM, specific activity 3.7 × 10^7 cpm/μmol
Glycogen synthase, 1 mg/ml
Histone, 20 mg/ml
Glycogen, 10%
EDTA, 0.25 M
UDP[U-^{14}C]Glucose, 50 mM (specific activity, 50,000 cpm/μmol)
ATP, 20 mM
Perchloric acid, 2.8 N, containing 2% phosphotungstic acid, 10 mM NaH$_2$PO$_4$, and 3 mM sodium pyrophosphate
Hydrochloric acid, 1 M
Sodium hydroxide, 1 M
Trichloroacetic acid, 10%

Procedure

Assay Based on Enzymic Activity. In a small test tube, mix 0.02 ml of Tris-HCl, 0.03 ml of sucrose, 0.25 ml of cysteine, 0.1 ml of MgCl$_2$, 0.1 ml of cyclic AMP, 0.1 ml of glycogen synthase,[6] and 0.18 ml of protein kinase plus water. The reaction mixture is incubated at 30° for 3 min, and the reaction is initiated by the addition of 0.02 ml of ATP. Aliquots of 0.2 ml are removed at 0, 3, and 6 min, and they are added to test tubes containing 0.03 ml of EDTA, 0.03 ml of glycogen, 0.03 ml of Tris-HCl, and 0.04 ml of distilled water to terminate the protein kinase reaction. Then 0.02 ml of UDP[U-^{14}C]glucose is added to these tubes to initiate the glycogen synthase reaction. These tubes are incubated at 37° for 10 min, and then 1 ml of trichloroacetic acid is added to stop the reaction and the resulting suspension is centrifuged to remove protein. Glycogen is precipitated from the supernatant solution by the addition of 4 ml of 95% ethanol. The turbid solution is centrifuged, and the pellet is dissolved in 1 ml of water. The glycogen is washed twice more by this procedure, and the final pellet is dissolved in 1 ml of water, transferred to a scintillation vial, and counted.[6] The amount of glucose transferred is determined by dividing the total counts incorporated by the specific activity of the UDP[U-^{14}C]glucose. When the total glycogen synthase activity is being measured, 0.04 ml of glucose 6-phosphate is added to the reaction mixture.

A unit of activity in this assay is defined as the amount of protein kinase required to convert 1 μmol of glycogen synthase to a form dependent on glucose 6-phosphate for activity per minute. Specific activity is expressed as units per milligram of protein. One micromole of kidney glycogen synthase will catalyze the transfer of about 5550 μmol (15

[6] H. A. Issa and J. Mendicino, *J. Biol. Chem.* **248**, 685 (1973).

μmol/min per milligram $\times$ 370 mg/μmol) of glucose for UDPglucose to glycogen per minute.

Assay Based on Transfer of ^{32}P. Histone is used as the protein substrate in the isotopic assay. In a small test tube mix 0.02 ml of Tris-HCl, 0.03 ml of sucrose, 0.25 ml of cysteine, 0.1 ml of cyclic AMP, 0.1 ml of histone, and 0.18 ml of protein kinase plus water. The mixture is incubated for 3 min at 30°, and then the reaction is initiated by the addition of 0.02 ml of [γ-^{32}P]ATP. Aliquots of 0.2 ml are removed at 0, 3, and 6 min and added to small test tubes containing 3 ml of 2.8 N perchloric acid containing 2% phosphotungstic acid, 10 mM NaH$_2$PO$_4$, 3 mM sodium pyrophosphate, and 1 mg of bovine serum albumin as carrier protein. The solution is mixed, and the precipitate is collected by centrifugation. The pellet is dissolved in 0.5 ml of 1 N NaOH. After the precipitate dissolves, 0.5 ml of water is added and the protein is reprecipitated by the addition of 3 ml of perchloric acid–phosphotungstic acid mixture. The precipitate is collected and washed twice more by this procedure. The final pellet is dissolved in 0.5 ml of 1 N NaOH and 0.5 ml of water. The sample is transferred to a vial containing 10 ml of scintillation solvent and 0.06 ml of HCl, then mixed and counted.[5] Only about 80% of the protein substrate is routinely recovered in this isolation procedure, and a correction can be made for this loss. A unit of activity is defined as the amount of protein kinase required to transfer 1 pmol of ^{32}P from [γ-^{32}P]ATP to protein substrate per minute at 30°. Specific activity is expressed as units per milligram of protein.[7]

Purification of the Cytosolic Catalytic Subunit of Protein Kinase

Fresh swine kidneys are obtained from a local slaughterhouse, cleaned, and stored at $-20°$ until used. All operations are carried out at 4°.

Step 1. Homogenization. Frozen swine kidneys (800 g) are cut into small pieces, suspended in 1.5 liters of 0.05 M Tris-HCl, pH 7.5, 20 mM 2-mercaptoethanol, 1 mM EDTA, and 0.1 mM phenylmethylsulfonyl fluoride and homogenized in a Waring blender for 2 min. The homogenate is centrifuged at 27,000 g for 20 min; the resultant precipitate is reextracted with 500 ml of the same buffer and centrifuged at 27,000 g for 15 min. The supernatants are combined and passed through cheesecloth to remove insoluble lipid material.

Step 2. Chromatography on Cellulose Phosphate. The extract is applied to a cellulose phosphate column (10 cm $\times$ 18 cm) that was previously equilibrated with 0.05 M Tris-HCl, pH 7.5, containing 10 mM

[7] M. M. Bradford, *Anal. Biochem.* **72**, 248 (1976).

2-mercaptoethanol and 1 mM EDTA. The column is washed with 400 ml of the same buffer, and the effluent and wash are combined.

Step 3. pH 5.5 Treatment. The pH of the solution is adjusted to 5.5 by the addition of 1 M acetic acid with continuous stirring. After stirring for 5 min the solution is centrifuged at 27,000 g for 20 min, the supernatant is collected, and the pH is immediately adjusted to 6.8 by the addition of 1 M Tris base. Glycogen synthase can be isolated from the precipitate obtained at this step.[1]

Step 4. Chromatography on DEAE-Cellulose. The supernatant from step 3 is applied to a DEAE-cellulose column (10 × 12 cm) previously equilibrated with 20 mM potassium phosphate buffer, pH 6.8, containing 10 mM 2-mercaptoethanol and 1 mM EDTA. The column is washed extensively with 12 liters of this buffer until the absorbance at 280 nm decreases to 0.05. The catalytic subunit is eluted with 0.1 mM cyclic AMP in the same buffer at a flow rate of 500 ml/hr. Fractions are collected until the absorbance at 254 nm reaches 1.0, to ensure the complete elution of the catalytic subunit. The fractions in a peak containing phosphotransferase activity were pooled (183 ml) and concentrated threefold by ultrafiltration using an Amicon PM-10 membrane.

Step 5. Chromatography on Blue Dextran–Sepharose 4B. Histone is added to the concentrated enzyme solution from step 4 to a final concentration of 0.1 mg/ml. The solution is then applied to a Blue Dextran–Sepharose 4B column (4.8 × 16 cm) previously equilibrated with 20 mM potassium phosphate, pH 6.8, containing 0.3 M sucrose, 10 mM 2-mercaptoethanol, and 1 mM EDTA. The column is washed with five bed volumes of the same buffer solution containing 30 μM cyclic AMP. The protein kinase is then eluted with 0.3 M potassium phosphate buffer containing 0.3 M sucrose, 10 mM 2-mercaptoethanol, 1 mM EDTA, and 0.1 mg of histone per milliliter. A symmetrical protein peak containing phosphotransferase activity is eluted from the column between 400 and 680 ml. The active fractions are pooled, concentrated by evaporation as described in step 4, and dialyzed overnight against three changes of 2 liters each of 20 mM potassium phosphate buffer, pH 6.8, containing 0.3 M sucrose, 10 mM mercaptoethanol, and 1 mM EDTA.

Step 6. Chromatography on Cellulose Phosphate. The dialyzed enzyme solution from step 5 is applied to a cellulose phosphate column (2.8 × 18 cm) previously equilibrated with 20 mM potassium phosphate, pH 6.8, containing 10 mM 2-mercaptoethanol, 1 mM EDTA, 0.3 M sucrose, and 30 μM cyclic AMP. The column is washed with 1 liter of this buffer. The phosphotransferase is then eluted with a linear gradient of 200 ml of 0.02 M potassium phosphate, pH 6.8, 0.3 M sucrose, 10 mM 2-mercaptoethanol, and 1 mM EDTA in the mixing chamber and 200 ml of the buffer with 0.3 M potassium phosphate, pH 6.8, in the reservoir. The

PURIFICATION OF PROTEIN KINASE FROM SWINE KIDNEY

Step	Volume (ml)	Protein (mg/ml)	Total activity (histone, μmol/min)	Specific activity (pmol/min/mg protein) Glycogen synthase	Histone	Yield (%)
Preparation of crude extract	2200	56	11.8	1.0	96	100
Chromatography on phosphocellulose	2000	52	10.6	1.2	102	94
Acid precipitation at pH 5.5	1700	46	8.1	1.8	103	69
Chromatography on DEAE-cellulose	62	11	4.6	35	6,740	39
Chromatography on Blue Dextran– Sepharose 4B	25	7	3.1	445	17,700	26
Chromatography on phosphocellulose	10	0.32	2.6	39,500	813,000	22
Chromatography on Sephadex G-100	6	0.17	2.4	105,000	2,300,000	20

kinase is eluted in a single protein peak that emerges after 100 ml of solution is collected. The active fractions are pooled, concentrated, and dialyzed as described previously.

Step 7. Gel Filtration on Sephadex G-100. The concentrated enzyme preparation from step 6 is applied to a Sephadex G-100 column (2.2 × 50 cm) previously equilibrated with 0.1 M potassium phosphate buffer, pH 6.8, containing 10 mM 2-mercaptoethanol, 0.3 M sucrose, and 1 mM EDTA. The column is eluted with the same buffer at a flow rate of 0.7 ml/min, and the effluent is collected in 5-ml fractions and assayed for phosphotransferase activity. The active fractions are combined and concentrated.

A summary of the purification procedure for isolation of the catalytic subunit from the soluble fraction of kidney homogenates is shown in the table. A homogeneous preparation of the enzyme with an overall purification of about 23,000-fold and a yield of 20% is obtained.

Solubilization and Isolation of the Catalytic Subunit of Protein Kinase from the Membrane Fraction

The particulate fraction obtained by centrifugation of the homogenate from 1000 g of kidney is dispersed in 1 liter of 0.02 M potassium phosphate, pH 6.8, containing 10 mM 2-mercaptoethanol, 0.8 M NaCl, and 1 mM EDTA. The pH of the suspension is adjusted to 4.9 by the addition of 1 M acetic acid. The resulting thick suspension is centrifuged. The supernatant is passed through cheesecloth and adjusted to pH 6.8 by the addition of 1 M Tris base. The precipitate is reextracted with 500 ml of the same buffer at pH 6.8. The extracts are combined and dialyzed overnight against 10 liters of 20 mM potassium phosphate, pH 6.8, 10 mM 2-mercaptoethanol and 1 mM EDTA. Purification is then carried out by the procedure described above for the isolation of the soluble catalytic subunit of protein kinase.

Properties

The purified soluble and particulate catalytic subunits of protein kinase were homogeneous as judged by sucrose density gradient centrifugation, gel filtration, and polyacrylamide gel electrophoresis in the presence of sodium dodecyl sulfate. The specific activity of the purified enzymes from at least 10 different preparations varied from 2.1 to 2.7 μmol/min/mg with histone as the substrate. The specific activity and amino acid composition of the soluble and particulate subunits were identical. In addition, they have the same molecular weight, 41,500, and show the same activity toward different enzyme and protein substrates.

Substrate Specificity. The apparent K_m of protein kinase for fructose-1,6-bisphosphatase, phosphofructokinase, glycogen synthase, and histone was 53 μM, 55 μM, 50 μM, and 63 μM, respectively. The apparent K_m for ATP is about 7.5 μM with all the protein substrates tested, and Mg^{2+} is required for maximum activity. The pH optimum of the catalytic subunit of protein kinase isolated from the soluble and membrane fractions is 6.5.

Section II

Aldolases and Transketolases

[32] Transketolase from Yeast, Rat Liver, and Pig Liver

By G. A. KOCHETOV

$$
\begin{array}{ccccc}
\text{CH}_2\text{OH} & & & & \text{CH}_2\text{OH} \\
| & & & & | \\
\text{CO} & & & & \text{CO} \\
| & & & & | \\
\text{HOCH} & + \text{ HCO} & \rightleftharpoons \text{ HCO} & + \text{ HOCH} \\
| & | & | & | \\
\text{HCOH} & \text{HCOH} & \text{HCOH} & \text{HCOH} \\
| & | & | & | \\
\text{R} & \text{R}' & \text{R} & \text{R}'
\end{array}
$$

Transketolase   (D-sedoheptulose-7-phosphate : D-glyceraldehyde-3-phosphate glycolaldehydetransferase, EC 2.2.1.1), an enzyme of the pentose pathway of carbohydrate metabolism, catalyzes the cleavage of the C—C bond in keto sugars and the subsequent transfer of a C—C residue (glycolic aldehyde) to aldo sugars.

With the exception of hydroxypyruvate, all the known donor substrates for transketolase have hydroxyl groups at C-3 and C-4 in the trans position. They include xylulose 5-phosphate, sedoheptulose 7-phosphate, fructose 6-phosphate, L-erythrulose, etc. D-Glyceraldehyde 3-phosphate, D-ribose 5-phosphate, D-erythrose 4-phosphate, glycolaldehyde, etc. may function as acceptor substrates.[1] The transketolase reaction is reversible except when hydroxypyruvate is used.

Transketolase is a thiamine enzyme and belongs to the transferases. It occurs in almost all animal and plant tissues as well as in microorganisms and localizes primarily in a soluble cell fraction.

Bakers' Yeast Transketolase

Assay Method A

Principle.[2] When xylulose 5-phosphate and ribose 5-phosphate are used as substrates, one of the products of the transketolase reaction is glyceraldehyde 3-phosphate. In the presence of glyceraldehyde-phosphate dehydrogenase (D-glyceraldehyde-3-phosphate : NAD^+ oxidoreductase (phosphorylating), EC 1.2.1.12) and NAD, the latter is reduced owing to

[1] E. Racker, *in* "The Enzymes" (P. D. Boyer, H. Lardy, and K. Myrbäck, eds.), Vol. 5, p. 397. Academic Press, New York, 1961.
[2] G. de la Haba, J. G. Leder, and E. Racker, *J. Biol. Chem.* **214**, 409 (1955).

oxidation of glyceraldehyde 3-phosphate (mole per mole). The amount of NADH is determined spectrophotometrically by the increase in absorbance at 340 nm. The molar extinction coefficient of NADH is 6.22×10^3 cm^{-1} M^{-1}.[3]

Reagents

Glycylglycine buffer, 0.15 M, pH 7.6
Cysteine, 0.15%, pH 7.6, prepared before use
MgCl$_2$, 0.06 M
Thiamine pyrophosphate, 3 mM; may be stored at 0–4° for several days
Sodium arsenate, 0.15 M
NAD, 7.5 mM, pH 6.0; at 0–4° may be stored for several days
Substrate solution: The potassium salt of the xylulose 5-phosphate and ribose 5-phosphate mixture[4] and of ribose 5-phosphate are prepared in a concentration of 37.5 mg/ml (calculated for the barium salt)[4]; pH 6.5. Then the ratio which is optimal for the transketolase activity assay is chosen. To this end, all the components needed for the transketolase assay (see below), except for substrates, are placed in a spectrophotometer cuvette, the concentration of transketolase being 1.5–2 times higher than normal. The solutions of phosphopentoses and ribose 5-phosphate are added in the ratio of 9 : 1, 8 : 2, 7 : 3, etc., respectively (the total added volume is to be 0.2 ml). The transketolase activity is assayed, and the ratio at which the activity is maximal is regarded as optimal. The phosphopentose and ribose 5-phosphate solutions are poured together in this ratio, and the resultant solution is used for routine work; it may be stored for several days at $-20°$. (Preparations of xylulose 5-phosphate and ribose 5-phosphate may be used as transketolase substrates. Their final concentration in the reaction mixture should be 0.9 mM and 1.2 mM, respectively.[5])
Rabbit muscle glyceraldehydephosphate dehydrogenase devoid of glycerol-3-phosphate dehydrogenase (L-glycerol-3-phosphate : NAD$^+$ 2-oxidoreductase, EC 1.1.1.8) and triosephosphate isomerase (D-glyceraldehyde-3-phosphate ketol-isomerase, EC 5.3.1.1); 50 units/ml.
Transketolase. The enzyme solution is diluted with 0.05 M glycylglycine buffer, pH 7.6, to a concentration of about 0.4 units/ml.
Procedure. Added to a cuvette are 1 ml of glycylglycine buffer, 1.1 ml of cysteine, 0.15 ml of MgCl$_2$, 0.1 ml of thiamine pyrophosphate, 0.1 ml of

[3] B. L. Horecker and A. Kornberg, *J. Biol. Chem.* **175**, 385 (1948).
[4] C. J. Gubler, L. R. Johnson, and J. H. Wittorf, this series, Vol. 18A [22].
[5] A. G. Datta and E. Racker, *J. Biol. Chem.* **236**, 617 (1961).

sodium arsenate, 0.1 ml of NAD, 0.2 ml of substrate solution, 0.2 ml of glyceraldehyde-phosphate dehydrogenase, and 0.05 ml of transketolase. The mixture is stirred, and the increase in absorbance at 340 nm is measured for 5 min.

Definition of Unit and Specific Activity. One unit is the amount of the enzyme that catalyzes the formation of 1 μmol of glyceraldehyde 3-phosphate per minute under the above conditions. Specific activity is expressed as units per milligram of protein. Protein can be determined by the method of Lowry *et al.*,[6] and also spectrophotometrically: for homogeneous transketolase at 280 nm $E_{1\,cm}^{1\%} = 14.5$.[7]

Assay Method B

Principle.[2] Alternatively, if the reaction mixture is supplemented with excess triosephosphate isomerase, glycerol-3-phosphate dehydrogenase, and NADH, glyceraldehyde 3-phosphate is isomerized to dihydroxyacetone phosphate, which is then reduced to α-glycerophosphate. Again the amount of glyceraldehyde 3-phosphate formed in the transketolase reaction is equal to the consumption of NADH.

Reagents

Glycylglycine buffer, 0.5 M, pH 7.6
NADH, 2 mM, pH 8.0; may be stored for several days at 0–4°
Thiamine pyrophosphate, 3 mM; may be stored for several days at 0–4°
MgCl$_2$, 90 mM
Ammonium sulfate fraction (2.03–2.5 M) from rabbit muscle containing triosephosphate isomerase and glycerol-3-phosphate dehydrogenase,[8] an aqueous solution, about 20 mg of protein per milliliter. It may be kept for several days at −16°.
Transketolase. The enzyme solution is diluted with 50 mM glycylglycine buffer, pH 7.6, to a concentration of about 0.4 unit/ml.
Substrate solution (see reagents for assay method A).

Procedure. A spectrophotometric cuvette is filled with distilled water, 2 ml; glycylglycine buffer, 0.3 ml; NADH, 0.15 ml; auxiliary enzymes, 0.1 ml; substrate solution, 0.2 ml; thiamine pyrophosphate, 0.1 ml; and MgCl$_2$, 0.1 ml. The mixture is stirred, and the absorbance is measured at 340 nm.

[6] O. H. Lowry, N. J. Rosebrough, A. L. Farr, and R. J. Randall, *J. Biol. Chem.* **193**, 265 (1951).
[7] C. P. Heinrich, K. Noack, and O. Wiss, *Biochem. Biophys. Res. Commun.* **49**, 1427 (1972).
[8] P. Srere, J. R. Cooper, M. Tabachnik, and E. Racker, *Arch. Biochem. Biophys.* **74**, 295 (1958).

Transketolase (0.05 ml) is added, the mixture is stirred, and the decrease in absorbance is measured for 5 min.

Assay Method C

Principle. If xylulose 5-phosphate and ribose 5-phosphate are used as transketolase substrates, one of the reaction products will be sedoheptulose 7-phosphate. Its amount is determined colorimetrically.[9,10]

Reagents

Glycylglycine buffer, 0.5 M, pH 7.6
$MgCl_2$, 24 mM
Thiamine pyrophosphate, 2 mM; may be stored for several days at 0–4°
Substrates solution, 50 mg/ml, calculated for barium salt (see reagents for assay method A)
Transketolase. The enzyme solution is diluted with 50 mM glycylglycine buffer, pH 7.6, to a concentration of about 1 unit/ml
Trichloroacetic acid, 5%

Procedure. Glycylglycine buffer (0.5 ml), $MgCl_2$ (0.1 ml), thiamine pyrophosphate (0.05 ml), and transketolase (0.05 ml) are poured into a glass tube in ice. The mixture is stirred, and after 5–10 min 0.1 ml of substrate is added. The reaction mixture is stirred. The tube with the reaction mixture is placed into a water bath at 37° and a 0.1-ml aliquot is transferred at 5, 10-, 15-, and 20-min intervals into 0.4 ml of 5% trichloroacetic acid. The content of sedoheptulose 7-phosphate is then determined.[9,10]

Definition of Unit and Specific Activity. A unit is the amount of enzyme catalyzing formation of 1 μmol of sedoheptulose 7-phosphate per minute under the above conditions. Specific activity is expresed as units per milligram of protein.

Purification Procedure

This work describes the procedure of Racker *et al.*[8] modified primarily in the ammonium sulfate fractionation step.[11] This allows removal of all the contaminating glyceraldehydephosphate dehydrogenase[12] and preparation of the homogeneous enzyme.

[9] J. J. Villafranca and B. Axelrod, *J. Biol. Chem.* **246**, 3126 (1971).
[10] T. Ozawa, S. Saiton, and T. Jomita, *Chem. Pharm. Bull.* **20**, 2715 (1972).
[11] L. E. Meshalkina and G. A. Kochetov, *Biochim. Biophys. Acta* **571**, 218 (1979).
[12] G. A. Kochetov, L. J. Nikitushkina, and N. N. Chernov, *Biochem. Biophys. Res. Commun.* **40**, 873 (1970).

All procedures are carried out at 0–4° unless otherwise stated.

Preparation of the Material. Commercial bakers' yeast is crushed by hand and dried in an air current (possibly warm) at room temperature. The yeast must be dried for approximately 1–1.5 days. The dried yeast may be stored at 0–4° in a stoppered flask for several months.

Step 1. Extraction. Dry yeast (300 g) is mixed with 900 ml of 66 mM Na_2HPO_4, placed in a water bath at 40°, and stirred until a homogeneous suspension is formed. After 2.5 hr the suspension is centrifuged for 30 min at 4° and 18,000 g; the sediment is discarded.

Step 2. Fractionation with Acetone. The yeast extract is placed in an acetone bath at −4° to 6° and cooled to 0°; 0.5 volume of cold acetone is added with stirring. At the beginning, the temperature of the yeast extract should not be higher than 0°; then, after addition of about 0.25 of the volume of acetone, it decreases as fast as possible to −2°. The duration of acetone addition is about 15 min.

The stirring is continued at −2° for another 1–2 min, then the mixture is centrifuged for 5 min at −2° and 18,000 g. The same amount of acetone is added to the centrifugate for 15 min at −2°. The mixture is stirred for another 1–2 min and centrifuged for 5 min at −2° and 18,000 g. The sediment is ground in a mortar with 100 ml of cold distilled water and dialyzed overnight against 50 volumes of distilled water. The mixture is centrifuged for 5 min at 18,000 g, and the sediment is discarded.

Step 3. Heating. The centrifugate is transferred, in 10- to 15-ml aliquots, into thin chemical glass tubes; the tubes are placed in water for 15 min at 55°, then chilled on ice. The mixture is centrifuged for 10 min at 18,000 g, the sediment is discarded (it can be washed with a small volume of cold distilled water, and the washing should be added to the centrifugate). The centrifugate is filtered through a porous-glass filter or through glass wool.

Step 4. Ethanol Fractionation. The filtrate obtained at step 3 is placed in an acetone bath and 0.5 of the volume of cold ethanol is added with stirring. The temperature of the filtrate prior to the addition of ethanol should be 0°; after adding 0.25 of the total amount of ethanol, it should be −6°. Ethanol is added for about 30 min; the stirring is continued for another 5 min at −6°, then the mixture is centrifuged for 5 min at the same temperature at 40,000 g. The sediment is dissolved at room temperature in 6–10 ml of distilled water. The part of the sediment that would not dissolve is separated by centrifugation for 15 min at 2–4° and 40,000 g and then discarded.

The sediment of the alcohol fraction suspended in water may be stored for 3–4 days at 0–4° or for 10–12 days at −16°.

Step 5. DEAE-Cellulose Chromatography. Part of the enzyme solution obtained at step 4 (400–500 mg of protein) is applied to a 5.5- × 3-cm DEAE-cellulose column equilibrated with 5 mM potassium phosphate buffer, pH 7.7. (A column of a larger diameter may be used for a greater amount of protein.) The same buffer is used for elution. The rate of elution is 1 ml/min. Fractions (3–5 ml) are collected and assayed for protein and transketolase activity (those with transketolase are, as a rule, pinkish). The extreme fractions of the main peak, which have low specific activity, are discarded.

The rest of the enzyme solution obtained at step 4 is treated analogously. Fractions with high specific activity eluted from all columns are pooled.

Step 6. Ammonium Sulfate Fractionation. To the pooled fractions from step 5 ammonium sulfate (29 g/100 ml) is added with stirring (about 30 min); the pH is adjusted to 7.6 with 1 M KOH, and the stirring is continued for another 15 min. If a precipitate is formed, it is removed by centrifugation for 5–10 min at 40,000 g. Ammonium sulfate (13 g per 100 ml) is added at room temperature (about 30 min) to the centrifugate. The mixture is stirred for 30–40 min and then centrifuged for 15 min at 16–18° and 40,000 g. The centrifugate is discarded, and the sediment is dissolved with 5 mM potassium phosphate buffer, pH 7.6, to a protein concentration of 2 mg/ml.

To the resultant solution 3.9 M ammonium sulfate solution, pH 7.6, is added in cold for about 30 min to 2.14 M. The stirring is continued for another 30 min, then the solution is centrifuged for 15 min at 40,000 g. To the centrifugate 3.9 M ammonium sulfate is added to 2.3 M. The precipitate is collected by centrifugation and dissolved in a minimal volume of 5 mM potassium phosphate buffer, pH 7.6.

In the same manner, but at room temperature, ammonium sulfate fractions (2.3–2.46, 2.46–2.54, 2.54–2.74 M) are obtained. They are collected by centrifugation at 16–18° and dissolved in a minimum volume of cooled 5 mM potassium phosphate buffer, pH 7.6.

The protein and transketolase are assayed. Most of the enzyme and the maximal specific activity are usually confined to the 2.3–2.46 M ammonium sulfate fraction.

At all stages of purification, including the DEAE-cellulose chromatography step, transketolase preparations display some glyceraldehyde-phosphate dehydrogenase activity. This enzyme is removed at the ammonium sulfate fractionation step. Otherwise the solutions of the precipitates possessing transketolase activity are pooled and solid ammonium sulfate is added in the cold to 2.74 M over a 30-min period. The solution is then immediately centrifuged for 20 min at 40,000 g. The sediment is dissolved in 1.6 M ammonium sulfate, pH 8.4 (final concentration of protein 1 mg/ml). After 48 hr the ammonium sulfate concentration is adjusted to 2.74 M by adding dry salt, as above, and the solution is immediately centrifuged for 20 min at 40,000 g. The sediment is dissolved in 5 mM potassium phosphate buffer, pH 7.6 (final concentration of protein 2 mg/ml) and refractionated with ammonium sulfate as described above.

Step 7. Crystallization. The enzyme solution obtained at step 6 is diluted with 5 mM glycylglycine buffer, pH 7.6, to a protein concentration of 5–10 mg/ml. At room temperature, 3.9 M ammonium sulfate, pH 7.6, is added gradually with constant stirring until turbidity appears; stirring is continued for another 30–40 min, then the solution is placed in a refrigerator. Transketolase can be stored under these conditions for several months. The enzyme is more or less evenly distributed between the crystals and the bulk solution. The specific activity of transketolase assayed at 25° by method A is about 20 units/mg.

Properties

Stability. Transketolase is sufficiently stable and does not lose activity for several weeks on storage in 8 mM glycylglycine buffer, pH 7.4, in a refrigerator.[2] When the enzyme is stored as a crystalline suspension in ammonium sulfate, its solubility in water and buffer solutions is much lower, but it can be fully dissolved by adding thiamine pyrophosphate and $MgCl_2$.[5]

pH Optimum. The enzyme has a maximal activity at pH 7.6.[5]

Homogeneity. Based on ultracentrifugation and disk electrophoresis in polyacrylamide gel, the enzyme is homogeneous.[13]

Molecular Properties. The molecular weight of transketolase is, according to the sedimentation equilibrium data, 158,000–159,000.[13,14] At a sufficiently low concentration the apoenzyme (but not holoenzyme) dissociates reversibly into two subunits of equal molecular weight.[13–15] Individual subunits are catalytically active and hardly differ from the dimer in specific activity.[16]

Cofactors. Thiamine pyrophosphate and bivalent cations, such as Ca^{2+}, Mg^{2+}, and some others, are transketolase cofactors.[17,18] Cations are indispensable for the binding of coenzyme with protein.[19] Partially purified enzyme preparations do not need any additional cofactors for

[13] R. Kh. Belyaeva, V. Ya. Chernyak, N. N. Magretova, and G. A. Kochetov, *Biokhimiya* **43**, 545 (1978).

[14] S. W. Cavalieri, K. E. Neet, and H. Z. Sable, *Arch. Biochem. Biophys.* **171**, 527 (1975).

[15] V. Ya. Chernyak, R. K. Belyaeva, N. N. Magretova, and G. A. Kochetov, *Biokhimiya* **42**, 159 (1977).

[16] G. A. Kochetov and O. N. Solovieva, *Biochem. Biophys. Res. Commun.* **84**, 515 (1978).

[17] E. Racker, G. de la Haba, and J. G. Leder, *J. Am. Chem. Soc.* **75**, 1010 (1953).

[18] P. C. Heinrich, H. Steffen, P. Janser, and O. Wiss, *Eur. J. Biochem.* **30**, 533 (1972).

[19] L. E. Meshalkina and G. A. Kochetov, *Dokl. Akad. Nauk SSSR* **248**, 1482 (1979).

TABLE I
AMINO ACID COMPOSITION OF BAKERS' YEAST
TRANSKETOLASE[a]

Amino acid	Time of hydrolysis (hr)		
	24	48	72
Tryptophan	14.6[b]	—	—
	13.3–14.7[c]	—	—
Lysine	76.9	76.0	78.3
Histidine	30.5	29.2	35.0
Arginine	39.6	38.7	38.2
Aspartic acid	115.1	120.6	108.8
Threonine	62.3	63.3	59.6
Serine	85.1	85.6	76.0
Glutamic acid	106.0	113.8	73.7
Proline	57.3	63.3	63.3
Glycine	88.3	101.9	91.9
Alanine	111.0	115.6	104.2
Half-cystine	16.8	—	—
Valine	59.6	69.6	72.8
Methionine	42.7	—	—
Isoleucine	54.6	58.7	60.5
Leucine	107.4	107.4	103.8
Tyrosine	43.2	41.4	40.0
Phenylalanine	46.4	46.0	44.1

[a] Residues per 140,000 g of protein.
[b] Spectrophotometric data.
[c] Colorimetric data.

maximum activity. The crystalline preparation is only partially active without cofactors. Complete removal of thiamine pyrophosphate (and partial removal of metal ions) occurs during dialysis against EDTA[17] or as a result of storage in an alkaline ammonium sulfate solution.[20]

Kinetic Constants. Transketolase has two active centers[21,22] of equal catalytic activity,[22] but with different affinities for the coenzyme. The dissociation constants (measured in the presence of calcium) are, respectively, $3.2 \times 10^{-8} M$ and $2.5 \times 10^{-7} M$.[23] The substrate with the highest affinity for the enzyme is xylulose 5-phosphate. The K_m (in the presence of ribose 5-phosphate as the second substrate) is $2.1 \times 10^{-4} M$. The K_m

[20] G. A. Kochetov and A. E. Isotova, *Biokhimiya* **35**, 1023 (1970).
[21] C. P. Heinrich, *Experientia* **29**, 1227 (1973).
[22] L. E. Meshalkina and G. A. Kochetov, *Biochim. Biophys. Acta* **51**, 218 (1979).
[23] G. A. Kochetov, N. K. Tikhomirova, and P. P. Philippov, *Biochem. Biophys. Res. Commun.* **63**, 924 (1975).

values for fructose 6-phosphate (in the presence of ribose 5-phosphate), erythrulose (in the presence of glyceraldehyde 3-phosphate), and ribose 5-phosphate (in the presence of xylulose 5-phosphate) are, respectively, $1.8 \times 10^{-3} M$, $4.9 \times 10^{-3} M$, and $4 \times 10^{-4} M$.[5]

Inhibitors.[5] Sulfate and phosphate ions at a concentration of 0.01–0.02 M halve the transketolase activity. Oxythiamine pyrophosphate is a potent competitive inhibitor of the coenzyme. Iodoacetate, p-chloromercuribenzoate, and N-ethylmaleimide produce no effect on activity.

Functional Groups. The transketolase active center contains histidine,[24,25] tryptophan,[6,26–29] and arginine[30,31] residues. The former two are necessary for the binding of the coenzyme. The function of arginine remains obscure.

Amino Acid Composition. The amino acid analysis data are given in Table I.[32]

Rat Liver Transketolase

Transketolase from rat liver was isolated and partially purified in 1953.[33] Later another method was reported[34] that yields a homogeneous enzyme. It is this method that we describe below.

Assay Method

The activity assay is identical to that for Bakers' yeast transketolase, except that neither thiamine pyrophosphate nor $MgCl_2$ is added.

Purification Procedure

Preparation of Adsorbents. DEAE-Sephadex A-25 is treated as follows: The ion exchanger in the Cl form is suspended in 10 volumes of distilled water, stirred for 24 hr and washed 5–10 times with distilled water to remove the fines. The exchanger is suspended in 0.2 M KCl and packed

[24] G. A. Kochetov and K. R. Kobylyanskaya, *Biokhimiya* **35**, 3 (1970).
[25] L. E. Meshalkina and G. A. Kochetov, *Dokl. Akad. Nauk SSSR* **246**, 228 (1979).
[26] G. A. Kochetov and R. A. Usmanov, *Biochem. Biophys. Res. Commun.* **41**, 1134 (1970).
[27] G. A. Kochetov, R. A. Usmanov, and V. P. Merslov, *FEBS Lett.* **9**, 265 (1970).
[28] C. P. Heinrich, K. Noack, and O. Wiss, *Biochem. Biophys. Res. Commun.* **44**, 275 (1971).
[29] R. A. Usmanov and G. A. Kochetov, *Biokhimiya* **43**, 1796 (1978).
[30] L. E. Meshalkina, O. E. Volkovitskaya, and G. A. Kochetov, *Dokl. Akad. Nauk SSSR* **245**, 1259 (1979).
[31] A. Kremer, R. M. Egan, and H. Z. Sable, *J. Biol. Chem.* **255**, 2405 (1980).
[32] G. A. Kochetov, K. R. Kobylyanskaya, and L. P. Belyanova, *Biokhimiya* **38**, 1303 (1973).
[33] B. L. Horecker, P. Z. Smyrniotis, and H. Klenov, *J. Biol. Chem.* **205**, 661 (1953).
[34] G. A. Kochetov and A. A. Minin, *Biokhimiya* **43**, 1631 (1978).

into a 25- × 5-cm column. The column is loaded with 5 liters of 0.1 M Tris-buffer, pH 9.2; 15 liters of 0.02 M Tris-buffer, pH 9.2, and 1 liter of 0.02 M KCl. The treatment of the ion exchanger and the packing of the column are performed 1–2 days before use. In this case application of the samples and elution of protein is from the bottom with upward flow. Hydroxyapatite is prepared by the method of Tiselius *et al.*[35]

The entire purification is carried out at 0–4° unless otherwise stated.

Acetone Powder Preparation. Fresh livers of white rats (200 g) are covered with 800 ml of acetone, cooled to −25°, and homogenized in a Waring blender for 60 sec. The homogenate is filtered on a Büchner funnel. The process is repeated with the residue; after a second filtration, the residue is dried with filter paper, crushed by hand, and dried at room temperature until there is no smell of acetone. The dried powder is sieved to remove large fragments and is stored in a refrigerator in a desiccator over $CaCl_2$. The acetone-dried powder can be stored for 3 months without loss of activity.

Step 1. Extraction. To 10 g of the acetone powder is added 116 ml of cold 15 mM K_2HPO_4. The mixture is stirred for 15 min, then centrifuged for 15 min at 20,000 g; the sediment is discarded.

Step 2. Ammonium Sulfate Fractionation. Solid ammonium sulfate is added slowly to the supernatant to 1.75 M with stirring. The stirring is continued for another 15 min, then the mixture is centrifuged for 10 min at 40,000 g, and the sediment is discarded. The supernatant is supplemented with solid ammonium sulfate to 2.93 M; 10 min later it is centrifuged for 20 min at 40,000 g; the sediment is dissolved in 10 ml of 0.1 M KCl solution.

Step 3. DEAE-Sephadex A-25 Chromatography.[36] The enzyme solution (not more than 700 mg of protein) is loaded onto a DEAE-Sephadex A-25 column; then the enzyme is eluted with 0.1 M KCl at a rate of 1 ml/min. The volume of each fraction is 5–8 ml. The fractions whose specific activity does not differ by more than 30% from the maximum value are pooled.

Transketolase is usually easy to separate from the protein impurities. However, one should monitor the elution rate and the pH of the buffer.

Step 4. Hydroxyapatite Chromatography. The pooled fractions from step 3 are loaded onto a 5- × 1.5-cm hydroxyapatite column equilibrated with 0.01 M potassium phosphate buffer, pH 6.8. For elution, a 0.01 to 0.5 M linear gradient of potassium phosphate buffer, pH 6.8, is used. The total volume is 200 ml. The elution rate is not higher than 0.2–0.25 ml/min. The volume of each fraction is 3 ml. Fractions of high specific activity are pooled and frozen. The results of a typical purification are summarized in Table II.

---

[35] A. Tiselius, S. Hjerten, and O. Levin, *Arch. Biochem. Biophys.* **65**, 132 (1956).
[36] L. H. Kirkegaard, *Biochemistry* **12**, 3627 (1973).

TABLE II
PURIFICATION PROCEDURE FOR RAT LIVER TRANSKETOLASE

Step	Total volume (ml)	Total protein (mg)	Total units[a]	Specific activity (units/mg protein)	Purification (fold)	Recovery (%)
Extraction	104	1350	10.6	0.008	—	—
Ammonium sulfate fractionation	15	630	7.6	0.012	1.5	71
DEAE-Sephadex chromatography	72	49.4	7.0[b]	0.14	17.5	66
Hydroxyapatite chromatography	14	3.9	5.9[b]	1.5	187.5	56

[a] The activity was assayed at 20° by NAD reduction.
[b] The activity was assayed after the removal of KCl (step 3) or phosphate (step 4), which inhibit the enzyme.

Properties

Homogeneity. By the criteria of ultracentrifugation and electrophoresis in polyacrylamide gel, the enzyme is homogeneous. $E_{1\,cm}^{1\%}$ at 280 nm = 13.0.[34]

Stability. The activity remains unaltered for at least a week when the enzyme is stored frozen.[37]

pH Optimum. The catalytic activity of transketolase is maximal at pH 7.6.[37]

Cofactors. Rat liver transketolase is isolated as the holoenzyme that contains 2 moles of thiamine pyrophosphate per mole of protein.[34] Both participate in catalysis[37]. At pH 5.0, the coenzyme dissociates with loss of catalytic activity. When thiamine pyrophosphate is added, activity is restored. The bivalent cations are without effect in this case. If holo-transketolase is kept at pH 5.0 in the presence of EDTA, the activity of such an enzyme, assayed in the presence of thiamine pyrophosphate and calcium, is higher than in the presence of thiamine pyrophosphate alone.[37]

Molecular Weight and Quaternary Structure. The molecular weight of transketolase determined by the sedimentation equilibrium method is 130,000. A molecule of the enzyme consists of two subunits of equal molecular weight.[34] The individual subunits are catalytically active and do not differ appreciably from the dimer in specific activity (the experiments were carried out with an immobilized enzyme).[37]

[37] A. A. Minin and G. A. Kochetov, *Biokhimiya* **46**, 195 (1981).

Transketolase from Pig Liver

Transketolase from pig liver was first isolated in a highly purified form in 1960.[38] Later a technique for preparation of the homogeneous enzyme was developed[39] as described below.

Assay Method

The enzymic activity assay is identical to that described for bakers' yeast transketolase, except that neither thiamine pyrophosphate nor $MgCl_2$ is added and the reaction is carried out at pH 7.8.

Purification Procedure

The entire work is performed at 0–4° unless otherwise stated.

Preparation of Acetone Powder. This step is the same as for rat liver transketolase except that tissue is homogenized for 2 min rather than for 60 sec. The acetone powder can be stored for several months at 0° in a vacuum desiccator without loss of activity.

Step 1. Extraction. Acetone powder (15–20 g) is suspended in 10 volumes of 15 mM Na_2CO_3, and the solution is mixed for 45 min at room temperature, then centrifuged in a refrigerator centrifuge for 15 min at 20,000 g; the sediment is discarded. The centrifugate can be stored overnight at 0–4° without loss of activity.

Step 2. Treatment with Protamine Sulfate. The centrifugate is mixed with dry protamine sulfate (1 mg per 20 mg of protein) for 5 min and centrifuged for 15 min at 40,000 g; the sediment is discarded.

Step 3. Treatment with DEAE-Cellulose. DEAE-cellulose is equilibrated with 2 mM potassium phosphate buffer, pH 7.6. Then the liquid is removed using a Büchner funnel and added to the centrifugate obtained in step 2 (600 mg of wet ion exchanger per 1 ml of enzyme solution). The resultant solution is mixed for 10 min and centrifuged for 5 min at 20,000 g. This procedure is repeated until the concentration of protein in the centrifugate decreases to 6–8 mg/ml.

Step 4. Fractionation with Ammonium Sulfate. The centrifugate obtained at step 3 is supplemented, with stirring, with dry ammonium sulfate to 1.95 M; it is mixed for another 10 min and centrifuged for 15 min at 40,000 g. The centrifugate is supplemented with dry ammonium sulfate to 2.93 M; 5 min later it is centrifuged for 20 min at 40,000 g. The centrifugate

[38] F. J. Simpson, *Can. J. Biochem.* **38**, 115 (1960).
[39] P. P. Philippov, I. K. Shestakova, N. K. Tikhomirova, and G. A. Kochetov, *Vestn. Mosk. Univ. Ser. Biol.* No. 3, p. 39 (1977).

TABLE III
PURIFICATION PROCEDURE FOR PIG LIVER TRANSKETOLASE

Step	Total volume (ml)	Total protein (mg)	Total units[a]	Specific activity (units/mg protein)	Purification (fold)	Recovery (%)
Extraction	150	6820	9.1	0.0013	—	100.0
Treatment with protamine sulfate	145	2889	8.7	0.003	2.3	95.6
Treatment with DEAE-cellulose	102	816	8.2	0.01	7.6	90.1
Ammonium sulfate fractionation	22	630	7.5	0.011	8.5	82.4
DEAE-Sephadex chromatography	45	25	6.3[b]	0.25	192.0	69.2
Hydroxyapatite chromatography	20	2.4	2.1[b]	0.88	677.0	23.1

[a] Activity was assayed at 20° by reduction of NAD.
[b] The activity was assayed after the removal of KCl (step 5) or phosphate (step 6), which inhibit the enzyme.

is discarded, and the sediment is dissolved in 20 ml of 50 mM Tris-HCl buffer, pH 8.5.

Step 5. DEAE-Sephadex A-25 Chromatography. This step is carried out as for rat liver transketolase.

Step 6. Hydroxyapatite Chromatography. The fractions collected from the DEAE-Sephadex A-25 column are applied to a 5- × 1.5-cm hydroxyapatite column equilibrated with 0.01 M potassium phosphate buffer, pH 6.5. For elution, a 0.01 to 0.5 M linear gradient of potassium phosphate buffer, pH 6.5, is used; the total volume is 200 ml. Fractions of 2 ml are collected at an elution rate of 0.5 ml/min. Fractions with high specific activity are pooled and stored at −20°. The activity of the enzyme remains unaltered for several days. The results of typical purification are summarized in Table III.

Properties

Homogeneity. According to the data from disc electrophoresis in polyacrylamide gel, isoelectrofocusing, and ultracentrifugation, the enzyme is homogeneous.[40]

[40] P. P. Philippov, I. K. Shestakova, N. K. Tikhomirova, and G. A. Kochetov, *Biochim. Biophys. Acta* **613**, 359 (1980).

TABLE IV
AMINO ACID COMPOSITION OF PIG LIVER
TRANSKETOLASE

Amino acid	Residues per 139,000 g of protein
Tryptophan	14
Lysine	89.6
Histidine	21.5
Arginine	39.9
Aspartic acid	108.8
Threonine	62.6
Serine	79.7
Glutamic acid	144.0
Proline	64.2
Glycine	133.7
Alanine	122.1
Cystine	—
Cysteine	—
Valine	87.4
Methionine	—
Isoleucine	75.0
Leucine	126.4
Tyrosine	27.4
Phenylalanine	42.0

pH Optimum of Activity and Stability.[40] The maximum activity in Tris-HCl and glycylglycine buffers is at pH 7.8–8.2. At 20° and pH 6.5–8.7, the enzyme does not inactivate during 20 min. At higher and lower pH values the enzyme is less stable. At 40° (pH 8.0) the activity hardly changes for 60 min, but it is reduced by half at 50° during 10 min. The stability increases significantly in the presence of 4 mM thiamine pyrophosphate.

Isoelectric Point. This point is in the pH range of 7.6–7.8.[40]

Cofactors. Transketolase is isolated from pig liver in the form of the holoenzyme. There are 2 mol of thiamine pyrophosphate per mole of protein.[40] Keeping the enzyme in acetate buffer, pH 5.0, results in the loss of the catalytic activity, which is restored, although not completely, when thiamine pyrophosphate and Mg^{2+} are added.[41] The native holoenzyme does not contain bivalent cations, and the catalytic activity of transketolase remains unchanged when bivalent cations are added.[40]

Molecular Weight and Quaternary Structure.[40] The molecular weight of transketolase, determined by the sedimentation equilibrium method, is

[41] J. Tomita, S. Saitou, and M. Ishikawa, *J. Nutr. Sci. Vitaminol.* **25,** 174 (1979).

138,000 ± 3000. The enzyme is a tetramer of the $\alpha_2\beta_2$ type. The molecular weights of the subunits determined by electrophoresis in polyacrylamide gel in the presence of sodium dodecyl sulfate are, respectively, 52,000–56,000 and 27,000–29,000.

Amino Acid Composition. The data for the amino acid analysis are given in Table IV.[42]

[42] J. K. Shestakova and G. A. Kochetov, *Vestn. Mosk. Univ., Ser. Biol.* No. 3, p. 66 (1979).

[33] Transketolase from Human Red Blood Cells[1]

By G. D. SCHELLENBERG, N. M. WILSON, B. R. COPELAND, and CLEMENT E. FURLONG

Ribose 5-phosphate + xylulose 5-phosphate $\rightleftarrows$ sedoheptulose
7-phosphate + glyceraldehyde 3-phosphate

Assay Method

Several assays for measuring transketolase (EC 2.2.1.1) are available.[2–6] The following method is used to measure transketolase activity during purification.

Principle. Transketolase activity is measured by a coupled assay based on the spectrophotometric measurement of changes in NADH concentration.[2] Xylulose 5-phosphate and ribose 5-phosphate are converted by transketolase to sedoheptulose 7-phosphate and glyceraldehyde 3-phosphate. In the coupled assay, triosephosphate isomerase (EC 5.3.1.1) converts the glyceraldehyde 3-phosphate to dihydroxyacetone phosphate, which is in turn converted to glycerol-3-phosphate by glycerol-3-phosphate dehydrogenase (EC 1.1.1.8) with the concomitant oxidation of NADH. The disappearance of NADH is measured by the

[1] Supported by Grant GM15253 and a Career Development Award GM00452 (to C.E.F.) from USPHS, and a grant from the Alcoholism and Drug Abuse Institute, University of Washington.
[2] E. H. J. Smeets, H. Muller and J. DeWael, *J. Clin. Chim. Acta* 33, 379 (1971).
[3] P. M. Dreyfus, *N. Engl. J. Med.* 267, 596 (1962).
[4] G. A. Kochetov, L. I. Nikitushkina, and N. N. Chernov, *Biochem. Biophys. Res. Commun.* 40, 873 (1970).
[5] G. A. Kochetov, R. A. Usmanov, and A. T. Mevkh, *Anal. Biochem.* 88, 296 (1978).
[6] G. L. A. Reinjnierse, A. R. VanDerHorst, K. De Kloet, and C. D. Voorhorst, *Clin. Chim. Acta* 90, 259 (1978).

decrease of A_{340}. Since transketolase activity is stimulated in some hemolysates by thiamine pyrophosphate, this cofactor is included in the assay. Mg^{2+}, which activates transketolase from some sources but not others,[2] is also included in the reaction mixture.

Reagents

Thiamine pyrophosphate chloride, 10 mM
$MgCl_2$, 120 mM
Tris-HCl, 1 M, pH 7.6
Xylulose 5-phosphate, sodium salt, 8 mM
Ribose 5-phosphate, sodium salt, 100 mM
NADH, 10 mM, freshly prepared
Mixture of triosephosphate isomerase (~1000 units/ml) and glycerol-3-phosphate dehydrogenase (~100 units/ml) type X (Sigma) in 10 mM Tris-HCl, pH 7.6

Procedure. Thiamine pyrophosphate (10 μl), $MgCl_2$ (10 μl), Tris-HCl (100 μl), NADH (10 μl), glycerol-3-phosphate dehydrogenase–triosephosphate isomerase (10 μl), and enzyme + water (660 μl) are mixed together in a quartz cuvette and placed in a temperature-controlled (37°) recording spectrophotometer. The mixture is allowed to come to temperature, and a steady baseline is established (340 nm). Xylulose 5-phosphate (100 μl) and ribose 5-phosphate (100 μl) are added to initiate the reaction, and the rate of absorbance change is measured over 5–10 min. Typical volumes of enzyme required are 5 μl for hemolysate, 50–200 μl for the DEAE-column fractions, and 20–100 μl for hydroxyapatite fractions.

Preparation of Hemolysates

Red cells from outdated blood bank blood or from freshly drawn heparinized blood are sedimented at 1750g for 10 min (4°). The serum and buffy coat are removed by aspiration. The cells are then washed 3–4 times by resuspension in 3 volumes of normal saline and centrifugation (1750g) with removal of the supernatant solution and buffy coat by aspiration after each centrifugation. The supernatant solution should be clear after the last wash. The packed cells are next vigorously shaken in 1.5 volumes of cold deionized water in a separatory funnel and allowed to stand for 30 min at 4°. The deionized water contains 1/60th volume of 100 mM phenylmethylsulfonyl fluoride in anhydrous isopropyl alcohol. The hemolysate is then vigorously mixed in the separatory funnel with a volume of toluene equal to 60% of the packed cell pellet volume. The toluenized preparation is transferred to a beaker and allowed to stand at 4° with occasional mixing for 20 min. The mixed preparation is transferred to polypropylene or polyethylene centrifuge tubes and sedimented at 1750 g for 20 min. The bulk of the toluene upper phase and membranes at the interface are discarded. The hemolysate (lower phase above the residual pellet) is removed carefully so as to avoid inclusion of any membranes or pellet. Sufficient 1 M Tris, pH 8.3 (23°) is added to achieve a final concentration of 10 mM. The hemolysate is then dialyzed overnight (4°) against 20–30 volumes of 10 mM Tris, pH 8.3.

Purification of Transketolase

Chromatography on DEAE-Cellulose. DEAE-cellulose (Whatman DE-23) is equilibrated with 0.5 N HCl for 30 min, collected on a sintered-glass funnel (coarse), and washed with distilled water until the pH of the wash solution is greater than 4. The resin is next equilibrated for 30 min with 0.5 N NaOH, collected on a sintered-glass funnel, and washed with distilled water until the pH of the wash solution is less than 8.0. The fines are removed by resuspension of the exchanger in 10–15 volumes of 10 mM Tris, pH 8.3, followed by aspiration of the supernatant solution and fines after the larger particles have settled. This process should be repeated 10 times to assure a good flow rate through the packed column.

A DEAE-cellulose column 8 times the volume of the original packed red cells is prepared with a height-to-diameter ratio of at least 7 : 1. It is equilibrated with 10 mM Tris, pH 8.3. The hemolysate is loaded on the column, which is then eluted (23°) with 10 mM Tris (pH 8.3, 23°)–10 mM NaCl. The transketolase should elute before any hemoglobin, as shown in Fig. 1. The fractions with transketolase activity are pooled and loaded directly on the hydroxyapatite column described below.

Chromatography on Hydroxyapatite. BioGel HT is *gently* defined by swirling in 5 mM KPO$_4$.[7] A column with a volume equal to one-seventh the volume of the original packed red cells is prepared and equilibrated with 20 column volumes of 5 mM phosphate (height to diameter, approximately 3 : 1). Column elution is most conveniently followed by continuous monitoring of absorbance at 206 nm. The pooled fractions from the DEAE-cellulose column are loaded directly on the column. The flow rate with a 1.5-meter pressure head should be 1–3 ml/min depending on the column volume and dimensions. For very small preparations, the flow rate is much less. The loaded column is first washed with 5 mM phosphate until the A_{206} of the eluent reaches a baseline level. The column is next washed with 100 mM phosphate until the A_{206} returns to the baseline level. An additional 3 column volumes of 100 mM phosphate buffer are then

---

[7] All phosphate buffers used for hydroxyapatite chromatography are prepared by diluting a 1 M stock of potassium phosphate, pH 6.9.

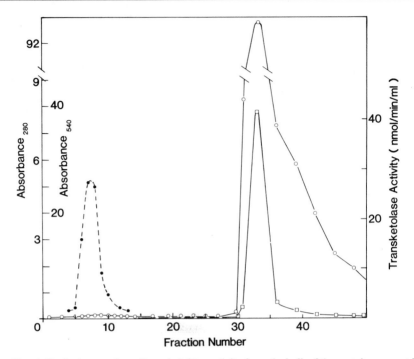

FIG. 1. Typical separation of transketolase activity from the bulk of the proteins present in crude red cell hemolysates by DEAE-cellulose chromatography. The column was prepared and eluted as described in the text. Hemoglobin and other more acidic proteins may be eluted with 1 M NaCl as indicated by the arrow; however, this is unnecessary since fresh DEAE-cellulose is used for each preparation. ●---●, Transketolase activity; ○——○, A_{280}; □——□, A_{540}.

washed through the column. This step is important in that the 100 mM wash removes most of the contaminating proteins. The column is then washed with 200 mM phosphate until the A_{206} reaches a baseline value. The column is finally eluted with 300 mM phosphate. The hydroxyapatite elution profile varies somewhat from batch to batch of BioGel HT. It may be necessary to adjust the concentrations of elution buffers in order to generate pure protein with this column. The activity usually elutes during the 200 mM phosphate wash (see Fig. 2). However, occasionally 300 mM phosphate is required to elute the transketolase activity from hydroxyapatite.

Fractions containing activity are analyzed by sodium dodecyl sulfate (SDS)–slab gel electrophoresis,[8] utilizing a sensitive silver staining procedure.[9] Bovine serum albumin is used as a molecular weight marker. The

[8] U. K. Laemmli, *Nature* (*London*) **227**, 680 (1970).
[9] B. R. Oakley, D. R. Kirsch, and N. R. Morris, *Anal. Biochem.* **105**, 361 (1980).

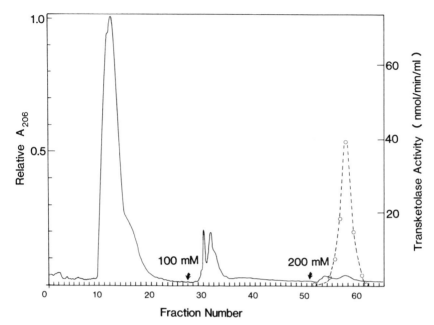

Fig. 2. Fractionation of the DEAE-cellulose-purified transketolase activity by hydroxyapatite chromatography. The column was prepared and eluted as described in the text. ---, Transketolase activity; ——, relative absorbance, A_{206}.

PURIFICATION OF HUMAN RED CELL TRANSKETOLASE

Step	Volume (ml)	Protein concentration (mg/ml)	Total protein (mg)	Total activity (nmol/min)	Specific activity (nmol/min/mg)	Yield (%)	Purification (fold)
Hemolysate	288	140.75	40,536	30,560	0.75	—	—
Fractionation on DEAE-cellulose	2060	0.016	33	9,604	291	31.4	388
Fractionation on hydroxyapatite	69.2	0.015	1.04	8,421	8097	27.5	10,796

transketolase subunit has a nearly identical relative mobility to bovine serum albumin. The fractions containing activity and a single protein band by SDS gel analysis[10] ($M_r = 70,000$) are pooled and dialyzed against 1 mM phosphate, pH 6.9. The pooled fractions may be concentrated by lyophilization and resuspension in distilled water following this step. The enzyme

[10] The silver stain often detects minor artifactual bands between 50,000 and 65,000 M_r. A lane containing sample buffer minus protein should be included to identify such bands.

is stable at $-20°$ when purified. A typical purification is summarized in the table. The overall yield varies from 15 to 40%.

As an alternative purification procedure, the outlet of the DEAE-cellulose column may be coupled directly to the inlet of the hydroxyapatite column with a three-way valve positioned between the two columns for sampling. As soon as the transketolase activity has eluted from the DEAE-cellulose column, the hydroxypatite column is eluted as described above.

[34] D-Tagatose-1,6-bisphosphate Aldolase (Class I) from *Staphylococcus aureus*

By RICHARD L. ANDERSON and DONALD L. BISSETT

D-Tagatose 1,6-bisphosphate $\rightleftharpoons$
dihydroxyacetone phosphate + D-glyceraldehyde 3-phosphate

This inducible class I aldolase functions in the catabolism of lactose and D-galactose in *Staphylococcus aureus*.[1-3] Although it can also catalyze the cleavage of D-fructose 1,6-bisphosphate and other D-2-ketohexose 1,6-bisphosphates,[3] it is distinct from the constitutive D-fructose-1,6-bisphosphate aldolase that occurs in the same organism.[1-4] In contrast, the D-tagatose-1,6-bisphosphate aldolase from *Klebsiella pneumoniae* is of class II and is relatively specific for D-tagatose 1,6-bisphosphate.[5]

Assay Method

Principle. The continuous spectrophotometric is based on the following sequence of reactions:

$$\text{D-Tagatose-6-P + MgATP} \xrightarrow[\text{kinase}]{\text{D-fructose-6-P}} \text{D-tagatose-1,6-P}_2 + \text{MgADP}$$

$$\text{D-Tagatose-1,6-P}_2 \xrightarrow[\text{aldolase}]{\text{D-tagatose-1,6-P}_2} \text{dihydroxyacetone-P + D-glyceraldehyde-3-P}$$

$$\text{D-Glyceraldehyde-3-P} \xrightarrow[\text{isomerase}]{\text{triose-P}} \text{dihydroxyacetone-P}$$

[1] D. L. Bissett and R. L. Anderson, *Biochem. Biophys. Res. Commun.* **52**, 641 (1973).
[2] D. L. Bissett and R. L. Anderson, *J. Bacteriol.* **119**, 698 (1974).
[3] D. L. Bissett and R. L. Anderson, *J. Biol. Chem.* **255**, 8750 (1980).
[4] F. Götz, S. Fischer, and K.-H. Schleifer, *Eur. J. Biochem.* **108**, 295 (1980).
[5] R. L. Anderson and J. P. Markwell, this volume [35].

$$2 \text{ Dihydroxyacetone-P} + 2 \text{ NADH} \xrightarrow[\text{dehydrogenase}]{\alpha\text{-glycerol-P}} 2 \text{ } \alpha\text{-glycerol-P} + 2 \text{ NAD}^+$$

D-Tagatose 6-phosphate, ATP, $MgCl_2$, and rabbit muscle D-fructose-6-phosphate kinase are included to effect the synthesis of the aldolase substrate, D-tagatose 1,6-bisphosphate. With D-fructose-6-phosphate kinase, triosephosphate isomerase, and α-glycerolphosphate dehydrogenase in excess, the rate of D-tagatose 1,6-bisphosphate cleavage is equal to one-half the rate of NADH oxidation, which is measured by the absorbance decrease at 340 nm.

Reagents

Sodium HEPES[6] buffer, pH 7.0, 0.25 M
$MgCl_2$, 0.2 M
ATP, 0.1 M
NADH, 0.01 M
D-Tagatose 6-phosphate, 0.05 M[7]
Crystalline D-fructose-6-phosphate kinase (rabbit muscle)
Crystalline triosephosphate isomerase–α-glycerolphosphate dehydrogenase mixture

Procedure. The following are added to a microcuvette with a 1.0-cm light path: 20 μl of HEPES buffer, 5 μl of $MgCl_2$, 5 μl of ATP, 5 μl of NADH, 5 μl of D-tagatose 6-phosphate, non-rate-limiting amounts of D-fructose-6-phosphate kinase (750 mU), triosephosphate isomerase (2000 mU), and α-glycerolphosphate dehydrogenase (200 mU), a rate-limiting amount of D-tagatose-1,6-bisphosphate aldolase, and water to a volume of 0.15 ml. The reaction is initiated by the addition of D-tagatose-1,6-bisphosphate aldolase. A control cuvette minus D-tagatose 6-phosphate measures NADH oxidase, which must be subtracted from the total rate. The rates are conveniently measured with a Gilford multiple-sample absorbance-recording spectrophotometer. The cuvette compartment should be thermostatted at 30°. Care should be taken to confirm that the rates are constant with time and proportional to the aldolase concentration.

Definition of Unit and Specific Activity. One unit is defined as the amount of enzyme that catalyzes the cleavage of 1 μmol of D-tagatose 1,6-bisphosphate per minute. Specific activity (units per milligram of protein) is based on protein determinations by the Lowry procedure.

Alternative Assay Procedure. If available, D-tagatose 1,6-bisphosphate may be substituted for D-tagatose 6-phosphate, ATP, $MgCl_2$, and D-fructose-6-phosphate kinase.

---

[6] 4-(2-Hydroxyethyl)-1-piperazineethanesulfonic acid.
[7] See R. L. Anderson, W. C. Wenger, and D. L. Bissett, this series, Vol. 89 [15].

PURIFICATION OF D-TAGATOSE-1,6-BISPHOSPHATE ALDOLASE FROM *Staphylococcus aureus*

Fraction	Volume (ml)	Total protein (mg)	Total activity (units)	Specific activity (units/mg protein)	Recovery (%)
Cell extract, pH 5.2	215	1092	138	0.126	(100)
Phosphocellulose I	150	27.2	118	4.34	85
Phosphocellulose II	72	4.52	57.9	12.8	42
DEAE-cellulose	8.4	2.52	49.1	19.5	36

Purification Procedure[3]

Organism and Growth Conditions. Staphylococcus aureus NCTC 8511 is grown at 37° in Fernbach flasks (1500 ml per flask) on a rotary shaker. The medium is the induction medium of McClatchy and Rosenblum[8] supplemented with 1% D-galactose (autoclaved separately). The inoculum is 7 ml of an overnight culture in the same medium except that no carbohydrate is added and the peptone concentration is increased to 2%. The cells are harvested by centrifugation 9 hr after inoculation, washed once by suspension in 0.85% (w/v) NaCl, and collected by centrifugation. The yield is about 4 g (wet weight) of cells per liter of medium.

Preparation of Cell Extracts. Cells (68 g) are suspended in 235 ml of 20 mM sodium acetate buffer (pH 5.2) containing 20% (v/v) glycerol and 0.2% (v/v) 2-thioethanol. The cells are broken by treating the suspension at 0 to 2° for 10 min (1 in 1-min bursts) with the 1.27-cm (diameter) horn of a Heat Systems-Ultrasonics W-185C sonifier at 100 W, in the presence of twice the packed-cell volume of glass beads (88–125 μm in diameter). The pH of the broken-cell suspension is 6.0; it is adjusted to 5.2 with 10% acetic acid prior to removal of the cellular debris and precipitated protein by centrifugation. The resulting supernatant fluid is designated the cell extract, although some purification of the aldolase had been achieved by the acid precipitation of other proteins.

General. The following procedures are performed at 0 to 4°. A summary of a typical purification is shown in the table. The success of this three-step scheme is due largely to the substrate-facilitated elution used in the second phosphocellulose chromatography run.

Phosphocellulose Chromatography I. A phosphocellulose column (3.1 × 22 cm) is equilibrated with the cell-extract buffer. The cell extract is applied to the column, which is then washed with 500 ml of the same buffer. The protein is eluted with a linear gradient (1660 ml; 90 ml/hr) of 0

[8] J. K. McClatchy and E. D. Rosenblum, *J. Bacteriol.* **86**, 1211 (1963).

to 0.4 M KCl in the same buffer. One hundred ten 15-ml fractions are collected, and those that contain most of the aldolase activity (fractions 41 through 50, about 0.20 M KCl) are combined.

Phosphocellulose Chromatography II. A phosphocellulose column (1.6 × 17 cm) is equilibrated with the same buffer. The combined fractions from the preceding step are diluted to 300 ml with the same buffer and applied to the column,which is then washed with 100 ml of the same buffer. The protein is eluted with a linear gradient (320 ml; 30 ml/hr) of 0 to 0.35 M KCl in the same buffer containing 1 mM D-fructose 1,6-bisphosphate. Forty-four 7.2-ml fractions are collected, and those that contain most of the aldolase activity (fractions 10 through 19, about 0.12 M KCl) are combined.

DEAE-Cellulose Chromatography. A DEAE-cellulose column (1.2 × 7.5 cm) is equilibrated with 20 mM potassium phosphate buffer (pH 7.5) containing 20% (v/v) glycerol and 0.2% (v/v) 2-thioethanol. The combined fractions from the preceding step are adjusted to pH 7.5 with 0.4 M K_2HPO_4, diluted to 300 ml with the pH 7.5 buffer, and applied to the column, which is then washed with 50 ml of the same buffer. The protein is eluted with a linear gradient (90 ml; 30 ml/hr) of 0 to 0.4 M KCl in the same buffer. Sixty-four 1.4-ml fractions are collected, and those that contain the highest specific activity of aldolase (fractions 36 through 41, which have equal specific activities) are combined. The enzyme is 155-fold purified with an overall recovery of 36%. It is free from the constitutive D-fructose-1,6-bisphosphate aldolase and is homogeneous when examined by electrophoresis on polyacrylamide gels under both native and denaturing conditions.

Properties[3]

Substrate Specificity. Both D-tagatose 1,6-bisphosphate (K_m = 1.5 mM) and D-fructose 1,6-bisphosphate (K_m = 2.5 mM) are cleaved, the latter at 47% the rate of the former at saturating concentrations. When the reaction was run in the reverse direction, the 1,6-bisphosphates of all four D-2-ketohexoses were formed, indicating that the aldolase is nonspecific with respect to carbon atoms 3 and 4 and also cleaves D-sorbose 1,6-bisphosphate and D-psicose 1,6-bisphosphate.

D-Fructose 1-phosphate (100 mM), and L-sorbose 1-phosphate (10 mM) are not cleaved (<1% of the rate with D-tagatose 1,6-bisphosphate).

Effect of Metal Ions, EDTA, and Borohydride; Reaction Mechanism. $MnCl_2$, $CoCl_2$, and $CaCl_2$ (1 mM) did not affect the reaction velocity in the alternative assay procedure, and 10 mM EDTA did not inhibit the enzyme.

NABH$_4$ (10 mM) caused 90% loss of activity in less than 5 min in the presence of substrate, but no loss in the absence of substrate. The lack of a divalent metal requirement and the inhibition of borohydride indicate that this enzyme is a class I (Schiff's base) aldolase.

pH Optimum. Activity as a function of pH is maximal at about pH 7.0 to 7.8, depending on the buffer.

Molecular Weight. D-Tagatose-1,6-bisphosphate aldolase is a monomeric enzyme with a molecular weight of about 37,000.

Stability. The DEAE-cellulose fractions were stable to storage at $-20°$ for several months.

[35] D-Tagatose-1,6-bisphosphate Aldolase (Class II) from *Klebsiella pneumoniae*

By RICHARD L. ANDERSON and JOHN P. MARKWELL

D-Tagatose 1,6-bisphosphate ⇌
   dihydroxyacetone phosphate + D-glyceraldehyde 3-phosphate

This inducible class II aldolase, which functions in the catabolism of galactitol but not of D-galactose or lactose in *Klebsiella pneumoniae,* is relatively specific for D-tagatose 1,6-bisphosphate.[1,2] In contrast, the D-tagatose-1,6-bisphosphate aldolase from *Staphylococcus aureus* is of class I and is nonspecific with respect to the D-2-ketohexose 1,6-bisphosphate substrate.[3]

Assay Method

The assay is similar to that described,[3] except that the buffer in the reaction mixture is 67 mM glycylglycine, pH 8.0.

Purification Procedure[2]

Organism and Growth Conditions. Klebsiella pneumoniae PRL-R3 (formerly designated *Aerobacter aerogenes* PRL-R3) is grown aerobically at 30° in a medium containing 0.15% KH$_2$PO$_4$, 0.71% Na$_2$HPO$_4$, 0.3%

[1] J. Markwell, G. T. Shimamoto, D. L. Bissett, and R. L. Anderson, *Biochem. Biophys. Res. Commun.* **71,** 221 (1976).

[2] J. P. Markwell and R. L. Anderson, in preparation.

[3] R. L. Anderson and D. L. Bissett, this volume [34].

PURIFICATION OF D-TAGATOSE-1,6-BISPHOSPHATE ALDOLASE FROM *Klebsiella pneumoniae*

Fraction	Volume (ml)	Total protein (mg)	Total activity (units)	Specific activity (units/mg protein)	Recovery (%)
Cell extract	43	653	41.6	0.0637	(100)
DEAE-cellulose I	44	60.3	46.5	0.770	112
Sephadex G-150	67	28.8	40.8	1.40	98
DEAE-cellulose II	35	2.35	22.6	3.07	54

$(NH_4)_2SO_4$, 0.01% $MgSO_4$, 0.0005% $FeSO_4 \cdot 7 H_2O$, and 0.5% galactitol (autoclaved separately). The cells are harvested by centrifugation and washed with 0.85% (w/v) NaCl.

Preparation of Cell Extracts. Pelleted cells are suspended in buffer A (20 mM sodium phosphate, pH 7.5, and 10%, v/v, glycerol) and broken by sonic treatment (10,000 Hz) in a Raytheon Model DF-101 (250-W) sonic oscillator cooled with circulating ice water. The broken-cell suspension is centrifuged at 12,000 g for 10 min, and the resulting supernatant fluid is designated the cell extract.

General. The following procedures are performed at 0 to 4°. A summary of a typical purification is shown in the table.

DEAE-Cellulose Chromatography I. This step separates D-tagatose-1,6-bisphosphate aldolase from the constitutive D-fructose-1,6-bisphosphate aldolase. The cell extract is applied to a column (1.2 × 10 cm) of DEAE-cellulose that has been equilibrated with buffer A. The column is washed with 100 ml of buffer A, and then the adsorbed protein is eluted with a 150-ml linear gradient of 0 to 0.45 M KCl in the same buffer. Fractions (1.4 ml) are collected, and those that contain most of the activity (fractions 35 to 66) are combined.

Sephadex G-150 Chromatography. The combined fractions from the preceding step are reduced in volume from 44 ml to 15 ml by pressure filtration, and the concentrate is applied to a column (3.2 × 81 cm) of Sephadex G-150 previously equilibrated with buffer A. Fractions (3.7 ml) are collected during elution with the same buffer. Fractions 50 through 67, which contain most of the aldolase activity, are combined.

DEAE-Cellulose Chromatography II. The combined fractions from the Sephadex G-150 step are applied to a column (1.2 × 5 cm) of DEAE-cellulose previously equilibrated with buffer A. The column is washed with 100 ml of buffer A, and then the absorbed protein is eluted with a 200-ml linear gradient of 0 to 0.4 M KCl in the same buffer. Fractions (1.8 ml) are collected, and those that contain the peak of the activity (fractions 57 through 75) are combined. D-Tagatose-1,6-bisphosphate aldolase is 48-

fold purified with an overall recovery of 54%. Electrophoresis of this preparation on 7.5% polyacrylamide gels, when stained for protein, reveals one major band at least four minor bands.

Properties[3]

Substrate Specificity. D-Tagatose 1,6-bisphosphate (K_m = 0.4 mM) is apparently the natural substrate. D-Fructose 1,6-bisphosphate (K_m = 0.9 mM) can also be cleaved, but at only 4% the rate of D-tagatose 1,6-bisphosphate at saturating concentrations. When the reaction was run in the reverse direction, only the bisphosphates of tagatose and fructose could be detected, indicating that sorbose 1,6-bisphosphate and psicose 1,6-bisphosphate are not substrates. D-Fructose 1-phosphate (50 mM) and L-sorbose 1-phosphate (10 mM) are also not cleaved (<0.1% of the rate with D-tagatose 1,6-bisphosphate).

Activation by Divalent and Monovalent Cations. Treatment of the enzyme with 10 mM EDTA followed by removal of the chelating agent by passage of the enzyme through a Sephadex G-25 column abolished the aldolase activity. The addition of metal salts (1 mM) to the assay resulted in the following relative specific activities (untreated enzyme = 100): $CdSO_4$, 315; $MnCl_2$, 272; $CoCl_2$, 254; $CaCl_2$, 104; $MgCl_2$, $ZnCl_2$, $FeCl_2$, $CuCl_2$, $NiCl_2$, and none, <3.

In an assay in which monovalent cations contributed by the assay reagents were Tris or cyclohexylammonium, the further addition of 33 mM KCl, NH_4Cl, or NaCl stimulated the activity severalfold.

Effect of Borohydride; Reaction Mechanism. Treatment of the enzyme for 15 min with $NaBH_4$ (10 mM or 100 mM) in the absence or the presence of substrate (0.5 mM D-tagatose 1,6-bisphosphate or 25 mM D-fructose 1,6-bisphosphate) resulted in no loss (<2%) of activity. This result and the divalent metal requirement indicate that the enzyme is a class II (non-Schiff's base) aldolase.

pH Optimum. Activity as a function of pH is maximal at pH 8.0 to 8.2 in glycylglycine buffer.

Molecular Weight. Estimation of the molecular weight by chromatography on a column of Sephadex G-150 yielded a value of 157,000.

Stability. The purified enzyme in 20 mM sodium phosphate buffer (pH 7.5) containing 10% (v/v) glycerol was stable (>90%) when stored at −20° for 3 months.

[36] Fructose-1,6-bisphosphate Aldolase from *Bacillus subtilis*

By SUSUMU UJITA and KINUKO KIMURA

Fructose 1,6-bisphosphate ⇌
    dihydroxyacetone phosphate + D-glyceraldehyde 3-phosphate

Assay Method

Aldolase activity can be determined by coupled spectrophotometric assay (methods A–C) and colorimetric assay.

Coupled Spectrophotometric Assay

Assay Method A. Aldolase activity can be determined by measuring the rate of glyceraldehyde 3-phosphate formation by aldolase from fructose 1,6-bisphosphate. The amount of glyceraldehyde 3-phosphate is measured based on the increase in absorbance at 340 nm due to the formation of NADH by coupled glyceraldehyde-3-phosphate dehydrogenase.

Assay Method B. Aldolase activity can be determined by measuring the rate of dihydroxyacetone phosphate produced by aldolase from fructose 1,6-bisphosphate. The amount of dihydroxyacetone phosphate is measured by determining the decrease in absorbance at 340 nm due to the disappearance of NADH by coupling with α-glycerolphosphate dehydrogenase.

Assay Method C. α-Glycerolphosphate dehydrogenase–triosephosphate isomerase mixture is used instead of α-glycerolphosphate dehydrogenase in assay B. In this assay system, glyceraldehyde 3-phosphate is isomerized to dihydroxyacetone phosphate. Therefore, aldolase activity is half of the observed value.

Reagents

Assay method A
    Tris (Cl⁻) buffer, 1 M, pH 7.4
    Fructose 1,6-bisphosphate, 170 mM
    NAD⁺, 1.7 mM
    Sodium arsenate, 170 mM
    Glyceraldehyde-3-phosphate dehydrogenase, 1 mg/ml
    Cysteine-HCl, 2 M
    Sodium hydroxide, 2 M

METHODS IN ENZYMOLOGY, VOL. 90

Assay method B
  Tris (Cl⁻) buffer, 1 M, pH 7.4
  Fructose 1,6-bisphosphate, 170 mM
  NADH, 1 mM
  α-Glycerolphosphate dehydrogenase, 1 mg/ml
  Cysteine-HCl, 2 M
  Sodium hydroxide, 2 M
Assay method C
  Tris (Cl⁻) buffer, 1 M, pH 7.4
  Fructose 1,6-bisphosphate, 170 mM
  NADH, 1 mM
  α-Glycerolphosphate dehydrogenase–triosephosphate isomerase
    mixture, 1 mg/ml
  Cysteine-HCl, 2 M
  Sodium hydroxide, 2 M

Procedure. Measurement is carried out in 1.0-cm light path cuvettes at 340 nm, using a recording spectrophotometer. The complete assay mixtures contain the following components at the final concentration: 100 mM Tris (Cl⁻) buffer, pH 7.4, 17 mM fructose 1,6-bisphosphate, 0.17 mM NAD⁺, 17 mM sodium arsenate, 0.1 mg of glyceraldehyde-3-phosphate dehydrogenase per milliliter, 200 mM cysteine-HCl, and 200 mM sodium hydroxide (assay A); 100 mM Tris (Cl⁻) buffer, pH 7.4, 17 mM fructose 1,6-bisphosphate, 0.1 mM NADH, 0.1 mg of α-glycerolphosphate dehydrogenase, 200 mM cysteine-HCl, and 200 mM sodium hydroxide (assay B); 100 mM Tris (Cl⁻) buffer, pH 7.4, 17 mM fructose 1,6-bisphosphate, 0.1 mM NADH, 0.1 mg/ml α-glycerolphosphate dehydrogenase–triosephosphate isomerase mixture, 200 mM cysteine-HCl, and 200 mM sodium hydroxide (assay C). All assay mixtures are incubated for 5 min at 25° before initiating the reaction by adding a suitable amount of aldolase. Assay A and assay B are convenient for routine use. Assay C is used for crude preparations. Cysteine used as an activator, and sodium hydroxide, used for the neutrization of cysteine hydrochloride salt, are sometimes omitted for kinetic studies. In substrate specificity studies, either assay A or assay B has to be selected, depending on the substrate.

Colorimetric Assay

A colorimetric assay with 2,4-dinitrophenylhydrazine, originally used by Blostein and Rutter[1] and later modified by Stribling and Perham,[2] is used.

[1] R. Blostein and W. J. Rutter, *J. Biol. Chem.* **238**, 3280 (1963).
[2] D. Stribling and R. N. Perham, *Biochem. J.* **131**, 833 (1973).

Reagents

Hydrazine hydrochloride, 0.56 M, neutralized to pH 7.5 with solid Tris

Fructose 1,6-bisphosphate, 0.05 M, neutralized to pH 7.0 with sodium hydroxide, and then finally in 0.05 M Tris (Cl$^-$) buffer, pH 7.5

Trichloroacetic acid, 10% (w/v)

Sodium hydroxide, 0.75 M

2,4-Dinitrophenylhydrazine, 1 mg/ml, in 2 M HCl

Procedure. To a test tube (a small tube is more convenient), is added 0.05 ml of hydrazine hydrochloride, 0.05 ml of fructose 1,6-bisphosphate, and 0.3 ml of distilled water (containing cysteine, metal, and other conditions in particular experiments). The reaction is started by addition of 0.1 ml of enzyme solution and incubated for 20 min at 37°. The reaction is stopped by addition of 0.5 ml of trichloroacetic acid. The formed triosephosphate hydrazones are assayed by addition of 1 ml, 0.75 M, sodium hydroxide. After 10 min, 1 ml of 2,4-dinitrophenylhydrazine is added, and the mixture is incubated at 37° for 10 min. The resulting colored solution is transferred to a larger tube; the small tube is washed with sodium hydroxide and the total reaction mixture is finally diluted with 7 ml of sodium hydroxide. The absorbance at 540 nm is measured in a spectrophotometer against a blank without enzyme. This method is suitable for the case where an added reagent interferes with the coupling enzyme(s) present.

Definition of Unit and Specific Activity. One unit of enzyme activity is defined as the amount of enzyme required to catalyze the cleavage of 1 μmol of fructose 1,6-bisphosphate per minute. Specific activity is expressed as units per milligram of protein. Protein concentrations are estimated by the biuret method for crude preparations and the spectrophotometric method of Warburg and Christian[2a] or the procedure of Lowry *et al.*[2b] for the purified material.

Cultivation and Harvest of Bacteria

Each *B. subtilis* PCI 219 (ATCC 6633) and *B. subtilis* 168 wild type (a gift from Professor H. Kadota, Kyoto University) is grown in a nutrient medium containing, per liter, glucose, 2 g; agar powder, 12 g; bouillon, 10 g at 37°. The media are sterilized by autoclaving 1 day before use. Vegetative cells are harvested at the late-exponential phase by centrifugation, and washed with 0.05 M Tris (Cl$^-$) buffer, pH 7.4, containing 1 mM

[2a] O. Warburg and W. Christian, *Biochem. Z.* **310**, 384 (1942).
[2b] O. H. Lowry, N. J. Rosebrough, A. L. Farr, and R. J. Randall, *J. Biol. Chem.* **193**, 265 (1951).

$MgSO_4$ and 1 mM 2-mercaptoethanol (buffer I). Spores are harvested after about 1–2 weeks' cultivation. Collected spores are subjected to sonic treatment for 9 min and centrifuged for 30 min at 44,000 g. Vegetative cells and mother cells of sporangia are removed by this treatment, if present. The precipitated spores are suspended in buffer I and stored at $-20°$ until use.

Purification Procedure

Unless otherwise specified, all operations are carried out at 0–5°, and centrifugation is performed at 20,000 g for 30 min. These purification procedures are summarized in Tables I and II.

Purification of Spore Enzyme

Step 1. Preparation of Crude Extract. Glass beads (mesh, 0.1 mm) are added to a stocked spore suspension in a 100-ml beaker of a vibrogen cell mill (Edmund Bühler, Tübingen) (4 volumes of beads to 1 volume of suspended spores), and the spores are disrupted at the maximum speed for 10 min with water cooling. The glass beads and spore debris are removed by decantation and centrifugation at 44,000 g for 30 min.

Step 2. Protamine Sulfate Treatment. A 1% solution of protamine sulfate (an amount equivalent to 5% of the total protein present in the crude

TABLE I
PURIFICATION OF SPORE AND VEGETATIVE CELL ALDOLASES FROM *Bacillus subtilis* PCI 219[a]

| | Spore[b] | | | | Vegetative cell[c] | | | |
Step	Protein (mg)	Total activity (units)	Specific activity (units/mg protein)	Purification (fold)	Protein (mg)	Total activity (units)	Specific activity (units/mg protein)	Puri facti (fol
Crude extract	4363	376	0.086	1	1402	110	0.078	1
Protamine sulfate	1964	405	0.206	2.4	1449	113	0.078	1
$(NH_4)_2SO_4$, 50–70%	160	101	0.625	7.3	853.3	82.1	0.096	1
DEAE-cellulose	12.9	76.5	5.95	69.1	96.1	78.0	0.81	10
DEAE-Sephadex A-50	1.7	54.7	32.3	375.1	14.7	57.8	3.93	50
Sephadex G-200	—	—	—	—	4.4	38.8	8.85	113

[a] The values of units from crude extract to ammonium sulfate fraction are based on assay C and others on assay A.
[b] Starting from 63.3 g of wet spores.
[c] Starting from 41 g of wet vegetative cells.

TABLE II
PURIFICATION OF VEGETATIVE CELL AND SPORE ALDOLASES FROM *Bacillus subtilis* 168

| | Spore[a] | | | | Vegetative cell[b] | | | |
Step	Protein (mg)	Total activity (units)	Specific activity (units/mg protein)	Purification (fold)	Protein (mg)	Total activity (units)	Specific activity (units/mg protein)	Purification (fold)
ude extract	356	13.7	0.038	1	1060	67.5	0.064	1
otamine sulfate	—[c]	—	—	—	1060	46.1	0.044	0.7
H$_4$)$_2$SO$_4$	—[c]	—	—	—	360	54.5	0.15	2.3
50–70%								
?AE-cellulose	43	8.1	0.19	5.0	79	42.6	0.54	8.4
?AE-Sephadex	4.1	8.5	2.1	55	5.2	46.5	8.9	139
A-50								
pharose 6B	0.6	3.1	5.3	140	1.7	22.6	13.2	206

[a] Starting from 20 g of wet spores.
[b] Starting from 52 g of wet vegetative cells.
[c] These steps are omitted, and the crude extract is directly subjected to DEAE-cellulose chromatography.

extract) is added to the crude extract, and the resulting precipitate is removed by centrifugation.

Step 3. Ammonium Sulfate Fractionation. To the protamine sulfate supernatant, 35 g of solid ammonium sulfate per 100 ml (50%) are added. The suspension is centrifuged after stirring for 30 min at 0°. Another 14 g of ammonium sulfate (70%) are added, and the centrifugation is repeated after stirring for 30 min.

Step 4. Chromatography on DEAE-Cellulose. The ammonium sulfate precipitate is dissolved in 20–30 ml of buffer I and desalted by passage through a Sephadex G-25 column (3.5 × 42 cm). After desalting, the protein solution is applied to a DEAE-cellulose column (2.5 × 23 cm) equilibrated with buffer I. The column is washed with 500 ml of buffer I and then with 500 ml of buffer I containing 0.1 M NaCl. Enzyme elution is performed with a linear gradient of 0.1 to 0.4 M NaCl in a total volume of 1000 ml of buffer I.

Step 5. Chromatography on DEAE-Sephadex A-50. The pooled fractions containing aldolase activity are combined with the same volume of buffer I to decrease the salt concentration to about 0.1 M and applied to a column (2 × 33 cm) of DEAE-Sephadex A-50 equilibrated with buffer I containing 0.1 M NaCl. The column is first eluted with 500 ml of buffer I containing 0.2 M NaCl and then with 700 ml of buffer I containing 0.225 M

NaCl. The fractions containing enzyme with high specific activity are collected. This purified enzyme is nearly homogeneous by SDS gel electrophoresis.

Purification of Vegetative Cell Enzyme

Step 1. Preparation of Crude Extract. The cell paste is suspended in 2.5 volumes of buffer I and disrupted by sonic treatment for periods of 3 min with 5 min intervals for a total treatment time of 9 min with ice cooling. Cell debris is removed by centrifugation.

Step 2–Step 5. From Protamine Sulfate Treatment to Chromatography on DEAE-Sephadex A-50. These steps are carried out as described above for the purification of spore enzyme. For DEAE-Sephadex A-50 column chromatography, however, enzyme elution is performed with a linear gradient of 0.1 to 0.4 M NaCl in a total volume of 700 ml of buffer I.

Step 6. Gel Filtration. The concentrated enzyme solution (0.5 ml) after DEAE-Sephadex A-50 chromatography is applied to a Sephadex G-200 superfine column (3.5 × 33 cm) and eluted with buffer I containing 0.1 M NaCl. Fractions of 2 ml are collected, and those containing aldolase activity are pooled. This enzyme solution is used in the following experiments unless otherwise stated.

Properties

K_m *Values and Optimum pH.* The K_m values of vegetative cell and spore enzymes are calculated to be 2 mM for fructose 1,6-bisphosphate by assay A and assay B. These enzymes have a maximum activity around pH 7.5 in Tris (Cl$^-$) buffer and phosphate buffer, but the enzyme activity is higher in Tris (Cl$^-$) buffer than in phosphate buffer in both assays A and B. This optimum pH is the same as that of a yeast-type aldolase.[3-6]

Inhibitors and Activators. Both cell and spore aldolases are inhibited more than 50% by chelators at the concentration of 0.2 mM supporting that these enzymes are metalloenzymes. Inhibition of these enzymes are also observed for sulfhydryl reagent and Ca^{2+}. As far as the metabolite effects are concerned, some α-keto acids, such as α-ketoglutarate, oxaloacetate, and pyruvate, inhibit this enzyme in the concentration range of 1–5 mM. On the contrary, K$^+$ and NH$_4^+$ ions activate these enzymes

[3] W. J. Rutter, *Fed. Proc., Fed. Am. Soc. Exp. Biol.* **23,** 1284 (1964).
[4] D. E. Morse and B. L. Horecker, *Adv. Enzymol.* **31,** 125 (1968).
[5] B. L. Horecker, O. Tsolas, and C. Y. Lai, *in* "The Enzymes" (P. D. Boyer, ed.), 3rd Ed., Vol. 7, p. 213. Academic Press, New York, 1972.
[6] W. J. Rutter, J. R. Hunsley, W. E. Groves, J. Calder, T. V. Rajkumar, and B. M. Woodfin, this series, Vol. 9, p. 479.

about twofold at the concentration of 50 mM as observed in other class II aldolases. These enzymes are also stimulated four- to fivefold by SH compounds such as cysteine and dithiothreitol (10–20 mM).

Substrate Specificity. These enzymes show a high degree of specificity for fructose 1,6-bisphosphate and show a trace of activity toward fructose 6-phosphate, deoxyribose 5-phosphate and ribose 5-phosphate, as do other class II aldolases.[3-6]

Effect of Sodium Borohydride on Enzyme Activity. These enzymes are not affected by incubation with borohydride in the presence of fructose 1,6-bisphosphate in contrast to the effects seen with rabbit muscle aldolase.[3-6]

Molecular Properties. The molecular weights of those enzymes are 150,000 ± 20,000 by Sephadex G-200 gel filtration and 120,000 ± 5000 by sucrose density gradient centrifugation. The molecular weights of subunits are 30,000 ± 2000 by SDS gel electrophoresis in both vegetative cell and spore aldolases. Hence, these enzymes are assumed to be tetramers.

Others. The properties of vegetative cell and spore aldolases from *B. subtilis* Marburg 168 and *B. licheniformis* NIH 168[7] are almost the same as those *B. subtilis* PCI 219.[8] But there are some differences between the aldolases from *B. subtilis* and *B. cereus* with regard to the molecular weight and electrophoretic mobilities. The molecular weight of *B. cereus* enzyme is 60,000.

[7] S. Ujita, T. Shiroza, and K. Kimura, *J. Biochem.* (*Tokyo*) **83**, 503 (1978).
[8] S. Ujita, *J. Biochem.* (*Tokyo*) **83**, 493 (1978).

[37] Fructose-bisphosphate Aldolases from Mycobacteria

By N. Jayanthi Bai, M. Ramachandra Pai, P. Suryanarayana Murthy, and T. A. Venkitasubramanian

Fructose 1,6-bisphosphate $\rightleftharpoons$
        dihydroxyacetone phosphate + D-glyceraldehyde 3-phosphate

Fructose-bisphosphate aldolase (fructose-1,6-bisphosphate D-glyceraldehyde-3-phosphate-lyase, EC 4.1.2.13) catalyzes the reversible cleavage of fructose bisphosphate (Fru-P$_2$) to glyceraldehyde 3-phosphate and dihydroxyacetone phosphate. The rabbit muscle type class I aldolase functions via the formation of a Schiff base intermediate between the substrate and a lysyl residue at the active site of the enzyme.

They are found in animals, protozoa, algae, and higher plants.[1-3] Class II aldolases do not form a Schiff base intermediate, but contain an essential divalent cation like Zn^{2+} and are found in bacteria, fungi, and blue-green algae.[1] Both these types of aldolases are present in *Euglena*,[4] *Chlamydomonas*,[5] and *Escherichia coli*.[6,7]

Aldolase of surface-grown cells of *M. tuberculosis* $H_{37}Rv$ is of class II type.[8] However, the same bacterium when grown in a fermentor acquires a mixture of both class I and class II aldolases in a proportion 90:10.[9] Contrary to earlier observations, class I aldolases are identified in microorganisms including *Peptococcus aerogenes*,[10] *Lactobacillus casei*,[11] *M. smegmatis*,[12,13] and species of *Staphylococcus* and *Peptococcus*.[14] The aldolase of *M. smegmatis* resembles rabbit muscle aldolase in many of its properties, such as formation of a Schiff base with dihydroxyacetone phosphate and photoinactivation in the presence of methylene blue. However, the *M. smegmatis* aldolase differs from other class I aldolases in that Fru-P_2 cleavage activity is insensitive to digestion with carboxypeptidase A and is unaffected by treatment with sulfhydryl reagents.[13] These differences in properties are found in the class I aldolase of *Peptococcus aerogenes* also.[10]

Assay Method

Principle. The assay for aldolase is based on the estimation of triose phosphate formed from Fru-P_2 and limiting amounts of the enzyme in presence of NADH, excess of triosephosphate isomerase and α-glycerolphosphate dehydrogenase. The triose phosphate formed is converted to α-glycerolphosphate with the concomitant oxidation of NADH to

[1] W. J. Rutter, *Fed. Proc., Fed. Am. Soc. Exp. Biol.* **23**, 1248 (1964).
[2] E. Grazi, T. Cheng, and B. L. Horecker, *Biochem. Biophys. Res. Commun.* **7**, 250 (1962).
[3] R. Fluri, T. Ramasarma, and B. L. Horecker, *Eur. J. Biochem.* **1**, 117 (1967).
[4] Yehchun Mo, B. G. Harris, and R. W. Grays, *Arch. Biochem. Biophys.* **157**, 580 (1973).
[5] G. K. Russell and M. Gibbs, *Biochim. Biophys. Acta* **132**, 145 (1967).
[6] S. A. Baldwin and R. N. Perham, *Biochem. J.* **169**, 643 (1978).
[7] S. A. Baldwin, R. N. Perham, and D. Stribling, *Biochem. J.* **169**, 633 (1978).
[8] N. J. Bai, M. R. Pai, P. S. Murthy, and T. A. V. Subramanian, *Indian J. Biochem. Biophys.* **12**, 181 (1975).
[9] N. J. Bai, M. R. Pai, P. S. Murthy, and T. A. V. Subramanian, *FEBS Lett.* **45**, 68 (1974).
[10] H. G. Lebherz and W. J. Rutter, this series, Vol. 42 [39].
[11] J. London, *J. Biol. Chem.* **249**, 7977 (1974).
[12] N. J. Bai, M. R. Pai, P. S. Murthy, and T. A. V. Subramanian, *Arch. Biochem. Biophys.* **168**, 230 (1975).
[13] N. J. Bai, M. R. Pai, P. S. Murthy, and T. A. V. Subramanian, *Arch. Biochem. Biophys.* **168**, 235 (1975).
[14] F. Goetz, E. Nuernberger, and K. H. Sehleifer, *FEMS Microbiol. Lett.* **5**, 253 (1979).

NAD, which is followed spectrophotometrically. Oxidation of 2 μmol of NADH is taken as equivalent to cleavage of 1 μmol of Fru-P$_2$. Only 1 μmol of NADH is oxidized for each micromole of fructose 1-phosphate (Fru-1-P) cleaved, since only 1 μmol of triose phosphate is formed in this reaction.

Reagents

Glycylglycine buffer, 0.5 M, pH 7.5
D-Fructose 1,6-bisphosphate sodium salt, 0.02 M, pH 7.5 (Sigma)
α-Glycerolphosphate dehydrogenase, 200 units/mg (Sigma)
Triosephosphate isomerase, 2400 units/mg (Sigma)
NADH, disodium salt (Sigma)

Procedure. The reaction mixture consists of 2 μmol of Fru-P$_2$, 0.3 μmol of NADH, 15 μg each of α-glycerolphosphate dehydrogenase and triosephosphate isomerase, 50 μmol of glycylglycine buffer pH 7.5, 0.003–0.03 unit of enzyme, and water in a final volume of 1 ml. The reaction is followed spectrophotometrically at 28° in a 1-ml cuvette having 1-cm light path. Change in absorbance at 340 nm with time is followed until a linear rate is obtained (3–5 min).

Crude extracts showed appreciable activity in the absence of substrate, due presumably to the presence of NADH oxidase activity. Correction for substrate-independent oxidation is necessary only when assaying crude enzyme preparations.

Definition of Unit and Specific Activity. A unit of aldolase activity is the amount of enzyme that catalyzes the cleavage of 1 μmol of Fru-P$_2$ under the conditions of assay. Specific activity is defined as units of activity per milligram of protein. The protein concentration of crude extract is determined by the procedure of Lowry *et al.*,[15] and that of purified fractions by the method of Warburg and Christian[16] or by measuring the absorbance at 280 nm using an extinction coefficient of 0.91 per milligram of protein per milliliter of solution.

Mycobacterium smegmatis Aldolase

Purification Procedure

Step 1. Growth and Harvest of Bacteria. Mycobacterium smegmatis, CDC No. 46, is grown in Youmans and Karlson[17] medium containing per

---

[15] O. H. Lowry, N. J. Rosebrough, A. L. Farr, and R. J. Randall, *J. Biol. Chem.* **193**, 265 (1951).

[16] O. Warburg and W. Christian, *Biochem. Z.* **310**, 384 (1941).

[17] G. P. Youmans and A. G. Karlson, *Am. Rev. Tuberc.* **55**, 529 (1947).

liter, 5 g of L-asparagine, 5.9 g of potassium dihydrogen phosphate, 0.5 g of potassium sulfate, 1.5 g of citric acid, 0.6 g of magnesium carbonate, and 20 ml of glycerol. These ingredients are dissolved in the order listed and made up to 1 liter with distilled water. The pH of the medium is adjusted to 7.0 with 40% potassium hydroxide.

The medium is distributed in 3-liter Haffkine flasks in lots of 1500 ml. Flasks are then plugged with cotton and sterilized at 20 psi for 15 min. The cultures are initiated in 250-ml conical flasks (80 ml of medium per flask) and, after 3 days growth, are transferred to the Haffkine flasks. Cells are grown as a surface culture at 37° and harvested at the mid-log phase (approximately 3 days) by centrifugation at 10,000 g, washed thoroughly with ice-cold water, and stored at $-20°$ until used. The average yield of M. smegmatis is about 10 g wet weight per liter of medium. When stored frozen, the cells retain the aldolase activity for at least 2 months.

Step 2. Crude Extract. Twenty grams of frozen cells are suspended in 100 ml of 0.05 M phosphate buffer containing 1 mM dithiothreitol and 1 mM EDTA. The cell suspension is sonicated using an MSE Ultrasonicator (6 amplitude setting) for 10 min maintaining the temperature at 6°. The solution is then centrifuged at 20,000 g at 0° for 30 min. The precipitate is discarded, and the supernatant is diluted to 100 ml with the buffer and used for further purification.

Step 3. Nucleic acids are removed from the crude cell-free extract by the addition of 1 mg of protamine sulfate for every 10 mg of protein in solution. Precipitated nucleic acids are centrifuged off at 10,000 g for 20 min.

Step 4. Ammonium Sulfate Fractionation. Solid ammonium sulfate (24.3 g/100 ml) is added to the above supernatant solution with stirring. The precipitate is removed by centrifugation at 20,000 g for 30 min. The pH of the supernatant fluid (115 ml) is adjusted to 4.5 by careful addition of 0.2 M acetate buffer, pH 4.0. The mixture is stirred for 10 min and centrifuged at 20,000 g for 20 min. The precipitate is removed, and the supernatant is used for further fractionation.

The supernatant (125 ml) is brought to 80% saturation by adding 25.8 g of ammonium sulfate per 100 ml of solution with constant stirring. The mixture is allowed to stand for 60 min and centrifuged at 20,000 g for 1 hr. The supernatant is discarded and the residue dissolved in 10 ml of 0.05 M phosphate buffer, pH 6.0, containing 1 mM EDTA and 1 mM dithiothreitol.

The enzyme solution is desalted by passing through a column of Sephadex G-25 (2 × 30 cm) previously equilibrated with the phosphate buffer. The enzyme at this stage can be stored at $-20°$ for 1 week without loss of activity.

TABLE I
PURIFICATION OF *Mycobacterium smegmatis* FRUCTOSE-BISPHOSPHATE ALDOLASE

Step	Total protein (mg)	Total activity (units)	Specific activity (units/mg protein)	Purification (fold)	Yield (%)
Crude extract	400	1480	3.7	—	100
Ammonium sulfate fractionation	47	1363	29.0	8	92
DEAE-cellulose chromatography	3.5	1062	303.4	82	72
Sephadex G-150 chromatography	1.15	966	840.0	227	65

Step 5. Chromatography on DEAE-Cellulose. A DEAE-cellulose column (2 × 30 cm) is prepared in 0.05 M phosphate buffer, pH 6.0, and equilibrated with the same buffer containing 0.1 M NaCl. The Sephadex G-25 eluate (12.5 ml) is layered on the column and then washed with 50 ml of the same buffer. Protein is eluted with a linear gradient of 0.1 to 0.5 M NaCl (total volume 400 ml). Fractions (5 ml) are collected at a flow rate of 30 ml/hr. Fractions with maximum activity (usually fractions 41–45) are pooled (25 ml).

Step 6. Chromatography on Sephadex G-150. The pooled fractions from the DEAE-cellulose column are concentrated using Aquacide II (Calbiochem) to a volume of 5 ml and applied on a Sephadex G-150 column (3 × 40 cm) previously equilibrated with 100 ml of 0.05 M phosphate buffer, pH 6.0. The enzyme is eluted with the same buffer at a flow rate of 18 ml/hr, and 2-ml fractions are collected. Fractions containing enzyme activity (usually fractions 77–85) are combined (18 ml) and concentrated (2 ml) using Aquacide II. The enzyme is stable for a period of 3 weeks when stored at −20°.

A typical purification is summarized in Table I.

Properties

Catalytic Specificity and Kinetic Properties. Mycobacterium smegmatis aldolase, like other class I aldolases, can use Fru-1-P as substrate. The K_m values for Fru-P$_2$ and Fru-1-P are calculated to be 2 × 10^{-6} M and 1 × 10^{-2} M, respectively, the corresponding V_{max} values being 14 μmol of Fru-P$_2$ cleaved per minute per milligram of protein and 0.68 μmol of Fru-1-P cleaved per minute per milligram of protein. The Fru-P$_2$: Fru-1-P cleavage activity ratio is 20. Fructose 6-phosphate is not acted upon significantly, and sedoheptulose diphosphate cleavage capacity is 2% of that of Fru-P$_2$

cleavage. Aldol synthesizing capacity from glyceraldehyde 3-phosphate and dihydroxyacetone phosphate is only 1.5% of Fru-P_2 cleavage.

The substrate (Fru-P_2) binding capacity of *M. smegmatis* aldolase (in moles per mole of tetramer) is 2, identical with that of rabbit muscle enzyme, and falls within the range of 1.6–3.8 reported for class I aldolases.

pH Optimum. *Mycobacterium smegmatis* aldolase has a broad pH optimum (pH 7–9). A broad pH optimum has been reported for class I aldolases from *Peptococcus aerogenes* and *E. coli*.

Activators and Inhibitors. Aldolase of *M. smegmatis* is not activated by K^+. The enzyme is also unaffected by Zn^{2+} and Mg^{2+}. Higher concentrations of Mn^{2+}, Co^{2+}, Cu^{2+}, and Fe^{2+} inhibit the enzyme activity.

o-Phenanthroline (0.25 mM) does not inhibit *M. smegmatis* aldolase, but completely inhibits rabbit muscle aldolase. Iodoacetamide, *N*-ethylmaleimide, and *p*-chloromercuribenzoate (5 × 10^{-4} M) also do not affect the enzyme. Dithiothreitol, 2-mercaptoethanol, glutathione, and cysteine (2 mM) have no effect on the enzyme activity.

Bacterial class II aldolases are sensitive to EDTA, but *M. smegmatis* aldolase is insensitive to 1 × 10^{-3} M EDTA.

Carboxypeptidase A treatment resulted in only 20% inactivation, and Fru-P_2 completely protected the enzyme against the inactivation.

Effect of DTNB. Aldolase of *M. smegmatis* possesses two thiol groups that could be modified by DTNB. However, such modification does not result in loss of enzyme activity, indicating the absence of thiol groups at the catalytic site.

Photooxidation for 20 min resulted in 50% inactivation of *M. smegmatis* aldolase.

Molecular Properties. The purified aldolase from *M. smegmatis* is found to be homogeneous by electrophoresis on polyacrylamide gel and cellulose acetate strip. A single symmetrical protein peak, which corresponds to the single activity peak, emerges from the Sephadex G-150 column. The molecular weight by Sephadex G-150 chromatography is calculated to be 158,000. The enzyme is tetrameric, with a subunit molecular weight of 40,000, and contains two types of subunits.

Fructose-bisphosphate Aldolase of *Mycobacterium tuberculosis,* Class II

The aldolase from *M. tuberculosis* $H_{37}Rv$ (NCTC, 7416) grown as surface culture is partially purified and characterized.

All precautionary measures are to be taken while handling the organism because of its pathogenic effects on man.

Purification Procedure

Step 1. Growth and Harvest of Bacteria. *Mycobacterium tuberculosis* $H_{37}Rv$ is grown as a surface culture on Youmans and Karlson medium for 3 weeks (other details are the same as described for *M. smegmatis*). Cells are collected by centrifugation at 10,000 *g*, washed thoroughly with ice-cold water, and stored at $-20°$ until used. Aldolase activity remains stable for 3 months in the frozen cells.

All subsequent steps are carried out at $0-4°$. Phosphate buffer, $0.05 M$, pH 7.0, containing 1 mM 2-mercaptoethanol is used throughout the purification procedure.

Step 2. Crude Extract. Cells (50 g) are suspended in the phosphate buffer and sonicated for 30 min in a Raytheon sonic oscillator, 9 kHz. Cell debris and unbroken cells are removed at 20,000 *g*, and the supernatant is used as crude enzyme.

Step 3. Nucleic acids are precipitated by adding 2% protamine sulfate solution (1 mg of protamine sulfate for every 10 mg of protein). The pH of the mixture is adjusted to 6.0 with 1 *M* ammonium hydroxide and stirred for 20 min; the precipitate is centrifuged off at 20,000 *g*.

Step 4. Ammonium Sulfate Fractionation. The step 3 supernatant is brought to pH 7.0 by the addition of 1 *M* ammonium hydroxide. Solid ammonium sulfate (24.3 g/100 ml) is added over a period of 30 min, during which time the pH is maintained at 7.0. The precipitate is sedimented at 20,000 *g* for 30 min and discarded. To the supernatant solution 25.8 g of ammonium sulfate is added per 100 ml and stirred for 30 min. The precipitate is collected by centrifugation at 20,000 *g* for 1 hr, dissolved in 8 ml of the phosphate buffer, and passed through a Sephadex G-25 column (1.5 × 30 cm). The protein eluate can be stored at $-20°$ without loss of enzyme activity for 4 weeks.

Step 5. Chromatography on DEAE-Cellulose. A DEAE-cellulose column (3 × 30 cm) is equilibrated with the phosphate buffer containing 0.1 *M* NaCl. The desalted protein solution (10 ml) is absorbed on the column and washed with the same buffer until the absorption of the eluate drops below 0.1 at 280 nm. No aldolase activity is detected in the effluent or wash. The enzyme is eluted using a linear gradient of NaCl, 0.1 to 0.5 *M* (500 ml total), at a flow rate of 24 ml/hr. Fractions (4 ml) are collected, and the most active ones (usually fractions 75–81, 28 ml) are concentrated to 5 ml using Aquacide II. Other methods of concentration including ammonium sulfate precipitation resulted in progressive inactivation of the enzyme.

Step 6. Chromatography on Sephadex G-200. The concentrated protein solution is layered on a Sephadex G-200 column (3 × 30 cm) and devel-

TABLE II
PURIFICATION OF SURFACE-GROWN *Mycobacterium tuberculosis* $H_{37}Rv$
FRUCTOSE-BISPHOSPHATE ALDOLASE

Step	Total protein (mg)	Total activity (units)	Specific activity (units/mg protein)	Yield (%)
Crude extract	787	2990	3.8	100
Ammonium sulfate fractionation	158.5	2092	13.2	70
DEAE-cellulose chromatography	9.8	1780	181.2	60
Sephadex G-200 chromatography	6.1	1480	243.1	50

oped with the phosphate buffer at the rate of 12 ml/hr. Three-milliliter fractions are collected; those with maximum enzyme activity (fractions 27–33, 21 ml) are combined and concentrated to 3 ml using Aquacide II. The enzyme is stored at $-20°$ without loss of activity for 4 weeks. A 64-fold purification with 50% recovery of the enzyme is thus achieved. Results of a typical enzyme preparation are summarized in Table II.

Properties

The partially purified aldolase is acid labile and loses 90% of the activity by exposure to pH 5.0 for 30 min. It has a sharp pH optimum of 8.6, a molecular weight greater than 200,000 and electrophoretic mobility only half that of *M. smegmatis* aldolase. K_m for Fru-P_2 cleavage is 0.8×10^{-4} M. The enzyme does not cleave Fru-1-P. o-Phenanthroline and EDTA ($1 \times 10^{-3} M$) completely inactivate the enzyme, but 8-hydroxyquinoline and α,α'-dipyridyl are less potent inhibitors. Sulfhydryl reagents, glutathione, cysteine, dithiothreitol, and 2-mercaptoethanol do not affect the enzyme activity, whereas K^+ elicits a stimulation of 1.7-fold. The aldolase activity is unaffected by treatment with $NaBH_4$ in the presence of dihydroxyacetone phosphate and by digestion with carboxypeptidase A. All these properties point out that aldolase of surface-grown *M. tuberculosis* is a typical bacterial Class II aldolase.

Fructose-bisphosphate Aldolase of *Mycobacterium tuberculosis,* Class I

The aldolase of *M. tuberculosis* $H_{37}Rv$ grown in a fermentor is also partially purified, and the properties are studied.

Purification Procedure

All steps of purification are carried out at 0–4°.

Step 1. Growth and Harvest of Bacteria. *Mycobacterium tuberculosis* $H_{37}Rv$ is grown in a fermentor (aeration 5–10 ml/min; rotation 300–500 rpm; silicone antifoam, 10–15 ml/liter) for 14 days and collected by centrifugation.

A cell-free extract is prepared by sonic disruption of a 20% cell suspension in 0.05 M phosphate buffer, pH 6.0, containing 1 mM 2-mercaptoethanol (other details are the same as for the surface-grown cells).

Step 2. Nucleic acids are precipitated with protamine sulfate (1 mg per 10 mg of protein) and removed.

Step 3. DEAE-Cellulose Chromatography. The step 2 supernatant (50 ml) is applied to a DEAE-cellulose column (3 × 45 cm) equilibrated with phosphate buffer, pH 6.0, containing 1 mM 2-mercaptoethanol. The column is washed with two column volumes of the equilibrating buffer, and the enzyme is eluted with a linear gradient of 0.1 to 0.5 M NaCl (total 800 ml) at a flow rate of 30 ml/hr. Three-milliliter fractions are collected and monitored for protein and aldolase activity. Two activity peaks are obtained; they are designated aldolase I and aldolase II. The fractions with high specific activity for aldolase I (fractions 115–126, 36 ml) and aldolase II (fractions 138–142, 15 ml) are pooled separately.

The pH of aldolase I solution is adjusted to 4.5 with 0.2 M acetate buffer, pH 4.0, and brought to 80% saturation with solid ammonium sulfate (25.8 g/100 ml). The precipitate is collected at 30,000 g for 1 hr, dissolved in 0.05 M phosphate buffer, pH 6.0, containing 1 mM 2-mercaptoethanol and 1 mM EDTA, and desalted by passing through Sephadex G-25.

Step 4. Sephadex Chromatography. The desalted enzyme solution is layered on a Sephadex G-150 column (2.5 × 30 cm) equilibrated with the phosphate buffer (0.05 M, pH 6.0) containing 1 mM EDTA and 1 mM 2-mercaptoethanol. The column is developed with the same buffer at a flow rate of 10 ml/hr. Fractions (2.5 ml) are collected and analyzed for aldolase activity and protein concentration. Fractions (usually 47–52) containing maximum activity are concentrated and stored at −20°. Aldolase I is thus purified by 250-fold with 48% recovery.

Pooled fractions of aldolase II are further purified by Sephadex G-200 sieving (1 × 40 cm) using 0.05 M phosphate containing 1 mM 2-mercaptoethanol. Fractions (1.5 ml) are collected and analyzed for enzyme activity and protein content; those with maximum activity (usually fractions 19–21) are combined.

TABLE III
PURIFICATION OF FERMENTOR-GROWN *Mycobacterium tuberculosis* $H_{37}Rv$
FRUCTOSE-BISPHOSPHATE ALDOLASES

	Aldolase I			Aldolase II	
Step	Total protein (mg)	Specific activity (units/mg protein)	Yield (%)	Total protein (mg)	Specific activity (units/mg protein)
Crude extract	227	2.41	100	ND[a]	ND
DEAE-cellulose chromatography	2	198.00	72	3.78	6.8
Sephadex chromatography	0.44	603.00	48	0.36	19.3

[a] ND, not detected.

Aldolase II of the fermentor-grown cells could be detected only after DEAE-cellulose chromatography.

Results of a typical preparation are summarized in Table III.

Properties

Properties of the aldolase I (major peak) are identical with those of class I aldolase of *M. smegmatis*. The aldolase II (minor peak) behaves like a typical bacterial class II aldolase and possesses properties identical with those of surface-grown cells of *M. tuberculosis*.

A switch from aerobiosis to anaerobiosis alters several of the glycolytic and tricarboxylic acid cycle enzymes. In surface-grown cells of *M. tuberculosis* glycolytic pathway is operative to the extent of 94%[18] and would necessitate the synthesis of a class II aldolase. Growth in a fermentor (aeration) would favor the synthesis of a class I aldolase to facilitate tricarboxylic acid cycle and gluconeogenesis.

[18] T. P. O'Barr and M. V. Rothlauf, *Am. Rev. Respir. Dis.* **101**, 964 (1970).

[38] Fructose-bisphosphate Aldolase from Human Erythrocytes[1]

By DON R. YELTMAN and BEN G. HARRIS

Fructose 1,6-bisphosphate $\rightleftharpoons$
  dihydroxyacetone phosphate + D-glyceraldehyde 3-phosphate

Assay Method

Principle. The assay is based on the oxidation of NADH by dihydroxyacetone phosphate in a coupled system containing triosephosphate isomerase and α-glycerophosphate dehydrogenase.[2]

Reagents

Assay mix: 100 mM Tris-HCl, pH 8.0, 1 mM fructose 1,6-bisphosphate (Fru-1,6-P$_2$), and 0.25 mM NADH
Coupling enzymes: triosephosphate isomerase and α-glycerophosphate dehydrogenase mix, 10 mg/ml

Procedure. One milliliter of assay mix and 10 μl of coupling enzyme mix are placed into a cuvette with a 1-cm light path and equilibrated to temperature in a spectrophotometer at 30°. The reaction is initiated by the addition of aldolase, and the rate of oxidation of NADH is monitored by a recording spectrophotometer at 340 nm. Two moles of NADH are oxidized by the conversion of 1 mol of Fru-1,6-P$_2$ to products.

Definition of Unit. One unit of aldolase activity is defined as the amount of enzyme required to catalyze the cleavage of 1 μmol of Fru-1,6-P$_2$ per minute at 30°. Specific activity is defined as units per milligram of protein.

Protein Determination. Protein concentrations are routinely monitored by the method of Bradford[3] using bovine serum albumin as standard.

Purification Procedure

The following operations are carried out at 4°.

Washing of Erythrocytes and Hemolysis. Recent outdated whole blood is sedimented at 1460 g for 10 min in a Sorvall GSA rotor, and the plasma and buffy coat are carefully removed by aspiration. The erythrocytes are resuspended in 4 volumes of 25 mM Tris-HCl, pH 7.8, and 0.1 M

---

[1] D-Fructose 1,6-bisphosphate: D-glyceraldehyde 3-phosphate-lyase, EC 4.1.2.13.
[2] R. Racker, *J. Biol. Chem.* **167**, 843 (1947).
[3] M. Bradford, *Anal. Biochem.* **72**, 248 (1976).

METHODS IN ENZYMOLOGY, VOL. 90

NaCl; the above process is repeated to remove any remaining buffy coat fragments. Two additional washes are performed with the same isotonic Tris buffer. The final centrifugation is carried out at 5000 g for increased packing of the erythrocytes.

Hemolysis is performed by resuspension of the packed cells in five volumes of 25 mM Tris-HCl, pH 7.8 and 10 mM 2-mercaptoethanol, and the suspension is stirred gently for 15 min.

Batch Process. Fresh hemolysate is mixed with phosphocellulose (exchange capacity = 0.95 meq/g) that has been equilibrated with the lysing buffer. A ratio of 1 g of damp phosphocellulose cake to 10 ml of hemolysate is usually used, although this value varies depending upon different phosphocellulose exchange capacities. The phosphocellulose–hemolysate suspension is gently stirred for 10 min and vacuum filtered; the filtrate is discarded. Essentially all the aldolase present in the hemolysate binds to phosphocellulose under these conditions.

Using the same process of resuspension, stirring, and filtration described above, the phosphocellulose is washed twice with 25 mM Tris-HCl, pH 7.8, 10 mM 2-mercaptoethanol; twice with 50 mM Tris-HCl, pH 7.8, 10 mM 2-mercaptoethanol; and twice with 50 mM Tris-HCl, pH 8.4, 10 mM 2-mercaptoethanol. This washing process removes approximately 98% of the hemoglobin while allowing most of the aldolase to remain bound to the cellulose ion exchanger. Between each of the above buffer changes, the cellulose is washed once with a 1:1 mixture of the two transition buffers. This gradual change in ionic strength and pH prevents release of aldolase that is caused by sudden changes in buffer pH or ionic strength.

A substrate elution of the aldolase is carried out by suspension of the phosphocellulose in the final wash buffer containing 2 mM Fru-1,6-P_2. The filtrate, which contains virtually all the aldolase that remains bound to the phosphocellulose, is dialyzed against saturated ammonium sulfate (3.94 M at 4°) in 50 mM triethanolamine-HCl, pH 7.3. The buffering of the ammonium sulfate solution is necessary because filtrate-induced pH changes approaching 8.0 causes aldolase denaturation during this step.

The precipitated protein is collected by centrifugation, resuspended, and dialyzed in 25 mM Tris-HCl, pH 7.3, 1 mM EDTA, 10 mM 2-mercaptoethanol.

Column Chromatography. The resuspended eluent from the batch process is applied to a phosphocellulose column (2.5 × 40 cm) equilibrated with 25 mM Tris-HCl, pH 7.8, 10 mM 2-mercaptoethanol. A wash process similar to that used in the batch phosphocellulose step is used to remove final traces of hemoglobin. Prior to the elution step, the column is washed with one volume of 1 mM NADH in 50 mM Tris-HCl, pH 8.4, without

ISOLATION OF FRUCTOSE-BISPHOSPHATE ALDOLASE FROM HUMAN ERYTHROCYTES

Fraction	Volume (ml)	Total units	Total protein (mg)	Specific activity (units/mg protein)	Yield (%)	Purification (fold)
Hemolysate	2000	240	116,000	0.002	(100)	(1)
Phosphocellulose batch	350	115	245	0.42	48	210
Phosphocellulose chromatography	10	80	6	11.4	33	5700
Reverse ammonium sulfate fractionation	3	48	3	16.1	20	8000

2-mercaptoethanol. This is effective in removing glyceraldehyde-3-phosphate dehydrogenase, which is a persistent contaminant until this wash is performed.

The aldolase is eluted with one column volume of 50 mM Tris-HCl, pH 8.4, 10 mM 2-mercaptoethanol, and 2 mM Fru-1,6-P$_2$. The eluted aldolase is again precipitated by dialysis against saturated ammonium sulfate at pH 7.3.

Reverse Ammonium Sulfate Fractionation. The final step involves fractionation[4] of 0.15 M decrements from 2.56 M to 1.27 M, with the bulk of the aldolase salting in at 1.92 M ammonium sulfate.

A typical purification procedure representing the washed erythrocytes from four units of whole blood is shown in the table. The process results in a 8000-fold purification and a specific activity of 16 units per milligram of protein. At this point the enzyme is stable for several months when stored as a crystalline suspension in ammonium sulfate at 4° (pH 7.3). Cell count, specific activity, and molecular weight parameters give an average value of 35,000 aldolase molecules per red blood cell.[5]

Properties of Human Erythrocyte Aldolase

Isoelectric focusing revealed the presence of one form of aldolase in the erythrocyte with an isoelectric point of 8.9.[5] The enzyme exhibits an $s_{20,w}^0$ value of 7.78 × 10^{-13} sec. Sedimentation equilibrium ultracentrifugation and gel filtration chromatography revealed a native molecular weight of 158,000, and ultracentrifugation in the presence of 6 M guanidinium chloride yielded a subunit molecular weight of 40,000; SDS-gel electrophoresis also gave a subunit molecular weight of 40,000. The enzyme,

---

[4] W. Jacoby, this series, Vol. 22 [23].

[5] D. R. Yeltman and B. G. Harris, *Biochim. Biophys. Acta* **484**, 188 (1977).

like other mammalian aldolases,[6] appears to be a tetramer with identical or near identical subunits. The enzyme has a Stokes' radius of 4.56 nm, a diffusion coefficient ($D_{20,w}$) of 4.68×10^{-7} cm^2/sec, and a frictional ratio of 1.27. The apparent K_m value for Fru-1,6-P$_2$ is 7.1×10^{-6} M, and for Fru-1-P it is 3.0×10^{-3} M; the V_{max} values for the two substrates are 16.1 and 0.4 units/mg, respectively. Therefore, the Fru-1,6-P$_2$: Fru-1-P ratio is about 40.

The amino acid composition[5] is very similar to that of the human heart[7] and rabbit muscle[8] enzymes. The human erythrocyte aldolase appears to be a type A isozyme or muscle-type aldolase.[5]

[6] K. Kawahara and C. Tanford, *Biochemistry* **5**, 1578 (1966).
[7] B. L. Allen, R. W. Gracy, and B. G. Harris, *Arch. Biochem. Biophys.* **155**, 325 (1973).
[8] C. Lai, N. Nakai, and D. Chang, *Science* **183**, 1204 (1974).

[39] Fructose-bisphosphate Aldolase from *Ascaris suum*

By MARIAN KOCHMAN and DANUTA KWIATKOWSKA

Fructose 1,6-bisphosphate $\rightleftharpoons$
dihydroxyacetone phosphate + D-glyceraldehyde 3-phosphate
Fructose 1-phosphate $\rightleftharpoons$ dihydroxyacetone phosphate + D-glyceraldehyde

Parasitic roundworms such as *Ascaris* have presumably evolved from free-living roundworms and have adapted to life in a rather anaerobic environment by developing a high glycolytic rate. Glycolytic enzyme activities and levels of glycolytic substrates have been measured in *Ascaris* tissues.[1,2] Aldolase from *Ascaris suum* muscle (EC 4.1.2.13) was first studied to determine whether the enzyme could induce protective antibody responses in swine against parasitic infection.[3,4] *Ascaris suum* aldolase has been purified to homogeneity and extensively characterized.[5,6] The procedures described below are based upon the study of Kochman and Kwiatkowska.[5]

[1] V. M. L. Srivastava, S. Ghatak, and C. R. Krishna Murti, *Parasitology* **60**, 157 (1970).
[2] J. Barrett and I. Beis, *Comp. Biochem. Physiol. B* **44**, 751 (1973).
[3] N. K. Mishra, Ph.D. Thesis, University of Nebraska, Lincoln (1970).
[4] N. K. Mishira and C. L. Marsh, *Exp. Parasitol.* **33**, 89 (1973).
[5] M. Kochman and D. Kwiatkowska, *Arch. Biochem. Biophys.* **152**, 856 (1972).
[6] J. R. Dedman, A. C. Lycan, R. W. Gracy, and B. G. Harris, *Comp. Biochem. Physiol. B* **44**, 291 (1973).

Assay Method

Principle. The assay is based on the oxidation of NADH by dihydroxyacetone phosphate (DHAP) in a coupled system containing triosephosphate isomerase and α-glycerophosphate dehydrogenase.[7]

Reagents

Tris-HCl buffer: A stock solution is prepared containing 50 mM Tris and 5 mM EDTA (adjusted to pH 7.5 with HCl).

Fructose bisphosphate, sodium salt (Sigma), 100 mM, pH 7.5

NADH (Sigma), 10 mg/ml in 1 mM NaOH, stored frozen at $-5°$

α-Glycerophosphate   dehydrogenase–triosephosphate   isomerase (Boehringer), 12 mg/ml crystalline suspension. For kinetic measurements a salt-free preparation is used.

Assay Mixture. To 9.5 ml Tris-HCl solution add 500 μl of fructose 1,6-bisphosphate (Fru-1,6-P$_2$) solution, 10 μl α-glycerophosphate dehydrogenase–triosephosphate isomerase suspension, and 100 μl of NADH (or 1 mg of solid NADH, to give $A_{340} \simeq 0.9$ to 1.0). This assay mixture is prepared fresh and kept at room temperature.

Procedure. One milliliter of assay mixture is measured into a 1.0-ml cuvette with a 1.0-cm light path. The enzyme is diluted in Tris-HCl solution to contain 0.3–0.9 unit/ml, and 10–20 μl of diluted enzyme is added to the cuvette and mixed rapidly. The cuvette is then placed in a spectrophotometer with a cuvette chamber thermostatted at 25°, and the change in absorbance at 340 nm is measured. The reaction rate is calculated from the linear portion of the curve. The reaction rate is expressed as micromoles of Fru-1,6-P$_2$ cleaved per minute, with 6.22×10^3 as the molar absorbance coefficient of NADH. In this assay each mole of Fru-1,6-P$_2$ cleaved results in the oxidation of 2 mol of NADH.

For kinetic measurements use 50 mM Tris-HCl buffer (pH 7.5) containing 0.6 mg of bovine serum albumin (Sigma) per milliliter. Salt-free aldolase solution is diluted in the above buffer (i.e., 3 μg/ml or 0.6 mg/ml for determination of K_m Fru-1,6-P$_2$ or K_m Fru-1-P, respectively), and then 20–30 μl of diluted enzyme is added to 2.97 ml of assay mixture containing albumin and variable amounts of substrate. The 0.05 or 0.1 absorbance scale is used. For estimation of the K_m Fru-1,6-P$_2$ value (which is rather low), the integrated method can be used.[8]

Definition of Unit and Specific Activity. A unit of activity is defined as the cleavage of 1 μmol of substrate per minute under the conditions de-

[7] R. Blostein and W. J. Rutter, *J. Biol. Chem.* **238**, 3280 (1963).
[8] A. A. Klesow and J. W. Beresin, *Biokhimiya* **37**, 170 (1972).

scribed above. Specific activity is defined as units of aldolase activity per milligram of protein. Aldolase concentration is estimated spectrophotometrically by using $E_{280}^{1\%} = 10.5$.[9]

Purification Procedure

Prepare 20 liters of stock buffer (10 mM Tris-HCl, 1 mM EDTA, pH 7.5 at 5°). All purification procedures are performed at 5° or in an ice bath.

Step 1. Extraction. Live ascarids are collected from an abattoir and washed in tap water. The animals can be kept overnight in buffered physiological saline containing 0.1% glucose.[6] The body walls of the worms are incised longitudinally, and the internal organs are discarded. The body walls, which contain cuticle and muscle, are washed with distilled water and collected in a beaker on ice. The tissue (450 g) is suspended in an equal volume of cold stock buffer containing 5 mM 2-mercaptoethanol and is homogenized in a blender for 1 min. The homogenate is centrifuged at 30,000 g for 40 min, and the precipitate is discarded. [Steps 1 and 2 should be performed in a hood in order to avoid developing hypersensitivity to volatile antigens present in ascarid tissue. Hands and eyes should be protected by gloves and goggles, respectively.]

Step 2. Ammonium Sulfate Precipitation. The opalescent supernatant solution is made 1.8 M with ammonium sulfate by the addition of 27.8 g of solid ammonium sulfate to 100 ml of extract over a 2-hr period. The pH is then adjusted to 7.5 with 6 M ammonium hydroxide. The solution is allowed to stand for 30 min and then centrifuged at 30,000 g for 30 min. To each 100 ml of supernatant fluid, 6.4 g of ammonium sulfate are added over a period of 1.5–2 hr to yield a final concentration of 2.2 M ammonium sulfate. The pH is again adjusted to 7.5, and the solution is allowed to stand overnight. The precipitate is collected by centrifugation at 30,000 g for 40 min. This precipitate is dissolved in stock buffer in order to obtain a final protein concentration between 15 and 20 mg/ml. Any insoluble material is removed by centrifugation. The clear supernatant solution is passed through a column of Sephadex G-25 coarse (4 × 100 cm) equilibrated with the same stock buffer.

Step 3. Substrate Elution from Phosphocellulose Column. The desalted solution is applied to a column of phosphocellulose (Sigma, 3 × 45 cm) equilibrated with stock buffer. The column is washed with 1.5–2 liters of the above buffer until the absorbance of the eluent is less than 0.05 at 280 nm. Substrate elution is then carried out with stock buffer containing 1 mM Fru-1,6-P$_2$. The aldolase is eluted in a single peak with a maximum

[9] J. R. Dedman, R. W. Gracy, and B. G. Harris, *Comp. Biochem. Physiol. B* **49**, 715 (1974).

PURIFICATION OF *Ascaris suum* MUSCLE FRUCTOSE-BISPHOSPHATE ALDOLASE[a]

Steps	Total protein (mg)	Total activity (units)	Specific activity (units/mg protein)	Purification (fold)	Recovery (%)
1. Crude extract	16,000	864	0.054	1	100
2. 1.8–2.2 M $(NH_4)_2SO_4$ precipitate	2,228	713	0.32	6	82
3. Phosphocellulose chromatography	37	377	10.2	188	43
4. DEAE-Sephadex A-50 chromatography	27	283	10.5	194	32
5. Crystallization	21	224	10.7	200	26

[a] Reproduced from Kochman and Kwiatkowska.[5] Ammonium sulfate concentrations have been converted to molarity.

specific activity of 8–10 μmol of Fru-1,6-P_2 cleaved per milligram per minute.

Step 4. DEAE-Sephadex Chromatography. The aldolase-containing fractions from the phosphocellulose column with specific activity greater than 2 are combined and applied to a column of DEAE-Sephadex A-50 (1.3 × 60 cm) previously equilibrated with the stock buffer. Elution is carried out with 600 ml of a linear NaCl gradient (0–0.4 M NaCl) in the same buffer. Two protein and aldolase activity peaks are eluted. The first narrow peak, which presumably contains the most homogeneous aldolase preparation (of specific activity in the range 10–11 units), is collected.

Step 5. Crystallization. The pooled aldolase fractions from the DEAE-Sephadex column are dialyzed against 2.5 M ammonium sulfate in stock buffer. The precipitate formed overnight is collected by centrifugation and then dissolved in a minimal amount of the buffer. Solid ammonium sulfate is added slowly until the solution becomes opalescent. It is immediately centrifuged, and the small precipitate is discarded. After standing overnight at 0–4°, crystals in the shape of small needles are formed. Sometimes amorphous material is formed, which slowly changes into crystalline form within a few days. Crystallization can be accelerated by the addition of ascarid aldolase crystals.

The purification procedure is summarized in the table.

Properties

Homogeneity. Crystalline ascarid muscle aldolase is homogeneous by polyacrylamide gel electrophoresis. Its electrophoretic mobility by cellulose acetate electrophoresis (pH 8.3) is slightly anodic, opposite to the mobility of vertebrate aldolases A or B. Vertebrate aldolase C exhibits almost double the anodic mobility rate of the ascarid enzyme. Sodium dodecyl sulfate–gel electrophoresis reveals one protein band. Electrophoresis of S-carboxymethylated ascarid aldolase (dissociated in 8 M urea) indicates that molecules of this enzyme are composed of two types of polypeptide chains. This heterogeneity, by analogy to rabbit muscle aldolase, may be due to deamidation of a single asparagine residue.[10]

Molecular Properties. The enzyme has a molecular weight of approximately 155,000–160,000 and is composed of four subunits of identical molecular weight.

Amino Acid Composition. Ascarid aldolase exhibits a high degree of homology in amino acid composition to vertebrate muscle aldolase.[9] Similarities between the ascarid, pig, and rabbit muscle enzymes are also apparent from tryptic peptide fingerprints.

Free-living nematode *Turbatrix aceti* aldolase differs in amino acid composition from *Ascaris suum* to the same extent as it differs from that of aldolases from mammals.[11] A convergence by the ascarid enzyme toward a structure similar to the mammalian form has been postulated.[9]

Functional Amino Acids. The enzyme has a C-terminal functional tyrosine residue and is inactivated by incubation with $NaBH_4$ in the presence of substrate, indicating a Schiff base formation between dihydroxyacetone phosphate and a lysyl residue.

Catalytic Properties. Ascarid aldolase has a specific activity for Fru-1,6-P_2 cleavage close to 11, which can be reduced by 95% after removal of C-terminal tyrosine residues. K_m values are $1 \times 10^{-6} M$ and $2 \times 10^{-3} M$ for Fru-1,6-P_2 and Fru-1-P, respectively. The Fru-1,6-P_2 : Fru-1-P cleavage activity ratio measured at 5 mM substrate concentration (which is nonsaturated for Fru-1-P) is 40.[5]

Immunological studies indicate lack of common immunological determinants between rabbit aldolases A and B and ascarid aldolase. A mixture of ascarid aldolase (As_4) and rabbit muscle aldolase (A_4) after acid or SDS dissociation and reconstitution results in one active hybrid As_2A_2 with electrophoretic mobility intermediate between those of the parental aldolases.

---

[10] C. Y. Lai and B. L. Horecker, *J. Cell. Physiol.* **76**, 381 (1970).
[11] A. Z. Reznick and D. Gershon, *Int. J. Biochem.* **8**, 53 (1977).

[40] Fructose-bisphosphate Aldolase from *Helix pomatia*

By MARIAN KOCHMAN, PAUL A. HARGRAVE, and JANINA BUCZYŁKO

Fructose 1,6-bisphosphate $\rightleftharpoons$
  dihydroxyacetone phosphate + D-glyceraldehyde 3-phosphate
Fructose 1-phosphate $\rightleftharpoons$ dihydroxyacetone phosphate + D-glyceraldehyde

Fructose-bisphosphate aldolase [EC 4.1.2.13] from molluscs has been purified and characterized only from calamary muscle[1] and from snail.[2,3] We present here the purification and properties of the enzyme from the commercially available snail *Helix pomatia*.[2,3]

Assay Method

Snail aldolase activity is measured at 25° as described for ascarid aldolase[4] except that 100 mM Tris-HCl buffer containing 1 mM EDTA, pH 7.5, is used. The final concentration of the substrate, fructose 1,6-bisphosphate (Fru-1,6-P_2), in the assay mixture is 2 mM. Aldolase concentration is measured assuming the same extinction coefficient as for rabbit aldolase A, $E_{280}^{1\%} = 9.38$.[5]

The kinetics are measured as described for ascarid aldolase.[4]

Purification Procedure

The isolation of snail aldolase is based on a general substrate elution procedure developed for Fru-1,6-P_2 aldolase purification.[6] All steps of the purification procedure are performed at 4–6°. The pH measurements are performed on cold (5°) solutions.

Step 1. Tissue Extraction. Foot muscles of fresh snails are cut off and immersed in liquid nitrogen. The muscles can be stored for months at −20° prior to use. Frozen muscles are thawed by immersion in stock

[1] J. C. Mareschal, E. Schonne, R. R. Crichton, and D. A. Strosberg, *FEBS Lett.* **54**, 97 (1975).

[2] J. Buczyłko, P. A. Hargrave, and M. Kochman, *Comp. Biochem. Physiol. B* **67**, 225 (1980).

[3] J. Buczyłko, P. A. Hargrave, and M. Kochman, *Comp. Biochem. Physiol. B* **67**, 233 (1980).

[4] See this volume [39].

[5] J. W. Donovan, *Biochemistry* **3**, 67 (1964).

[6] E. E. Penhoet, M. Kochman, and W. J. Rutter, *Biochemistry* **8**, 4391 (1969).

buffer (10 mM Tris-HCl, pH 7.5, containing 1 mM EDTA). Muscles (600–700 g) free of mucous foam are ground in meat grinder. The minced muscle is suspended in 2 volumes of stock buffer containing 10 mM 2-mercaptoethanol. After 20 min of extraction, the suspension is centrifuged at 11,000 g for 1 hr.

Step 2. Ammonium Sulfate Fractionation. The opalescent supernatant is made 1.2 M with ammonium sulfate by addition of 175 g of solid ammonium sulfate to 1 liter of solution over a period of 1 hr. The pH is then adjusted to 7.5 with 6 M ammonium hydroxide, and the solution is slowly stirred for 30 min to complete protein precipitation, followed by centrifugation at 11,000 g for 1 hr. The precipitate is discarded, and the resulting supernatant is brought to 2.2 M in ammonium sulfate (by the addition of 160 g of this reagent per 1 liter of supernatant) over a period of 90 min. The pH is adjusted to 7.5, and the solution is allowed to stand overnight. The precipitate is then collected by centrifugation at 11,000 g for 50 min. The 1.2–2.2 M ammonium sulfate precipitate is dissolved in a small volume of stock buffer containing 5 mM 2-mercaptoethanol. Approximately 100 ml of the above solution is dialyzed against 3–4 liters of this buffer, which is changed five times over a 24-hr period.

Step 3. DEAE-Cellulose Chromatography. The desalted viscous solution is applied to a column of DEAE-32 Cellulose (Whatman, 4 × 50 cm) equilibrated with 10 mM Tris-HCl, pH 7.8, containing 1 mM EDTA. The column is first washed with 800 ml of Tris-HCl buffer until the A_{280} of the effluent drops below 0.5. Elution is carried out with 900 ml of a linear NaCl gradient (0 to 0.4 M NaCl) in the above buffer. Aldolase activity appears in the ascending part of the first broad protein peak. The fractions with activity more than 0.4 unit/ml are collected (200–300 ml) and immediately passed through a column of Sephadex G-25 coarse (6 × 100 cm) equilibrated with stock buffer.

Step 4. Phosphocellulose Chromatography. Desalted fractions from the DEAE-cellulose column are applied to a phosphocellulose column (1.8 × 35 cm) equilibrated with the stock buffer. The column is washed with 700 ml of the same buffer until the A_{280} of the effluent drops below 0.1. Substrate elution is then carried out with 100 ml of stock buffer containing 2.5 mM Fru-1,6-P$_2$. Aldolase is eluted as a single sharp peak.

Step 5. DEAE-Sephadex Chromatography. The aldolase-containing fractions from the phosphocellulose column with specific activities greater than 7 units/mg are combined and applied to a column of Sephadex A-50 (0.8 × 60 cm) previously equilibrated with the stock buffer containing 5 mM 2-mercaptoethanol. Elution is carried out with 180 ml of a linear NaCl gradient (0 to 0.4 M NaCl) in the same buffer. A small amount of aldolase appears in the breakthrough peak or at the beginning of the gradient, and

PURIFICATION OF *Helix pomatia* MUSCLE ALDOLASE[a]

Step	Total protein (mg)	Specific activity (units/mg protein)	Total activity (units)	Purifi- cation (fold)	Recovery (%)
1. 11,000 g supernatant of crude extract (from 700 g of muscle)	24,200	0.036	880	1	100
2. 1.2–2.2 M (NH$_4$)$_2$SO$_4$ precipitate	21,150	0.04	846	1.1	96
3. DEAE-cellulose chromatography	3,160	0.14	450	3.9	51
4. Phosphocellulose chromatography	19	12.1	230	336	26
5. DEAE-Sephadex A-50 chromatography	10	15.2	152	442	17

[a] Reproduced from Buczyłko et al.[2] with permission of the publishers. Ammonium sulfate concentrations have been converted to molarity.

this rather diffuse peak can be discarded. The major sharp protein and aldolase activity peak appears at 0.09 M NaCl. Aldolase fractions from this peak with specific activity greater than 12 units/mg (presumably containing the most homogeneous material) are combined and then dialyzed against a solution of 2.74 M ammonium sulfate in stock buffer containing 10 mM 2-mercaptoethanol. The precipitated aldolase with a specific activity of 13.5–16 units/mg can be stored as an ammonium sulfate suspension at 4° for months without loss of activity.

A typical purification is summarized in the table.

Notes on the Purification Procedure.[7] The DEAE-cellulose step is essential in order to separate the greater part of the mucous material from aldolase. In the presence of this mucous material, aldolase does not bind to phosphocellulose. After phosphocellulose chromatography it is best to limit the time of exposure of aldolase to substrate, since during prolonged incubation a slow inactivation of aldolase can be observed.

[7] The purification procedure may be used successfully for purification of bovine or porcine aldolase C with only minor modifications: in step 1, the fresh tissue is homogenized in 1.5 volumes (v/w) of stock buffer containing 0.1 M phenylmethanesulfonyl fluoride; in step 2, aldolase is precipitated between 1.8 and 2.43 M (NH$_4$)$_2$SO$_4$; in step 3, the DEAE-cellulose column is washed with 1.5 liters of Tris-HCl–EDTA buffer until the red band is eluted and A_{280} of the eluate drops below 0.1; and in step 4 the phosphocellulose column is equilibrated with stock buffer at pH 7.3. This procedure yields aldolase C with specific activity of 9 units/mg and containing 0.92 C-terminal tyrosines per subunit.[7a]

[7a] J. Buczyłko, K. Palczewski, and M. Kochman, *Int. J. Biochem.*, in press.

Properties

Homogeneity. The purified aldolase is homogeneous by poly-acrylamide gel electrophoresis, cellulose acetate electrophoresis, or chromatography on Sephadex G-200.

Molecular Properties. Helix pomatia aldolase is a tetramer of 40,000 subunit molecular weight. It exhibits a stronger tendency for disulfide cross-linking in the absence of reducing agents (2-mercaptoethanol) than does rabbit aldolase A,[8] yielding less active material of 320,000 molecular weight. Like all class I aldolases,[9] the snail aldolase is inactivated by incubation with $NaBH_4$ in the presence of substrate, thus suggesting the presence of a lysine residue in the active site. Snail aldolase contains more glycine, phenylalanine, and dicarboxylic amino acids than rabbit muscle aldolase A. Comparison of amino acid composition based upon composition coefficient (CC)[10] suggests that snail aldolase is related to rabbit aldolase C > A > B. Among invertebrates, aldolase from *Turbatrix aceti*[11] exhibits a high CC to the snail enzyme, whereas class I bacterial aldolases[12–14] and class II metalloaldolases[15,16] exhibit poor similarities. Glycine rather than proline is the N terminus of snail aldolase. It has a C-terminal functional tyrosine residue. Digestion of snail aldolase with carboxypeptidase A results in loss of activity toward Fru-1,6-P$_2$ and fructose 1-phosphate (Fru-1-P) by 95 and 55%, respectively. In contrast to rabbit aldolases A[17] and C,[18] tyrosine is the only amino acid that is released. There is no linear relationship between C-terminal tyrosine digestion and activity loss, analogous to behavior of the lobster enzyme,[19] indicating that snail aldolase subunits are not functionally equivalent.

Snail muscle aldolase forms active interspecies hybrids with mammalian aldolases A, B, and C after acid dissociation and reassociation of the mixture of two parental homomeric forms. The distribution of reconstituted tetramers is not statistical. Hybridization of snail muscle aldolase (Sn_4) and rabbit liver aldolase (B_4) yields homomers Sn_4, $B_4 >$ symmetri-

[8] M. T. Mas, J. Buczyłko, and M. Kochman, *Eur. J. Biochem.* **100**, 393 (1979).
[9] W. J. Rutter, *Fed. Proc., Fed. Am. Soc. Exp. Biol.* **23**, 1248 (1964).
[10] J. R. Dedman, R. W. Gracy, and B. G. Harris, *Comp. Biochem. Physiol. B* **49**, 715 (1974).
[11] A. Z. Reznick and D. Gershon, *Int. J. Biochem.* **8**, 53 (1977).
[12] H. G. Lebherz, R. A. Bradshaw, and W. J. Rutter, *J. Biol. Chem.* **248**, 1660 (1973).
[13] S. A. Baldwin and R. N. Perham, *Biochem. J.* **169**, 643 (1978).
[14] D. Stribling and R. N. Perham, *Biochem. J.* **131**, 833 (1973).
[15] C. E. Harris, R. D. Kobes, D. C. Teller, and W. J. Rutter, *Biochemistry* **8**, 2442 (1969).
[16] B. L. Horecker, O. Tsolas, and C. Y. Lai, *in* "The Enzymes" (P. D. Boyer, ed.), 3rd Ed., Vol. 7, p. 242. Academic Press, New York, 1972.
[17] J. A. Winstead and F. Wold, *J. Biol. Chem.* **239**, 4212 (1964).
[18] E. E. Penhoet, M. Kochman, and W. J. Rutter, *Biochemistry* **8**, 4396 (1969).
[19] A. Guha, this series, Vol. 42 [35].

cal heteromer $Sn_2B_2 \geqslant$ asymmetrical heteromers Sn_3B, SnB_3, indicating that intersubunit contacts are not equivalent and have changed during evolution. In contrast, a pattern of reconstituted molecules approaching statistical distribution is observed for hybridization of rabbit aldolase A with snail muscle aldolase.[20]

Catalytic Properties. *Helix pomatia* aldolase has specific activity similar to that of rabbit muscle aldolase. V_{max}(Fru-1,6-P$_2$) = 2500 mol of Fru-1,6-P$_2$ cleaved per mole of enzyme per minute at 25°. Its K_m value is very low: K_m(Fru-1,6-P$_2$) = 0.3 μM, K_m(Fru-1-P) = 1 mM, and the Fru-1,6-P$_2$: Fru-1-P activity ratio = 20 at 5 mM substrate concentration, which is nonsaturated for Fru-1-P.

Adenine nucleotides are competitive inhibitors of snail aldolase: K_I(ATP) = 0.2 mM, K_I(ADP) = 0.5 mM, K_I(AMP) = 1.1 mM.[20] These values differ markedly from that found for rabbit aldolase B[21] > aldolase A[22] and are more similar to those for aldolase C.[7]

Spectroscopic Properties. The snail aldolase ultraviolet absorption spectrum exhibits a maximum at 278 nm and a minimum at 250 nm, with $A_{280} : A_{250} = 2.6$. The comparable value for aldolase A is $A_{280} : A_{250} = 3.2$.[23]

There is no major difference in secondary structure of snail and rabbit muscle aldolase judged by their optical rotatory dispersion and circular dichroism spectra.[3,24,25] *Helix* seems to be a dominant structural element in both aldolases; however, the amount of α and β structure seems to be slightly lower in the enzyme from snail than from rabbit muscle.[3]

[20] J. Buczyłko and M. Kochman, unpublished results.
[21] A. A. Kasprzak and M. Kochman, *J. Biol. Chem.* **256**, 6127 (1981).
[22] A. A. Kasprzak and M. Kochman, *Eur. J. Biochem.* **104**, 443 (1980).
[23] T. Baranowski, *J. Biol. Chem.* **180**, 535 (1949).
[24] L. S. Hsu and K. E. Neet, *J. Mol. Biol.* **97**, 351 (1975).
[25] M. E. Magar, *J. Biol. Chem.* **242**, 2517 (1967).

[41] 2-Keto-3-deoxy-galactonate-6-phosphate Aldolase from *Pseudomonas saccharophila*

By H. PAUL MELOCHE and EDWARD L. O'CONNELL

2-Keto-3-deoxy-6-phosphogalactonate $\rightleftharpoons$ pyruvate + D-glyceraldehyde 3-phosphate

2-Keto-3-deoxy-galactonate-6-phosphate (KDPGal) aldolase (EC 4.1.2.21) is found in extracts of galactose-grown *Pseudomonas saccharophila* along with 2-keto-3-deoxy-gluconate-6-phosphate (KDPG) al-

METHODS IN ENZYMOLOGY, VOL. 90

dolase (EC 4.1.2.14). Previously published methods[1] for the separation of the two aldolases from this organism have been difficult to reproduce in this and other laboratories. Consequently the problem was restudied. As reported here, the two enzymes can be separated by chromatography on DEAE-cellulose under conditions wherein the bulk of KDPGal aldolase activity is found in the column breakthrough while much of the activity toward KDPG is retarded. Further separation is achieved by taking advantage of the instability of KDPG aldolase under mildly alkaline conditions. In contrast, KDPGal aldolase is completely stable under the conditions employed.

Assay

Reagents

Disodium NADH (Sigma Chemical Co., St. Louis, Missouri)
Imidazole buffer, 1 M, pH 7
Rabbit muscle lactate dehydrogenase
2-Keto-3-deoxy-6-phospho-D-galactonate   (KDPGal)   sodium   or lithium salt, 25 mM

KDPGal is synthesized by the net enzyme-catalyzed condensation of pyruvate and D-glyceraldehyde 3-phosphate adopting the previously reported method for 2-keto-3-deoxy-6-phosphogluconate,[2] KDPG was synthesized using 6-phosphogluconate dehydratase.[3]

Procedure. A stock solution is prepared by dissolving 10 mg of disodium NADH in 0.5 ml of imidazole buffer and 9.5 ml of distilled water. This solution is then used to dissolve ca. 1.5 mg (1200 IU) of rabbit muscle lactate dehydrogenase that had been pelleted by centrifugation to remove the ammonium sulfate component of the crystalline suspension.

For the assay, 50 μl of the above stock solution is combined with 100 μl of water and 10 μl of 25 mM KDPGal in a 0.5-ml microcuvette of 1-cm path length. The assay is initiated by addition of the aldolase, and the reaction velocity is followed at 28° in an absorbance-recording spectrophotometer at 340 nm as previously reported.[4]

KDPG aldolase was assayed as reported previously.[5] Using an absorbance change of 1.0 per minute as one unit, with a reaction volume of 0.16 ml, this unit is equivalent to 0.0257 μmol of KDPGal cleaved per minute.

[1] C. W. Shuster, this series, Vol. 9, p. 254.
[2] H. P. Meloche and W. A. Wood, this series, Vol. 9, p. 51.
[3] E. L. O'Connell and H. P. Meloche, this series, Vol. 89 [16].
[4] H. P. Meloche, J. M. Ingram, and W. A. Wood, this series, Vol. 9, p. 520.
[5] R. A. Hammerstedt, H. Möhler, K. A. Decker, D. Ersfeld, and W. A. Wood, this series, Vol. 42, p. 258.

Protein was determined by the ratio of absorbances at 280 nm and 260 nm.[6]

Cell Growth

Pseudomonas saccharophila (ATCC 15946) is grown in submerged culture in a fermentor in a medium composed (per liter) of KH_2PO_4 (1.5 g), Na_2HPO_4 (7.15 g), $(NH_4)_2SO_4$ (3.0 g), $MgSO_4 \cdot 7 \ H_2O$ (100 mg), $FeSO_4 \cdot H_2O$ (5 mg), and galactose (10–12.5 g).

Growth is carried out at 30° with moderate stirring and aeration. The aeration requirement is readily satisfied by maintaining the system at 1 atm air backpressure; this also serves to reduce foaming. The air throughput rate is approximately 0.1 volume per minute. Silicone antifoam can be added. Generation time is approximately 3 hr, and the final pH of the system remains higher than 6.8. If automatic pH control is available, a second portion of 1% galactose solution can be added to the fermentor as the cells approach the stationary phase of growth. In this case pH is maintained between the limits 6.8 and 7.2 using ammonia gas. Under such conditions, we have routinely isolated ~30 pounds of cell paste from a 400-liter fermentor employing a total of 2.5% galactose (w/v), using the pilot-plant facilities at the Oak Ridge National Laboratories, Biology Division. This result, obtained in collaboration with E. F. Phares, represents a cell yield ranging to 34 g/liter for the two-stage growth process.

The wild-type organism is exquisitely sensitive to glucose. This is reflected as very slow growth on galactose, initially, presumably due to traces of glucose present. However, growth is rapidly initiated on galactonate or gluconate. These cells, then, are adapted for facile growth on galactose, while remaining intolerant to glucose.

Enzyme Purification

Crude Extract and Heat Step. Cell paste (420 g) is suspended in 1 liter of sodium phosphate buffer, 50 mM, pH 6.5, containing EDTA, 5 mM (buffer A) and is disrupted in a Manton–Gaulin mill (Gaulin Co., Everett, Massachusetts). The cells can also be ruptured using a high-energy output sonic oscillator. The disrupted cell suspension is then heated to 60°. After cooling in an ice bath, the preparation is centrifuged at 5°, and the supernatant is dialyzed overnight at room temperature against 5 liters of buffer A.

DEAE-Cellulose Chromatography. Unless otherwise stated, all further steps are carried out at room temperature. The sample is then passed

---

[6] E. Layne, this series, Vol. 3, p. 447.

PURIFICATION OF 2-KETO-3-DEOXY-GALACTONATE-6-PHOSPHATE (KDPGal) ALDOLASE

| | Total units | | | Specific activity |
Step	KDPGal aldolase	KDPG[a] aldolase	Activity ratio	(KDPGal aldolase) (units/mg protein)
Heated crude extract	550,000	325,000	1.7	24
DEAE-cellulose	289,000	29,000	10	26
Dialysis	270,000	15,000	18	—
DEAE-Sephadex chromatography I	147,000	3,060	48	1071
DEAE-Sephadex chromatography II	114,300	~1,000	100+	2270
Sephacryl S-200 chromatography I	86,000	—	—	3500
Sephacryl S-200 chromatography II	75,000	—	—	5000+

[a] KDPG aldolase, 2-keto-3-deoxy-gluconate-6-phosphate aldolase.

through a 5- × 75-cm bed of DEAE-cellulose equilibrated against buffer A, and the breakthrough volume is collected. The column is washed with an additional 2 liters of buffer A, and the washings and breakthrough are combined. As shown in the table, the sample applied to the column contains 550,000 units of KDPGal aldolase at a specific activity of 24 units per milligram of protein and a KDPGal : KDPG aldolase activity ratio of 1.7 : 1. The combined breakthrough and wash exhibited a specific activity of 26 and an activity ratio of 10 : 1 with the recovery of 289,000 units.

Dialysis. Protein is isolated from the pool fractions by addition of 500 mg of $(NH_4)_2SO_4$ per milliliter and centrifugation at 5°. The pellet is dissolved in 100 ml of 100 mM ammonium bicarbonate and dialyzed against 5 volumes of the same buffer. Subsequent to dialysis, 270,000 units of KDPGal aldolase are recovered and the activity ratio increases to 18 : 1.

First Chromatography on DEAE-Sephadex. The above preparation is applied to a 2.5- × 20-cm bed of DEAE-Sephadex equilibrated against 100 mM ammonium bicarbonate. After application of the sample, the column is washed with 100 mM ammonium bicarbonate until the absorbance of the effluent at 280 nm returns to background. In this step about two-thirds of the activity is adsorbed on the column. Proteins are eluted in a single peak using 250 mM NaCl in 100 mM $(NH_4)_2CO_3$. The ratio of activities in the pool is 48 : 1, and the specific activity of the KDPGal aldolase is approximately 1000. Proteins are recovered by dissolving 500 mg of $(NH_4)_2SO_4$ per milliliter in the pooled protein sample followed by centrifugation. The pellet is dissolved in a small volume of 10 mM $(NH_4)_2SO_4$ and dialyzed against 10–20 volumes of the same buffer.

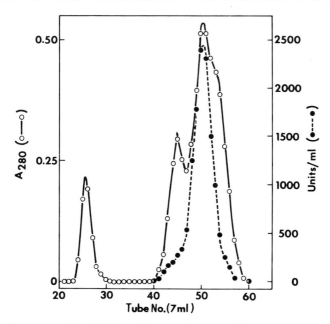

FIG. 1. Chromatography of 2-keto-3-deoxy-galactonate-6-phosphate aldolase on Sephacryl S-200.

Second Chromatography on DEAE-Sephadex. The sample from above is applied to a 2.5- × 20-cm bed of DEAE-Sephadex equilibrated against 10 mM (NH$_4$)$_2$CO$_3$. Once again the column is washed with the above buffer until the absorbance at 280 nm decreases to background. Elution is achieved by the dropwise addition of 300 ml of 250 mM NaCl in 10 mM (NH$_4$)$_2$SO$_4$ to the head volume of the column (about 50 ml). During this procedure considerable shrinking of the DEAE-Sephadex bed occurs. Fractions are collected. At the void volume is found a major protein peak containing 10% of the total activity applied. A gradual elution of protein peaking at 2.7 void volumes is also evident. The pooled peak fractions contain the bulk of the aldolase at a specific activity in excess of 2000. The sample is concentrated to a small volume, ca. 5 ml, using ultrafiltration under N$_2$ pressure.

Chromatography on Sephacryl S-200. The sample from the preceding step is applied to a 2.5- × 80-cm bed of Sephacryl S-200 (Superfine) equilibrated against 50 mM Tris-5 mM EDTA, pH 7, and eluted with the same buffer. A series of proteins elutes, one of which is enriched for the aldolase. A typical result is shown in Fig. 1. Fractions 47–53 are pooled yielding 86,000 units at a specific activity of 3500 units per milligram of protein. This is concentrated by ultrafiltration and rechromatographed on the same column. The overall recovery is excellent and the specific activity of the peak tubes ranges to 6500 units per milligram of protein.

General Comments

Growth of *Pseudomonas saccharophila* on galactose as sole carbon source produces both KDPGal and KDPG aldolases. It has proved to be infeasible to isolate each enzyme, free of the other, from extracts of these cells. It is evident from the table that while the initial DEAE-cellulose step partially resolved the two activities, each enzyme is retained to some degree by the column. In a typical case, elution of the DEAE column with higher ionic strength phosphate buffer after collecting both the breakthrough and wash gives KDPG and KDPGal aldolases at an activity ratio of 3 : 1. The activities are not resolved by gel filtration, nor can significant separation be achieved on DEAE-Sephadex at a lower pH, where KDPG aldolase survives. Consequently, KDPG aldolase must be purified from extracts of this organism grown on gluconate that does not induce KDPGal aldolase.

It is of interest that a specific activity of 6000 units per milligram of protein for KDPGal aldolase is about half that of *Pseudomonas putida*[2] KDPG aldolase, which has been studied in this and other laboratories.[5] KDPGal aldolase reductively binds 1 μmol of pyruvate per 168,000 units of activity lost.[7] This value compares to about 346,700 units inactivated per mole of pyruvate bound for KDPG aldolase from *P. putida*,[8] a three-subunit enzyme. This consistency in relationship of specific activities and stoichiometry of reductive inactivation suggests that KDPGal aldolase also is a three-subunit enzyme. The K_m of KDPG aldolase from *P. putida* is 5×10^{-5} mM[7] for KDPG cleavage, with an absolute specificity for the open-chain form, 9%, of the total substrate[8] in solution. This gives a V/K value of 9.44×10^8 mol^{-1} sec^{-1}, showing that turnover of the condensation product by KDPG aldolase is diffusion limited. The reported K_m of KDPGal aldolase is 5×10^{-4} mM,[1] while our data gives a K_m in the range of 5×10^{-5} mM.[7] Thirteen percent of the substrate is in the open-chain form.[8] If one assumes the enzyme to be specific for the keto form of KDPGal, the magnitude of the K_m value allows anticipating a V/K in the range of 10^7 to 10^8 mol^{-1} sec^{-1}, again supporting diffusion-limited catalysis for C—C bond turnover by KDPGal aldolase.

Evidence on hand suggests that KDPGal aldolase at a specific activity of about 6000 units per milligram of protein is homogeneous. In addition KDPG aldolase from *P. saccharophila*, the same organism, appears to

[7] H. P. Meloche and E. L. O'Connell, unreported observation.
[8] C. H. Midelfort, R. J. Gupta, and H. P. Meloche, *J. Biol. Chem.* **252**, 3486 (1977).

have a similar specific activity. As mentioned above, the two activities are not well resolved by gel filtration, suggesting similar molecular weights. KDPGal aldolase elutes just behind, but not resolved from, KDPG aldolase of *P. putida,* which has a molecular weight of 73,000. Catalysis by both KDPG and KDPGal aldolases of *P. saccharophila* are Schiff's-base mediated, and their sole difference in overall stereomechanism is in the orientation of that carbonyl face of D-glyceraldehyde-3-phosphate presented to the pyruvyleneamine during the condensation–cleavage reaction.[9] Consequently, this pair of activities represents two enzymes from a single source, but under separate genetic control, having similar turnover numbers, but mechanistically differing by a single stereochemical constraint.

[9] H. P. Meloche and C. T. Monti, *J. Biol. Chem.* **250,** 6875 (1975).

Acknowledgment

This work was supported in part by grants from the National Science Foundation, PCM 79-11565, and the National Institutes of Health, GM24926 and RR05690.

[42] 2-Keto-3-deoxy-D-xylonate Aldolase (3-Deoxy-D-pentulosonic Acid Aldolase)

By A. STEPHEN DAHMS and ALAN DONALD

2-Keto-3-deoxy-D-xylonate → pyruvate + glycolaldehyde

2-Keto-3-deoxy-D-xylonate aldolase (EC 4.1.2.18) functions in the metabolism of D-xylose and, presumably, other D-pentoses in a pseudomonad designated MSU-1.[1,2]

Assay Method

Principle. The continuous spectrophotometric assay measures the rate of pyruvate formation by coupling the reaction to lactate dehydrogenase. With the coupling enzyme in excess, the rate of 2-keto-3-deoxy-D-xylonate cleavage is equal to the rate of NADH oxidation as monitored at 340 nm.

[1] A. S. Dahms, *Biochem. Biophys. Res. Commun.* **60,** 1933 (1974).
[2] A. S. Dahms, *Fed. Proc., Fed. Am. Soc. Exp. Biol.* **33,** 1900 (1974).

Reagents

N,N-Bis(2-hydroxyethyl)glycine (Bicine) buffer, 0.5 M, pH 8.0
2-Keto-3-deoxy-D-xylonate,[3] 0.20 M
$MnCl_2$, 0.20 M
NADH, 5 mM
Crystalline lactate dehydrogenase

Procedure. The following are added to microcuvettes with a 1.0-cm light path: 10 μl of buffer, 10 μl of 2-keto-3-deoxy-D-xylonate, 10 μl of $MnCl_2$, 10 μl of NADH, a nonlimiting amount of lactate dehydrogenase, a limiting amount of 2-keto-3-deoxy-D-xylonate aldolase, and water to a final volume of 0.10 ml. The reaction is initiated by the addition of the aldolase. A control cuvette minus substrate measures NADH oxidase activity, which must be subtracted from the total rate. The cuvette compartment is thermostatted at 25°. Care should be taken that the rates are proportional to the aldolase activity.

Definition of Unit and Specific Activity. One unit is defined as the amount of enzyme that catalyzes the cleavage of 1 μmol of 2-keto-3-deoxy-D-xylonate per minute. Specific activity is in terms of units per milligram of protein. Protein was determined by the method of Lowry[4] with bovine serum albumin as a standard.

Purification Procedure

The enzyme is purified from pseudomonad MSU-1 (ATCC 27855). Growth and preparation of cell extracts are as described elsewhere.[5] All steps except for the heat step are performed at 0–4°. A summary of the purification procedure is given in the table.

Protamine Sulfate Treatment. To a cell extract containing 0.2 M ammonium sulfate is added an amount of 2% (w/v) protamine sulfate solution (pH 7.0) to give a concentration of 0.33%. After 30 min, the precipitate is removed by centrifugation and discarded.

Ammonium Sulfate Fractionation. The protein in the supernatant from the protamine sulfate step is fractionated by the addition of crystalline

---

[3] 2-Keto-3-deoxy-D-xylonate can be conveniently prepared enzymatically as described by A. S. Dahms and R. L. Anderson [*J. Biol. Chem.* **247**, 2233 (1972)], employing D-xylonate and D-*xylo*-aldonate dehydratase (A. S. Dahms and A. Donald, this volume [49]). Chemical synthesis of the mixture of the enantiomeric 3-deoxypentulosonic acids can also be carried out using glycolaldehyde and a pyruvyl carbanion generated by alkaline β-decarboxylation of oxaloacetate [A. S. Dahms and R. L. Anderson, *J. Biol. Chem.* **247**, 2233 (1972)].

[4] O. H. Lowry, N. J. Rosebrough, A. L. Farr, and R. J. Randall, *J. Biol. Chem.* **193**, 264 (1951).

[5] A. S. Dahms, D. Sibley, W. Huisman, and A. Donald, this volume [47].

PURIFICATION OF 2-KETO-3-DEOXY-D-XYLONATE ALDOLASE

Fraction	Volume (ml)	Total protein (mg)	Total activity (units)	Specific activity (units/mg protein)
Cell extract	175	4100	279	0.068
Protamine sulfate supernatant	310	3805	298	0.078
$(NH_4)_2SO_4$ precipitate	25	875	255	0.291
Sephadex G-200	40	176	205	1.16

ammonium sulfate. The protein precipitating between 1.42 M and 2.14 M is collected and dissolved in 0.1 M Bicine–0.1 mM 2-mercaptoethanol (pH 7.4).

Chromatography on Sephadex G-200. The above fraction is chromatographed on a column (4 × 45 cm) of Sephadex G-200 equilibrated with 50 mM Bicine–0.1 mM 2-mercaptoethanol (pH 7.4). Fractions (10 ml) are collected, and those with the highest specific activity are combined. 2-Keto-3-deoxy-D-xylonate aldolase is purified 17-fold and is free of 2-keto-3-deoxy-L-arabonate (3-deoxy-L-pentulosonic acid) aldolase.[6]

Properties

Substrate Specificity. Of several 2-keto-3-deoxyaldonic acids tested, only 2-keto-3-deoxy-D-xylonate (K_m = 0.97 mM) served as a substrate. Compounds (5 mM) that were not cleared (<1% of the rate) were: 2-keto-3-deoxy-L-arabonate (3-deoxy-L-pentulosonic acid), 2-keto-3-deoxy-D-fuconate, 2-keto-3-deoxy-D-gluconate and its 6-phosphoester, and 2-keto-4-hydroxy-DL-glutarate.

pH Optimum. Aldolase activity as a function of pH is maximal in the pH range of 7.4–8.2.

Metal Requirement. The aldolase requires a divalent cation for activity when assayed in the presence of EDTA. In the presence of 0.3 mM EDTA, the relative activities of various 1 mM metal salts were as follows: $MnCl_2$, 100; $CoCl_2$, 76; $MgCl_2$ or $MgSO_4$, 33; $ZnCl_2$, 15; $CaCl_2$, 4; $NiCl_2$, $CuCl_2$, and $FeCl_2$, 0. The aldolase is totally insensitive to borohydride reduction whether in the presence of pyruvate, glycolaldehyde, or 2-keto-3-deoxy-D-xylonate.

Equilibrium Constant. The equilibrium constant for the reaction at pH 8.0 and 30° was determined 0.46 mM in the cleavage reaction and 0.36 mM in the condensation reaction. Condensation is stereospecific, since

[6] R. L. Anderson and A. S. Dahms, this series, Vol. 42 [42].

the product does not serve as a substrate for 2-keto-3-deoxy-L-arabonate (3-deoxy-L-pentulosonic acid) aldolase.

Stability. The partially purified enzyme lost 50% of its activity in 2 months when stored at $-20°$. The half-life of the aldolase is 2 min at 55°, about one-sixth of the half-life of 2-keto-3-deoxy-L-arabonate aldolase under exactly identical conditions. In crude extracts containing both aldolases, the half-lives at 53° were 30 min and 3 hr, respectively.

Comparison between the 3-Deoxy- D- *and the 3-Deoxy-* L-*pentulosonic Acid Aldolases.* Both enzymes function in pentose biodegradation in this organism and cleave their respective substrates to pyruvate and glycolaldehyde.[1,6] The 3-deoxy-L-pentulosonic acid aldolase functions in the metabolism of L-arabinose and D-fucose.[7] The 3-deoxy-D-pentulosonic acid aldolase is induced by growth on xylose and probably functions in the biodegradation of D-pentoses in addition to D-xylose.[1] The aldolases are specific for their respective enantiomeric 3-deoxypentulosonic acid substrates, have similar pH optima and equilibrium constants, and are unaffected by borohydride in the presence of either substrate or individual cleavage products. The enzymes, however, possess different metal requirements, molecular weights, and thermal sensitivities.

Acknowledgment

This research was supported by United States Public Health Research Grant GM-22197, National Science Foundation Grant GB-38671, and the California Metabolic Research Foundation.

[7] A. S. Dahms and R. L. Anderson, *J. Biol. Chem.* **247**, 2238 (1972).

[43] 4-Hydroxy-4-methyl-2-ketoglutarate Aldolase from *Pseudomonas putida*

By STANLEY DAGLEY

4-Hydroxy-4-methyl-2-ketoglutarate → 2 pyruvate
4-Carboxy-4-hydroxy-2-ketoadipate → pyruvate + oxaloacetate

The tricarboxylic acid 4-carboxy-4-hydroxy-2-ketoadipate (CHA) is an intermediate in the degradation of gallate (3,4,5-trihydroxybenzoate) by *Pseudomonas putida*[1] and of protocatechuate (3,4-dihydroxybenzoate)

[1] B. F. Tack, P. J. Chapman, and S. Dagley, *J. Biol. Chem.* **247**, 6444 (1972).

by *Pseudomonas testosteroni*.[2,3] The same aldolase that degrades CHA to pyruvate and oxaloacetate also readily forms two molecules of pyruvate from one molecule of 4-hydroxy-4-methyl-2-ketoglutarate (HMG). This compound is easier to synthesize than CHA and is more convenient to use in assays, since it is stable in neutral solutions. Further, before the investigations with bacteria began, a similar enzyme had been partially purified from germinating peanut cotyledons.[4] Since the name HMG aldolase had already been given to the plant enzyme, this was retained for the enzyme later purified to homogeneity from *P. putida*,[1] although the "natural" substrate in bacterial systems is CHA.

Assay Method

Principle. Enzyme activity is assayed by spectrophotometric measurement of the rate of formation of pyruvate from HMG. A solution of HMG is prepared by mild alkaline hydrolysis[4] of pyruvic aldol (4-carboxy-2-ketovalerolactone), which is obtained by acid-catalyzed condensation of pyruvic acid.[4,5] Pyruvic aldol is sometimes formed spontaneously from pyruvic acid when stored and can be collected as a white deposit.

Reagents

Tris-HCl buffer, 50 mM, pH 8.0
MgCl$_2$, 10 mM
4-Hydroxy-4-methyl-3-ketoglutarate (HMG), 10 mM
NADH (Sigma)
Lactate dehydrogenase

Procedure. The following are added to a cuvette with a 1.0-cm light path: 2.5 ml of buffer, 0.4 ml of HMG, 0.4 ml of MgCl$_2$, 0.41 mg of NADH, 25 units of lactate dehydrogenase, and a limiting amount of aldolase to give a total volume of 3.0 ml. Measurements of decrease in absorbance at 340 nm must be corrected for any oxidation of NADH that occurs in a reaction mixture from which substrate is omitted.

Definition of Unit and Specific Activity. One unit of HMG aldolase is defined as the amount of enzyme that catalyzes the disappearance of 1 μmol of substrate per minute at 20°. Specific activity is defined as units of enzyme activity per milligram of protein. Under the conditions of the assay, only one enantiomer of the synthetic racemic mixture used is at-

[2] D. A. Dennis, P. J. Chapman, and S. Dagley, *J. Bacteriol.* **113**, 521 (1973).
[3] V. L. Sparnins and S. Dagley, *J. Bacteriol.* **124**, 1374 (1975).
[4] L. M. Shannon and A. Marcus, *J. Biol. Chem.* **237**, 3342 (1962).
[5] A. W. K. DeJong, *Recl. Trav. Chim. Pays-Bas* **20**, 81 (1901).

tacked[1]; evidence has been presented that this is (R)-HMG.[6] Since two molecules of pyruvate are formed from each molecule of substrate cleaved, the specific activities given in the purification procedure are one-half of those previously based on product formation.[1]

Purification Procedure

Bacterial. Syringic (4-hydroxy-3,5-dimethoxybenzoic),[7] 3,4,5-trimethoxybenzoic,[8] and 3,4,5-trimethoxycinnamic[9] acids are plant constituents that support growth of widely distributed soil bacteria. Strains of the fluorescent *Pseudomonas putida* can be isolated without difficulty from soil, using these aromatic acids as sole carbon sources for elective culture. Since CHA is formed during catabolism,[9] the aldolase is fully induced in bacteria when grown with any of these compounds, which, since they are less susceptible to spontaneous oxidation, are more convenient for use as growth substrates than gallic acid.[3] Further, nonfluorescent pseudomonads that oxidize 3,4-dihydroxybenzoate (protocatechuate)[10] also elaborate HMG aldolase; the enzyme is partially purified from 4-hydroxybenzoate-grown *Pseudomonas testosteroni* (British NCIB 8893).[6] The source of enzyme for the procedure used here is a strain of *P. putida* isolated from soil in St. Paul, Minnesota, by elective culture in media containing syringic acid. Cells are grown with aeration[7] in a medium containing, in grams per liter: $KHPO_4$, 2.65; Na_2HPO_4, 4.33; NH_4Cl, 2.0; Bacto-Peptone (Difco Lab., Detroit, Michigan), 0.2; $MgSO_4$, 0.4; and sodium syringate, 0.5. The last two compounds are omitted during autoclaving and are added, as sterile solutions, after cooling. Sodium syringate solutions are sterilized by Millipore filtration. Cultures are grown at 30°.

Step 1. Preparation of Cell Extract. A cell paste (140 g) is suspended in 520 ml of buffer and sonicated, in portions of 170 ml, for 15 min in an ice–salt bath with a Branson sonifier (Branson Instruments, Inc., Stamford, CT). These organisms, and other pseudomonads, are also conveniently broken in a Hughes bacterial press.[3,11] A clear cell extract is obtained by centrifugation at 18,000 g for 30 min.

Step 2. Heat Treatment. Portions (100 ml) of crude cell extract are held at 60° for 7 min and centrifuged; the precipitate is discarded.

Step 3. Ammonium Sulfate Fractionation. Crystalline ammonium sul-

---

[6] C. S. Ritter, P. J. Chapman, and S. Dagley, *J. Bacteriol.* **113**, 1064 (1973).
[7] B. F. Tack, P. J. Chapman, and S. Dagley, *J. Biol. Chem.* **247**, 6438 (1972).
[8] M. I. Donnelly and S. Dagley, *J. Bacteriol.* **142**, 916 (1980).
[9] M. I. Donnelly and S. Dagley, *J. Bacteriol.* **147**, 471 (1981).
[10] R. Y. Stanier, N. J. Palleroni, and M. Doudoroff, *J. Gen. Microbiol.* **43**, 159 (1966).
[11] S. Dagley and D. T. Gibson, *Biochem. J.* **95**, 466 (1965).

fate (143 g) is added with stirring to the combined supernatants and allowed to stand for 20 min. The precipitate is collected by centrifugation, dissolved in 150 ml of buffer, and dialyzed for 18 hr against five changes of the same buffer containing 10 mM NaCl. The solution is then clarified by centrifugation at 105,000 g for 2.5 hr.

Step 4. TEAE-Cellulose Chromatography. The extract is applied to a TEAE-cellulose column (5.5 × 50 cm) previously equilibrated with buffer containing 10 mM NaCl, and is then eluted with a linear gradient of 0.01 M to 0.70 M NaCl. With a flow rate of 1.5 ml/min, maximum activity is usually eluted in fractions (15 ml) 50–59; these are pooled and precipitated by addition of 35.2 g of $(NH_4)_2SO_4$. The precipitate is taken up in 12.5 ml of buffer and dialyzed overnight against two changes of 5 liters of the same buffer.

Step 5. Preparative Disc Electrophoresis. This procedure uses a Canalco Prep-Disc model, using 0.37 M Tris-HCl of pH 8.9 containing 2.4 mM N,N,N',N'-tetramethylethylenediamine for elution of enzyme, as described in the manual supplied by Canalco Industrial Corporation, Rockville, MD. The amount of cell extract from step 4 is sufficient for two runs, each with 55 mg of protein. When the bromophenol blue used as tracking dye has moved half-way into the separating gel, with a current of 8 mA and a flow rate of 0.6 ml of eluting buffer per minute, collection of 10-ml fractions is started; aldolase activity is found only in fractions 77–89, which elute 4–6 hr after the tracking dye. These fractions from the two runs are pooled, and all the protein is precipitated by adding sufficient powdered $(NH_4)_2SO_4$. After dissolving the precipitate in buffer, the solution is dialyzed against the same buffer until free from $(NH_4)_2SO_4$. At this stage the preparation contains only a trace of contaminating protein when examined by analytical gel electrophoresis or ultracentrifugation. It may be assumed that another design of preparative electrophoresis apparatus could be used at this stage of purification, with suitable modifications of procedure.

Step 6. Sephadex G-200 Chromatography. To the enzyme (4 ml) is added a sufficient amount of each solid to give a solution containing 0.1 M NaCl and 30% sucrose; the solution is then layered onto a column (3 × 83 cm) of Sephadex G-200. The enzyme is eluted by maintaining a constant flow rate (0.2 ml/min) of 50 mM Tris-HCl buffer, pH 8.0, containing 0.1 M NaCl; fractions of 6.5 ml are collected. The aldolase appears in fractions collected after about 21 hr, and at this stage the elution profiles for activity and protein are coincident. Fractions containing activity are pooled, protein is precipitated with $(NH_4)_2SO_4$, and the solution, dissolved in buffer, is dialyzed. One band is now shown in analytical disc electrophoresis, and one single symmetrical peak is observed upon ultracentrifugation.

Results of a typical purification are summarized in the table.

PURIFICATION OF 4-HYDROXY-4-METHYL-2-KETOGLUTARATE ALDOLASE[a]

Fraction	Total protein (mg)	Enzyme (total units)	Specific activity (units/mg protein)	Yield (%)	Purification (fold)
1. Crude extract	14,250	8415	0.6	100	1.0
2. Heat treatment	5,000	5050	1.0	60	1.7
3. Ammonium sulfate fractionation	1,100	3580	3.3	43	5.5
4. TEAE-cellulose	110	1530	13.9	24	18
5. Preparative disc electrophoresis	13.6	774	57	9.2	30
6. Sephadex G-200	8.5	570	67	7	112

[a] Modified from Tack et al.[1]

Properties

Substrate Specificity. Substrates include HMG (K_m = 0.29 mM) and CHA (K_m = 0.067 mM), but 4-hydroxy-2-ketovalerate and 4-hydroxy-2-ketocaproate, which undergo aldol cleavage as catabolites of the meta-fission degradative pathways of catechol and 4-methylcatechol, respectively, are not attacked. Both enantiomers of racemic CHA are decomposed, but only one isomer of HMG is attacked unless the enzyme concentration is high, when slow removal of the second enantiomer can be detected.[1]

Effect of pH. Maximum activity is at pH 8.0. As the pH is raised above 8.5, both HMG and CHA increasingly undergo nonenzymic fission.

Metal Ions. The enzyme requires Mg^{2+} (K_m = 0.17 mM) or Mn^{2+}; Ca^{2+} and Zn^{2+} give no response. Co^{2+} activates, but at 0.7 mM $CoSO_4$ the rate of nonenzymic reaction approaches that for the enzyme.

Molecular Properties. From the amino acid composition, the apparent partial specific volume $\bar{v}$ = 0.742 ml/g; the sedimentation coefficient $s_{20,w}^0$ = 7.98 × 10^{-13} S; the fractional ratio $f:f_0$ = 1.15, indicative of an approximately spherical molecule. From these data and from sedimentation equilibrium determinations, the approximate molecular weight is 150,000. The low mobility of the enzyme in sodium dodecyl sulfate electrophoresis is greatly increased upon reduction with mercaptoethanol; one band with the mobility expected for a protein of molecular weight about 27,000 is then observed. It appears, therefore, that before reduction six polypeptide chains of similar size are joined through disulfide linkages.

[44] 4-Hydroxy-2-ketopimelate Aldolase

By STANLEY DAGLEY

4-Hydroxy-2-ketopimelate → pyruvate + succinic semialdehyde

4-Hydroxy-2-ketopimelate (HKP) aldolase is a widely distributed bacterial enzyme that functions in all the pathways of aromatic catabolism, so far investigated, for which homoprotocatechuate (3,4-dihydroxyphenylacetate) serves as substrate for a benzene ring-fission dioxygenase.[1]

Assay Method

Principle. Enzyme activity is assayed by spectrophotometric measurement of the rate of formation of pyruvate from HKP.

Reagents

Tris-HCl buffer, 50 mM, pH 8.0
MgCl$_2$, 0.1 M
4-Hydroxy-2-ketopimelate,[2] 10 mM
NADH (Sigma)
Lactate dehydrogenase

Procedure. The following are added to a cuvette with a 1.0-cm light path: 0.8 ml of buffer, 0.1 ml of MgCl$_2$, 0.1 μmol of NADH, 4 units of lactate dehydrogenase, and 0.25 ml of HKP. When a limiting amount of aldolase is added, together with any additional buffer required, the total volume is 1.5 ml. Before adding enzyme, 1 min is allowed to elapse in order to observe any decrease in absorbance at 340 nm that might occur, due to traces of pyruvate initially present. The decrease in absorbance upon addition of enzyme must also be corrected for any enzymic oxidation of NADH occurring in the absence of HKP during the first steps of enzyme purification.

Definition of Unit and Specific Activity. One unit of HKP aldolase is defined as the amount of enzyme that catalyzes the disappearance of 1 μmol of substrate per minute. Specific activity is defined as units of en-

[1] S. Dagley, *in* "The Bacteria" (N. L. Ornston and J. R. Sokatch, eds.), Vol. 6, p. 328. Academic Press, New York, 1978.

[2] The preparation of a solution of 4-hydroxy-2-ketopimelate, of known concentration and free from pyruvate, by treating succinic semialdehyde with oxaloacetate, is described by P.-K. Leung, P. J. Chapman, and S. Dagley [*J. Bacteriol.* **120**, 168 (1974)].

METHODS IN ENZYMOLOGY, VOL. 90

zyme activity per milligram of protein. Both enantiomers of synthetic HKP are attacked.[2]

Purification Procedure

Bacteria. HKP aldolase is elaborated by bacteria that are induced to oxidize homoprotocatechuate after growth at the expense of L-tyrosine or 4-hydroxyphenylacetate. Gram-positive bacteria in this category that grow with tyrosine[3] include various strains of *Bacillus* and also *Micrococcus lysodeikticus* (British NCIB 9278). Gram-negative strains that grow with 4-hydroxyphenylacetate and can then oxidize homoprotocatechuate, but not homogentisate, include *Pseudomonas putida* strains NCIB 10015[4] and NCIB 9865[5] and also *Acinetobacter*[6]; some strains of *Escherichia coli* also exhibit these characteristics.[7] The purification procedure described here was developed for *Acinetobacter*.[2] Cells are grown in a mineral salts medium[6] with 4-hydroxyphenylacetate as sole carbon source, using forced aeration at 30°. To furnish sufficient material for enzyme purification, cultures of 16 liters are grown in 20-liter bottles supplied with aeration; after overnight growth, 8 g of sodium 4-hydroxyphenylacetate is added, and the culture is harvested 2 hr later (yield, 40–50 g, wet weight, of cells).

Step 1. Preparation of Cell Extract. The procedure for preparing 360 ml of crude cell extract is described in this volume.[8]

Step 2. Heat Treatment. The cell extract is divided into three portions; each is held at 70° for 6 min, and the precipitates collected at that time are discarded.

Step 3. Ammonium Sulfate Fractionation. To the solution (305 ml) at 4° is added, slowly with stirring, 53.3 g of powdered ammonium sulfate; after standing for 1 hr, the precipitate is removed by centrifugation and discarded. A second addition of 42 g of ammonium sulfate gives a precipitate that is collected on the centrifuge, dissolved in 8 ml of buffer, and dialyzed for 19 hr against four changes of the same buffer. The solution is then clarified by centrifugation and made up to 14 ml with buffer.

Step 4. DEAE-Cellulose Chromatography. The extract is applied to a DEAE-cellulose column (4.0 × 37 cm) previously equilibrated with buffer

---

[3] V. L. Sparnins and P. J. Chapman, *J. Bacteriol.* **127**, 362 (1976).
[4] Y-L. Lee and S. Dagley, *J. Bacteriol.* **131**, 1016 (1977).
[5] M. G. Barbour and R. C. Bayly, *J. Bacteriol.* **142**, 480 (1980).
[6] V. L. Sparnins, P. J. Chapman, and S. Dagley, *J. Bacteriol.* **120**, 159 (1974).
[7] Some strains of *E. coli* that grow at the expense of 4-hydroxyphenylacetate and oxidize homoprotocatechuate are listed by R. A. Cooper and M. A. Skinner [*J. Bacteriol.* **143**, 302 (1980)].
[8] S. Dagley, this volume [43].

PURIFICATION OF 4-HYDROXY-2-KETOPIMELATE ALDOLASE[a]

Fraction	Total protein (mg)	Enzyme (total units)	Specific activity (units/mg protein)	Yield (%)	Purification (fold)
1. Crude extract	4860	1240	0.255	100	1
2. Heat treatment	780	1050	1.36	85	5.4
3. Ammonium sulfate fractionation	127	706	5.6	56	22
4. DEAE-cellulose	5.7	242	41.0	19	174
5. Sephadex G-150	3.86	164	42.5	13	181

[a] From Leung et al.[2]

containing 0.05 M NaCl and is eluted with a linear gradient of 0.05 M to 0.4 M NaCl in a total volume of 4 liters of buffer. With a flow rate of 2.2 ml/min, maximum activity is usually eluted in fractions (20 ml) 105–111 at 0.2 to 0.25 M NaCl; these are pooled, concentrated by pressure dialysis with a Diaflo PM-30 membrane (Amicon Corporation, Lexington, Massachusetts) and dialyzed against buffer.

Step 5. Sephadex G-150 Chromatograhy. The enzyme (1 ml) is layered onto a column (1.7 × 96 cm) of Sephadex G-150 and eluted with buffer at a flow rate of 0.15 ml/min. Aldolase activity appears around fractions (3 ml) 30–35, and at this stage activity and protein profiles coincide. Samples show one band upon disc electrophoresis in acrylamide gels and also in urea–acetic acid acrylamide gel.

Results of a typical purification are summarized in the table.

Properties

Substrate Specificity. The aldolase attacks both enantiomers of HKP (K_m = 0.067 mM). 4-Hydroxy-2-ketovalerate, which is the aldolase substrate in the meta-fission degradation of catechol, is attacked at 3% of the rate for HKP. 4-Hydroxy-4-methyl-2-ketoglutarate and 4-carboxy-4-hydroxy-2-ketoadipate, substrates for HMG aldolase,[8] are not attacked by HKP aldolase.

Effect of pH. Maximum activity is at pH 8.0. At pH 7.0 activity is 70% of maximum, and at pH 8.8 rapid nonenzymic fission of HKP occurs.

Metal Ions. The pure enzyme cleaves HKP without additions, but 1 mM MgCl$_2$ or MnCl$_2$ stimulate activity twofold.

Molecular Properties. From the amino acid composition the apparent partial specific volume $\bar{v}$ = 0.751; the sedimentation coefficient $s_{20,w}^0$ = 7.99 × 10^{-13} S; the frictional ratio $f{:}f_0$ = 1.21, which indicates a roughly spherical shape for the protein. From sedimentation equilibrium measurements a molecular weight of 158,000 is calculated. Sodium dodecyl sulfate electrophoresis gives a major band corresponding to a molecular weight of 25,700; there is a trace of a component of molecular weight 57,800, which disappears upon pretreatment with mercaptoethanol. HKP aldolase is therefore very similar in physical properties to HMG aldolase[8]; they are approximately spherical proteins of about the same molecular weight, apparently consisting of six similar subunits, with extensive disulfide linkages between these subunits only in the case of HMG aldolase. Despite these similarities, the enzymes differ sharply in substrate specificities and in the absolute requirement of HMG aldolase for divalent metal ions.[8]

Section III

Dehydratases

[45] D-Gluconate Dehydratase from *Clostridium pasteurianum*

By Gerhard Gottschalk and Rudolf Bender

$$\text{D-Gluconate} \xrightarrow{\text{Fe}^{2+}} \text{2-keto-3-deoxy-D-gluconate} + H_2O$$

Assay Method

Principle. D-Gluconate dehydratase (D-gluconate hydrolyase, EC 4.2.1.39) catalyzes the irreversible dehydration of D-gluconate to form 2-keto-3-deoxy-D-gluconate (KDG). The enzyme from *C. pasteurianum* is inactivated by exposure to oxygen, but the activity can be completely restored by incubation with sulfhydryl compounds and ferrous ions.[1] The assay involves three steps: (*a*) the activation of D-gluconate dehydratase; (*b*) the incubation of enzyme with D-gluconate; and (*c*) the colorimetric determination of KDG.

Reagents

Tris(hydroxymethyl)aminomethane (Tris)-HCl buffer, pH 8.0, 100 mM
2-Mercaptoethanol
Ferrous ammonium sulfate, 100 mM
Sodium D-gluconate, 50 mM, in 100 mM Tris-HCl buffer, pH 7.6
EDTA, 20 mM, pH 7.0
Sodium periodate, 25 mM, in 0.125 N H$_2$SO$_4$
Sodium arsenite, 2% (w/v), in 0.5 N HCl
2-Thiobarbituric acid, pH 2.0, 0.3% (w/v)

Procedure. The activation of D-gluconate dehydratase is done in small glass tubes of a total volume of 1.5 ml; 5–200 μl of enzyme solution are mixed with 100 mM Tris-HCl buffer, pH 8.0, to give a final volume of 1 ml. Seven microliters of 2-mercaptoethanol and 10 μl of 100 mM ferrous ammonium sulfate solution are added. The components are mixed, and the tubes are flushed with nitrogen gas for 2 min, then closed with rubber stoppers and incubated for 45 min at 50°. When kept anaerobic the activated enzyme solution can be stored for several weeks at 4° without loss of activity.

The D-gluconate dehydratase reaction is performed anaerobically in stoppered glass tubes at 25°. Into glass tubes 2.95 ml of 100 mM Tris-HCl buffer, pH 7.6, containing 50 mM sodium gluconate are pipetted and made

[1] R. Bender and G. Gottschalk, *Eur. J. Biochem.* **40**, 309 (1973).

METHODS IN ENZYMOLOGY, VOL. 90

anaerobic with a stream of nitrogen gas; 0.05 ml of an anaerobic 100 mM ferrous ammonium sulfate solution is added. After adjustment of the temperature to 25°, the reaction is started by the addition of 5–20 μl of activated enzyme solution. Aliquots (0.5 ml) are withdrawn after 0, 15, and 30 min, treated each with 0.5 ml of 20 mM neutralized EDTA, and analyzed for KDG.

The amount of KDG formed is determined using the thiobarbituric acid assay of Weissbach and Hurwitz.[2] Aliquots (0.2 ml) of the above solutions are added to 0.25 ml of 25 mM NaIO$_4$ in 0.125 N H$_2$SO$_4$. After incubation for 20 min at room temperature, 0.5 ml of a 2% (w/v) solution of sodium arsenite in 0.5 N HCl is added. After 2 min this is followed by the addition of 2.0 ml of 0.3% (w/v) 2-thiobarbituric acid (pH 2.0). The solution is incubated for 10 min at 100° (boiling water) and cooled to room temperature. The absorption at 549 nm is measured against a blank. An absorption coefficient of $\epsilon = 67.8 \times 10^3 M^{-1}$ cm^{-1} is used to calculate the amount of KDG formed.[1]

For the determination of KDG formed, the semicarbacide method[3] as modified by Kersters et al.[4] can also be used. For kinetic experiments it is advisable partially to purify a mixture of 2-keto-3-deoxygluconate kinase and 2-keto-3-deoxy-6-phosphogluconate aldolase and to do coupled optical assays. A convenient purification method for these enzymes from C. pasteurianum has been described.[1]

Definition of Unit and Specific Activity. One unit of enzyme activity is defined as the quantity of enzyme that yields 1 μmol of KDG per minute under the above conditions. Specific activity is expressed as units per milligram of protein.

Purification Procedure

Clostridium pasteurianum (ATCC 6013; DSM 525) is grown anaerobically in 20-liter carboys at 37°. The modified synthetic medium of Carnahan and Castle[5] contains the following components per liter: 20 g of sodium gluconate, 0.95 g of KH$_2$PO$_4$, 5.75 g of K$_2$HPO$_4$, 1 g of (NH$_4$)$_2$SO$_4$, 0.1 g of NaCl, 0.1 g of CaCO$_3$, 0.25 g of MgSO$_4$ · 7 H$_2$O, 2 mg of MnCl$_2$ · 4 H$_2$O, 12 mg of Na$_2$MoO$_4$ · 2 H$_2$O, 5.6 mg of FeSO$_4$ · 7 H$_2$O, 1 μg of biotin, and 5 μg of p-aminobenzoic acid. Sodium gluconate and the phosphates are autoclaved separately. The pH value of the gluconate–phosphate solution is adjusted to 7.8 with 5 N KOH before autoclaving.

[2] A. Weissbach and J. Hurwitz, J. Biol. Chem. **234**, 705 (1959).
[3] J. MacGee and M. Doudoroff, J. Biol. Chem. **210**, 617 (1954).
[4] K. Kersters, J. Khan-Matsubara, L. Nelen, and J. De Ley, Antonie van Leeuwenhoek **37**, 233 (1971).
[5] J. E. Carnahan and J. E. Castle, J. Bacteriol. **75**, 121 (1958).

Prior to inoculation the medium is made anaerobic by passing nitrogen gas through a sterile cotton–wool filter and then through the medium for 15 min. A 10% (v/v) inoculum is used. When the cultures have reached an optical density of 4–5 (at 546 nm against water), cells are harvested and stored at $-20°$.

Step 1. Crude Extract. Frozen cells (500 g) of *C. pasteurianum* are thawed in 500 ml of 50 mM Tris-HCl buffer, pH 8.0. The cell suspension is homogenized using an Ultra-Turrax homogenizer (Janke and Kunkel KG, Staufen) and sonicated with a 600-W ultrasonic disintegrator (Schoeller and Co., Frankfurt) for 4 hr in a continuous-flow cell. The temperature is kept below 8° by cooling the cell with ethanol of $-5°$. Cell debris is removed by centrifugation at 10,000 g for 30 min at 4°, and the pH of the supernatant is adjusted to 6.9. The crude extract can be frozen at $-20°$ if the pH has been adjusted to 7.4–8.0. Under these conditions the enzyme is stable for several weeks.

The following steps are performed at 0–4° (unless otherwise mentioned).

Step 2. Cetyltrimethylammonium bromide (CTAB) Treatment. Add 0.18 volume of a 2% (w/v) solution of CTAB slowly to the crude extract (20–25 mg of protein per milliliter) with stirring. The precipitate is removed by centrifugation at 10,000 g for 10 min.

Step 3. Heat Treatment. 2-Mercaptoethanol and ferrous ammonium sulfate are added to the supernatant of step 2 to give final concentrations of 100 and 2.5 mM, respectively. The pH is adjusted with 1 N acetic acid to 6.7. The solution is heated in a 80° water bath to 65° and kept at this temperature for 15 min. It is then cooled in an ice bath and centrifuged at 8000 g for 10 min.

Step 4. Ammonium Sulfate Fractionation. The pH of the solution is adjusted to 7.0 with 1 N KOH, and pulverized ammonium sulfate is added over a period of 45 min with stirring to give a concentration of 45.2 g/100 ml. After centrifugation at 5000 g for 5 min, the supernatant is discarded and the pellet is dissolved in 15 ml of 50 mM Tris-HCl buffer, pH 6.0. Centrifugation at higher speeds results in a significant loss of enzyme activity. The protein solution is dialyzed against 100 volumes of 50 mM Tris-HCl buffer, pH 6.0, for 8 hr and against 50 mM Tris-HCl buffer, pH 7.4, containing 100 mM KCl, for another 8 hr.

Step 5. DEAE-Sephadex Chromatography. The protein solution is allowed to flow into a column (1.5-cm diameter × 20 cm) of DEAE-Sephadex A-50, equilibrated against the second dialysis buffer. The column is washed with 200 ml of the same buffer and is developed with a 420-ml linear gradient from 50 mM Tris-HCl buffer, pH 7.4, containing 100 mM KCl to the same buffer with 300 mM KCl. Fractions of approxi-

PURIFICATION OF GLUCONATE DEHYDRATASE FROM *Clostridium pasteurianum*[a]

Step[b]	Volume (ml)	Total activity (units)	Total protein (mg)	Specific activity (units/mg protein)	Yield (%)
1. Crude extract	400	34,400	8600	4.0	100
2. CTAB[c] supernatant	410	29,900	4750	6.3	87
3. Heat treatment	380	26,900	1630	16.5	78
4. Ammonium sulfate fractionation	30.5	25,000	790	31.6	73
5. DEAE-Sephadex eluate	2.5	16,400	177	92.7	48
6. First Sephadex G-200	2.0	9,000	77	117	26
Second Sephadex G-200	1.5	5,550	42	132	16

[a] From Bender and Gottschalk.[1]
[b] In step 4 the data are given after dialysis; in steps 5 and 6, after concentration.
[c] CTAB, cetyltrimethylammonium bromide.

mately 8.3 ml are collected at an average flow rate of 15 ml/hr. Fractions 10–20 are combined and concentrated by ultrafiltration using the Diaflo cell 54 (Amicon Corp., Lexington, Massachusetts) with the membrane UM-20E at a pressure of 3 bar (300 KPa).

Step 6. Sephadex G-200 Chromatography. First, the solution from step 5 is added to a column (1.5-cm diameter × 100 cm) of Sephadex G-200 that has been equilibrated against 50 mM Tris-HCl buffer, pH 7.5. The column is developed with the same buffer and the eluate is directly passed into a second column of the same dimensions and further fractionated by ascending chromatography. Fractions of 3 ml are collected at an average flow rate of 4 ml/hr. Second, after concentration of the dehydratase-containing fractions, the chromatography on Sephadex G-200 is repeated. A typical purification is summarized in the table.

Properties

Purity and Stability. Analytical polyacrylamide gel electrophoresis of the purified enzyme shows that it is approximately 80% pure. Homogeneous enzyme preparations can be obtained by preparative polyacrylamide gel electrophoresis. The enzyme preparation is free of KDG kinase and 2-keto-3-deoxy-6-phosphogluconate aldolase. When stored at −20° in 50% glycerol, the D-gluconate dehydratase loses approximately 30% of its activity within 5 months.

Molecular Weight. The molecular weight of the enzyme has been found to be 130,000. The enzyme consists of two subunits of identical size.

Activation and Cofactor Requirements. Activation of the dehydratase requires the presence of both 2-mercaptoethanol (or dithiothreitol or other sulfhydryl compounds) and ferrous ions. Since the activation of the enzyme at room temperature takes more than 2 hr, it is conveniently done at 50°. Divalent metal ions are also required for dehydratase activity. Ferrous ions (2 mM) are most effective but can be partially replaced by Mn^{2+}, Co^{2+}, or Mg^{2+}. Sulfhydryl compounds are inhibitory for activity. At concentrations of 10 mM the activity decreases 56% with 2-mercaptoethanol, 35% with L-cysteine, and 63% with dithiothreitol. Therefore, the activated enzyme solutions have to be diluted as in the procedure described so that this inhibition becomes negligible.

The pH optimum of the dehydratase reaction is pH 7.3 to pH 7.8, and the K_m value for D-gluconate amounts to 5.5 mM.

Substrate Specificity. D-Gluconate dehydratase exhibits a high substrate specificity. 6-Phospho-D-gluconate, D-arabonate, D-xylonate, L-mannonate, D-galactonate, D-galactarate, D-galacturonate, and D-glucoheptonate do not serve as substrates. The enzyme can be used for the synthesis of KDG from D-gluconate.[6,7]

Distribution. The enzyme has been detected in *Rhodopseudomonas sphaeroides*,[8] in several clostridia,[9,10] in a subgroup of the *Achromobacter–Alcaligenes* group of bacteria,[11] in *Aspergillus niger*,[12] and in extremely halophilic bacteria.[13] The purification of D-gluconate dehydratase from *Alcaligenes* has been described.[14] This enzyme requires Mg^{2+} for activity, is apparently not inactivated by oxygen, and has a broader substrate specificity and a considerably lower specific activity than the enzyme described here.

[6] K. Kersters and J. De Ley, this series, Vol. 41 [22].
[7] R. Bender and G. Gottschalk, *Anal. Biochem.* **61,** 275 (1974).
[8] M. Szymona and M. Doudoroff, *J. Gen. Microbiol.* **22,** 167 (1960).
[9] J. R. Andreesen and G. Gottschalk, *Arch. Mikrobiol.* **69,** 160 (1969).
[10] R. Bender, J. R. Andreesen, and G. Gottschalk, *J. Bacteriol.* **107,** 570 (1971).
[11] J. De Ley, K. Kersters, J. Khan-Matsubara, and J. R. Shewan, *Antonie van Leeuwenhoek* **36,** 193 (1970).
[12] T. A. Elzainy, M. M. Hassan, and A. M. Allam, *J. Bacteriol.* **114,** 457 (1973).
[13] G. A. Tomlinson, T. K. Koch, and L. I. Hochstein, *Can. J. Microbiol.* **20,** 1085 (1974).
[14] K. Kersters and J. De Ley, this series, Vol. 42 [48].

[46] D-Mannonate and D-Altronate Dehydratases of *Escherichia coli* K12

By J. ROBERT-BAUDOUY, J. JIMENO-ABENDANO, and F. STOEBER

$$\text{D-Mannonate} \rightarrow \text{2-keto-3-deoxy-D-gluconate} + H_2O \quad (1)$$
$$\text{D-Altronate} \rightarrow \text{2-keto-3-deoxy-D-gluconate} + H_2O \quad (2)$$

In *Escherichia coli*, D-mannonate dehydratase (D-mannonate hydro-lyase, EC 4.2.1.8) and D-altronate dehydratase (D-altronate hydro-lyase, EC 4.2.1.7) act, respectively, on D-mannonate, the intermediate aldonate of the glucuronate branch [Eq. (1)], and on D-altronate, the intermediate aldonate of the galacturonate branch [Eq. (2)] of the hexuronate pathway,[1,2] yielding 2-keto-3-deoxy-D-gluconate.

Assay Method

Principle. Mannonate and altronate dehydratase activities are estimated by formation of 2-keto-3-deoxy-D-gluconate (KDG) from D-mannonate and D-altronate, respectively, under the conditions established by Waravdekar and Saslaw[3] modified by Weissbach and Hurwitz.[4] According to these methods, determination of KDG complexed with thiobarbituric acid[3,4] provides a sensitive measure of the reaction rate.

Reagents

Glycylglycine buffer, 400 mM, pH 8.3
2-Mercaptoethanol, 1 M
FeSO$_4$, 16 mM
Potassium D-mannonate, 0.6 M[5,6]
or
Potassium D-altronate, 0.25 M[7,8]
D-Mannonate or D-altronate dehydratase diluted so as to contain about 20 units/ml

[1] G. Ashwell, A. J. Wahba, and J. Hickman, *J. Biol. Chem.* **235**, 1559 (1960).
[2] F. Stoeber, A. Lagarde, G. Nemoz, G. Novel, M. Novel, R. Portalier, J. Pouysségur, and J. Robert-Baudouy, *Biochimie* **56**, 199 (1974).
[3] V. S. Waravdekar and L. D. Saslaw, *Biochim. Biophys. Acta* **24**, 439 (1957).
[4] A. Weissbach and J. Hurwitz, *J. Biol. Chem.* **234**, 705 (1959).
[5] C. S. Hudson and H. S. Isbell, *J. Am. Chem. Soc.* **56**, 2480 (1929).
[6] J. Robert-Baudouy, Thesis, Université Claude Bernard, Lyon (1971).
[7] J. W. Pratt and N. K. Richtmyer, *J. Am. Chem. Soc.* **77**, 1906 (1955).
[8] J. Jimeno-Abendano, Thesis, Université Claude Bernard, Lyon (1968).

METHODS IN ENZYMOLOGY, VOL. 90

Trichloroacetic acid, 10% (w/v) containing 20 mM $HgCl_2$
Periodic acid, 0.25 M, dissolved in 0.125 N H_2SO_4
Sodium arsenite, 2% (w/v) dissolved in 0.5 N HCl
Thiobarbituric acid, 0.3% (w/v), freshly prepared and supplemented with 1% of 0.5 N HCl

Procedure. A stock solution is prepared that contains 10 ml of glycylglycine buffer, 400 mM; 1 ml of 2-mercaptoethanol, 1 M; 1 ml of $FeSO_4$, 16 mM; and 0.24 ml of potassium D-mannonate, 0.6 M, for mannonate dehydratase assay or 0.30 ml of potassium D-altronate, 0.25 M, for altronate dehydratase assay. A 0.75-ml aliquot of stock solution is incubated at 37° for 10 min. Separately, 0.75 ml of the crude extract or purified fraction (see below) is brought to room temperature. The reaction is started by the addition of 0.75 ml of crude extract or purified fraction to 0.75 ml of the above reaction mixture. The reaction is stopped by addition of 0.5 ml of 10% trichloroacetic acid containing 20 mM $HgCl_2$. The precipitate is removed by centrifugation at 6000 g during 15 min in a Martin Christ centrifuge. The supernatnt solution can be assayed directly for the formation of KDG or after several days of storage at $+4°$ as follows. An aliquot of this supernatant estimated to contain 0.002 to 0.03 μmol of KDG is made up to 0.20 ml with 10% trichloroacetic acid and incubated with 0.25 ml of periodic acid for 20 min at room temperature. The oxidation is stopped by addition of 0.5 ml of sodium arsenite, and the tube is shaken for a few seconds until the transitory brown color disappears; 2.0 ml of thiobarbituric acid reagent are added. The mixture is placed in a boiling-water bath for 10 min and then removed and cooled at room temperature for 20 min. The pink color formed has a maximum extinction at 549 nm. Under these conditions, when read in a cuvette with a 1-cm light path, 0.01 μmol of KDG gives an absorbance reading of 0.290.

Definition of Unit and Specific Activity. One unit of enzyme forms 1 μmol of KDG in 10 min under the conditions of the assay. Specific activity is defined as the number of enzyme units per milligram of protein. Protein is determined by the method of Lowry *et al.*[9] or Warburg and Christian.[10]

Purification Procedure[11,12]

For the preparation of both altronate and mannonate dehydratases, extracts of *E. coli* grown on D-glucuronate are used.

[9] O. H. Lowry, N. J. Rosebrough, A. L. Farr, and R. J. Randall, *J. Biol. Chem.* **193**, 265 (1951).
[10] O. Warburg and W. Christian, *Biochem. Z.* **310**, 384 (1941).
[11] J. Robert-Baudouy and F. Stoeber, *Biochim. Biophys. Acta* **309**, 473 (1973).
[12] J. Robert-Baudouy, J. Jimeno-Abendano, and F. Stoeber, *Biochimie* **57**, 1 (1975).

TABLE I
PURIFICATION OF D-MANNONATE DEHYDRATASE

Step	Volume (ml)	Total protein (mg)	Total activity (units)	Recovery (%)	Specific activity (mU/mg protein)	Purification (fold)
Crude extract	100	1680	483	100	287	1
Protamine sulfate supernatant	115.5	524	465	96	887	3.1
Dialyzate, 1.56–2.93 M $(NH_4)_2SO_4$	14.4	357	345	71	966	3.4
DEAE-cellulose	100	70	250	52	3,570	12.4
Sephadex G-200 gel filtration	46	9.2	163	34	17,600	61

TABLE II
PURIFICATION OF D-ALTRONATE DEHYDRATASE

Step	Volume (ml)	Total protein (mg)	Total activity (units)	Recovery (%)	Specific activity (mU/mg protein)	Purification (fold)
Crude extract	100	1680	1206	100	717	1
Protamine sulfate supernatant	115.5	524	1182	98	2,255	3
Dialyzate 1.56–2.93 M $(NH_4)_2SO_4$	14.4	357	965	80	2,703	3.7
DEAE-cellulose	90	59	712	59	12,067	17
Sephadex G-200 gel filtration	70	21	470	39	22,380	31

Each step of purification is summarized in tables (Table I for mannonate dehydratase; Table II for altronate dehydratase). Procedural details
(the same for both dehydratases) are given only for mannonate dehydratase purification.

All steps of this purification are performed at +4°. All centrifugation
operations are for 30 min at 12,000 g. The crude extract is prepared from 3
liters of culture (1.7 g of bacteria dry weight).

Preparation of Crude Extract. *Escherichia coli* K12 wild-type cells[13] are
grown aerobically on a minimum salt medium containing D-glucuronate or
D-galacturonate (hexuronates) potassium salt as sole carbon source
(glucuronate induces the synthesis of both dehydratases; and galacturonate, the synthesis of altronate dehydratase only). The growth medium
contains potassium dihydrogen phosphate, 1.32%; ammonium sulfate,
0.2%; magnesium sulfate, 0.02%; calcium chloride, 0.001%; ferric sulfate,

---

[13] J. Robert-Baudouy, R. C. Portalier, and F. Stoeber, *Mol. Gen. Genet.* **118**, 351 (1972).

0.00005%; and potassium hexuronate, 0.97% (w/v). The final pH of the medium is adjusted to 7.2.

Four or 12 flasks containing 250 ml of growth medium are inoculated with *E. coli* K12 and placed on a shaker overnight at 37°. The cells are harvested by centrifugation at 4° and 6000 g and washed twice with Tris buffer (10 mM, pH 7.0). The cells are suspended in the same buffer so as to obtain 1.8×10^{10} to 4.5×10^{10} bacteria per milliliter (12–30 mg of bacterial dry weight per milliliter). The crude extract is prepared by disrupting the cell membranes in a 10-kc Raytheon sonicator for 20 min at 1°. The cell fragments are removed by centrifugation and discarded. Crude extract may be stored at $-20°$ for a few weeks.

Nucleic Acid Precipitation. Crude extract (100 ml) is precipitated by 21 ml of protamine sulfate (20 g/liter). After incubation with the protamine for 12 hr and removal of the precipitate, the supernatant contains 6.5% of nucleic acids in relation to the total absorbing material.[10]

Ammonium Sulfate Fractionation. Ammonium sulfate is added to obtain a precipitate between 1.56 M and 2.93 M ammonium sulfate. The precipitate is dissolved in 12 ml of 10 mM mono-disodium phosphate buffer, pH 7.0 (buffer I), and the solution is stored overnight at 4°. The fraction containing the bulk of the activity is dialyzed twice for 15 hr against 500 ml of buffer I; this fraction contained less than 0.5% of nucleic acids.[10]

DEAE-Cellulose Chromatography. The above dialyzed fraction is absorbed on a column (2 × 45 cm) of DEAE-cellulose (Serva: 0.81 capacity) equilibrated with buffer I containing 10 mM 2-mercaptoethanol (buffer II). The proteins are eluted with a linear gradient of 0 to 0.4 M NaCl in buffer II at a rate of 15 ml hr^{-1} cm^{-2}; 1.8-ml fractions are collected. Fractions 200–400 (Fig. 1), containing most of the mannonate dehydratase activity, are pooled and treated with ammonium sulfate between 2.54 M and 3.51 M. The precipitate is dissolved in buffer II and dialyzed twice for 10 hr against the same buffer (500 ml).

Sephadex Gel Filtration. The above dialyzed fraction is absorbed on a column (2.5 × 39 cm) of Sephadex G-200 gel (Pharmacia) equilibrated in buffer II containing 100 mM NaCl and eluted with the same buffer at a rate of 2.5 ml hr^{-1} cm^{-2}; 2-ml fractions are collected. Fractions 27–50 contain all the enzyme activity. At this step mannonate dehydratase is purified 61-fold, and 34% of original activity is recovered.

Properties[11,12]

Stability. Crude extracts of mannonate and altronate dehydratases in Tris buffer, 10 mM, pH 7.0 (buffer III) retain 90% of the initial activity for 1 month at 4°.

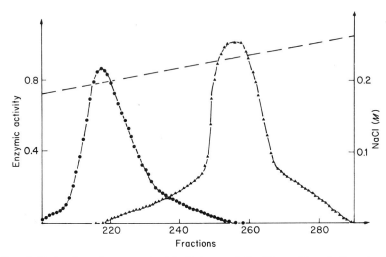

FIG. 1. Separation of mannonate dehydratase activity (●——●) and altronate dehydratase activity (▲——▲) on DEAE-cellulose; - - - -, NaCl gradient. Enzymic activities are given in arbitrary units. For details see Tables I and II.

Heat Stability. At 52° the half-life of mannonate dehydratase in buffer III is 2 min. At 59° the half-life of altronate dehydratase in the same buffer is 4 min.

At 59° the mannonate dehydratase in the crude extract is very rapidly inactivated; after 1 min of treatment only 10% of the activity remains. Under the same conditions, 90% of altronate dehydratase activity is recovered.

Effect of pH and Ionic Strength. The mannonate dehydratase exhibits maximum activity and stability between pH 7.0 and 8.5; the altronate dehydratase exhibits maximum activity at pH 8.3. The assay of the two enzymes is performed at pH 8.3 in 400 mM glycylglycine buffer. Variation of buffer concentration between 100 mM and 400 mM has no effect on enzyme activities.

Specificity. The mannonate dehydratase and the altronate dehydratase appear to be highly specific for D-mannonate and D-altronate, respectively. None of the following compounds is dehydrated by the mannonate dehydratase: D-arabono-1,4-lactone, L-gulonolactone, D-gulono-1,4-lactone, D-galactonolactone, D-talonolactone, D-gluconate, D-1,5-gluconolactone, D-glucuronate, D-glucuronamide, D-galacturonate, D-mannonic amide, or D-altronate.

None of the following compounds is a substrate for altronate dehydratase: D-arabonate, D-glucurate, galactarate, D-gluconate, D-galactonate,

TABLE III
COMPARATIVE ACTION OF SOME EFFECTORS OF MANNONATE AND
ALTRONATE DEHYDRATASES

Effector	Mannonate dehydratase		Altronate dehydratase	
	Effector concentration (mM)	Relative activity	Effector concentration (mM)	Relative activity
Ions				
Fe^{2+}	0.8	100	0.8	100
Mn^{2+}	0.8	90	0.8	20
Co^{2+}	0.8	40	0.8	10
Ni^{2+}	0.8	40	0.8	5
Cu^{2+}	0.8	32	0.8	0
Zn^{2+}	0.8	25	0.8	0
Sugars				
D-Glucuronate	75	190	40	85
D-Galacturonate	50	150	40	77
D-Altronate	50	170	1	100[a]
D-Mannonate	6	100[a]	40	98
D-Arabonate	50	100	40	15
D-Glucarate	50	100	40	68
D-Mannonic amide	0.05	0	0.05	100

[a] Activity in the presence of only the substrate of the enzyme (without another sugar).

D-gulonate, L-gulonate, D-ribonate, D-galacturonate, D-glucuronate, or D-mannonate.

Kinetic Properties. Kinetic studies show that the mannonate and altronate dehydratases are consistent with Henri–Michaelis laws.

The K_m for D-mannonate and D-altronate is estimated to be 20 mM (± 10) for mannonate dehydratase and 0.8 mM for altronate dehydratase.

Effectors and Inhibitors. The effect of different ions is summarized in Table III. The best activation for the two enzymes is obtained by Fe^{2+} at 0.8 mM. The mannonate dehydratase is inhibited by D-gluconate ($K_i = 20$ mM), D-mannitol ($K_i = 30$ mM), D-sorbitol ($K_i = 9$ mM); these inhibitions are competitive. On the other hand, D-mannonic amide ($K_i = 0.02$ mM) and D-glucuronic amide ($K_i = 20$ mM) exert a noncompetitive inhibition. For the altronate dehydratase only D-arabonate exhibits a competitive inhibition ($K_i = 5$ mM) and D-glucarate inhibits weakly the same enzyme. In Table III, comparative action of some sugars shows that the two dehydratases are not sensitive to the same inhibitors and that some sugars are activators of mannonate dehydratase.

Sulfhydryl Requirement. Sulfhydryl compounds are necessary for maintenance of activity, stability of reaction rate, and optimum activation of the enzyme. Addition of 50 mM 2-mercaptoethanol in the presence of 0.8 mM FeSO$_4$ is optimal.

Genetics. Mannonate and altronate dehydratase genes are located at min 97 and 66, respectively, on the *E. coli* K12 chromosome map.[13,14]

[14] R. C. Portalier, J. Robert-Baudouy, and F. Stoeber, *Mol. Gen. Genet.* **118**, 335 (1972).

[47] D-Galactonate Dehydratase

By A. Stephen Dahms, David Sibley, William Huisman, and Alan Donald

Galactonate dehydratase (D-galactonate hydro-lyase, EC 4.2.1.6) functions in the metabolism of D-galactose and D-galactonic acid in a pseudomonad designated MSU-1.[1]

Assay Method[1]

Principle. The α-keto acid formed during the reaction is measured as its semicarbazone by the procedure of MacGee and Doudoroff.[2]

Reagents

Glycylglycine buffer, 0.40 M, pH 7.0
MgCl$_2$, 50 mM
D-Galactonate, 50 mM
Semicarbazide reagent: 1% semicarbazide hydrochloride and 1.5% sodium acetate trihydrate in water

Procedure. The reaction mixture consists of 25 μl of glycylglycine buffer, 5 μl of MgCl$_2$, 100 μl of D-galactonate, D-galactonate dehydratase, and water to a volume of 0.20 ml. The reaction is initiated by the addition of the enzyme. The reaction mixture is incubated at 25° for 30 min and is then quenched by the addition of 1.0 ml of the semicarbazide reagent. After a 15-min incubation at 25°, the mixture is diluted to 5.0 ml with water, and the absorbance is read at 250 nm in a 1-cm quartz cuvette against a diluted reagent blank. Controls are necessary to correct for the absorbance of protein and other reaction components.

[1] A. Donald, D. Sibley, D. E. Lyons, and A. S. Dahms, *J. Biol. Chem.* **254**, 132 (1979).
[2] J. MacGee and M. Doudoroff, *J. Biol. Chem.* **210**, 617 (1954).

Definition of Unit and Specific Activity. One unit is defined as the amount of enzyme that catalyzes the formation of 1 μmol of α-keto acid per minute, based on a molar absorptivity of 10,200.[2] Specific activity is in terms of units per milligram of protein. Protein concentration is conveniently measured by the Lowry procedure[3] using bovine serum albumin as a standard.

Qualitative Assay. When a rapid procedure for detection of 2-keto-3-deoxyaldonic acid is desired (as in scanning fractions from chromatographic columns), the periodate 2-thiobarbituric acid procedure of Weissbach and Hurwitz[4] may be used. The molarity absorptivity of 2-keto-3-deoxygalactonic acid is 60,200 at 550 nm.[1]

Purification Procedure[1]

Growth of Organism. The enzyme is isolated from pseudomonad MSU-1 (ATCC 27855). The organism is grown aerobically in Fernbach flasks containing 1 liter of medium. The flasks are agitated on a rotary shaker at 32°. The medium consists of 1.35% $Na_2HPO_4 \cdot 7\ H_2O$. 0.15% KH_2PO_4, 0.3% $(NH_4)_2SO_4$, 0.2% $MgSO_4 \cdot 7\ H_2O$, 0.0005% $FeSO_4$, and 0.5% D-galactose. The carbohydrate is autoclaved separately and added aseptically to the mineral medium. The cells are harvested 36 hr after inoculation.

Preparation of Cell Extracts. Cells are harvested by centrifugation, washed once by resuspension in 0.01 M Tris buffer, pH 7.1, and finally resuspended in the same buffer to a concentration of 1 g wet weight per 16 ml of buffer. Aliquots (35 ml) of the suspension are added to a jacketed cell cooled with circulating ice water and are exposed to sonic oscillations at 200 W for 10 min from a Biosonik Model III (Bronwill Scientific Instruments, Rochester, New York) equipped with a ¾-inch step horn, during which time the temperature is maintained below 8°. The treated material is pooled and centrifuged at 42,000 g for 10 min at 2°. The supernatant fluid is used as the cell extract.

General. The following procedures were performed at 0° to 4°. A typical purification is summarized in the table.

Protamine Sulfate Treatment. To a cell extract containing 0.2 M ammonium sulfate is added an amount of 2% (w/v) protamine sulfate solution (pH 7.0, 0.01 M Tris, 25°) to give a final concentration of 0.33%. After 20 min, the precipitate is removed by centrifugation and discarded.

Ammonium Sulfate Fractionation. The protein in the supernatant from the protamine sulfate step is fractionated by the addition of crystalline

[3] O. H. Lowry, N. J. Rosebrough, A. L. Farr, and R. J. Randall, *J. Biol. Chem.* **193**, 265 (1951).

[4] A. Weissbach and J. Hurwitz, *J. Biol. Chem.* **234**, 705 (1959).

PURIFICATION OF D-GALACTONATE DEHYDRATASE

Fraction	Total units	Specific activity (units/mg protein)	Protein (mg/ml)	Purification (fold)
Cell extract	46.2	0.103	15.0	1
Protamine sulfate supernatant	48.4	0.133	11.1	1.3
(NH₄)₂SO₄ precipitate	39.3	0.34	36	3.3
Sephadex G-200 concentrate	33.3	2.48	11.2	24
Calcium phosphate gel	28.7	11.8	1.22	115
DEAE-cellulose	11.4	38.0	0.1	369

ammonium sulfate. The protein precipitating between 1.20 M and 2.66 M is collected by centrifugation, and dissolved in 0.05 M glycine–0.01 M sodium cacodylate (pH 8.8).

Sephadex G-200. The above fraction is chromatographed on a column (4 × 45 cm) of Sephadex G-200 equilibrated with the glycine–cacodylate buffer. Fractions (3 ml) are collected during elution with the same buffer, and those with the highest specific activity were combined.

Calcium Phosphate Gel. The above fraction was concentrated to 12 ml with a Diaflo ultrafiltration cell (Amicon Corp., Lexington, Massachusetts) fitted with a UM-10 membrane. The concentrated protein (containing xylonate, gluconate, L-arabonate, and D-galactonate dehydratase activities) is treated with 20% (v/v) calcium phosphate gel (Bio-Rad; dry weight 40.5 mg/ml). The gel suspension is centrifuged for 1 min in a Beckman microfuge, and the centrifugate is successively eluted with 2 ml each of 10, 20, 30, 40, 50, and 100 mM sodium phosphate buffer (pH 7.0). About 87% of the activity is recovered in the 30–40 mM range.

DEAE-Cellulose. The above fraction is diluted 1:3 with distilled water and adsorbed onto a DEAE-cellulose column, 2 × 3 cm (Sigma Chemical Co., St. Louis, Missouri; exchange capacity 0.9 meq/g) pretreated as recommended by Sober et al.[5] and equilibrated with 10 mM potassium phosphate buffer, pH 7.0. The column is then eluted with a stepwise gradient of 20-ml volumes of the same buffer containing 0, 0.1, 0.2, 0.3, 0.4, and 0.8 M NaCl. D-Galactonate dehydratase elution is initiated with 0.2 M NaCl. Fractions containing the highest specific activity are combined. The fraction is purified 369-fold with a 25% overall recovery. The fraction is free of D-gluconate,[6] D-xylonate,[6] and L-arabonate-D-fuconate

[5] H. A. Sober, F. J. Gutter, M. Wyckoff, and E. A. Peterson, *J. Am. Chem. Soc.* **78,** 756 (1956).

[6] A. S. Dahms, *Biochem. Biophys. Res. Commun.* **60,** 1433 (1974).

dehydratase[7] activities and contains a single polypeptide on electrophoresis in 7% polyacrylamide gels stainable with either Coomassie
Blue or Amido Black.

Properties[1]

Substrate Specificity. The enzyme is specific for D-galactonate. The
following substances do not serve as substrates: D-galactarate, 6-iodo-
6-deoxy-D-galactonate, 6-deoxy-D-galactonate (D-fuconate), 6-deoxy-
L-galactonate, L-arabonate, L-galactonate, L-lyxonate, D-arabonate,
D-galacturonate, D-gluconate, L-gluconate, D-glucarate, D-glucuronate,
D-mannonate, L-rhamnonate, D-xylonate, D-lyxonate, D-ribonate, 2-
ketogluconate, 3-deoxygalactonate, cellobionate, lactobionate, and 4,6-
dideoxy-L-*xylo-*, -DL-*ribo-*, -L-*ribo-*, -L-*arabino-*, and -D-*xylo*-hexonic
acids.

The apparent K_m value for D-galactonate is 1.14 mM.

pH Optimum. The optimal pH was 7.1 in glycylglycine buffer.

Metal Requirement. Dehydratase activity is absent in the presence of
EDTA (1 mM). The further addition of various salts (10 mM) to the
enzyme pretreated with 0.5 mM EDTA resulted in the following rates:
MgCl$_2$ or MgSO$_4$, 100; MnCl$_2$, 84; CoCl$_2$, 28; NiCl$_2$, 7; ZnSO$_4$ and FeSO$_4$,
0. The optimal Mg^{2+} concentration was about 1 mM.

Stability. The purified enzyme was stable to storage at 4° for 6 months
or at −20° for 1 year in 0.05 M glycine–0.01 M sodium cacodylate (pH 8.8).
The enzyme could also be stored for up to 6 months without loss of
activity at −20° after lyophilization in the above buffer. The enzyme is
quite stable to thermal treatment and possesses a half-life of over 5 min at
90°. Mg^{2+}, substrate, and dithiothreitol increased the rate of thermal inactivation 2.7-, 4-, and 8-fold, respectively, at 95°.

Molecular Properties. Purified D-galactonate dehydratase migrates as a
single component upon electrophoresis in nondenaturing 7% polyacrylamide gels; the position of the stained protein and the dehydratase activity are coincident. The native protein has a molecular weight of 240,000 to
245,000 as demonstrated by gel filtration and sedimentation on sucrose
density gradients. Sodium dodecyl sulfate–polyacrylamide gel electrophoresis yielded a subunit molecular weight of 57,000, indicating that
galactonate dehydratase is a tetramer; cross-linking studies with dimethyl
suberimidate and dimethyl adipamidate also demonstrated that the enzyme exists as a tetramer. The enzyme exhibits a pI of 4.4–4.6 pH units

[7] R. L. Anderson and A. S. Dahms, this series, Vol. 42 [49].

by isoelectric focusing. The protein contains 6.0 and 14.5 tryptophan residues per tetramer as determined by N-bromosuccinimide oxidation[8] in 0.1 M acetate buffer in the absence and in the presence of 8 M urea, respectively.

Cofactors. The enzyme contains no detectable amount of vitamin B_{12} derivatives, and dialysis, charcoal treatment, gel filtration, or ambient light are without effect on activity.

Inhibitors. Iodoacetamide or p-mercuribenzoate at 3.4 mM produced 90 and 97% inhibition, respectively. Dithiothreitol, 2-mercaptoethanol, and Ellman's reagent are without effect on dehydratase activity.

Mechanism. During the dehydration of galactonate in HTO, 1 solvent proton was incorporated into the product, consistent with a mechanism involving an α,β-elimination of water and ketonization of an intermediate enol.[1] A back incorporation of tritium from HTO into galactonate was observed. In addition, incubation of the dehydratase with 2-keto-3-deoxy-D-galactonate in HTO did not lead to tritium exchange. The data are consistent with the immediate dehydration product being the enol that undergoes spontaneous ketonization. The enzyme does not possess 2-keto-3-deoxygalactonate aldolase, 2-ketodeoxygluconate aldolase, 2-keto-3-deoxygalactonate kinase, or 2-keto-3-deoxy-6-phosphogalactonate aldolase activity.

Induction. The enzyme is induced by growth on D-galactose, D-galactonate, D-fucose, and L-arabinose 44-, 45-, 38-, and 11-fold, respectively, over the levels present in glucose-grown cells. D-Fucose (D-fuconate) and L-arabinose (L-arabonate) are gratuitous inducers of the enzyme.

Acknowledgments

This research was supported by United States Public Health Research Grant GM-22197, National Science Foundation Grant GB-38671, and the California Metabolic Research Foundation.

[8] T. F. Spande and B. Witkep, this series, Vol. 11 [61].

[48] D-Galactonate/D-Fuconate Dehydratase

By RICHARD L. ANDERSON and CHARLES L. HAUSWALD

D-Galactonate → 2-keto-3-deoxy-D-galactonate + H_2O
D-Fuconate → 2-keto-3-deoxy-D-fuconate + H_2O

This enzyme functions in the catabolism of D-galactonate in wild-type *Klebsiella pneumoniae*, wherein it is induced by D-galactonate but not by D-fuconate.[1,2] It also functions in the catabolism of D-fuconate in a mutant strain that is constitutive for the D-galactonate operon.[1,2]

Assay Method[2]

Principle. The assay is a modification of the spectrophotometric semicarbazide method of MacGee and Doudoroff.[3] After a timed interval, the semicarbazone of the α-keto acid product is prepared and determined by measuring its absorbance at 250 nm.

Reagents

PIPES[4] buffer, pH 7.0, 0.3 M
EDTA, 6 mM
$MgCl_2$, 60 mM
Potassium D-galactonate, 0.12 M
Semicarbazide reagent (1.0% semicarbazide-HCl in 1.5% sodium acetate)

Procedure. The following constituents are added to a 1.3- × 10.0-cm test tube: 50 μl of buffer, 50 μl of EDTA, 50 μl of $MgCl_2$, 50 μl of D-galactonate, enzyme, and water to a total volume of 0.30 ml. The reaction is initiated by the addition of D-galactonate/D-fuconate dehydratase. After incubation at 30° for a measured time period (usually 20 min), the reaction is terminated by the addition of 1.0 ml of the semicarbazide reagent. After an additional 20 min, the mixture is diluted to 5.0 ml with distilled water and centrifuged to remove any precipitate. The absorbance is then determined at 250 nm against a diluted reagent blank. The semicarbazone may be quantitated using a molar absorption coefficient of 10,200.[3]

Definition of Unit and Specific Activity. One unit of dehydratase is defined as the amount that converts 1 μmol of D-galactonate to product in

---

[1] C. L. Hauswald and R. L. Anderson, in preparation.
[2] C. L. Hauswald and R. L. Anderson, in preparation.
[3] J. MacGee and M. Doudoroff, *J. Biol. Chem.* **210**, 617 (1954).
[4] Piperazine-N,N'-bis(2-ethanesulfonic acid).

METHODS IN ENZYMOLOGY, VOL. 90

PURIFICATION OF D-GALACTONATE/D-FUCONATE DEHYDRATASE FROM
Klebsiella pneumoniae

Fraction	Volume (ml)	Total protein (mg)	Total activity (units)	Specific activity (units/mg protein)	Yield (%)
Cell extract	485	7830	2470	0.315	(100)
Protamine sulfate precipitate	908	6670	2380	0.357	96
(NH₄)₂SO₄ precipitate	800	3910	2120	0.542	86
DEAE-cellulose I	835	668	1340	2.01	54
Sepharose A-5M	224	243	993	4.09	40
Hydroxyapatite	470	155	950	6.13	38
DEAE-cellulose II	75	98	782	7.98	32
Sephadex G-200	23	84	657	7.82	27

1 min. Specific activity (units per milligram of protein) is based on protein determinations by the Lowry procedure.

Purification Procedure[2]

Organism and Growth Conditions. The common strain of *Klebsiella pneumoniae* PRL-R3 (formerly designated *Aerobacter aerogenes* PRL-R3) may be used, although the procedure described here was done with a D-fuconate-positive mutant (CH-101) of a uracil auxotroph.[1] The organism is cultured aerobically in 20 liters of medium in a New Brunswick Microferm fermentor and harvested by centrifugation. The medium consists of 0.15% KH_2PO_4, 0.71% Na_2HPO_4, 0.3% $(NH_4)_2SO_4$, 0.01% $MgSO_4$, 0.0005% $FeSO_4 \cdot 7 H_2O$, 0.005% uracil, and 0.5% D-galactonate (autoclaved separately).

Preparation of Cell Extracts. Cells (111.5 g) are slurried in 0.05 M potassium phosphate buffer (pH 7.0) to give about 500 ml of a thick suspension. The cells are broken by sonic vibration (10,000 Hz) with a Raytheon sonic oscillator, Model DF-101, cooled with circulating ice water. The broken-cell suspension is centrifuged at 12,000 g, and the pellet is discarded; the supernatant solution is designated the cell extract.

General. The following procedures are performed at 0 to 4°. A summary of the purification procedure is shown in the table.

Protamine Sulfate Treatment. The cell extract is diluted with 0.05 M potassium phosphate buffer (pH 7.0) to give a protein concentration of 10 mg/ml, and $(NH_4)_2SO_4$ is added to 2.6% (w/v). Then 20% by volume of a 2% (w/v) protamine sulfate solution (pH 5.0) is added slowly with stirring.

After 30 min, the precipitate is sedimented by centrifugation, and the supernatant solution is collected by decantation.

Ammonium Sulfate Precipitation. Ammonium sulfate (201 g/liter) is added slowly with stirring to 908 ml of supernatant from the protamine sulfate step. The mixture is stirred for 30 min and then clarified by centrifugation. Additional ammonium sulfate (189 g/liter) is added to the 955 ml of supernatant solution. After 30 min of constant stirring, the precipitate is collected by centrifugation and dissolved in 0.05 M potassium phosphate buffer (pH 7.0) to give a final volume of 800 ml.

DEAE-Cellulose Chromatography I. The $(NH_4)_2SO_4$ fraction is applied to a column (4.0 × 18.5 cm) of DEAE-cellulose that has been equilibrated with the same buffer, and the protein that adsorbs is eluted with a 3.6-liter linear gradient of 0 to 0.35 M KCl in 0.05 M potassium phosphate buffer (pH 7.0). Fractions (20 ml) are collected at a flow rate of 2.0 ml/min, and those that contain most of the activity (0.085 to 0.16 M KCl) are combined.

Sepharose A-5M Chromatography. The above fraction is concentrated to 34 ml by pressure dialysis, and two equal aliquots are chromatographed separately on a column (2.6 × 80 cm) of Sepharose A-5M equilibrated with 0.05 M potassium phosphate buffer (pH 7.0). The protein is eluted with the same buffer, and fractions (4.0 ml) are collected at a flow rate of 40 ml/hr. Fractions 52–76, which contain most of the activity, are combined from both runs.

Hydroxyapatite Chromatography. The above pooled fraction (224 ml) was diluted 8-fold with 1 mM potassium phosphate buffer (pH 6.8) and applied to a column (7.9 × 2.0 cm) of hydroxyapatite (Bio-Rad Laboratories, Richmond, California) equilibrated with the same buffer. The protein is eluted with a 2.4-liter linear gradient of 4 to 400 mM potassium phosphate buffer, pH 6.8. Fractions (22 ml) are collected at a flow rate of 200 ml/hr, and those that contain most of the activity (40 to 120 mM potassium phosphate) are combined.

DEAE-Cellulose Chromatography II. The above fraction is applied directly to a column (1.5 × 15.5 cm) of DEAE-cellulose equilibrated with 0.05 M potassium phosphate buffer (pH 6.0). The protein is eluted with a 400-ml linear gradient of 0.0 to 0.35 M KCl in the same buffer. Fractions (4.0 ml) are collected at a flow rate of 90 ml/hr, and those that contain the peak of the activity (120 to 165 mM KCl) are combined.

Sephadex G-200 Chromatography. The above fraction is concentrated to 26 ml by pressure dialysis and applied to a column (4.0 × 95 cm) of Sephadex G-200 equilibrated with 0.05 M potassium phosphate buffer, pH 7.0. The protein is eluted with the same buffer, and fractions (4.0 ml) are collected at a flow rate of 90 ml/hr. Fractions 52–70, which contain most of the activity, are combined. The dehydratase is purified 25-fold with an overall recovery of 27%. Despite the low-fold purification, the enzyme is >95% pure when examined by electrophoresis on polyacrylamide gels under both native and denaturing conditions.

Properties[2]

Substrate Specificity and Kinetic Constants. Of nine aldonic and aldaric acids tested, only D-galactonate ($K_m = 0.9$ mM) and D-fuconate ($K_m = 1.3$ mM) served as substrates. The V_{max} value with D-fuconate as the substrate is 30% of that with D-galactonate. Compounds that did not serve as substrates (<1% of the rate with D-galactonate) at 40 mM were D-gluconate, D-glucarate, D-mannonate, D-xylonate, D-lyxonate, D-arabonate, and L-arabonate.

Effect of EDTA and $MgCl_2$. EDTA (1.0 mM) in the assay mixture abolished the dehydratase activity in the absence of $MgCl_2$ (10 mM). Both EDTA and $MgCl_2$ are required in the assay mixture to assure maximal activity and proportionality with enzyme concentration.

Molecular Weight. The molecular weight is 290,000 ± 20,000. The subunit molecular weight is 46,000, indicating that the enzyme is hexameric.

pH Optimum. The pH optimum varies from 6.5 to 7.5, depending on the buffer.

Stability. The purified enzyme in 0.05 M potassium phosphate buffer (pH 7.0) is stable for at least a month at either 4° or −20°.

[49] D-*xylo*-Aldonate Dehydratase

By A. Stephen Dahms and Alan Donald

D-Xylonate → 2-keto-3-deoxy-D-xylonate (D-pentulosonic acid)

This enzyme functions in the metabolism of D-xylose in the pseudomonad designated MSU-1.[1–3]

Assay Method

Principle. The keto acid formed during the reaction is measured as its semicarbazone by the method of MacGee and Doudoroff.[4]

[1] A. S. Dahms, *Biochem. Biophys. Res. Commun.* **60**, 1433 (1979).
[2] A. S. Dahms, *Fed. Proc., Fed. Am. Soc. Exp. Biol.* **33**, 1900 (1974).
[3] A. Donald, D. Sibley, D. E. Lyons, and A. S. Dahms, *J. Biol. Chem.* **254**, 2132 (1979).
[4] J. MacGee and M. Doudoroff, *J. Biol. Chem.* **210**, 617 (1954).

Reagents

N,N-Bis(2-hydroxyethyl)glycine (Bicine) buffer, 0.1 M
D-Xylonate,[5] 0.10 M
$MgCl_2$, 0.10 M
Semicarbazide reagent: 1% semicarbazide hydrochloride and 1.5% sodium acetate · 3 H_2O in water

Procedure. The reaction mixture consists of 0.05 ml of buffer, 0.02 ml of $MgCl_2$, 0.02 ml of D-xylonate, D-*xylo*-aldonate dehydratase, and water to a volume of 0.15 ml. The reaction is initiated by the addition of enzyme. The reaction mixture is incubated at 25° for 30 min and is then quenched by the addition of 1.0 ml of the semicarbazide reagent. After a 15-min incubation at 30°, the mixture is diluted to 5.0 ml with H_2O. The absorbance is read at 250 nm in a 1.0-cm quartz cuvette against a diluted reagent blank. Controls are necessary to correct for the absorbance of protein and other reaction components.

Definition of Unit and Specific Activity. One unit is defined as the amount of enzyme that catalyzes the formation of 1 μmol of α-keto acid per minute, calculated on the basis of a molar absorption coefficient of 10,200.[4] Specific activity is in terms of units per milligram of protein. Protein is conveniently measured by the method of Lowry[6] using bovine serum as a standard.

Qualitative Assay.[1] When a rapid assay for 2-keto-3-deoxyaldonic acid formation is desired (as in scanning fractions from chromatographic columns), the periodate-2-thiobarbituric acid procedure of Weissbach and Hurwitz[7] may be used. The molar absorptivity of 2-keto-3-deoxy-D-xylonate was determined to be 54,500 at 551 nm.

Purification Procedure

The enzyme is purified from pseudomonad MSU-1. The organism is grown as described elsewhere.[8] The preparation of cell extracts is also as described there, except that the cells are suspended in 0.1 M Bicine buffer, pH 8.0. The following procedures are performed at 0–4°.

Protamine Sulfate Treatment. To a cell extract containing 0.2 M ammonium sulfate is added an amount of 2% (w/v) protamine sulfate solution (pH 7.0 in 0.10 M Bicine buffer) to give a final concentration of 0.33%. After 30 min, the precipitate is removed by centrifugation and discarded.

[5] D-Xylonate can be prepared as the K^+ salt from D-xylose by the method of S. Moore and K. P. Link, *J. Biol. Chem.* **133**, 293 (1940).
[6] O. H. Lowry, N. J. Rosebrough, A. L. Farr, and R. J. Randall, *J. Biol. Chem.* **193**, 265 (1951).
[7] A. Weissbach and J. Hurwitz, *J. Biol. Chem.* **234**, 705 (1959).
[8] A. S. Dahms, D. Sibley, W. Huisman, and A. Donald, this volume [47].

PURIFICATION OF D-*xylo*-ALDONATE *Dehydratase*

Fraction	Volume (ml)	Total protein (mg)	Total activity (units)	Specific activity (units/mg protein)
Cell extract	175	4100	270	0.066
Protamine sulfate supernatant	310	3805	258	0.068
$(NH_4)_2SO_4$ precipitate	20	750	153	0.204
Sephadex G-200	32	155	102	0.653
Heat step	30.5	70	97	1.38

Ammonium Sulfate Fractionation. The protein in the supernatant from the protamine sulfate step is fractionated by the addition of crystalline ammonium sulfate. The enzyme precipitating between 0 and 1.66 M is collected by centrifugation and dissolved in 0.10 M Bicine (pH 8.0).

Chromatography on Sephadex G-200. The above fraction is chromatographed on a 4- × 50-cm column of Sephadex G-200 equilibrated with 0.10 M Bicine buffer, pH 8.0. Fractions (8 ml) are collected during elution with the same buffer, and those with the highest specific activity are pooled.

Heat Step. The pooled Sephadex fractions are concentrated by pressure dialysis (Amicon, UM-10 membrane) to a protein concentration of 5.2 mg/ml. The solution is heated at 55° for 10 min and then quickly cooled and clarified by centrifugation. The enzyme is stable under the above conditions but must be freed of $(NH_4)_2SO_4$ prior to the heat step.

The thiobarbituric acid procedure must be employed with caution. 2-Keto-3-deoxyxylonate aldolase copurifies with the dehydrase and results in the net conversion of D-xylonate to pyruvate and glycolaldehyde.[1,9] The aldolase is inactivated totally during the heat step. This problem can also be circumvented by the use of D-gluconate as substrate.

The above procedures resulted in 21-fold purification of D-*xylo*-aldonate dehydratase with an overall recovery of 36%. A typical purification is summarized in the table.

Properties

Substrate Specificity. Of 18 aldonic, uronic, and aldaric acids tested, only D-xylonate and D-gluconate served as substrates (<5% of the rate with D-xylonate). Among those compounds that were inactive at 25 mM concentrations were L-arabonate, D-arabonate, D-fuconate, L-fuconate, 6-iodo-6-deoxygalactonate, D-mannonate, L-mannonate, L-galactonate,

[9] A. S. Dahms and A. Donald, this volume [42].

D-galactonate, L-gluconate, D-ribonate, D-glucuronate, D-galacturonate, and D-galactarate. The K_m and relative V_{max} values were 2.98 mM (100%) and 2.77 mM (55%) for D-xylonate and D-gluconate, respectively.

All data are consistent with one enzyme functioning on either D-xylonate or D-gluconate. Column elution profiles and pH optimum with both substrates were superimposable. Reaction rates with mixed substrates were not additive, and thermal inactivation profiles with both substrates were superimposable and exhibited first-order kinetics. The enzyme can be localized on 7% polyacrylamide gels by employing a modified thiobarbituric acid procedure or with TTC[10]; one protein band is associated with dehydratase activity on both xylonate and gluconate.

pH Optimum. Dehydratase activity as a function of pH was optional at 8.0 in 50 mM Bicine.

Metal Requirement. The enzyme exhibits an absolute requirement for a divalent cation. The addition of various salts at 5 mM to the EDTA-inhibited enzyme resulted in the following relative rates (percentage of original activity): $MgCl_2$ or $MgSO_4$ (100), $FeSO_4$ (75), $PbCl_2$ (50), Mn (13). $CuSO_4$, $CoCl_2$, $ZnSO_4$, and $NiCl_2$ were inactive.

Stability. The enzyme was stable in the frozen state for several months. The half-life at 60° and 70° was 14 and 2 min, respectively. The enzyme is not affected by thiols, iodoacetic acid, *p*-mercuribenzoate, or *N*-ethylmaleimide at 2 mM.

Induction. The dehydratase is induced by growth on D-xylose, but not by D-gluconate, D-glucose, L-arabinose, or D-fucose. Although the enzyme exhibits D-gluconate dehydratase activity, it is not induced by growth on glucose or gluconate, in contrast to the induction of 6-phosphogluconate dehydratase by the same sugars.[1,3] The D-*xylo*-aldonate dehydratase is probably not operative in glucose (gluconate) metabolism in this organism.

[10] O. Gabriel and S. Wang, *Anal. Biochem.* **27**, 545 (1969).

Acknowledgments

This research was supported by United States Public Health Research Grant GM-22197, National Science Foundation Grant GB-38671 and the California Metabolic Research Foundation.

Section IV

Synthases

[50] L-*myo*-Inositol-1-phosphate Synthase from Bovine Testis[1,2]

By Yun-Hua H. Wong, Linda A. Mauck, and William R. Sherman

$$\text{D-Glucose-6-P} \xrightarrow{\text{NAD}^+} \text{L-}myo\text{-inositol-1-P}$$

L-*myo*-Inositol-1-phosphate synthase (EC 5.5.1.4) appears to be widely distributed. In all the instances thus far reported, the enzyme is NAD^+ dependent, and in each case NADH is not a reaction product; i.e., the enzyme is an oxidoreductase. There are a number of studies that support the mechanism shown in Fig. 1, however, the actual intermediates have never been isolated. In the natural course of the reaction, the product, L-*myo*-inositol 1-phosphate, is hydrolyzed to *myo*-inositol by L-*myo*-inositol-1-phosphatase, thus giving the only known de novo source of this inositol.

Assay Method

Principle. The assay is a modification of the colorimetric method of Barnett *et al.*[3] wherein inorganic phosphate released from *myo*-inositol 1-phosphate by periodate oxidation is measured using a malachite green-enhanced molybdate method.[4]

Reagents

Standard buffer: 50 mM Tris-HCl, pH 7.4, 5% ethanol, 0.02% NaN_3
Glucose 6-phosphate, 20 mM in standard buffer
NAD^+, 10 mM in standard buffer
Dithiothreitol (DTT), 10 mM in standard buffer
Bovine serum albumin, 10 mg/ml in standard buffer
Trichloroacetic acid, 10% (w/v)
Sodium periodate, 0.3 M
Sodium sulfite, 2 M
Malachite green reagent: A—malachite green 0.05% in distilled water; B—ammonium molybdate 4.2% in 50% HCl. Mix 3 volumes of A and 1 volume of B.
Tween 20, 0.6%

[1] We wish to acknowledge the help of Allison Swift in this work. We also acknowledge support from NIH Grants NS-05159, RR-00954, and AM-20579.
[2] L. A. Mauck, Y.-H. Wong, and W. R. Sherman, *Biochemistry* **19**, 3623 (1980).
[3] J. E. G. Barnett, R. E. Brice, and D. L. Corina, *Biochem. J.* **119**, 183 (1970).
[4] K. Itaye and M. Ui, *Clin. Chim. Acta* **14**, 361 (1966).

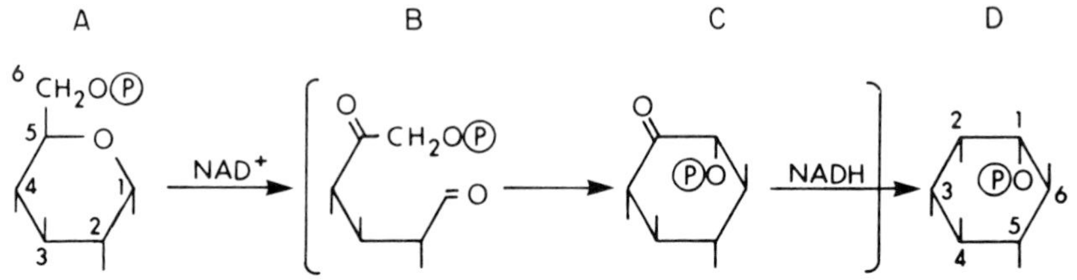

FIG. 1. Mechanism for the reaction of L-*myo*-inositol-1-phosphate synthase. (A) D-Glucose 6-phosphate, (B) D-5-ketoglucose 6-phosphate, (C) L-*myo*-inosose 2-1-phosphate, (D) L-*myo*-inositol 1-phosphate.

Procedure. The assay mixture contains glucose 6-phosphate, 0.4 μmol; NAD^+, 0.2 μmol; dithiothreitol, 0.1 μmol; 0.01 ml of ethanol; Tris-HCl, pH 7.4, 10 μmol; NaN_3, 40 μg; and 0.2 unit of L-*myo*-inositol-1-phosphate synthase in a final volume of 0.2 ml. In order to stabilize the enzyme, 0.1 mg of bovine serum albumin is added to the assay mixture at the BioGel A-1.5m and hydroxyapatite stages. After a 1-hr incubation at 37°, 20 μl of 10% trichloroacetic acid is added, the suspension is mixed and then centrifuged. An aliquot of 0.1 ml of the supernatant is incubated with 50 μl of 0.5 M sodium periodate for 1 hr at 37°. At the end of the incubation, 50 μl of 2 M sodium sulfite is added followed by 1 ml of malachite green reagent and 0.1 ml of 0.6% Tween-20. The samples are then measured spectrophotometrically at 660 nm.

Definition of Unit and Specific Activity. One unit is defined as the amount of enzyme required for the synthesis of 1 nmol of L-*myo*-inositol 1-phosphate per minute under the standard assay conditions. Specific activity is expressed as units per milligram of protein. Protein is determined by the method of Warburg and Christian[5] or by the dye-binding technique of Bradford[6] using protein standards as noted.

Purification Procedure

Purification without Heating

Step 1. Homogenization and Centrifugation. One hundred grams of fresh or frozen testis is decapsulated and homogenized in 200 ml of 0.154 M KCl containing 0.2 mM dithiothreitol and 5% ethanol. The homogenate is centrifuged at 100,000 g for 1 hr in the Beckman L5-65 ultracentrifuge.

Step 2. Celite Column. To the supernatant is added 10 g of Celite-545 with stirring followed by solid $(NH_4)_2SO_4$ to 1.58 M at 0°. The mixture is

[5] A. Warburg and W. Christian, *Biochem. Z.* **310**, 384 (1941).
[6] M. Bradford, *Anal. Biochem.* **72**, 248 (1976).

stirred for 1 hr, then the Celite suspension is filtered and washed with 50 ml of $1.84 M$ $(NH_4)_2SO_4$ solution in standard buffer. The Celite preparation is resuspended in the $1.84 M$ $(NH_4)_2SO_4$ solution and poured into a 2.5- × 35-cm column. After settling, and decanting the supernatant, the column is eluted at 15–25 ml/hr with a 400-ml linear gradient from $1.84 M$ to $0 M$ $(NH_4)_2SO_4$ in standard buffer. Fractions of about 4 ml volume are collected and analyzed colorimetrically.

After the active Celite fractions are pooled, they are dialyzed against standard buffer, then concentrated to about 6 mg/ml by pressure dialysis using an Amicon XM-100A filter.

Step 3. Chromatography on DEAE-Cellulose. The Celite concentrate is introduced onto a Whatman DE-52 column at a loading of 6 mg of protein per milliliter of DE-52. The column is eluted with a 6-column volume linear gradient from 0 to $0.4 M$ KCl in standard buffer. Active fractions are pooled, concentrated, and dialyzed overnight against $0.2 M$ KCl in standard buffer.

Step 4. Gel Filtration. The dialyzate is placed on a BioGel A-1.5m column (5 × 100 cm) that had previously been equilibrated with standard buffer containing $0.2 M$ KCl. The column is eluted at 50 ml/hr, and 10-ml fractions are collected. The active fractions are pooled, concentrated by Amicon ultrafiltration, and dialyzed against 5 mM Tris-HCl, pH 7.0, containing 0.2 mM DTT and 0.02% NaN_3. Analysis is carried out in the presence of bovine serum albumin.

Step 5. Chromatography on Hydroxyapatite. The enzyme is placed on a BioGel HT hydroxyapatite column (2.5 mg of protein per milliliter of the absorbent), which had been previously equilibrated with 5 mM, pH 7.0, Tris-HCl buffer. The homogeneous enzyme is eluted from the column with a linear gradient in 5 mM Tris-HCl, pH 7.0, and 0.02% NaN_3 of 0 to 50 mM potassium phosphate. Each fraction is dialyzed against standard buffer and then analyzed in the presence of bovine serum albumin.

A typical purification is summarized in the table.

Alternative Purification with Heating

Step 1A. Heating Followed by $(NH_4)_2SO_4$ Precipitation. Following the original procedure of Eisenberg[7] for partial purification of this enzyme from rat testis, the supernatant from step 1 is heated at 60° for 2 min and then centrifuged at 40,000 g for 30 min. Solid $(NH_4)_2SO_4$ is added to the supernatant to a concentration of $1.58 M$. The suspension is allowed to stand overnight at 4°, and the precipitate is collected by centrifugation at 40,000 g for 30 min. The precipitate is then dissolved in the minimum

[7] F. Eisenberg, Jr., *J. Biol. Chem.* **242**, 1375 (1967).

PURIFICATION OF BOVINE TESTIS *myo*-INOSITOL-1-PHOSPHATE SYNTHASE TO HOMOGENEITY

Step	Total protein (mg)[a]	Yield (% total activity)	Specific activity (nmol/min/mg protein)			Purification (fold)
			BSA[b]	Igs[b]	$A_{280} : A_{260}$[a]	
Celite–ammonium sulfate fractionation	261	100	7	2	3	1
DEAE-cellulose chromatography	55	85	24	8	12	4
BioGel A-1.5m filtration	10.3	47	72	23	36	12
Hydroxyapatite chromatography	1.6	37	528	156	181	60

[a] Using the spectrophotometric method ($A_{280} : A_{260}$) of Warburg and Christian[5]; see also E Layne, this series, Vol. 3, p. 451.

[b] Using bovine serum albumin or γ-globulins as a standard and the method of Bradford.[6]

amount of standard buffer, and the resulting solution is dialyzed against the same buffer overnight. The specific activity of these preparations is typically 12 nmol/min per milligram of protein based on $A_{280} : A_{260}$.

Properties of the Enzyme

Homogeneity. After the hydroxyapatite stage, the enzyme preparation appears as a single band on 7% polyacrylamide disc gel electrophoresis. The locations of the activity extracted from the gel and the stained band coincide.

Several hydroxyapatite fractions have been examined by sedimentation equilibrium in an analytical ultracentrifuge. These fractions give linear plots of log C vs r^2. Linearity of this plot is a necessary condition for homogeneity. Schlieren patterns from other sedimentation studies appear to be symmetrical, with occasional evidence of heterogeneity that may be due to aggregation.

Molecular Weight. The molecular weight of the synthase as determined by sedimentation equilibrium ($v = 0.725$, assumed) is 209,000.

Cofactor Requirement. The enzyme has a requirement for NAD^+, regardless of the stage of purification or whether it is heated in the initial step, or not. The K_m of NAD^+ is 0.011 mM.

Stability. The Celite and DEAE-cellulose enzyme preparations are stable for several months when stored at $-70°$ in the presence of standard buffer containing 0.2 mM DTT. However, the enzyme preparations from the agarose gel filtration and hydroxyapatite stages are unstable toward freezing and thawing.

Substrate Specificity. The enzyme is most active with D-glucose 6-phosphate, but a number of C-2 analogs of D-glucose 6-phosphate, namely, D-mannose 6-phosphate,[8] D-2-deoxyglucose 6-phosphate (11.5% the rate of glucose 6-phosphate),[9] and D-2-deoxy-2-fluoroglucose 6-phosphate (0.2% the rate of glucose 6-phosphate)[9] are also substrates. Neither D-allose 6-phosphate nor D-galactose 6-phosphate are converted to product by the enzyme.[9]

Inhibitors.[9] D-Mannose 6-phosphate, D-2-deoxyglucose 6-phosphate, and D-2-deoxy 2-fluoroglucose 6-phosphate are competitive inhibitors of the enzyme. The presence of 5 mM D-6-phosphogluconic acid, 5 mM D-fructose 6-phosphate, or 5 mM D-allose 6-phosphate results in 90%, 76%, and 6% inhibition, respectively. D-Galactose 6-phosphate and L-α-glycerophosphate are noninhibitory at 12 mM.

Effect of Monovalent Cations and Ethanol.[2] The enzyme is 2-fold stimulated by either 50 mM NH$_4^+$ or K$^+$, an effect that is not additive. Sodium ions inhibit the enzyme by 78% at 153 mM. The effect of Na$^+$ and K$^+$ on the enzyme is not on K_m but on V_{max}. Ethanol stimulates the enzyme 2-fold, and with added K$^+$, 2.5-fold. The effect of ethanol on the enzyme is on K_m. In the presence of ethanol, the effect of salt on V_{max} disappears.

Effect of Divalent Cations and EDTA.[10] The enzyme, at the DEAE-cellulose stage of purification, is uninhibited by EDTA to a concentration of 50 mM. Mg^{2+} and Mn^{2+} activation occurs up to a concentration of 10 mM and 0.5 mM, respectively. Co^{2+} and Ca^{2+} were found to be inactive and Cd^{2+}, Cu^{2+}, Zn^{2+} were inhibitory. There is no metal requirement for basal activity. From the evidence[10–12] that has thus far accrued, the bovine synthase appears to have neither a class I nor a class II aldolase mechanism at the ring-closure step of the reaction. It is our belief that the cyclization is simply a base-catalyzed aldol condensation with product specificity given by the enzyme.

Other Purifications to Homogeneity. Maeda and Eisenberg[12] have reported the purification of rat testis synthase to homogeneity by a procedure similar to that reported here (and by Mauck *et al.*[2]) except that two additional steps are employed: prior to the ammonium sulfate step the enzyme is heated, and before the hydroxyapatite column a glucose 6-phosphate affinity gel is employed. It is reported that bovine testis[13] and

[8] W. R. Sherman, S. L. Goodwin, and K. D. Gunnell, *Biochemistry* **10**, 3491 (1971).

[9] Y.-H. H. Wong and L. A. Mauck, *Fed. Proc., Fed. Am. Soc. Exp. Biol.* **40**, 1866 (1981).

[10] W. R. Sherman, M. W. Loewus, M. Z. Pina, and Y.-H. H. Wong, *Biochim. Biophys. Acta* **660**, ·299 (1981).

[11] W. R. Sherman, A. Rasheed, L. A. Mauck, and J. Wiecko, *J. Biol. Chem.* **252**, 5672 (1977).

[12] T. Maeda and F. Eisenberg, Jr., *J. Biol. Chem.* **255**, 8458 (1980).

[13] F. Pittner and O. Hoffmann-Ostenhof, *Monatsh. Chem.* **107**, 793 (1976).

rat testis[14] L-*myo*-inositol-1-phosphate synthase can be purified to homogeneity in a single step from an $(NH_4)_2SO_4$ fraction using an NAD^+-Sepharose. We were unable to reproduce this work[2,15] (our enzyme preparation did not bind to the NAD^+-Sepharose). Evidence has been presented,[12] the implication of which is that the synthase purified by the NAD^+-Sepharose method[14] may not have been homogeneous.

[14] F. Pittner, W. Fried, and O. Hoffmann-Ostenhof, *Hoppe-Seyler's Z. Physiol. Chem.* **355**, 222 (1974).
[15] W. R. Sherman, P. P. Hipps, L. A. Mauck, and A. Rasheed, *in* "Cyclitols and Phosphoinositides" (W. W. Wells and F. Eisenberg, Jr., eds.), p. 279. Academic Press, New York, 1978.

[51] 3-Hexulose-6-phosphate Synthase from *Methylomonas* (*Methylococcus*) *capsulatus*

By J. RODNEY QUAYLE

D-Ribulose 5-phosphate + formaldehyde $\rightleftharpoons$ D-*arabino*-3-ketohexulose 6-phosphate

The enzyme catalyzes a key step in the assimilation of carbon during growth of some bacteria on reduced C_1 compounds. The purification described below has been the subject of a previous report.[1] A similar enzyme from *Methylomonas aminofaciens* 77a has been purified to homogeneity.[2]

Assay Methods

Principle. Two assay methods can be used. Method 1 involves discontinuous colorimetric measurement of the rate of ribulose 5-phosphate-dependent removal of formaldehyde. Method 2, which is a modification of that used by van Dijken *et al.*,[3] involves continuous measurement of the rate of ribulose 5-phosphate- and formaldehyde-dependent formation of hexose phosphate. Method 2 is more convenient to use, but it depends on having previously purified phosphohexuloisomerase as a coupling enzyme.[4]

[1] T. Ferenci, T. Strøm, and J. R. Quayle, *Biochem. J.* **144**, 477 (1974).
[2] N. Kato, H. Ohashi, Y. Tani, and K. Ogata, *Biochim. Biophys. Acta* **523**, 236 (1978).
[3] J. P. van Dijken, W. Harder, A. J. Beardsmore, and J. R. Quayle, *FEMS Microbiol. Lett.* **4**, 97 (1978).
[4] J. R. Quayle, this series, Vol. 89 [96].

Reagents for Method 1

Sodium potassium phosphate buffer, 100 mM, pH 7.0
Magnesium chloride, 50 mM
Formaldehyde, 50 mM, prepared by heating 300 mg of paraformal-
dehyde in 10 ml of water overnight in a sealed tube at 100°. The
strength (approximately 1 M) should be checked by the procedure
of Nash,[5] and the solution diluted accordingly.
D-Ribose 5-phosphate, sodium salt, 50 mM
Phosphoriboisomerase (type I, from spinach; Sigma Chemical Co.,
London)

Reagents for Method 2

Sodium potassium phosphate buffer, 100 mM, pH 7.0
Magnesium chloride, 25 mM
NADP, 10 mM
Formaldehyde, 50 mM (prepared from paraformaldehyde, see
Method 1)
D-Ribose 5-phosphate, sodium salt, 50 mM
Glucose-6-phosphate dehydrogenase [grade II, from yeast;
Boehringer Corporation (London) Ltd.]
Phosphoglucoisomerase [from yeast; Boehringer Corporation (Lon-
don) Ltd.]
Phosphoriboisomerase (type I, from spinach; Sigma Chemical Co.,
London)
Phosphohexuloisomerase (purified from *M. capsulatus*)[4]

Procedure for Method 1. The assay mixture contains, in a final volume
of 1 ml, sodium potassium phosphate buffer, 0.5 ml; magnesium chloride,
0.1 ml; formaldehyde, 0.1 ml; ribose 5-phosphate, 0.1 ml; phosphoriboisomerase, 1.75 units. This is preincubated for 15 min at 37° to
ensure formation of an equilibrium mixture of ribose 5-phosphate and
ribulose 5-phosphate. After this preincubation, extract containing up to
0.04 unit of hexulose-phosphate synthase activity is added, and formaldehyde consumption is followed at 37° by taking timed samples (0.1 ml)
from the assay mixture over a period of up to 10 min. The samples are
added to 1 ml of 10% (w/v) trichloroacetic acid solution, and the concentration of formaldehyde remaining is determined colorimetrically by the
method of Nash.[5] Formaldehyde disappearance not dependent on
hexulose-phosphate synthase activity is corrected for, when necessary,
by running parallel assays lacking ribose 5-phosphate. It should be noted
that D-ribulose 5-phosphate can be used in the above assay in place of
ribose 5-phosphate and phosphoriboisomerase; however, apart from the

[5] T. Nash, *Biochem. J.* **55**, 416 (1953).

increased cost, commercial ribulose 5-phosphate usually contains an inhibitor of hexulose-phosphate synthase.[1]

Procedure for Method 2. The assay mixture in a cuvette (10-mm light path) contains, in a final volume of 1 ml; sodium potassium phosphate buffer, 0.5 ml; magnesium chloride, 0.1 ml; NADP, 25 μl; ribose 5-phosphate, 0.1 ml; glucose-phosphate dehydrogenase, 0.7 unit; phosphoglucoisomerase, 0.7 unit; phosphoriboisomerase, 1.75 units; phosphohexuloisomerase, 1 unit. The mixture is equilibrated at 37° for several minutes, and then extract containing the hexulose-phosphate synthase is added. After a suitable time interval, formaldehyde (0.1 ml) is added and the rate of absorbance increase is followed at 340 nm. The resulting rate minus the formaldehyde-independent rate is taken to be due to hexulose-phosphate synthase activity.

Units. One unit of hexulose-phosphate synthase is defined as that amount of enzyme catalyzing the formation of 1 μmol of hexulose phosphate per minute at 37°. Specific activity is expressed as units per milligram of protein.

Purification Procedure

Growth of Organism. The strain of *Methylomonas capsulatus* ATCC No. 19069 (*Methylococcus capsulatus*) used, originated from Foster and Davis.[6] It is grown at 37° in a fermentor, sparged with methane–air (1 : 20, v/v), in a mineral salts medium containing per liter: Na_2HPO_4, 0.6 g; KH_2PO_4, 0.4 g; NH_4Cl, 0.4 g; $MgSO_4 \cdot 7 H_2O$, 0.2 g; $FeCl_3 \cdot 6 H_2O$, 16.7 mg; $CaCl_2 \cdot 2 H_2O$, 0.66 mg; $ZnSO_4 \cdot 7 H_2O$, 0.18 mg; $CuSO_4 \cdot 5 H_2O$, 0.16 mg; $MnSO_4 \cdot 4 H_2O$, 0.15 mg; $CoCl_2 \cdot 6 H_2O$, 0.18 mg; H_3BO_3, 0.1 mg; $Na_2MoO_4 \cdot 2 H_2O$, 0.3 mg. *It should be noted that methane–air mixtures containing more than 5% methane are explosive.* The actively growing culture is harvested by centrifugation at 4500 g for 15 min, and the cell paste is stored at −15°. It should be noted that the supernatant from step 2, described below, can be used for the purification of phosphohexuloisomerase.[4]

Step 1. Preparation of Cell-Free Extract. Cell paste (24 g wet weight) is suspended in 75 ml of 20 mM sodium potassium phosphate buffer, pH 7.0, containing 5 mM $MgCl_2$. The suspension is disrupted by ultrasonication for 6 min at 0–4° in an MSE ultrasonic disintegrator (150 W) and centrifuged at 6000 g for 10 min, yielding the cell-free extract as supernatant.

Step 2. Preparation of Particulate Fraction. The extract is further centrifuged at 38,000 g for 60 min, the pellet obtained is used for the purification of hexulose-phosphate synthase, and the supernatant can be used for

[6] J. W. Foster and R. H. Davis, *J. Bacteriol.* **91**, 1924 (1966).

the purification of phosphohexuloisomerase.[4] The pellet obtained from the centrifugation of the cell-free extract is resuspended with the aid of a glass homogenizer in 40 ml of 20 mM sodium potassium phosphate buffer, pH 7.0, containing 5 mM MgCl$_2$ and 1 M NaCl.

Step 3. Solubilization with 1 M NaCl. After 1 hr at 0° the resuspended particulate fraction is centrifuged at 38,000 g for 60 min; the pellet obtained is subjected to two further extractions and centrifugations. The supernatants from the three extractions are combined.

Step 4. Ammonium Sulfate Fractionation. To the combined supernatants from step 3, solid (NH$_4$)$_2$SO$_4$ is added with stirring. The protein precipitating between (NH$_4$)$_2$SO$_4$ concentrations of 1.025 M and 2.05 M is collected by centrifugation at 25,000 g for 15 min and dissolved in 10 mM sodium potassium phosphate buffer, pH 7.4, containing 2.5 mM MgCl$_2$.

Step 5. DEAE-Cellulose Chromatography. The protein solution from step 4 is applied to a column (19 × 5 cm) of DEAE-cellulose (Whatman DE-52) previously equilibrated with 10 mM sodium potassium phosphate buffer, pH 7.4, containing 2.5 mM MgCl$_2$. Hexulose-phosphate synthase activity is eluted from the column with the equilibration buffer, close to the void volume of the column. Fractions containing synthase activity are concentrated to 5.6 ml by ultrafiltration through a Diaflo PM-30 membrane filter (Amicon Corp., Lexington, Massachusetts).

Step 6. Calcium Phosphate Gel Treatment. The concentrated enzyme from step 5 is further purified by adsorption on calcium phosphate gel, prepared by the method of Keilin and Hartree.[7]

The protein solution from step 5 is treated with the gel, at a protein : dry calcium phosphate ratio of 1 : 1 (w/w), the gel is washed with 10 ml of 10 mM sodium potassium phosphate buffer, pH 7.4, containing 2.5 mM MgCl$_2$, and the hexulose-phosphate synthase is extracted by treating the gel twice with 3-ml volumes of 200 mM sodium potassium phosphate buffer, pH 7.4, containing 2.5 mM MgCl$_2$. The two extracts are combined, and the hexulose-phosphate synthase contained therein is stored at − 15°.

A typical purification is summarized in the table.

Properties

Purity. The enzyme preparation is not homogeneous; analysis of 50 μg by polyacrylamide gel electrophoresis either in the absence or the presence of sodium dodecyl sulfate showed four protein bands on staining with Coomassie Blue. The enzyme preparation does not contain detectable activities of phosphoriboisomerase, ribulose-5-phosphate 3-epimerase, or phosphohexuloisomerase.

[7] D. Keilin and E. F. Hartree, *Proc. R. Soc. London, Ser. B* **124**, 397 (1938).

PURIFICATION OF HEXULOSE-PHOSPHATE SYNTHASE FROM *Methylomonas capsulatus*

Step	Fraction	Total volume (ml)	Total protein (mg)	Total activity[a] (units)	Specific activity (units/mg protein)	Yield (%)
1	Cell-free extract	91	1940	3310	1.71	100
2	Resuspended particulate fraction	50	1350	2750	2.04	83
3	Solubilized particulate fraction	125	414	1720	4.17	52
4	$(NH_4)_2SO_4$ fraction (1.025–2.05 M)	20.5	194	1560	8.03	47
5	Pooled, concentrated, fractions after DEAE-cellulose chromatography	5.6	14.0	823	58.9	25
6	Calcium phosphate gel eluate	6.0	7.50	517	69.0	15

[a] Results obtained by Assay Method 1.

Substrate Specificity. No formaldehyde fixation can be detected in the presence of enzyme and the following compounds: ribose 5-phosphate, xylulose 5-phosphate, allulose 6-phosphate, fructose 6-phosphate, glyceraldehyde 3-phosphate, and dihydroxyacetone phosphate.

Effect of Metal Ions. Hexulose-phosphate synthase has an absolute requirement for a bivalent metal ion, Mg^{2+} and Mn^{2+} being most effective; Co^{2+} and Zn^{2+} are much less effective. Mg^{2+} or Mn^{2+} are required not only for enzyme activity, but also for enzyme stability. The enzyme rapidly, and apparently irreversibly, is denatured in the absence of one of these metal ions. Ni^{2+}, Ca^{2+}, and Cu^{2+} at 1 mM concentration are inhibitory to the enzyme (64, 49, and 24% inhibition, respectively).

Stability. The purified enzyme, in the presence of 2.5 mM $MgCl_2$, is stable for 6 months at $-15°$, but is rapidly inactivated at elevated temperatures; e.g., activity is totally lost within 5 min at 60°.

pH Optimum. The enzyme has optimum activity, in the synthetic direction, at pH 7.0.

Michaelis Constants. The apparent K_m for ribulose 5-phosphate (determined at 4 mM formaldehyde concentration) is $8.3 \times 10^{-5} M$; for formaldehyde (determined at 0.57 mM ribulose 5-phosphate concentration) it is $4.9 \times 10^{-4} M$. In the reverse direction, the apparent K_m for D-arabino-3-hexulose 6-phosphate is $7.5 \times 10^{-5} M$.

Molecular Weight and Subunit Structure. The molecular weight of hexulose-phosphate synthase in 50 mM sodium potassium phosphate buffer, pH 7.4, containing 2.5 mM $MgCl_2$, as measured by gel permeation through a column of Sephadex G-200,[8] corresponds to 310,000. If a similar

[8] P. Andrews, *Biochem. J.* **96**, 595 (1965).

buffer at pH 4.6 is used, a single species is eluted from the column at a position corresponding to a molecular weight of approximately 49,000. This suggests that the enzyme can dissociate into six subunits. Partial dissociation is also observed during gel filtration in buffer of low phosphate concentration (10 mM sodium potassium phosphate, pH 7.4, containing 2.5 mM MgCl$_2$).

Equilibrium Constant. The equlibrium constant of the reaction, at pH 7.0 and 30°, as determined in the synthetic direction is $(4.0 \pm 0.7) \times 10^{-5}$ (mean of six determinations ± standard deviation).

[52] 3-Hexulose-phosphate Synthase from *Methylomonas* M15

By HERMANN SAHM, HORST SCHÜTTE, and MARIA-REGINA KULA

D-Ribulose 5-phosphate + formaldehyde $\rightleftharpoons$ D-arabino-3-hexulose 6-phosphate

The first step in the ribulose monophosphate cycle of formaldehyde fixation in some methylotrophic bacteria is the condensation of formaldehyde and D-ribulose 5-phosphate to give D-*arabino*-3-hexulose 6-phosphate.[1,2] This reaction is catalyzed by the enzyme 3-hexulose-phosphate synthase. So far this enzyme has been purified from a methane-utilizing bacterium, *Methylococcus capsulatus*,[3] and the two methanol-utilizing bacteria, *Methylomonas* M15[4] and *Methylomonas aminofaciens*.[5]

Assay Method

Principle. The enzyme activity is measured by determining the incorporation of [14C]formaldehyde into sugar phosphates according to the method of Lawrence *et al.*[6] Because of the known inhibitory effect of commercial ribulose 5-phosphate,[3] this sugar is generated from ribose 5-phosphate by ribose-5-phosphate isomerase.

[1] M. B. Kemp, *Biochem. J.* **139,** 129 (1974).
[2] T. Strøm, T. Ferenci, and J. R. Quayle, *Biochem. J.* **144,** 465 (1974).
[3] T. Ferenci, T. Strøm, and J. R. Quayle, *Biochem. J.* **144,** 477 (1974).
[4] H. Sahm, H. Schütte, and M. R. Kula, *Eur. J. Biochem.* **66,** 591 (1976).
[5] N. Kato, H. Ohashi, Y. Tani, and K. Ogata, *Biochim. Biophys. Acta* **523,** 236 (1978).
[6] A. J. Lawrence, M. B. Kemp, and J. R. Quayle, *Biochem. J.* **116,** 631 (1970).

Reagents

Tris-HCl buffer, 300 mM, adjusted to pH 7.5 (A)
D-Ribose 5-phosphate (sodium salt), 100 mM
[^{14}C]Formaldehyde (2 μCi/ml), 80 mM
MgCl$_2$, 100 mM
D-Ribose-5-phosphate isomerase (EC 5.3.1.6), 100 units per milliliter
    of Tris-HCl buffer (A)
Ethanol
Barium acetate, 5%, w/v

Procedure. The reaction mixture consists of 0.05 ml of Tris-HCl buffer, 0.1 ml of D-ribose-5-phosphate, 0.05 ml of [^{14}C]formaldehyde, 0.05 ml of MgCl$_2$, 0.05 ml of D-ribose-5-phosphate isomerase, and limiting amounts of the enzyme in a final volume of 0.4 ml. After a 5-min preincubation period at 30°, the reaction is started by the addition of the enzyme and incubation is continued for 5 min at 30°. The fixation of [^{14}C]formaldehyde is linear for 8 min and up to 1 μg of enzyme protein. The reaction is stopped by adding 1.5 ml of ethanol and 0.1 ml of barium acetate solution to precipitate the sugar phosphates. The mixture is left for 10 min at 0° and then filtered under reduced pressure through a circle (2.5 cm^2) of glass fiber paper (Whatman GF/A). The precipitate is washed twice with 2 ml of ethanol–formaldehyde (19:1, v/v) and once with 2 ml of ethanol, sucked dry on the pump, and then further dried under a radiant lamp for 5 min. Each disk is placed in a scintillation vial and assayed for radioactivity by liquid-scintillation counting. When defined starting concentrations of ribulose 5-phosphate are required in the assay, ribulose 5-phosphate is previously synthesized with ribose-5-phosphate isomerase. The ribulose 5-phosphate concentration is determined by the cysteine-carbazole method,[7] and known amounts are then used in the assay instead of ribose 5-phosphate and ribose-5-phosphate isomerase.

Definition of Enzyme Unit and Specific Activity. One unit of enzyme activity is defined as the amount of enzyme that catalyzes the incorporation of 1 μmol of formaldehyde into sugar phosphates per minute under these assay conditions. Specific activity is defined as the number of units per milligram of protein.

Production of 3-Hexulosephosphate Synthase

For enzyme preparations *Methylomonas* M15 (DSM 580) is grown in fermentors on a basal medium with 1% (v/v) methanol as the sole carbon and energy source as described earlier.[8] Cells are harvested by centrifuga-

---

[7] C. Ashwell and J. Hickman, *J. Biol. Chem.* **266**, 65 (1957).
[8] R. A. Steinbach, H. Schütte, and H. Sahm, this series, Vol. 89 [46].

tion at the end of the exponential growth phase in a high-speed continuous-flow centrifuge and stored at $-20°$ until used.

Purification Procedure

All steps are carried out at 4°, and all buffers contain 5 mM MgCl$_2$, 1 mM EDTA, and 0.5 mM phenylmethanesulfonyl fluoride unless otherwise stated.

, *Step 1. Preparation of Crude Extract.* Frozen cells (300 g) are softened overnight in the cold room and suspended with 1 liter of 0.02 M Tris-HCl buffer (pH 7.5) using a Waring blender. The cells are broken by two passages through a Manton–Gaulin homogenizer at 600 kp/cm^2. The suspension is cooled in ice after each run. Cell debris are removed by centrifugation at 40,000 g for 2 hr. The enzyme remains in the supernatant in contrast to the 3-hexulose-phosphate synthase from *Methylococcus capsulatus,* which would sediment at this step.[3]

Step 2. Chromatography on DEAE-Cellulose. A glass column (10 × 90 cm) is packed with DEAE-cellulose (Whatman DE-52) under slight pressure and equilibrated with 0.02 M Tris-HCl buffer (pH 7.5). The clear crude extract obtained in step 1 is applied to the column with a flow rate of 500 ml/hr. The column is washed first with 5 liters of equilibration buffer. The enzyme is then eluted with a linear gradient produced from 10 liters of 0.02 M and 10 liters of 0.2 M Tris-HCl buffer (pH 7.5). Fractions of 200 ml are collected and tested for 3-hexulose-phosphate synthase activity. The enzyme is eluted at approximately 0.1 M Tris-HCl concentration. The peak fractions are combined and concentrated to about 25 ml by ultrafiltration using an Amicon hollow-fiber H1P10 and a PM-10 filter membrane successively.

Step 3. Gel Filtration on Sephadex G-75. A column (5 × 90 cm) is packed with Sephadex G-75 and equilibrated with 0.02 M Tris-HCl buffer (pH 7.5). The concentrated enzyme solution obtained at step 2 is applied to the column, and elution is carried out with the same buffer at a flow rate of 80 ml/hr. The effluent is collected in fractions of 15 ml, and those exhibiting the highest enzyme activities are pooled and concentrated by ultrafiltration using an Amicon PM-10 membrane.

Step 4. Chromatography on DEAE-Sephadex A-50. The enzyme solution obtained at step 3 is applied to a column (5 × 50 cm) packed with DEAE-Sephadex A-50 and equilibrated with 0.02 M Tris-HCl buffer (pH 7.5). After sample application the column is washed with 1 liter of the same buffer and the protein is eluted with a linear gradient between 2 liters of 0.02 M and 2 liters of 0.2 M Tris-HCl buffer (pH 7.5). The flow rate is 42 ml/hr, and fractions of 10 ml are collected. The enzyme activity is eluted

PURIFICATION OF 3-HEXULOSE-PHOSPHATE SYNTHASE FROM *Methylomonas* M15

Enzyme fraction	Volume (ml)	Total protein (mg)	Total activity (units × 10³)	Specific activity (units/mg protein)	Yield (%)	Purification (fold)
Crude extract	970	14,065	63.0	4.5	100	1
DEAE-cellulose	1800	540	31.0	58	49	12.9
Gel filtration (Sephadex G-75)	220	253	16.0	62.5	25	13.9
DEAE-Sephadex A-50	590	190	12.5	66.5	20	14.8

with about 0.1 M Tris-HCl buffer. Active fractions are combined, concentrated by ultrafiltration, and stored at $-20°$.

A summary of a typical purification is given in the table.

Remarks. If a large column is not available, the first DEAE-cellulose chromatography may be replaced by a batchwise adsorption and desorption procedure. In this case the order of chromatographic steps should be reversed and gel filtration be performed last. Losses of activity in the permeate of PM-10 membranes or hollow-fiber H1P10 cartridges are negligible. The protease inhibitor phenylmethanesulfonyl fluoride is included in all buffers to prevent proteolytic degradation.

Properties

Molecular Weight and Subunit Structure. The molecular weight of the 3-hexulose-phosphate synthase was determined by gel filtration on Sephadex G-100 to be approximately 40,000. Furthermore, by the sedimentation equilibrium method a molecular weight of 43,000 ± 3000 was calculated. Polyacrylamide gel electrophoresis in the presence of sodium dodecyl sulfate indicated that the enzyme is a dimer composed of two probably identical subunits with a molecular weight of 22,000.

Specificity. It seems that the enzyme is highly specific for D-ribulose 5-phosphate within the range of formaldehyde acceptors tested. Studies with the purified enzyme by measuring the incorporation of formaldehyde into sugar phosphates showed no detectable activity with the following compounds: D-fructose 6-phosphate, D-fructose 1,6-bisphosphate, D-glucose 1-phosphate, D-glucose 6-phosphate, lithium hydroxypyruvate, glyceraldehyde 3-phosphate, dihydroxyacetone phosphate, D-ribose 5-phosphate, and D-xylose 5-phosphate. These results confirm the finding of Kemp[1] and Ferenci et al.[3] that ribulose 5-phosphate is the acceptor molecule for formaldehyde in the aldol condensation catalyzed by the

3-hexulose-phosphate synthase. The Michaelis constants for formaldehyde and D-ribulose 5-phosphate were calculated to be 1.1 mM and 1.6 mM, respectively.

Effect of pH. The enzyme exhibits an optimum for activity at 7.5–8.0 and is stable at the pH range of 6.0–8.0 at 30° for 1 hr. However, the enzyme stability decreases at pH values below 6.0.

Metal Ion Requirement. The enzyme has a requirement for Mg^{2+} (5 mM). The apparent K_m value for MgCl$_2$ was found to be 0.25 mM. Instead of Mg^{2+}, Mn^{2+} is also very effective for promoting enzyme activity, whereas Cd^{2+}, Co^{2+}, and Zn^{2+} activate the enzyme only partially. The metal ions Ca^{2+}, Cu^{2+}, Hg^{2+}, and Ni^{2+} inhibit the enzyme.

Section V

Phosphatases

[53] Fructose-bisphosphatase, Zinc-Free, from Rabbit Liver

By MARGARET M. DeMAINE, CAROL A. CAPERELLI, and
STEPHEN J. BENKOVIC

D-Fructose 1,6-bisphosphate + H_2O → D-fructose 6-phosphate + P_i

In an earlier volume[1] there appears a procedure for the purification
from rabbit liver of fructose-1,6-bisphosphatase (FBPase) (EC 3.1.3.11)
with high specific activity at neutral pH. As this enzyme has a high affinity
for Zn^{2+} ($K_{dissociation} \simeq 0.1 \ \mu M$),[2,3] which occurs in trace quantities in most
chemicals, the homogeneous purified enzyme still contains ca. 2 mol of
Zn^{2+} per mole (143,00 daltons). It is possible that the native enzyme
contains Zn^{2+} under certain cellular conditions. A technique to obtain
enzyme virtually free of zinc will be described.

Assay Method

The enzyme is assayed spectrophotometrically by following the rate of
NADPH accumulation at 340 nm in the presence of excess glucose-6-
phosphate dehydrogenase and glucose-6-phosphate isomerase. $(NH_4)_2SO_4$
is removed from solutions of auxiliary enzymes by dialysis at pH 7.5
versus 10 mM Tris-HCl or 5 mM N-2-hydroxyethylpiperazine-N'-2-
ethanesulfonic acid (HEPES)-KOH.

The routine assay mixture at 25° contains 50 mM Tris-HCl, pH 7.5;
5 mM $MgCl_2$; 0.1 mM EDTA; 0.2 mM $NADP^+$; 3.5 units of glucose-6-
phosphate isomerase; 0.7 unit of glucose-6-phosphate isomerase, FBPase,
and (to initiate reaction) 0.1 mM fructose 1,6-bisphosphate.

FBPase Purification

Neutral FBPase is purified from frozen livers of young, 24-hr-fasted
rabbits by the published procedure[1] with the following modifications.[4] The

[1] E. H. Ulm, B. M. Pogell, M. M. de Maine, C. B. Libby, and S. J. Benkovic, this series,
Vol. 42 [60].
[2] F. O. Pedrosa, S. Pontremoli, and B. L. Horecker, *Proc. Natl. Acad. Sci. U.S.A.* **74**, 2742
(1977).
[3] P. A. Benkovic, C. A. Caperelli, M. M. de Maine, and S. J. Benkovic, *Proc. Natl. Acad.
Sci. U.S.A.* **75**, 2185 (1978).
[4] S. J. Benkovic, W. A. Frey, C. B. Libby, and J. J. Villafranca, *Biochem. Biophys. Res.
Commun.* **57**, 196 (1974).

10-min heat treatment at 65° is usually carried out on small portions of the homogenate in thin-wall glass Erlenmeyer flasks that are shaken manually in a large water-bath at 72°. After rapid chilling, pH adjustment from 7.0 to 6.0, and centrifugation, the supernatant is dialyzed overnight against 16 volumes of doubly distilled, deionized water. The dialyzed fraction is centrifuged at 13,000 g for 40 min and then added dropwise to the slurry of CM-cellulose. Washing and elution are carried out as described.[1] Eluted enzyme is concentrated by dialysis against saturated $(NH_4)_2SO_4$ solution adjusted to pH 7.0. The centrifuged precipitate is redissolved in 50 mM Tris-HCl, pH 7.5 (recrystallized Tris), and dialyzed against twelve changes of 10 mM buffer to remove any remaining $(NH_4)_2SO_4$.

This purified protein shows one sharp band on disc gel electrophoresis in the absence and in the presence of sodium dodecyl sulfate. At 280 nm the absorbance of a solution containing 1.0 mg of neutral FBPase per milliliter in a 1.0-cm light path is 0.71.

Zn Analysis

Atomic absorption spectroscopy is employed to measure the concentration of the stock Zn^{2+} solution, and the levels of Zn^{2+} (and Mn^{2+}) in buffer solutions, water, and FBPase samples. Measurements are made with a Perkin-Elmer 305-A spectrophotometer, equipped with a HGA 2000 graphite furnace and controller. Zn is measured at 213.9 nm, and Mn is measured at 279.5 nm.

Removal of Zn^{2+} from FBPase

Glassware used in preparation and storage of the samples is soaked in aqua regia (concentrated HCl and HNO_3, 3 : 1) and then rinsed well with doubly distilled deionized water. Plastic ware is soaked in Micro cleaner (International Products Corp., Box 118, Trenton, New Jersey), rinsed with doubly distilled deionized water, soaked in boiling 0.1 M EDTA, and then rinsed thoroughly with doubly distilled deionized water. Plastic containers, thus treated, are regarded as being freer of heavy-metal contaminants than the treated glassware.

Buffer is prepared from Tris base that had been recrystallized from 95% ethanol containing 0.0001% EDTA. Stock solutions of tris buffer are extracted with 0.001% dithizone in CCl_4. HEPES is recrystallized from 90% ethanol containing 0.005% EDTA. For work with ultrapure enzyme, $MgCl_2$ is prepared from MgO (99.5%, Alfa Inorganics, Ventron Corp., Danvers, Massachusetts).

Dialysis against doubly distilled deionized water for 12 days (30 changes) at 4° seemed to effect consistently the most complete removal of

QUANTITATIVE SPECTROSCOPIC ANALYSES OF Zn^{2+} AND Mn^{2+}

Treatment	Mol M^{2+}/mol FBPase		Concentration (M)	
	Zn^{2+}	Mn^{2+}	Zn^{2+}	Mn^{2+}
FBPase purified to homogeneity	1.9, 2.2	—	—	—
FBPase after 12-day dialysis vs H_2O	0.032, 0.020	0.134	—	—
FBPase after Chelex columns and dialysis vs buffer	1.69, 1.99	2.58	—	—
HEPES, 10 mM, pH 7.5	—	—	37.1 nM	<20 nM
Doubly distilled deionized H_2O	—	—	43.8 nM	<20 nM

Zn^{2+} (and Mn^{2+}) (see the table) from concentrated (5–12 mg/ml) FBPase purified to homogeneity as described above. Surprisingly, a procedure involving passage of FBPase through columns of Chelex-100 and dialysis against buffer effected little removal of Zn^{2+} (see the table). Specifically, FBPase (9 mg/ml, 1.5 ml) was passed through three 0.9- × 16-cm columns of Chelex-100, with concentration (collodion bag apparatus, Schleicher & Schuell) after elution from each column, followed by dialysis against 10 mM HEPES, pH 7.5, for 24 hr at 4°.

Removal of the tightly bound Zn^{2+} apparently did not irreversibly alter the FBPase structure. Incubation of aliquots of Zn^{2+}-free FBPase (0.1 mg/ml) with 10 μM Zn^{2+}, 0.5 mM Mn^{2+}, or 5 mM Mg^{2+} (the optimum concentrations of the required metal cofactor) for intervals ranging from 1 min to 24 hr, followed by assays, showed activities 50 to 100% the values observed for enzyme from which the trace Zn had not been removed.[3]

However, experimental studies of other properties of zinc-free FBPase were technically difficult because exposure of the enzyme to buffer or other reagents immediately resulted in binding of the adventitious Zn^{2+} by FBPase. The dissociation constant of Zn^{2+} from the first site of rabbit liver FBPase is estimated at <0.1 μM.[3] Binding constants for the second, third, and fourth sites are 9.99×10^6 M^{-1}, 1.55×10^6 M^{-1}, and 1.49×10^6 M^{-1}, respectively. Substrate binding measurements have been carried out in the presence of 50 mM EDTA to maintain "zinc-free" enzyme.[5]

Acknowledgment

Financial support for this research was provided by Grant GM 13306 from the U.S. Public Health Service. Margaret de Maine is an awardee of an Eloise Gerry Fellowship from Sigma Delta Epsilon.

[5] M. M. de Maine and S. J. Benkovic, *Arch. Biochem. Biophys.* **205**, 308 (1980).

[54] Fructose-bisphosphatase from Ox Liver

By H. G. Nimmo and K. F. Tipton

Fructose 1,6-bisphosphate + H_2O → fructose 6-phosphate + P_i

Assay Method

Fructose-1,6-bisphosphatase (EC 3.1.3.11) is assayed by coupling the production of fructose 6-phosphate to the reduction of $NADP^+$ in the presence of excess glucosephosphate isomerase (EC 5.3.1.9) and glucose-6-phosphate dehydrogenase (EC 1.1.1.49).[1]

Reagents

Triethanolamine hydrochloride–KOH, 100 mM, pH 7.4
Glycine–KOH, 100 mM, pH 9.6
KCl, 1.0 M
$MgSO_4$, 210 mM
EDTA, 10 mM
Fructose 1,6-bisphosphate, 10 mM
$NADP^+$, 15 mM
Glucosephosphate isomerase from yeast, 10 mg/ml
Glucose-6-phosphate dehydrogenase from yeast, 5 mg/ml
Adenosine 5'-monophosphate, 15 mM

Procedure. Each cuvette contains 0.5 ml of pH 7.4 or pH 9.6 buffer, 0.1 ml of KCl, 0.01 ml of each of $MgSO_4$, EDTA, fructose bisphosphate, and $NADP^+$, 0.002 ml of glucose phosphate isomerase (7 units), 0.003 ml of glucose-6-phosphate dehydrogenase (3 units) and enzyme in a total volume of 1.0 ml. To monitor the sensitivity of the enzyme to inhibition by AMP (see below), 0.01 ml of AMP is included in the assay. The reduction of $NADP^+$ is measured at 340 nm and 30°.

The final pH values in the cuvettes for the two buffers are 7.2 and 9.5, respectively. The reaction is usually started by addition of the enzyme solution. However, when the enzyme is assayed after step 5 (substrate elution from CM-cellulose) this procedure gives rise to slowly accelerating rates. At this step, therefore, the reaction is started by addition of the fructose bisphosphate.

With the stated amounts of the coupling enzymes, there is a lag of 30–40 sec before a linear rate is established. Rate is proportional to fructose-bisphosphatase concentration up to about 0.2 absorbance unit/min.

[1] E. Racker and E. A. R. Schroeder, *Arch. Biochem. Biophys.* **74**, 326 (1958).

METHODS IN ENZYMOLOGY, VOL. 90

The assay can be used for the crude extract, but when the activity of this fraction is measured in the presence of AMP, accelerating rates are sometimes obtained. This apparently results from destruction of AMP in the cuvette. One unit of activity is the amount required to catalyze the formation of 1 μmol of product per minute.

Purification

Each step is carried out as rapidly as possible at 0–4°.

Step 1. Extraction. Ox liver is obtained from a freshly slaughtered animal, cooled in ice, and transported to the laboratory. Five 200-g portions are homogenized in 4 volumes of 0.15 M KCl for 15 sec at full speed in a Kenwood blender. The homogenate is strained to remove any large lumps of tissue and centrifuged for $6 \times 10^5 g \cdot$ min.

Step 2. pH Treatment. The supernatant is carefully decanted and then adjusted to pH 4.55 ± 0.05 by the dropwise addition of cold 5 M acetic acid, with stirring. The pH is immediately adjusted to 6.5 ± 0.1 by the dropwise addition of cold 5 M NaOH, with stirring. The resulting suspension is centrifuged for $1 \times 10^5 g \cdot$ min, and the supernatant is decanted.

Step 3. $(NH_4)_2SO_4$ Fractionation. To each liter of supernatant 327 g of solid $(NH_4)_2SO_4$ are added slowly. The material is stirred for 30 min and centrifuged for $2 \times 10^5 g \cdot$ min. The precipitate is discarded, and a further 66 g of $(NH_4)_2SO_4$ are added per liter of supernatant. The material is stirred and centrifuged as before, and the precipitate is dissolved in 250 ml of 5 mM malonic acid–NaOH, pH 6.0 (adjusted at 4°) (buffer A). The solution is desalted by passage through a column of Sephadex G-25 (bed volume 2 liters) equilibrated with buffer A.

Step 4. Gradient Elution from CM-Cellulose. The desalted solution is loaded onto a 5- × 10-cm column of CM-cellulose equilibrated in buffer A. The column is washed with 1 liter of buffer A and then developed with a 2-liter linear gradient of 5 to 30 mM malonic acid–NaOH, pH 6.0. Fructose-bisphosphatase is eluted in a single sharp peak at 18–20 mM buffer. Fractions with activity greater than 0.5 unit/ml are pooled and desalted into 5 mM malonic acid–NaOH, pH 6.5 (adjusted at 4°) (buffer B) on a column of Sephadex G-25.

Step 5. Substrate Elution from CM-Cellulose. The desalted material from step 4 is loaded onto a 2- × 10-cm column of CM-cellulose equilibrated in buffer B. The column is washed with buffer B until the effluent is protein-free as judged by its absorbance at 280 nm. The column is then developed with a 0.1 mM solution of fructose bisphosphate in buffer B adjusted to pH 6.5 at 4° with NaOH. Fructose-bisphosphatase is eluted in a sharp peak with the fructose bisphosphate front. Fractions with specific

PURIFICATION OF OX LIVER FRUCTOSE-BISPHOSPHATASE[a]

Fraction	Volume (ml)	Enzyme activity at pH 7.2 (units/ml)	Protein concentration (mg/ml)	Specific activity at pH 7.2 (units/mg protein)	"pH ratio"	Percentage inhibition by 15 μM AMP	Purification (fold)	Yield (%)
Step 1	3600	1.1	108	0.01	2.10	54.3	1	100
Step 2	2550	1.2	34	0.04	2.05	55.9	4	79
Step 3	127	12.0	26	0.46	1.95	51.7	46	40
Step 4	330	4.1	1.8	2.3	2.03	53.6	230	35
Step 5	69	15.2	1.0	15.2	2.06	56.1	1520	27
Step 6	8.8	106	4.6	22.5	2.04	54.9	2250	25

[a] Reproduced from Nimmo and Tipton[2] by permission of the *Biochemical Journal*.

activity in excess of 10 units/mg are pooled, dialyzed into 20 mM triethanolamine hydrochloride–KOH buffer, pH 7.5, containing 2 mM MgSO$_4$ and 1 mM EDTA (buffer C), and concentrated by ultrafiltration to 4–6 mg/ml.

Step 6. Gel Filtration on Sephadex G-200. The concentrated material is loaded onto a 3- × 60-cm column of Sephadex G-200 equilibrated in buffer C. The column is developed with buffer C at 10 ml/hr, and 3-ml fractions are collected. The fructose-bisphosphatase is eluted as a single symmetrical peak with the specific activity constant across the peak. For storage the enzyme is concentrated by ultrafiltration to 3–4 mg/ml, (NH$_4$)$_2$SO$_4$ is added (0.5 g/ml), and the suspension is kept at 4°. Under these conditions the enzyme loses less than 2% activity per month.

Comments on the Purification. The results of a typical purification are shown in the table. The procedure has been completed successfully by several workers in different laboratories. The overall purification is normally 2000- to 2500-fold, the final specific activity and the overall yield being 21–25 units/mg and 18–25%, respectively. The final material is homogeneous by the criteria of gel electrophoresis under both denaturing and nondenaturing conditions and analytical ultracentrifugation.[2]

The malonate buffer for step 5 must be adjusted to the correct pH at 4°. If the buffer is made up at room temperature, the pH will be too low and the fructose-bisphosphatase may be eluted from CM-cellulose as a very dilute solution rather than as a sharp peak. The enzyme loses activity very rapidly in these circumstances. This loss of activity can be minimized by using plastic test tubes at this step to avoid surface adsorption and denaturation and by concentrating the enzyme solution as rapidly as possible.

[2] H. G. Nimmo and K. F. Tipton, *Biochem. J.* **145**, 323 (1975).

Properties of the Enzyme

Assessment of Proteolysis. It is well known that mammalian liver fructose-bisphosphatases are very susceptible to proteolysis during the isolation procedure.[3] The most obvious kinetic effects of proteolysis are a change in the pH optimum of the enzyme from neutral to alkaline pH values and a decrease in its sensitivity to inhibition by AMP.[3] Neither the "pH ratio" (the ratio of the activity of the enzyme at pH 7.2 to its activity at pH 9.5) of ox liver fructose-bisphosphatase nor its sensitivity to inhibition by 15 μM AMP change during the isolation (see the table). Furthermore, the enzyme shows only one band on sodium dodecyl sulfate (SDS)–polyacrylamide gels, and its native molecular weight, as judged by gel filtration, is not altered by the purification.[2] These pieces of evidence indicate that the purified enzyme is undegraded.

Tryptophan Content. The enzyme contains one residue of tryptophan per subunit as judged spectrophotometrically[2] and by amino acid analysis after hydrolysis with methanesulfonic acid (H. G. Nimmo, unpublished observations). The fluorescence spectrum of the enzyme is typical of a tryptophan-containing protein.[2] By contrast, the native fructose-bisphosphatases from rabbit, rat, and human liver contain no tryptophan.[4-6] This appears to be caused simply by a species difference.

Subunit Composition. The molecular weight of the enzyme measured by analytical ultracentrifugation is 140,000 ± 3000.[2] The subunit molecular weight estimated by SDS gel electrophoresis is 35,500 ± 1500.[2] Peptide mapping suggests that only one type of subunit is present, so the enzyme must be a tetramer composed of identical subunits.[2]

Kinetics. The enzyme requires divalent cations for activity; Mg^{2+} ions are normally used. The K_m for fructose bisphosphate (2 μM at pH 7.2, 6 μM at pH 9.5) is not greatly affected by the level of Mg^{2+} ions. The enzyme responds hyperbolically to Mg^{2+} ions at pH 9.5, with a K_m of 0.15 mM, but sigmoidally at pH 7.2.[7] AMP is well known to be an allosteric inhibitor of mammalian fructose-bisphosphatases[3]; for the ox liver enzyme the sensitivity to inhibition by AMP is controlled by the concentration of Mg^{2+} ions.[8] For example, the apparent K_i for AMP is 4 μM at 0.25 mM Mg^{2+} but is 25 μM at 3.0 mM Mg^{2+}.[8]

Activation by Chelating Agents. Ox liver fructose-bisphosphatase is

[3] B. L. Horecker, E. Melloni, and S. Pontremoli, *Adv. Enzymol.* **42**, 193 (1975).
[4] S. Pontremoli, E. Melloni, F. Salamino, M. Michetti, L. H. Botelho, H. A. El-Dorry, D. K. Chu, C. E. Isaacs, and B. L. Horecker, *Arch. Biochem. Biophys.* **191**, 825 (1978).
[5] G. A. Tejwani, F. O. Pedrosa, S. Pontremoli, and B. L. Horecker, *Arch. Biochem. Biophys.* **177**, 253 (1976).
[6] A. Dzugaj and M. Kochman, *Biochim. Biophys. Acta* **614**, 407 (1980).
[7] H. G. Nimmo and K. F. Tipton, *Eur. J. Biochem.* **58**, 567 (1975).
[8] H. G. Nimmo and K. F. Tipton, *Eur. J. Biochem.* **58**, 575 (1975).

markedly stimulated at neutral pH by EDTA and other chelating agents.[2] The requirement for a chelating agent can be removed by treating all the buffers used for storage and assay of the enzyme with Chelex resin.[2] Moreover, EDTA does not bind directly to the enzyme.[2] It is reasonable to conclude that the effect of chelating agents is to remove inhibitory heavy-metal ions from the enzyme.

[55] Fructose-1,6-bisphosphatase from Turkey Liver

By PETER F. HAN and JOE JOHNSON, JR.

$$\text{Fructose 1,6-bisphosphate} + H_2O \xrightarrow{\text{FBPase}} \text{fructose 6-phosphate} + P_i$$

Assay Method

Principle. For the assay of fructose-1,6-bisphosphatase (FBPase; EC 3.1.3.11), the formation of fructose 6-phosphate is measured spectrophotometrically by following the reduction of NADP at 340 nm in a coupled reaction containing excess of phosphoglucose isomerase and glucose-6-phosphate dehydrogenase. Alternatively, the P_i formed can be determined by the method of Tashima and Yoshimura.[1]

Reagents for Spectrophotometric Method

Buffer: 60 mM diethanolamine–60 mM triethanolamine, pH 7.5 or 9.2
MgCl$_2$, 0.15 M
KCl, 10 M
EDTA, 10 mM, neutralized to pH 7.5 with 1 N NaOH
NADP, 10 mM, sodium salt
Fructose 1,6-bisphosphate (FBP), 5 mM, sodium salt
Phosphoglucose isomerase, 1 mg/ml (Sigma)
Glucose-6-phosphate dehydrogenase, 1 mg/ml (Sigma)

Procedure. In a cuvette having a 1-cm light path, place 0.5 ml of buffer, 0.01 ml of MgCl$_2$, 0.01 ml of KCl, 0.01 ml of EDTA, 0.01 ml of NADP, 0.005 ml each of phosphoglucose isomerase and glucose-6-phosphate dehydrogenase, and the appropriate amount of FBPase and water to a total volume of 0.99 ml. After incubation for 5 min at 25°, start the reaction by adding 0.01 ml of FBP. In monitoring the FBPase activity during isolation, 1 unit of 5′-nucleotidase (Sigma) is additionally included in each assay

---

[1] Y. Tashima and N. Yoshimura, *J. Biochem. (Tokyo)* **78**, 1161 (1975).

PURIFICATION OF TURKEY LIVER FRUCTOSE-1,6-BISPHOSPHATASE

Step and fraction	Total units[a]	Specific activity[a] (units/mg protein)	Yield (%)	Activity ratio pH 7.5:9.2[b]
1. Crude extract	1085	0.09	100	3.5
2. Heat fraction	1041	0.32	96	3.6
3. Phosphocellulose-treated fraction	1031	0.91	95	3.6
4. Phosphocellulose eluate	749	25.1	69	3.7
5. Blue Sepharose fraction	662	25.3	61	3.7

[a] Calculated on the basis of enzyme activity at pH 7.5.

[b] Since the inhibition of enzyme activity by high fructose 1,6-bisphosphate (FBP) concentration is nearly absent at pH 9.2, this ratio varies greatly with the concentration of FBP used in the assay.[3]

mixture to hydrolyze 5'-AMP, which is a potent inhibitor of FBPase and is present in enzyme solution throughout the course of purification.

Unit and Specific Activity. One unit of enzyme activity is defined as the amount that causes the formation of 1 μmol of fructose 6-phosphate per minute at 25°. Specific activity is defined as units per milligram of protein. Protein concentration is measured by the method of Lowry *et al.*[2] or calculated from the absorbance at 280 nm. The absorbance of a 0.5 N NaCl solution containing 1 mg of purified FBPase (dry weight) per milliliter in 1-cm light path is 0.705 at 280 nm.

Purification Procedure

All operations are carried out at 2–4° unless otherwise stated. Results of a typical purification are summarized in the table.

Step 1. Preparation of Crude Extract. Freshly removed young turkey livers (200 g) are cut into small pieces and homogenized for 2 min at medium speed in a Waring blender with 600 ml of ice-cold 40 mM Tris-HCl buffer (pH 8.0) containing 1 mM EDTA, 10 mM 2-mercaptoethanol, 0.5 mM 5'-AMP, and 3 mM FBP. The homogenate is centrifuged for 40 min at 32,000 g, and the precipitate is discarded (crude extract, 590 ml).

Step 2. Heat Treatment. The crude extract (pH 7.6) is transferred to two 500-ml beakers and heated with constant stirring in a water bath maintained at 85°. When the temperature of the enzyme solution reaches 68°, the mixture is rapidly cooled in an ice bath and centrifuged to remove the precipitate (heat fraction, 492 ml).

[2] O. H. Lowry, N. J. Rosebrough, A. L. Farr, and R. J. Randall, *J. Biol. Chem.* **193**, 265 (1951).

Step 3. Removal of Unwanted Proteins by Phosphocellulose. The pH of the heat fraction is adjusted to 5.6, and the treated phosphocellulose,[3] corresponding to 30 g of dry weight, is added with constant stirring. During the addition, the pH of the solution is kept constant at 5.6. The deep red phosphocellulose is then removed by vacuum filtration. To the filtrate, another 30 g of treated phosphocellulose is added, maintaining the pH at 5.6. The phosphocellulose is again removed by filtration. At this point, about 60% of the proteins in the heat fraction is removed, leaving all the FBPase activity unabsorbed (phosphocellulose-treated fraction, 490 ml).

Step 4. Phosphocellulose Absorption and Chromatography. In order to remove AMP and FBP from the enzyme solution, the phosphocellulose-treated fraction is dialyzed for 6 hr at 2° against three changes of 5 liters of 40 mM Tris-HCl buffer (pH 7.5) containing 0.1 mM EDTA and 1.5 mM MgCl$_2$. It is again dialyzed for 1 hr at 25° against 5 liters of the same buffer to hydrolyze the residual FBP remaining in the enzyme solution. The pH of the enzyme solution is then adjusted to 5.6, and treated phosphocellulose, previously equilibrated with 30 mM sodium acetate buffer (pH 5.6), is added to absorb all the FBPase activity. In this experiment, phosphocellulose corresponding to 20 g of dry powder is added. The enzyme-exchanger suspension is then poured into a 600-ml coarse sintered-glass funnel and washed with 5 liters of 0.2 mM sodium acetate buffer (pH 6.3) containing 0.1 mM EDTA. The slurry is poured into a glass column (2.5 × 45 cm) and washed again with the same washing buffer until all the unabsorbed proteins are removed ($A_{280 \text{ nm}} = 0$). The FBPase is then eluted with the same washing buffer additionally containing 0.1 mM each of 5'-AMP and FBP at a flow rate of 40 ml/hr. Fractions of 6 ml are collected, and the enzyme activity appears as a sharp peak between fractions 15 and 22. These fractions are pooled, and (NH$_4$)$_2$SO$_4$ is added slowly to 80% saturation. The protein is collected by centrifugation, dissolved in 20 ml of 10 mM Tris-HCl buffer (pH 7.5) containing 0.1 mM EDTA, and dialyzed against 3 liters of the same buffer for 2 hr.

Step 5. Blue Sepharose Chromatography. To the dialyzed phosphocellulose eluate, about 32 g of moist Blue Sepharose CL-6B (Pharmacia), previously washed with 3 N KCl and then equilibrated with 10 mM Tris-HCl buffer (pH 7.5), is added slowly with stirring to absorb all the FBPase. The enzyme-exchanger suspension is then poured into a column (2.2 × 15 cm) and washed with 3 liters of 10 mM Tris-HCl buffer (pH 7.5) containing 0.1 mM EDTA. At this point no protein is detected in the filtrate. The FBPase is finally eluted with the same washing buffer plus 0.5 mM of 5'-AMP at the flow rate of about 40 ml/hr. Fractions of 3.5 ml are collected and FBPase activity appears as a sharp peak with constant

---

[3] P. F. Han, G. S. Owen, and J. Johnson, Jr., *Arch. Biochem. Biophys.* **168,** 171 (1975).

specific activity in peak fractions. Fractions containing FBPase activity are pooled and dialyzed for 6 hr against two changes of 50 volumes of 80% saturated $(NH_4)_2SO_4$. FBPase is then collected by centrifugation, suspended in a small volume of 80% saturated $(NH_4)_2SO_4$, and stored at 3–4°. It is stable for at least 3 months.

Comments. Turkey liver FBPase can be satisfactorily isolated from freshly removed livers or frozen livers. The freezing of livers is best carried out at $-20°$ immediately after removal from turkey and washing with 1 mM EDTA. Since FBPase is susceptible to modification by proteolytic enzymes,[4] the processes of extraction and heating should be performed between pH 7.5 and 8.0. The inclusion of 5'-AMP and FBP in grinding buffer has several advantages.[5] They can synergistically increase the resistance of FBPase to proteolytic modification. They can also significantly protect the enzyme against heat inactivation so that the enzyme solution can be heated without loss of FBPase activity to 68–70° to denature potential proteolytic enzymes and other unwanted proteins (proteolytic modification is not observed after the heat step). Moreover, these two compounds can specifically prevent the absorption of FBPase by phosphocellulose so that an excess of exchanger can be added to remove more unwanted proteins. Elution of FBPase from phosphocellulose column should employ a combination of 5'-AMP and FBP, both at lower concentration, since trace amounts of impurities are often eluted with FBPase when high FBP concentration (2 mM) is used. The FBPase in phosphocellulose eluate is usually homogeneous. The chromatography on Blue Sepharose serves to ensure the high purity of the enzyme. This procedure has been used to purify many enzymes, including FBPase.[6-8]

Properties

Homogeneity, Molecular Weight, and Subunit Structure. The purified enzyme is homogeneous in disc gel electrophoresis and shows a sharp peak in sucrose density gradient centrifugation. The molecular weight is approximately 144,000, as calculated by sedimentation analysis in a sucrose density gradient. Analysis in sodium dodecyl sulfate (SDS) disc-gel electrophoresis shows the presence of a single protein band corresponding to the molecular weight of 36,000. The carboxy-terminal amino acid has been found to be lysine.

[4] S. Pontremoli, E. Melloni, F. Balestrero, A. T. Franzi, A. DeFlora, and B. L. Horecker, *Proc. Natl. Acad. Sci. U.S.A.* **70**, 303 (1973).
[5] P. F. Han, V. V. Murthy, and J. Johnson, Jr., *Arch. Biochem. Biophys.* **173**, 293 (1976).
[6] S. T. Thompson, K. H. Cass, and E. Stellwagen, *Proc. Natl. Acad. Sci. U.S.A.*, **72**, 669 (1975).
[7] Y. Tashima, H. Mizunuma, and D. D. Hasegawa, *J. Biochem. (Tokyo)* **86**, 1089 (1979).
[8] H. Kido, A. Vita, and B. L. Horecker, *Anal. Biochem.* **106**, 450 (1980).

Effect of pH and Substrate Concentration. Without chelating agents the enzyme has optimum activity at pH 8.4–8.5. Addition of chelating agents or treatment of assay system with Chelax 100^9 shifts the pH optima to 7.4–7.6. The enzyme has high affinity for FBP. At pH 7.5 the value of K_m is estimated to be 5.3 μM and the maximum activity is attained at about 10 μM. Concentrations of FBP greater than 10 μM are inhibitory. The degree of inhibition decreases with increasing pH or when Mg^{2+} is replaced by Mn^{2+} as metal cofactor.[3]

Effect of Divalent Cations. This enzyme absolutely requires divalent cations (Mg^{2+} or Mn^{2+}) for activity. The optimum concentrations for these two cations at pH 7.5 are 1–1.5 mM with Mg^{2+} and 0.05–0.07 mM with Mn^{2+}. Above optimum concentrations, both cations inhibit enzyme activity, greater inhibition being observed with Mn^{2+}. Unlike FBPases from other sources,[10,11] Mn^{2+} is more effective than Mg^{2+} as the essential metal cofactor for turkey liver FBPase.[3] The maximum activity with Mg^{2+} is about 60% of the maximum activity observed with Mn^{2+}.

Inhibition by 5'-AMP. 5'-AMP is a specific allosteric inhibitor of this enzyme with K_i estimated to be about 10 μM at pH 7.5 and 30°. Inhibition of this enzyme is not observed with the following 5'-AMP analogs up to 0.8 mM concentration: 5'-IMP, 5'-GMP, 5'-UMP, 5'-CMP, ADP, ATP, 2'-AMP, 3'-AMP, cyclic 3',5'-AMP, 8-azido-5'-AMP, and adenosine. It should be pointed out that the commercial preparations of these compounds are frequently contaminated with 5'-AMP. It is therefore essential to ensure that they are free of 5'-AMP before testing. This can be conveniently carried out by treating these compounds with 5'-AMP deaminase followed by removal of 5'-AMP deaminase by ultrafiltration.[12] The sensitivity of this enzyme to inhibition by 5'-AMP decreases greatly with increasing pH or temperature.[3]

Selective Desensitization to Inhibition by 5'-AMP or High FBP Concentration. As observed with chicken liver FBPase,[13] acetylation of turkey liver FBPase with aspirin results in irreversible desensitization to inhibition by 5'-AMP and high FBP concentration. The desensitization to 5'-AMP is specifically prevented when treatment with aspirin is carried out in the presence of 5'-AMP, while FBP can selectively prevent the

[9] H. G. Nimmo and K. F. Tipton, *Biochem. J.* **145**, 323 (1975).
[10] S. Traniello, E. Melloni, S. Pontremoli, C. L. Sia, and B. L. Horecker, *Arch. Biochem. Biophys.* **149**, 222 (1972).
[11] O. M. Rosen, *Arch. Biochem. Biophys.* **114**, 31 (1966).
[12] G. Y. Han, H. C. McBay, P. F. Han, G. S. Owen, and J. Johnson, Jr., *Anal. Biochem.* **83**, 326 (1977).
[13] P. F. Han, G. Y. Han, H. C. McBay, and J. Johnson, Jr., *Biochem. Biophys. Res. Commun.* **85**, 747 (1978).

diminished sensitivity to high FBP concentration. The ability of the enzyme to bind radioactively labeled 5'-AMP disappears almost completely after the enzyme is desensitized to 5'-AMP inhibition by aspirin.

Activation by Monovalent Cation. Like other FBPases,[14] the enzyme from turkey liver is activated by some monovalent cations such as K^+, NH_4^+, and Tl^+. This characteristic is, however, readily abolished after converting the enzyme to the "alkaline" form with papain or subtilisin, or by treatment with aspirin or pyridoxal phosphate followed by reduction with $NaBH_4$. Li^+ is an inhibitor with K_i estimated to be 0.9 mM at pH 7.5.

Inhibition by Zn^{2+} and Its Reversal by Chelating Agents. Zn^{2+} is a potent inhibitor of FBPase.[9,15] The activity of turkey liver FBPase at pH 7.5 is reduced about 50% or 92% in the presence of 0.4 μM or 1.5 μM Zn^{2+}, respectively. The sensitivity to Zn^{2+} inhibition decreases greatly with increasing pH. The inhibitory effects of Zn^{2+} and 5'-AMP are synergistic. Inhibition of the enzyme activity by 1.5 μM Zn^{2+} is completely reversed by the following natural chelating agents: imidazole pyruvate (0.2 mM),[16] histidine (1.2 mM), cysteine (0.3 mM), mercaptoacetate (0.3 mM), mercaptopyruvate (0.3 mM), and glutathione (2 mM). The ability of chelators to reverse Zn^{2+} inhibition decreases greatly if FBP is first bound to the enzyme. If 5'-AMP is also present, then chelators become almost completely incapable of reversing Zn^{2+} inhibition when added to the enzyme after substrate.[17] This indicates that the prior binding of FBP to FBPase hinders the removal of Zn^{2+} from the inhibitory sites of the enzyme by chelators, especially when 5'-AMP is also present.

Immobilization. Turkey liver FBPase can readily be immobilized on CNBr-activated Sepharose 4B (Sigma) as follows: moist activated Sepharose (equivalent to 1 g dry weight), previously treated with 1 mM HCl and then thoroughly washed with 0.1 N $NaHCO_3$ solution (pH 7.9) containing 0.5 N NaCl and 0.2 mM EDTA, is added to 5 ml of the same washing solution additionally containing 4 mM FBP and 2 mg of purified FBPase. The mixture is incubated at 25° with gentle shaking until about 90% of the FBPase is coupled by Sepharose (usually less than 30 min). The enzyme-exchanger is then thoroughly washed on a sintered-glass funnel with 0.5 N NaCl solution containing 0.2 mM EDTA. The immobilized

[14] E. Hubert, J. Villanueva, A. M. Gonzalez, and F. Marcus, *Arch. Biochem. Biophys.* **138**, 590 (1970).

[15] G. A. Tejwani, F. O. Pedrosa, S. Pontremoli, and B. L. Horecker, *Proc. Natl. Acad. Sci. U.S.A.* **73**, 2692 (1976).

[16] P. F. Han, G. Y. Han, T. W. Cole, G. S. Owen, and J. Johnson, Jr., *Experientia* **34**, 704 (1978).

[17] P. F. Han, G. Y. Han, H. C. McBay, and J. Johnson, Jr., *Biochem. Biophys. Res. Commun.* **93**, 558 (1980).

enzyme retains about 43% of the specific activity of the native enzyme when assayed at pH 7.5 by measuring the release of P_i according to the method of Tashima and Yoshimura.[1]

Interaction with Turkey Liver Aldolase. As observed with rabbit liver FBPase and aldolase,[18] the ability of turkey liver FBPase or turkey liver aldolase to penetrate into the gel phase of Ultrogel AcA-34 (LKB) decreases significantly when these two enzymes are simultaneously present in the reaction mixture. This phenomenon is, however, not observed when 5'-AMP is also present in the mixture. This suggests that turkey liver FBPase is incapable of forming a complex with turkey liver aldolase in the presence of 5'-AMP.

[18] J. S. MacGregor, V. N. Singh, S. Davoust, E. Melloni, S. Pontremoli, and B. L. Horecker, *Proc. Natl. Acad. Sci. U.S.A.* **77**, 3889 (1980).

[56] Fructose-1,6-bisphosphatase from Chicken and Rabbit Muscle

By John S. MacGregor, A. E. Annamalai, A. van Tol, W. J. Black,[1] and B. L. Horecker

$$\text{Fructose 1,6-bisphosphate} + H_2O \rightarrow \text{fructose 6-phosphate} + P_i$$

Assay Method

Principle. Fructose 6-phosphate produced in the reaction is converted to glucose 6-phosphate by phosphohexoisomerase. In the presence of glucose-6-phosphate dehydrogenase this product reduces NADP. The production of NADPH is measured spectrophotometrically at 340 nm.[2]

Reagents

Sodium fructose 1,6-bisphosphate, 5.0 mM
Diethanolamine, 0.2 M–triethanolamine, 0.2 M, adjusted to pH 7.5 with HCl (DEA–TEA buffer)
$MgCl_2$, 0.5 M
Ammonium sulfate [$(NH_4)_2SO_4$], 0.4 M
EDTA, 10 mM, adjusted to pH 7.5 with NaOH

[1] Deceased.
[2] B. L. Horecker and A. Kornberg, *J. Biol. Chem.* **175**, 385 (1948).

NADP, 2 mM

Glucose-6-phosphate dehydrogenase (Boehringer-Mannheim Corp.), crystalline suspension in $(NH_4)_2SO_4$, 5 mg/ml, diluted 1 : 10 in DEA–TEA buffer

Phosphohexoisomerase (Boehringer-Mannheim Corp.), crystalline suspension in $(NH_4)_2SO_4$, 2 mg/ml, diluted 1 : 5 in DEA–TEA buffer

Phosphoenolpyruvate, 50 mM

ATP, 50 mM

Adenylate kinase (Boehringer-Mannheim Corp.), 0.5 mg/ml

Pyruvate kinase (Boehringer-Mannheim Corp.), 1.0 mg/ml

Procedure. Place into a quartz cell (1-cm light path) in a final volume of 1 ml: 0.1 ml of TEA–DEA buffer, 0.01 ml of $MgCl_2$, 0.1 ml of EDTA, 0.01 ml of NADP, 0.01 ml each of glucose-6-phosphate dehydrogenase and phosphohexose isomerase, and the aliquot of enzyme to be measured. Record the absorbance at 340 nm at 1-min intervals until no further increase occurs. Add fructose bisphosphate (0.02 ml), and record the absorbance at 340 nm.

In the initial purification steps for muscle fructose-1,6-bisphosphatases (Fru-P₂ases) it is necessary to include an AMP-removing system in the assay. This consists of 0.01 ml each of phosphoenolpyruvate, ATP, adenylate kinase, and pyruvate kinase.

Definition of Unit and Specific Activity. A unit of enzyme is defined as the amount that converts 1 μmol of fructose bisphosphate to fructose 6-phosphate in 1 min at room temperature. Specific activity is expressed as units per milligram of protein. Protein is determined by the method of Bücher[3] standardized with rabbit muscle aldolase. For the last steps protein is determined by analysis with fluorescamine after alkaline hydrolysis[4] with bovine serum albumin as the standard, or by dividing the absorbance at 280 nm by 0.81 or 0.70 for chicken muscle or rabbit muscle Fru-P₂ases, respectively.

Purification Procedure, Chicken Muscle Fru-P₂ase

All operations are carried out at 0–4° unless otherwise specified. The results of a typical purification procedure in which the enzyme is assayed in the presence of either Mg^{2+} or Mn^{2+} are presented in Table I.

Extract of Fresh Muscle. Two Cornish Cross hens are killed by exsanguination, and their breast muscles are rapidly excised, placed on ice, and processed immediately. The muscle (248 g) is rinsed in ice-cold

[3] T. Bücher, *Biochim. Biophys. Acta* **1**, 292 (1947).
[4] N. Nakai, C. Y. Lai, and B. L. Horecker, *Anal. Biochem.* **58**, 563 (1974).

TABLE I
PURIFICATION OF CHICKEN BREAST MUSCLE FRUCTOSE-1,6-BISPHOSPHATASE

Fraction	Volume (ml)	Total activity (units)[a] 0.2 mM MnCl$_2$	Total activity (units)[a] 5 mM MgCl$_2$	Specific activity[b] 0.2 mM MnCl$_2$	Specific activity[b] 5 mM MgCl$_2$	Recovery (%)
Extract	949	2031	1034	0.29	0.15	100
Heated fraction	852	1823	818	3.6	1.6	90
Blue Dextran–Sepharose eluate	12	1358	611	46.0	22.4	67

[a] The enzyme was assayed spectrophotometrically at pH 7.5 in the presence of 40 mM $(NH_4)_2SO_4$ as described in Assay Methods.
[b] Protein was determined by the turbidity method of Bücher[3] except in the last step, where fluorescamine was used after alkaline hydrolysis.[4]

homogenizing buffer (10 mM Tris-HCl, pH 7.2, containing 1 mM EDTA), cut into small pieces, and extracted with four volumes of homogenizing buffer in a large Waring blender at low speed for three 15-sec intervals. The homogenate is centrifuged at 20,000 g for 40 min, and the supernatant solution is filtered through four layers of cheesecloth (Extract).

Heat Treatment. The extract is adjusted to pH 7.2 with 2 N NaOH, transferred to a 3-liter Erlenmeyer flask, and heated with constant stirring in a water bath initially at 85°. The temperature of the solution should reach 67° within approximately 5 min and is maintained at that temperature for an additional 3 min, after which it is cooled in a salt–ice bath to 15° within 6 min. The heat-treated extract is centrifuged at 40,000 g for 30 min, and the supernatant solution is collected by filtration through glass wool (Heated fraction).

Chromatography on Blue Dextran–Sepharose. The heated fraction is dialyzed overnight against four volumes of saturated $(NH_4)_2SO_4$ solution. The precipitated protein is collected by centrifugation, dissolved in 25 ml of 10 mM Tris-HCl, pH 7.5, containing 1 mM EDTA and dialyzed for a total of 2 hr against 1 liter of the same buffer, replaced after 1 hr with 1 liter of fresh buffer. The enzyme solution, diluted to 500 ml with 10 mM Tris buffer, pH 7.5, is treated with 100 g (equivalent dry weight) of Blue Dextran–Sepharose (Blue Sepharose CL-6B, Pharmacia Fine Chemicals) that has been equilibrated with the same buffer. The solution is stirred in the cold for 30 min, at which time assay of an aliquot of the supernatant should show that >90% of the enzyme is bound to the adsorbent. The suspension is filtered through a coarse sintered-glass funnel using a slight vacuum, and the adsorbent is washed on the funnel with 4 liters of 10 mM

TABLE II
PURIFICATION OF RABBIT MUSCLE FRUCTOSE-1,6-BISPHOSPHATASE

Step	Volume (ml)	Total activity (units)	Specific activity (units/ml)	Yield (%)
Extract	2500	1300	0.06[a]	100
Heated fraction	2440	1098	0.68	84
Phosphocellulose eluate	40	513	22.1	40

[a] See Table I.

Tris-HCl buffer, pH 7.5, until the absorbance of the filtrate measured at 280 nm is less than 0.01. The Sepharose is then transferred to a column, 3.3 × 16 cm, and washed with 500 ml of 10 mM Tris-HCl, pH 7.5, containing 1 mM AMP. The enzyme is eluted with the same buffer, containing 1 mM AMP and 1 mM Fru-P$_2$, at a flow rate of 15 ml/hr (Blue Dextran–Sepharose eluate).

Ammonium Sulfate Precipitation. Pooled fractions containing the activity are dialyzed overnight against four volumes of saturated $(NH_4)_2SO_4$ solution. The precipitated protein is stored at 4° as a suspension in 80% saturated $(NH_4)_2SO_4$.

Homogeneity. The homogeneity of the enzyme is based on the following criteria. It emerges from the Blue Dextran column as a sharp peak with constant specific activity; a single band is observed in SDS-slab gel electrophoresis; the specific activity is high, comparable to that reported by Annamalai *et al.*[5] for the homogeneous chicken enzyme isolated by chromatography on a column of phosphocellulose P-11. The enzyme can be crystallized from $(NH_4)_2SO_4$ solutions, but the specific activity is not significantly altered.[5]

Purification Procedure, Rabbit Muscle Fru-P$_2$ase

All procedures were performed at 0–4° unless otherwise stated. The results of a typical purification procedure are presented in Table II.

Extraction. A rabbit is killed by exsanguination, and the hind leg and back muscles (652 g) are removed, cut into small pieces, and homogenized in 2500 ml of 10 mM Tris-HCl, pH 7.5, containing 1 mM EDTA (homogenization buffer) in a large Waring blender run at high speed for two 1-min intervals. The homogenate is centrifuged at 20,000 g for 40 min, and the residue is discarded. The supernatant solution, pH 6.5, is adjusted to pH 7.1 with 13.8 ml of 2 N KOH (Extract).

[5] A. E. Annamalai, O. Tsolas, and B. L. Horecker, *Arch. Biochem. Biophys.* **183**, 48 (1977).

Heat Treatment. The extract is transferred to four 2-liter Erlenmeyer flasks and heated with constant stirring in a water bath initially at 85°. After approximately 6 min, when the temperature of the enzyme solution has reached 67°, the flasks are removed and rapidly cooled in an ice bath to 15°. The suspension is centrifuged at 20,000 *g* for 30 min, and the supernatant solution is collected by filtration through glass wool (Heated fraction).

Chromatography on Phosphocellulose. Phosphocellulose (Whatman, P-11) is prepared by removing the fines by decantation from water, followed by alternate washings with 0.5 *N* NaOH and 0.5 *N* HCl, then with water until neutral, followed by equilibration in 0.1 *M* sodium malonate buffer, pH 5.8, containing 1 m*M* EDTA. After equilibration in malonate buffer the exchanger is filtered and stored as a moist cake at 4°.

The heated fraction is diluted fivefold in ice-cold 1 m*M* EDTA, and the pH is adjusted to 5.8 by the addition of 1 *M* malonic acid. Phosphocellulose preequilibrated in malonate buffer is added (125 g of moist cake) to the diluted enzyme solution with constant stirring until more than 90% of the Fru-P$_2$ase activity is adsorbed. The pH of the slurry is maintained at 5.8 during the addition of phosphocellulose by the addition of 1 *M* malonic acid. The resin is collected on a Büchner funnel using a gentle vacuum and washed with an additional 2 liters of malonate buffer at a flow rate of about 100 ml/hr, until the absorbance of the effluent at 280 nm is less than 0.02. Fru-P$_2$ase activity is then eluted at a flow rate of 30 ml/hr with 0.1 *M* sodium malonate, pH 5.8, containing 1 m*M* EDTA and 2 m*M* Fru-P$_2$. Fractions containing enzyme with high specific activity are pooled and precipitated by dialysis against (NH$_4$)$_2$SO$_4$, as previously described for the chicken muscle Fru-P$_2$ase.

Properties

Molecular Weight and Subunit Structure. The chicken and rabbit muscle Fru-P$_2$ases were found to have similar molecular weights of 143,000–144,000, and each contains four apparently identical subunits, $M_r = 36,000$ each.[5,6] The NH$_2$ terminus is blocked.[7]

Catalytic Properties. In the presence of 0.2 m*M* Mn^{2+}, and 0.1 m*M* EDTA or a mixture of 1 m*M* citrate plus 1 m*M* histidine, the pH optimum is 7.1. In the absence of chelators, the optimum is shifted to pH 8.0.[5,6]

[6] W. J. Black, A. van Tol, J. Fernando, and B. L. Horecker, *Arch. Biochem. Biophys.* **151,** 591 (1972).

[7] B. L. Horecker, J. S. MacGregor, and S. Pontremoli, *in* "Science and Scientists, Essays by Biochemists, Biologists and Chemists" (M. Kageyama, ed.), p. 107. Jpn. Sci. Soc. Press, Tokyo, 1981.

Rabbit and chicken muscle Fru-P$_2$ases are very sensitive to inhibition by AMP with K_i values of 0.45 and 0.32 mM, respectively.[5,6] The activities are increased severalfold by the addition of 40 mM (NH$_4$)$_2$SO$_4$, and both enzymes are more active when assayed with 0.2 mM MnCl$_2$, as compared to 5 mM MgCl$_2$. In the presence of 40 mM NH$_4{}^+$, the substrate-concentration curve is hyperbolic with $K_m \simeq$ 10 mM.[5,6]

[57] Fructose-1,6-bisphosphatase from Rabbit Liver (Neutral Form)

By B. L. Horecker, W. C. McGregor, S. Traniello, E. Melloni, and S. Pontremoli

$$\text{Fructose 1,6-bisphosphate} + \text{H}_2\text{O} \rightarrow \text{D-fructose 6-phosphate} + \text{P}_i$$

Assay Method

Principle. Fructose 6-phosphate produced in the reaction is converted to glucose 6-phosphate by phosphohexoisomerase. In the presence of glucose-6-phosphate dehydrogenase this product reduces NADP. The production of NADPH is measured spectrophotometrically at 340 nm.

Reagents

Sodium fructose 1,6-bisphosphate, 0.01 M
Diethanolamine, 0.2 M-triethanolamine, 0.2 M, adjusted to pH 7.5 with HCl (DEA–TEA buffer)
MgCl$_2$, 0.2 M
NADP, 0.02 M
Glucose-6-phosphate dehydrogenase (Boehringer-Mannheim Corp.), crystalline suspension in (NH$_4$)$_2$SO$_4$, 5 mg/ml
Phosphohexoisomerase (Boehringer-Mannheim Corp.), crystalline suspension in (NH$_4$)$_2$SO$_4$, 10 mg/ml
EDTA, 10 mM, adjusted to pH 7.5 with NaOH
(NH$_4$)$_2$SO$_4$, 1 M

Procedure. Place in a quartz cell (1-cm light path) in a final volume of 1 ml: 0.1 ml of DEA–TEA buffer, 0.01 ml of MgCl$_2$, 0.01 ml of EDTA, 0.04 ml of (NH$_4$)$_2$SO$_4$, 0.01 ml of NADP, 1 μl each of glucose-6-phosphate dehydrogenase and phosphohexoisomerase, and the aliquot of enzyme to be measured. Record the absorbance at 340 nm for several minutes until no further increase occurs. Add fructose bisphosphate (0.01 ml), and record the absorbance at 340 nm. Use the interval between 2 and 10 min to calculate the rate.

Definition of Unit and Specific Activity. A unit of enzyme is defined as the amount that converts 1 μmol of fructose bisphosphate to fructose 6-phosphate in 1 min at room temperature (21–23°), calculated from the rate of NADPH formation in the spectrophotometric assay. Specific activity is expressed as units per milligram of protein. Protein is determined by measurement of absorbance at 280 nm, using a value of 0.63 for a solution containing 1 mg of purified enzyme per milliliter. In the initial steps of purification, the turbidometric method of Bücher[1] is employed.

Purification Procedure[2]

Extraction of Fresh Livers. Freshly collected rabbit livers (10 kg) are homogenized in a Waring CB-6 commercial blender with an equal volume of cold 0.25 M sucrose adjusted to pH 7.5 and containing 0.5 mM EDTA. The homogenate is brought to a total volume of about 30 liters with sucrose solution and mixed for about 20 min at 4°. The suspension is centrifuged in Sorvall GS-3 rotors at 8500 rpm for 40 min. The supernatant is decanted through polyester fiber and adjusted to pH 7.3 with 1 N NaOH.

Heat Precipitation. The solution is heated to 63° and maintained at that temperature for 3 min. It is then rapidly cooled to <4° in an ice bath and with the aid of a cooling coil. The precipitate is removed by continuous centrifugation in an RK ultracentrifuge (Electronucleonic Corp., Fairfield, New Jersey) with RK 10 core, 35,000 rpm, ca. 225 ml/min.

Phosphocellulose Treatment, Negative Adsorption. Phosphocellulose P-11 purchased from Whatman Biochemicals Ltd., Kent, England, is prepared as follows: 500 g (dry weight) of exchanger is soaked overnight in H_2O, and the fines are removed by decantation. The exchanger is transferred to 5 liters of 0.5 N NaOH, stirred for 10 min, washed with H_2O to neutrality on a coarse sintered-glass funnel, stirred for 15 min in 8 liters of 0.5 N HCl, washed with H_2O to neutrality, and finally equilibrated with the appropriate buffer. The exchanger is stored in the cold as a moist cake. The quantities of phosphocellulose required at pH 6.3 to remove the red color while adsorbing less than 10% of the activity are determined in a small pilot experiment using 50-ml aliquots of the heat-treated supernatant solution. To the bulk of the heat-treated supernatant solution add the

---

[1] T. Bücher, *Biochim. Biophys. Acta* **1**, 292 (1947).

[2] The method described is appropriate for large-scale preparations and has been carried out many times in the Biopolymer Laboratory of Hoffmann-La Roche Inc. It can also be adapted to laboratory-scale preparations carried out with 500 g of liver.

proportional quantity (ca. 100 g per liter of supernatant solution), maintaining the pH at 6.3 with 1 M NaOH or 1 M malonic acid as required, during the addition. Mix at 2° with stirring for 30 min. Filter and retain the P-11 paste for recycling and re-use (P-11, negative adsorption).

Adsorption on Phosphocellulose and Elution. Dilute the filtrate from the negative adsorption with an equal volume of 0.5 mM EDTA. With 50-ml aliquots, determine the quantity of P-11 paste required at pH 5.6 to adsorb >90% of the activity. Add the proportional amount (ca. 20 g per liter of solution) to the bulk of the solution from the negative adsorption. During the addition, the pH is maintained at 5.6 with 1 M NaOH or 1 M malonic acid, as required. The suspension is stirred for about 30 min, after which a small sample is centrifuged and the supernatant solution is assayed. If it contains less than 10% of the original activity, the entire suspension is centrifuged in a basket centrifuge (Komline Sanderson, Peapack, New Jersey) and washed by continuous flow with 4 × 10-liter washes of 0.12 M NaOAc, containing 0.5 mM EDTA, pH 5.6. Follow with 5 × 10 liters of 0.18 M NaOAc, containing 0.5 mM EDTA, pH 5.75. Pour the suspension on a coarse sintered-glass funnel and continue to wash with suction until the A_{280} of the effluent is less than 0.03. Elute the activity with 0.18 NaOAc, pH 6.3, containing 0.5 mM EDTA and 1 mM fructose 1,6-bisphosphate by gravity filtration. Use 8 × 500-ml elutions and collect separately. Assay and dialyze those fractions with activity overnight against 4 volumes of cold saturated $(NH_4)_2SO_4$.

Ammonium Sulfate Fractionation. This step is required if the specific activity of the eluted fraction is below 15 units/mg. Collect the precipitates in the dialysis bags by centrifugation. Suspend each precipitate in approximately 30 ml of 60% cold saturated $(NH_4)_2SO_4$ (prepared by appropriate dilution of an ammonium sulfate solution saturated at 4°). Centrifuge and assay for fructose-1,6-bisphosphatase activity. Repeat the extraction once with 60% saturated $(NH_4)_2SO_4$ and then several times with the same volumes of 50% saturated $(NH_4)_2SO_4$. Assay and combine those fractions with specific activity >20 units/mg. Dialyze the combined fractions overnight against 10 volumes of 80% saturated $(NH_4)_2SO_4$ solution. Remove from the dialysis bags and store at 4° in 80% saturated $(NH_4)_2SO_4$ solution.

Results of a typical purification are summarized in the table.

Properties

Activators and Inhibitors. The enzyme is activated by Mg^{2+} or Mn^{2+}. When both metal cofactors are present, the enzyme shows properties seen with Mn^{2+} alone. It is also activated by monovalent cations such as NH_4^+

PURIFICATION OF NEUTRAL RABBIT LIVER FRUCTOSE-1,6-BISPHOSPHATASE

Fraction	Volume (ml)	Total units	Recovery (%)	Specific activity (units/mg protein)
1. Extract	31,000	55,900	—	0.094
2. Heat-treated fraction	30,000	55,300	99	0.33
3. P-11, negative adsorption	29,600	45,500	81	0.43
4. P-11 eluates	3,130	30,000	53	9.2
5. $(NH_4)_2SO_4$	158	28,700	51	23.0

and K^+ and inhibited by Na^+.[3] Zn^{2+} is a strong inhibitor at low concentration and an activator at higher concentrations.[4] AMP is an allosteric inhibitor.[3,5]

pH Optimum. Under the assay conditions employed in the purification, the pH optimum is close to 7.5.[6]

Homogeneity and Molecular Weight. The enzyme contains four identical subunits, each with $M_r = 35,000$. The molecular weight of the native enzyme is 143,000. Purified preparations yield a single band in disc gel or sodium dodecyl sulfate gel electrophoresis.[6] The preparations are sufficiently homogeneous for sequence analysis. High-quality crystals have been obtained for X-ray analysis.[7]

Stability. The precipitated enzyme suspended in 80% $(NH_4)_2SO_4$ and stored in the refrigerator is stable for several years.

---

[3] A. Vita, H. Kido, S. Pontremoli, and B. L. Horecker, *Arch. Biochem. Biophys.* **209**, 598 (1981).

[4] S. Pontremoli, E. Melloni, F. Salamino, B. Sparatore, and B. L. Horecker, *Arch. Biochem. Biophys.* **188**, 90 (1978).

[5] T. Taketa and B. M. Pogell, *J. Biol. Chem.* **240**, 651 (1965).

[6] S. Tranillo, S. Pontremoli, Y. Tashima, and B. L. Horecker, *Arch. Biochem. Biophys.* **146**, 161 (1971).

[7] B. Soloway and A. McPherson, *J. Biol. Chem.* **253**, 2461 (1978).

[58] Fructose-1,6-bisphosphatase from Snake Muscle (*Zaocys dhumnades*)

By GEN-JUN XU, JIAN-PING SHI, and YING-LAI WANG

$$\text{Fructose-1,6-bisphosphatase} + H_2O \xrightarrow{\text{Fru-P}_2\text{ase}} \text{fructose 6-phosphate} + P_i$$

Assay Method

Principle. The assay was based on the estimation of the inorganic phosphate formed by the hydrolysis of fructose 1,6-bisphosphate (Fru-P_2)[1] using a modification of the method of Fiske and Subbarow.[2,3]

Reagents

HEPES, 0.2 M, pH 7.5, or glycine buffer, 0.2 M, pH 9.2
Trichloroacetic acid, 10% (w/v)
KCl, 2 M
EDTA, 0.1 M
MgSO$_4$, 0.5 M
H$_2$SO$_4$, 9%
Ammonium molybdate, 6.6%
FeSO$_4$, 8%
Fructose 1,6-bisphosphate, sodium salt, 0.75 mM (Fru-P$_2$)

Procedure. Prepare a stock solution (assay mixture) containing 2.5 ml of HEPES buffer (for assay at pH 7.5) or glycine buffer (for assay at pH 9.2), 2.0 ml of KCl, 0.1 ml of EDTA, 0.2 ml of MgSO$_4$, and 0.2 ml of H$_2$O. For each assay, 0.5 ml of this assay mixture is required.

The reaction is carried out at 37° in a 10-ml conical centrifuge tube containing 0.5 ml of assay mixture, 0.2 ml of Fru-P$_2$, and 0.2 ml of H$_2$O. The reaction is begun by the addition of 0.1 ml of appropriately diluted enzyme solution to the mixture which has been preincubated at 37° for 10 min, and terminated by the addition of 0.5 ml of 10% trichloroacetic acid. If a precipitate forms, it is removed by centrifugation. An aliquot of the supernatant solution is added to 3 ml of H$_2$O, followed by 1 ml of molybdate ammonium and 0.5 ml of FeSO$_4$. The solution is allowed to stand for 5 min, and the absorbance is read at 660 nm. The results are expressed as micromoles of inorganic phosphate formed, using a potassium diphos-

---

[1] Abbreviations: Fru-P$_2$, fructose 1,6-bisphosphate; Fru-P$_2$ase, fructose 1,6-bisphosphatase.
[2] C. H. Fiske and Y. SubbaRow, *J. Biol. Chem.* **66**, 375 (1925).
[3] K. Lohmann and L. Jenrassik, *Biochem. J.* **178**, 419 (1926).

phate solution as the standard. One unit of activity is defined as the quantity required to generate 1 μmol of inorganic phosphate per minute at 37° under above conditions. The specific activity is expressed as units per milligram of protein. Protein concentration is determined by the method of Lowry et al.[4] with bovine serum albumin as standard.

Purification Procedure[5,6]

All procedures are carried out at 4°, unless otherwise stated.

Extraction. Freshly collected snake (*Zaocys dhumnades*) muscle (9.5 kg) is cut into small pieces and homogenized in small portions in a Waring blender with three w/v portions of a solution containing 10 mM Tris, 1 mM EDTA, 0.2 mM phenylmethylsulfonyl fluoride, pH 7.5, for 1.5 min at high speed. The homogenate is centrifuged at 10,000 g for 20 min (crude extract).

Heat Treatment. The supernatant solution is rapidly heated (within 5 min) to 67° in a stainless steel container, using a water bath at 80°, and maintained at this temperature for 1 min. The resulting solution is then cooled to 10° in an ice bath with stirring and centrifuged at 3000 g for 20 min. The supernatant solution is filtered through glass wool (heated fraction).

Ammonium Sulfate Fractionation. To the heated fraction $(NH_4)_2SO_4$ is added to 0.55 saturation (351 g/liter); after 1 hr at 4° the precipitate is removed with a fluted filter. The clear solution is treated with additional ammonium sulfate to 0.85 saturation (221.5 g/liter) and kept overnight. It is then centrifuged at 10,000 g for 30 min, and the precipitate is dissolved in 200 ml of 0.12 M acetate buffer, pH 5.8, and dialyzed against the same buffer.

Phosphocellulose Absorption (Ammonium Sulfate Fraction). The dialyzed ammonium sulfate fraction (ca. 300 ml) is applied to a preequilibrated phosphocellulose column (P-11, Whatman, 4.2 × 18 cm)[7] at a flow-rate of 20 ml/hr. The column is washed with about 1.5 liters of 0.12 M acetate, pH 5.8, at the rate of 80 ml/hr until $A_{280} \leq 0.05$. Finally the column is washed with 0.23 M acetate buffer, pH 6.7, containing 0.2 M KCl, until

---

[4] O. H. Lowry, N. J. Rosebrough, A. L. Farr, and R. J. Randall, *J. Biol. Chem.* **193**, 265 (1951).

[5] B. M. Pogell, *Biochem. Biophys. Res. Commun.* **7**, 325 (1962).

[6] S. Pontremoli, E. Melloni, F. Salamino, M. Michetti, L. H. Botelho, H. A. El-Dorry, D. K. Chu, C. I. Isaacs, and B. L. Horecker, *Arch. Biochem. Biophys.* **191**, 825 (1978).

[7] P-11 phosphocellulose is prepared by alternate washings with 0.5 N NaOH, H_2O, 0.5 N HCl, and, finally, with water until the pH is 5.6.

PURIFICATION OF SNAKE MUSCLE FRUCTOSE-1,6-BISPHOSPHATASE

Fraction	Volume (ml)	Total activity (units)	Specific activity (units/mg protein)	Recovery (%)
Crude extract	7200	7695	0.078	100
Heated fraction	6350	4799	0.21	62.4
Ammonium sulfate fraction	300	4533	0.37	58.9
Phosphate-cellulose eluate	375	2688	48	34.9

$A_{280} \leq 0.02$. Fructose-1,6-bisphosphatase is recovered by the elution with 0.23 M acetate buffer, pH 6.7, containing 0.2 M KCl and 2 mM Fru-P$_2$ase at a flow-rate of 30 ml/hr. The eluate is collected in 8-ml fractions. The tubes containing Fru-P$_2$ase activity are pooled and precipitated with ammonium sulfate at 0.85 saturation and stored in refrigerator. Results of a typical purification are summarized in the table.

Properties

The purified enzyme shows a single band in sodium dodecyl sulfate–gel electrophoresis corresponding to a subunit weight of 35,000, and a molecular weight of 140,000 is estimated by gel filtration.[8] The maximum activity is in the neutral range and the ratio of activity measured at pH 7.5 to that measured at pH 9.2 is in excess of 2.5. The enzyme is sensitive to inhibition by AMP with K_i approximately 1.4×10^{-6} M. Snake muscle Fru-P$_2$ase shows an absolute requirement for divalent metal cations, which can be satisfied by Mg^{2+}, Mn^{2+}, or Zn^{2+}. Activity is highest with 20 mM MgCl$_2$. K$^+$, NH$_4^+$, and Cs$^+$, but not Na$^+$ and Li$^+$, enhance the activity. In the absence of KCl the pH optimum for the snake muscle enzyme is at pH 6.6; in the presence of 0.4 M KCl, the optimum shifts to pH 7.0. Snake muscle Fru-P$_2$ase can be modified by subtilisin, which increases the activity at pH 9.2 about 4.5-fold, while decreasing slightly the activity at pH 7.5.

[8] J.-P. Shi and G.-J. Xu, *Acta Biochim. Biophys. Sin.* **13**, 89 (1981).

[59] Fructose-1,6-bisphosphatase from Rat Liver

By Frank Marcus, Judith Rittenhouse, Tapati Chatterjee, and M. Marlene Hosey

Fructose 1,6-bisphosphate + H_2O → fructose 6-phosphate + P_i

Assay Method

Principle. Fructose-1,6-bisphosphatase catalyzes the hydrolysis of fructose 1,6-bisphosphate to fructose 6-phosphate and P_i. Enzyme activity can be determined spectrophotometrically by following the rate of formation of NADPH at 340 nm in the presence of excess glucosephosphate isomerase and glucose-6-phosphate dehydrogenase.[1] Alternatively, the inorganic phosphate formed by hydrolysis of fructose 1,6-bisphosphate can be determined colorimetrically.[2]

Reagents for the Spectrophotometric Assay

Triethanolamine, 0.1 M, adjusted to pH 7.5 with diethanolamine-HCl, 0.1 M

$MgCl_2$, 0.1 M

Ammonium sulfate, 1 M

D-Fructose 1,6-bisphosphate, tetrasodium salt (98–100% pure, Sigma 750-1), 2 mM

NADP (98–100% pure, Sigma), 3 mM

EDTA, disodium salt, 10 mM

Glucose-6-phosphate dehydrogenase, and glucosephosphate isomerase (Boehringer-Mannheim) diluted to 14 units/ml each with 5 mM Tris-HCl, pH 7.5

Fructose-1,6-bisphosphatase diluted to a concentration of 0.2–2.0 units/ml with 20 mM potassium phosphate buffer, pH 7.4, containing 0.1 mM EDTA, and 0.1 mM dithiothreitol

Procedure. The incubation mixture (1.0 ml) contains 20 mM triethanolamine–diethanolamine buffer, pH 7.5, 2 mM $MgCl_2$, 40 mM ammonium sulfate, 0.15 mM fructose 1,6-bisphosphate, 0.3 mM NADP, 0.1 mM EDTA, 0.7 unit each of glucose-6-phosphate dehydrogenase and glucosephosphate isomerase. The reaction is initiated by the addition of

[1] G. A. Tejwani, F. O. Pedrosa, S. Pontremoli, and B. L. Horecker, *Arch. Biochem. Biophys.* **177**, 255 (1976).

[2] F. Marcus, *Arch. Biochem. Biophys.* **122**, 393 (1967).

10–20 μl of fructose-1,6-bisphosphatase and NADPH formation is followed spectrophotometrically at 340 nm.

Definition of Unit and Specific Activity. A unit of fructose-1,6-bisphosphatase activity is defined as the amount of enzyme that catalyzes the formation of 1 μmol of fructose 6-phosphate per minute at 30° under the conditions described above. Specific activity is expressed in terms of units per milligram of protein. Protein concentration is determined by the dye-binding assay of Bradford.[3]

Purification Procedure

All procedures are performed at 0–4° unless otherwise stated.

Step 1. Extraction. Sprague–Dawley rats (250–300 g) are killed by cervical dislocation, and the livers are rapidly removed and chilled in ice. The livers from 10–14 rats are added to 2 volumes (w/v) of 0.25 M sucrose–0.1 mM dithiothreitol–1 mM EDTA adjusted to pH 8 with KOH, and the mixture is homogenized for 40 sec in a Waring blender. The homogenate is centrifuged at 100,000 g for 45 min, and the supernatant solution is filtered through glass wool. If required, the crude supernatant may be rapidly frozen in a Dry Ice–ethanol bath and stored for at least 2 weeks at $-90°$.

Step 2. Heat Treatment. The extract is transferred to a 1-liter Erlenmeyer flask and heated with constant swirling in a water bath maintained at 90–95°. When the temperature of the solution reaches 69° (within approximately 3 min), the flask is placed in a water bath at 69–70° for 3 min. The solution is then cooled in an ice bath and centrifuged at 17,000 g for 30 min; the supernatant is collected.

Step 3. Phosphocellulose AffiGel Blue Chromatography. The heat-treated fraction is passed through a set of two columns connected in series. The first is a 3.8- × 5.0-cm column of phosphocellulose (Whatman P-11) and the second column (2.8 × 18.5 cm) is packed with AffiGel Blue, 50–100 mesh (Bio-Rad 153-7301). Both columns are equilibrated with 10 mM potassium malonate buffer (pH 6.0)–0.1 mM EDTA. Under these conditions, fructose-1,6-bisphosphatase is not retained by the columns and appears in the effluent front diluted about 1.2-fold.

Step 4. Ammonium Sulfate Fractionation. To the column effluent, solid ammonium sulfate (34 g per 100 ml of solution) is added with stirring. After 1 hr, the precipitate is collected by centrifugation, dissolved in a minimum volume of 10 mM potassium malonate buffer (pH 6.0) containing 0.1 mM EDTA and 0.1 mM dithiothreitol, and dialyzed overnight against 1 liter of the above pH 6.0 buffer.

[3] M. Bradford, *Anal. Biochem.* **72**, 248 (1976).

PURIFICATION OF FRUCTOSE-1,6-BISPHOSPHATASE FROM RAT LIVER[a]

Step and fraction	Total volume (ml)	Units/ml	Total units	Protein (mg/ml)	Specific activity (units/mg protein)	Recovery (%)
1. Crude extract	290	7.0	2030	49	0.14	100
2. Heat treated	230	6.5	1495	9.3	0.7	74
3. Phosphocellulose AffiGel Blue	290	5.0	1450	2.0	2.5	71
4. Ammonium sulfate	17	79.5	1352	11.6	6.9	67
5. AffiGel Blue	34	30.6	1040	0.9	34.0	51

[a] From 169 g of rat liver.

Step 5. AffiGel Blue Chromatography. The protein solution is applied to an AffiGel Blue column (1.5 × 13 cm) equilibrated with 10 mM potassium malonate (pH 6.0)–0.1 mM EDTA. For optimal performance of this step, the application of the sample should be performed at a slow flow rate (20 ml/hr) and the amount of protein applied to the column should be approximately 5 mg of protein per milliliter of bed volume of AffiGel Blue. After application of the sample, the column is washed at a fast flow rate (ca. 7 ml/min) with 10 mM potassium malonate, pH 6.0, containing 0.1 mM EDTA, 0.1 mM dithiothreitol, and 0.6 mM K phosphate. The washing is continued until the absorbance of the effluent at 280 nm decreases to less than 0.01. Approximately 800 ml of buffer solution are required, and it is usual for a very small amount of enzyme activity (less than 0.2 unit/ml) to be present in this effluent. At this stage, 5 ml of 10 mM potassium malonate, pH 6.0, containing 0.1 mM EDTA and 0.1 mM dithiothreitol are passed through the column. The enzyme is then eluted at a flow rate of 20 ml/hr with 10 mM potassium malonate(pH 6.0)–0.1 mM EDTA–0.1 mM dithiothreitol, also containing 0.25 mM AMP and 0.1 mM fructose 1,6-bisphosphate. Fractions of 3 ml are collected, and those fractions having more than 3 units of fructose-1,6-bisphosphatase per milliliter are pooled.

The purification procedure is summarized in the table. The pure enzyme obtained by the above procedure may be stored at −20°, after prior addition of 0.5 volume of glycerol. If stored at −20° without glycerol, the enzyme frequently becomes insoluble. The insoluble material can be redissolved by the addition of mercaptoethanol (to a final concentration of about 50 mM) and stirring for 30–60 min at 4°. Rat liver fructose-1,6-bisphosphatase can also be stored (for at least 2 weeks at 4°) as a 65% ammonium sulfate suspension, after ammonium sulfate precipitation of the enzyme in the presence of 20 mM potassium phosphate, pH 7.4.

Properties

Rat liver fructose-1,6-bisphosphatase prepared as described above exhibits a single protein band in sodium dodecyl sulfate–polyacrylamide gel electrophoresis[4] with a subunit molecular weight of 42,000. In our experience, the enzyme prepared by two other procedures[1,5] shows gel patterns consisting of several closely migrating bands (40,000–42,000 molecular weight) that may represent the native enzyme and products of limited proteolytic digestion.[6] A number of different subunit molecular weights for rat liver fructose-1,6-bisphosphatase ranging from 36,000 to 45,000, and values ranging from 140,000 to 176,000 for the native tetramer, are reported in the literature.[1,5–9] A subunit molecular weight of 35,000–38,000 has been reported consistently for neutral pH optimum mammalian fructose-1,6-bisphosphatases,[10] with the exception of the rat liver enzyme.

Purified rat liver fructose-1,6-bisphosphatase,[1,7] as well as the enzyme present in crude extracts,[11–13] shows maximum activity at neutral pH (7.0 to 7.5). The enzyme has a very low K_m for fructose 1,6-bisphosphatase (1–5 μM), and excess substrate inhibition is observed at concentrations greater than 50 μM.[1,5,7,11–13] Fructose-1,6-bisphosphatase requires a divalent cation for activity, and this requirement can be fulfilled either by Mg^{2+} or Mn^{2+}.[1] Zn^{2+}, Fe^{2+}, and Fe^{3+} are strongly inhibitory.[12] The inhibition by Zn^{2+} is due to binding to a noncatalytic divalent metal binding site,[14] and the activation of the enzyme by EDTA or histidine appears to be due to removal of contaminating Zn^{2+} by these chelating agents.[1,15] Rat liver fructose-1,6-bisphosphatase, like other fructose-1,6-bisphosphatases,[16,17] is activated by monovalent cations.[1] Activation is observed with K^+ or NH_4^+. Na^+ is relatively neutral, and Li^+ is strongly inhibitory.[18]

[4] U. K. Laemmli, *Nature (London)* **227**, 680 (1970).
[5] J.-P. Riou, T. H. Claus, D. A. Flockhart, J. D. Corbin, and S. Pilkis, *Proc. Natl. Acad. Sci. U.S.A.* **74**, 4615 (1977).
[6] M. M. Hosey and F. Marcus, *Proc. Natl. Acad. Sci. U.S.A.* **78**, 91 (1981).
[7] S. Traniello, *Biochim. Biophys. Acta* **341**, 129 (1974).
[8] A. Orengo and D. M. Patenia, *Comp. Biochem. Physiol. B* **55**, 283 (1976).
[9] J. G. Zalitis and H. C. Pitot, *Arch. Biochem. Biophys.* **194**, 620 (1979).
[10] S. J. Benkovic and M. M. deMaine, *Adv. Enzymol. Relat. Areas Mol. Biol.* **53**, 45 (1982).
[11] K. Taketa and B. M. Pogell, *J. Biol. Chem.* **240**, 651 (1965).
[12] A. H. Underwood and E. A. Newsholme, *Biochem. J.* **95**, 767 (1965).
[13] K. Sato and S. Tsuiki, *Biochim. Biophys. Acta* **159**, 130 (1968).
[14] F. O. Pedrosa, S. Pontremoli, and B. L. Horecker, *Proc. Natl. Acad. Sci. U.S.A.* **74**, 2742 (1977).
[15] H. G. Nimmo and K. I. Tipton, *Biochem. J.* **145**, 323 (1975).
[16] E. Hubert, J. Villanueva, A. M. Gonzalez, and F. Marcus, *Arch. Biochem. Biophys.* **138**, 590 (1970).
[17] F. Marcus and M. M. Hosey, *J. Biol. Chem.* **255**, 2481 (1980).
[18] J. Rittenhouse and F. Marcus, unpublished data (1981).

Rat liver fructose-1,6-bisphosphatase is allosterically inhibited by adenosine 5'-monophosphate (AMP).[11] This highly specific inhibition is temperature dependent. Lowering of temperature increases AMP inhibition.[11] K_i values for AMP inhibition of about 20 μM and $n = 2$ are obtained at 25°.[1] The inhibition of fructose-1,6-bisphosphatases by AMP is considered to be important for the control of gluconeogenesis.[19-21] Work with fructose 2,6-bisphosphate,[22,23] an activator of phosphofructokinase, has shown that fructose 2,6-bisphosphate is a potent inhibitor of rat liver fructose-1,6-bisphosphatase and that it also enhances inhibition of the enzyme by AMP. The effects of fructose 2,6-bisphosphate on phosphofructokinase and fructose-1,6-bisphosphatase appear to be related to the stimulation of gluconeogenesis by glucagon.[24,25]

Rat liver fructose-1,6-bisphosphatase can be phosphorylated *in vitro* by the catalytic subunit of cyclic AMP-dependent protein kinase, and about 4 mol of phosphate are incorporated per mole of enzyme tetramer.[5] The phosphorylation of fructose-1,6-bisphosphatase in rat hepatocytes is stimulated by glucagon, but no changes in enzyme activity are observed.[26] Thus, the physiological role of phosphorylation of rat liver fructose-1,6-bisphosphatase remains to be determined. The amino acid sequence of a peptide from the phosphorylation site of rat liver fructose-1,6-bisphosphatase has been determined in two laboratories.[27,28] The amino acid sequence Ser-Arg-Pro-Ser (P)-Leu-Pro-Leu-Pro was reported by Pilkis *et al.*[27] The same sequence, but with a tyrosine residue instead of proline at position 3, was reported by Humble *et al.*[28] The cyclic AMP-dependent phosphorylation site is located near the COOH terminus of rat liver fructose-1,6-bisphosphatase.[6] In contrast to the results obtained with the rat liver enzyme, homogeneous preparation of mouse liver, rabbit liver, and pig kidney fructose-1,6-bisphosphatase

[19] H. A. Krebs, *Proc. R. Soc. London, Ser. B* **159**, 545 (1964).
[20] E. A. Newsholme and C. Start, *in* "Regulation in Metabolism," p. 278. Wiley, New York, 1973.
[21] F. Marcus, *in* "The Regulation of Carbohydrate Formation and Utilization in Mammals" (C. M. Veneziale, ed.), p. 269. University Park Press, Baltimore, Maryland, 1981.
[22] S. J. Pilkis, R. El-Maghrabi, J. Pilkis, and T. Claus, *J. Biol. Chem.* **256**, 3619 (1981).
[23] E. Van Schaftingen and H. G. Hers, *Proc. Natl. Acad. Sci. U.S.A.* **78**, 2861 (1981).
[24] E. Van Schaftingen, M. F. Jett, L. Hue, and H. G. Hers, *Proc. Natl. Acad. Sci. U.S.A.* **78**, 3483 (1981).
[25] K. Uyeda, E. Furuya, and L. J. Luby, *J. Biol. Chem.* **256**, 8394 (1981).
[26] T. H. Claus, J. Schlumpf, M. R. El-Maghrabi, M. McGrane, and S. J. Pilkis, *Biochem. Biophys. Res. Commun.* **100**, 716 (1981).
[27] S. J. Pilkis, M. R. El-Maghrabi, B. Coven, T. H. Claus, H. S. Tager, D. F. Steiner, P. S. Keim, and R. L. Heinrikson, *J. Biol. Chem.* **255**, 2770 (1980).
[28] E. Humble, U. Dahlquist-Edberg, P. Ekman, E. Netzel, U. Ragnarsson, and L. Engström, *Biochem. Biophys. Res. Commun.* **90**, 1064 (1979).

are not phosphorylated by cyclic AMP-dependent protein kinase because they lack the phosphorylatable peptide in their COOH terminus.[6] Cleavage of phosphorylated rat liver fructose-1,6-bisphosphatase with cyanogen bromide and purification of the phosphorylated CNBr fragment has located the phosphorylation site in a peptide of about 50 amino acids.[29] The partial determination of the amino acid sequence of this peptide by automated Edman degradation reveals the sequence Gly-Ser-Thr-Glu-Asp-Val-Glx-Glu-Phe-Leu-Glu-Ile-Tyr-Asn-Lys-Asp-Lys-Ala-Lys-Ser-Arg-Pro-Ser-Leu-Pro-Leu-Pro. This sequence can be placed as beginning 17–19 residues distant from the COOH terminus of both pig kidney[29] and rabbit liver[30] fructose-1,6-bisphosphatase. However, the rat liver enzyme extends beyond the COOH terminal amino acid of the two other fructose-1,6-bisphosphatases. The proline-rich extension of rat liver fructose-1,6-bisphosphatase contains the cyclic AMP-dependent phosphorylation site.[31] It remains to be determined whether one or more peptides, accounting for about 30 amino acids, are removed from the nonphosphorylatable fructose 1,6-bisphosphatases by endogenous proteases during the enzyme isolation procedures.

Acknowledgment

This work was supported by National Institutes of Health Research Grant AM 21167.

[29] F. Marcus, J. Rittenhouse, T. Chatterjee, I. Reardon, and R. L. Heinrikson, unpublished data (1981).
[30] S. C. Sun, A. G. Datta, E. Hannappel, V. N. Singh, O. Tsolas, E. Melloni, S. Pontremoli, and B. L. Horecker, *Arch. Biochem. Biophys.* **206**, 265 (1981).
[31] F. Marcus, J. Rittenhouse, T. Chatterjee, I. Reardon, and R. L. Heinrikson, *Fed. Proc.*, *Fed. Am. Soc. Exp. Biol.* **41**, 1136 (1982).

[60] Fructose-1,6-bisphosphatase from Mouse and Rabbit Intestinal Mucosa

By YOHTALOU TASHIMA and HIDEO MIZUNUMA

$$\text{D-Fructose 1,6-bisphosphate} + H_2O \rightarrow \text{D-fructose 6-phosphate} + P_i$$

The small intestinal mucosa contains fructose-1,6-bisphosphatase (D-fructose-1,6-bisphosphate 1-phosphohydrolase, EC 3.1.3.11) in a relatively high concentration.[1,2] The purified enzyme has similar properties to those of the known liver,[3-6] kidney,[7] and muscle[8] fructose-1,6-bisphosphatases, such as molecular and subunit molecular weights, AMP inhibition, pH optimum, and the metal requirement. Rabbit intestinal fructose-1,6-bisphosphatase is very similar to the liver enzyme. On the other hand, mouse intestinal enzyme is different from the liver enzyme.

Assay Method

Spectrophotometric Assay

The production of fructose 6-phosphate is measured spectrophotometrically by following the reduction of $NADP^+$ with glucose-6-phosphate isomerase and glucose-6-phosphate dehydrogenase.

Reagents

Tris-HCl, 0.2 M, pH 7.5 or pH 9.1
$MgCl_2$, 20 mM
EDTA, 10 mM
$NADP^+$, 15 mM
Fructose 1,6-bisphosphate, sodium salt, 1 mM
Glucose-6-phosphate isomerase, 700 units/ml
Glucose-6-phosphate dehydrogenase, 350 units/ml

Procedures. In a 1.0-ml cuvette having a 1-cm light path, place 0.25 ml of buffer, 0.1 ml of $MgCl_2$, 0.1 ml of EDTA, 0.01 ml of $NADP^+$, 0.001 ml each of auxiliary enzymes, the amount of enzyme to be assayed, and water to a total volume of 0.9 ml. After a 4-min incubation without substrate, start the reaction by the addition of 0.1 ml of 1 mM fructose 1,6-bisphosphate and read the absorbance at 340 nm for 10 min. The assay is performed at 30°. The formation of fructose 6-phosphate results in the reduction of an equivalent amount of $NADP^+$.

[1] H. Mizunuma and Y. Tashima, *J. Biochem.* (*Tokyo*) **84**, 327 (1978).
[2] H. Mizunuma, M. Hasegawa, and Y. Tashima, *Arch. Biochem. Biophys.* **201**, 296 (1980).
[3] Y. Tashima, H. Mizunuma, and M. Hasegawa, *J. Biochem.* (*Tokyo*) **86**, 1089 (1979).
[4] S. Pontremoli and E. Melloni, this series, Vol. 42 [57].
[5] E. H. Ulm, B. M. Pogell, M. M. De Maine, C. B. Libby, and S. J. Benkovic, this series, Vol. 42 [60].
[6] G. A. Tejwani, F. O. Pedrosa, S. Pontremoli, and B. L. Horecker, *Arch. Biochem. Biophys.* **177**, 255 (1976).
[7] Y. Tashima, G. Tholey, G. Drummond, H. Bertrand, J. S. Rosenberg, and B. L. Horecker, *Arch. Biochem. Biophys.* **149**, 118 (1972).
[8] W. J. Black, A. van Tol, J. Fernando, and B. L. Horecker, *Arch. Biochem. Biophys.* **151**, 576 (1972).

Phosphate Release Assay

Fructose-1,6-bisphosphatase is also measured by the release of inorganic phosphate. The method has the advantage that no other additions are required other than those essential for the hydrolysis of fructose 1,6-bisphosphate.

Reagents

Fructose 1,6-bisphosphate, sodium salt, 1 mM
Ribulose 1,5-bisphosphate, sodium salt, 1 mM
Sedoheptulose 1,7-bisphosphate, sodium salt, 1 mM
Tris-HCl, 0.2 M, pH 7.5
$MgCl_2$, 20 mM
EDTA, 10 mM

Procedure. Mix 0.25 ml of buffer, 0.1 ml of $MgCl_2$, 0.1 ml of EDTA, 0.1 ml of one of the substrates, and water in a final volume of 0.98 ml. Start the reaction by addition of 0.02 ml of enzyme solution at 30° and incubate for 10–15 min.

Estimation of Inorganic Phosphate

MODIFICATION[9] OF THE METHOD OF ITAYA AND UI[10]

Color-Developing Reagent. Mix slowly 1 volume of 3% $(NH_4)_6Mo_7O_{24} \cdot 4 H_2O$ in water with 1 volume of 0.06% Malachite Green in 6 N HCl. After 1 day, the mixture is filtered and stored at room temperature. The mixed solution deteriorates slowly with slight precipitation, and can be used for 1 week. The precipitate is filtered off before use.

Procedure. Add 0.5 ml of the color-developing reagent to 1 ml of a test solution and mix. After 15 min at room temperature, the absorbance at 650 nm is measured. The intensity of the color is stable for 1 hr after the 15 min of incubation. The calibration curve is linear in the region of 0–300 ng of P_i per milliliter. At 100 ng/ml, P_i gives an absorbance of 0.160 cm^{-1}. The background absorbance is very small as compared with that of the original method. Substances in the assay mixture for fructose-1,6-bisphosphatase do not interfere, though sulfate compounds give a strong color.

MODIFICATION OF THE METHOD OF FISKE AND SUBBAROW[11]

Reagents. Concentrated molybdate–sulfuric acid reagent is used to increase the sensitivity. Dissolve 6 g of $(NH_4)_6Mo_7O_{24} \cdot 4 H_2O$ in 400 ml of water, and mix with 33 ml of 95–96% H_2SO_4. This solution can be stored

[9] Y. Tashima and N. Yoshimura, *J. Biochem. (Tokyo)* **78**, 1161 (1975).
[10] K. Itaya and M. Ui, *Clin. Chim. Acta* **14**, 361 (1966).
[11] C. H. Fiske and Y. SubbaRow, *J. Biol. Chem.* **66**, 375 (1925).

at 4° for at least 6 months without deterioration. Reducing reagent is made as described in the original method.

Procedures. Add 0.05 ml of 10% sodium dodecyl sulfate to 1 ml of the assay mixture to stop the enzyme reaction. Proteins in the sample are solubilized, and the deproteinization step is eliminated.[12] Add 0.2 ml of the molybdate–sulfuric acid solution to the sample solution, and 0.025 ml of the reducing reagent solution. The color development is measured at 660 nm after 10 min at 30°. At 3 μg/ml, P_i gives an absorbance of 0.30 cm^{-1}. This modified method is suitable for the measurement of phosphorus in the concentration range from 0.2 to 3 μg/ml.

Definition of Unit and Specific Activity

One unit of enzyme activity is defined as the amount that catalyzes the hydrolysis of 1 μmol of fructose 1,6-bisphosphate per minute at 30°. Specific activity is defined as units per milligram of protein. Protein content is determined by the method of Lowry *et al.*[13] using bovine serum albumin as a standard.

Purification Procedure

All operations are performed at 0–5° unless otherwise stated.

Purification from Mouse Small Intestinal Mucosa

Step 1. Extraction. Ten male white mice, 20–30 g body weight, fed on laboratory chow *ad libitum* are sacrificed. The small intestines are quickly removed and slit longitudinally on ice. The fresh small intestines are washed with ice-cold homogenizing medium (containing 0.15 *M* KCl in 10 m*M* sodium phosphate buffer, pH 7.4). The intestinal mucosa is scraped out with a glass microscope slide on ice, and immediately homogenized in a glass homogenizer with a Teflon pestle with 2 volumes of the homogenizing medium. The homogenate is centrifuged at 40,000 *g* for 30 min, and the supernatant solution is carefully collected.

Step 2. Heat Treatment. 2-Mercaptoethanol and EDTA are mixed with the crude extract at final concentrations of 10 m*M* and 1 m*M*, respectively. The mixture is heated to 58° in a water bath (65°) and then chilled rapidly in an ice bath. After centrifugation at 40,000 *g* for 20 min, the supernatant solution is collected.

Step 3. Phosphocellulose Treatment. The heated fraction is dialyzed twice against 1 liter of 10 m*M* malonate buffer (pH 6.7) containing 1 m*M*

---

[12] Y. Tashima, *Anal. Biochem.* **69**, 410 (1975).
[13] O. H. Lowry, N. J. Rosebrough, A. L. Farr, and R. J. Randall, *J. Biol. Chem.* **193**, 265 (1951).

TABLE I

PURIFICATION OF FRUCTOSE-1,6-BISPHOSPHATASE FROM MOUSE
SMALL INTESTINAL MUCOSA

Step	Total protein (mg)	Total activity pH 7.5 (units)	Specific activity pH 7.5 (units/mg protein)	Activity ratio pH 7.5 : pH 9.1	Yield (%)
Crude extract	325	12.2	0.038	1.62	(100)
Heat treatment	152	15.3	0.10	1.67	125
Phosphocellulose treatment at pH 6.7	82.7	11.0	0.13	2.71	90
Phosphocellulose chromatography at pH 5.8	0.82	9.0	10.9	2.57	73
Blue Sepharose treatment	0.38	8.4	22.2	2.60	69

EDTA for 5 hr. The fraction is adjusted to pH 6.7 with 10 mM malonic acid and added to 10 g of phosphocellulose slurry preequilibrated with the same solution for dialysis. The suspension is stirred continuously for 30 min. The mixture is filtered with suction through a sintered-glass filter, and the filtrate is retained.

Step 4. Phosphocellulose Chromatography. The phosphocellulose-treated fraction is adjusted to pH 5.8 with 10 mM malonic acid and mixed with 15 g of the phosphocellulose slurry preequilibrated with 10 mM malonate buffer (pH 5.8) containing 1 mM EDTA. Fructose-1,6-bisphosphatase is completely adsorbed during constant stirring. The exchanger is washed 5 times with the same buffer (pH 5.8) with suction on a sintered-glass funnel. The slurry of washed exchanger is poured into a glass column, and the column is washed for 1 hr with the pH 5.8 buffer. The enzyme is eluted with the same buffer containing 1 mM fructose 1,6-bisphosphate, 20 μM AMP, and 50 mM NaCl. Fractions containing fructose-1,6-bisphosphatase are collected and dialyzed twice against 1 liter of the pH 5.8 buffer for 4 hr.

Step 5. Blue Sepharose Treatment. The phosphocellulose eluate is adjusted to pH 6.1 with 10 mM NaOH. The eluate is poured into a column packed with Blue Sepharose CL-6B gel (bed volume, 80 ml) preequilibrated with 10 mM malonate buffer (pH 6.1) containing 1 mM EDTA. The enzyme passed through the column is recovered and concentrated by ultrafiltration. The concentrated fructose-1,6-bisphosphatase is dialyzed against an appropriate medium overnight.

The final preparation is electrophoretically homogeneous. The results of a typical purification are summarized in Table I.

Purification from Rabbit Small Intestinal Mucosa

Step 1. Extraction. Crude extract is obtained from the intestinal mucosa of a male white rabbit, 3 kg body weight and fasted overnight, by the method described for the purification of mouse intestinal fructose-1,6-bisphosphatase except that the scraped mucosa is homogenized with 4 volumes of the homogenizing solution.

Step 2. Heat Treatment. 2-Mercaptoethanol and EDTA are added to the crude extract to final concentrations of 10 and 1 mM, respectively. The mixture is heated to 60° in a water bath (65°) and then chilled rapidly in an ice bath. After centrifugation at 40,000 g for 20 min, the supernatant is collected.

Step 3. Ammonium Sulfate Fractionation. Finely ground ammonium sulfate is slowly added to the heated fraction with constant stirring to a concentration of 2.1 M. After gentle stirring for 15 min, the fraction is centrifuged at 20,000 g for 10 min. The pellet is discarded, and the supernatant is adjusted to 3.1 M with ammonium sulfate, stirred, and centrifuged as above. The pellet is dissolved in a minimum volume of 10 mM malonate buffer (pH 6.2) containing 1 mM EDTA and dialyzed overnight against two changes of 1 liter of the same buffer containing 1 mM EDTA and 0.05% 2-mercaptoethanol.

Step 4. Phosphocellulose Chromatography. The dialyzate is centrifuged at 20,000 g for 10 min. The supernatant solution is adjusted to pH 6.2 with 0.1 M NaOH and loaded onto a column of 5 g of phosphocellulose slurry preequilibrated with 10 mM malonate buffer (pH 6.2) containing 1 mM EDTA. The column is washed with 50 ml of 10 mM malonate buffer (pH 6.2) containing 50 mM NaCl and 1 mM EDTA. Fructose-1,6-bisphosphatase is eluted with 2 mM fructose 1,6-bisphosphate–150 mM NaCl–1 mM EDTA–10 mM malonate buffer (pH 6.2). Fractions having the enzyme activity are pooled and dialyzed for 6 hr against two changes of 1 liter of 10 mM malonate buffer (pH 6.5) containing 1 mM EDTA and 0.05% 2-mercaptoethanol.

Step 5. Blue Sepharose Chromatography. The dialyzate is poured into a column packed with Blue Sepharose CL-6B gel (bed volume, 80 ml) preequilibrated with 10 mM malonate buffer (pH 6.5) containing 1 mM EDTA. The column is washed with the same buffer for 1 hr. The elution medium consists of 0.1 mM AMP, 0.05 mM fructose 1,6-bisphosphate, 1 mM EDTA, and 10 mM malonate buffer (pH 6.5). Fructose-1,6-bisphosphatase is eluted in a sharp peak. Fractions containing the enzyme activity are collected and concentrated by ultrafiltration. The concentrated enzyme is dialyzed against an appropriate medium overnight.

Table II summarizes the results obtained from a typical purification.

TABLE II
PURIFICATION OF FRUCTOSE-1,6-BISPHOSPHATASE FROM RABBIT
SMALL INTESTINAL MUCOSA[a]

Step	Total protein (mg)	Total activity pH 7.5 (units)	Specific activity pH 7.5 (units/mg protein)	Activity ratio pH 7.5 : pH 9.1	Yield (%)
Crude extract	2564	134.2	0.052	2.94	(100)
Heat treatment	897	133.8	0.14	3.25	100
Ammonium sulfate fractionation	307	136.3	0.44	3.31	102
Phosphocellulose elution	30	94.1	3.1	2.50	70
Blue Sepharose elution	2.2[b]	108.1	50.6	2.91	81

[a] Reproduced from Mizunuma et al.[2]
[b] The quantity of purified enzyme was estimated from the absorbance at 280 nm. The absorbance of fructose-1,6-bisphosphatase at 1 mg of protein per milliliter in a 1.0-cm light path is 0.73 at 280 nm [S. Traniello, S. Pontremoli, Y. Tashima, and B. L. Horecker, Arch. Biochem. Biophys. 146, 161 (1971)].

Comments

Intestinal homogenates contain a proteolytic activity that catalyzes the conversion of fructose-1,6-bisphosphatase to a form having the increased activity at alkaline pH. The heat treatment inactivates the proteolytic activity, and there is no evidence that the enzyme is modified after this step.

The electrophoretically homogeneous preparation can be obtained by the phosphocellulose chromatography step when NaCl is omitted from the elution buffer, but this procedure is not reproducible and gives a poor recovery of the enzyme. High ionic strength in the elution buffer makes the recovery high, and then Blue Sepharose chromatography is required.

Rabbit intestinal enzyme is well adsorbed on Blue Sepharose gel and is completely eluted at high purity from the gel with the low concentration of AMP. Fructose 1,6-bisphosphate in combination with AMP causes the enzyme to elute very sharply. Fructose 1,6-bisphosphate at more than 0.5 mM concentration coelutes a small amount of aldolase. Therefore, it is necessary to use a low concentration of fructose 1,6-bisphosphate, 0.05 mM and also to remove aldolase at previous steps as much as possible. In the present method, most of the aldolase is removed by the ammonium sulfate fraction.

Mouse intestinal enzyme is poorly adsorbed on Blue Sepharose gel

TABLE III
PROPERTIES OF INTESTINAL FRUCTOSE-1,6-BISPHOSPHATASE

Property	Mouse[1]		Rabbit[2]	
	Value	Method	Value	Method
Molecular weight	140,000	Gel filtration	145,000	Low-speed sedimentation equilibrium
Sedimentation coefficient ($s_{20,w}$)			7.2	Moving-boundary method
Partial specific volume			0.742	Calculated from the amino acid composition
Subunit molecular weight	37,500	SDS–gel electrophoresis	37,000	SDS–gel electrophoresis
K_m for fructose-1,6-P_2	<1 μM		2.6 μM	
K_m for sedoheptulose-1,7-P_2	16 μM		8.0 μM	
K_m for ribulose-1,5-P_2	3.1 μM		21 μM	
K_i for AMP	1 μM		17 μM	
pH optimum	7.3		7.5	

around pH 6 under the present conditions. The contaminants in the phosphocellulose eluate are successfully removed by use of Blue Sepharose gel.

Properties

Some of the molecular and kinetic properties of mouse and rabbit intestinal fructose-1,6-bisphosphatases are listed in Table III.

Substrates. High concentrations of fructose 1,6-bisphosphate are inhibitory. Both enzymes hydrolyze ribulose 1,5-bisphosphate and sedoheptulose 1,7-bisphosphate in addition to fructose 1,6-bisphosphate.

Metal Cations. Mg^{2+} is essential for both intestinal enzymes. Ca^{2+} is inhibitory. NH_4^+ and K^+ activate the activity of mouse enzyme approximately 1.5-fold, but Na^+ is inhibitory.

Activators and Inhibitors. EDTA activates both intestinal enzymes in the neutral pH region. 5'-AMP is a powerful inhibitor of both intestinal enzymes.

Isozymic Type. Mouse intestinal enzyme is electrophoretically distinct from the liver enzyme,[1] whereas rabbit intestinal enzyme does not separate from the liver enzyme.[2] In mouse the intestinal enzyme is more sensitive to AMP inhibition than is the liver enzyme, and it is distinguishable from the liver enzyme by the antiserum raised against mouse liver enzyme.[7] In rabbit, the K_i value for AMP of the intestinal enzyme is identical to that of the liver enzyme, and is not distinguishable from the liver enzyme by the antiserum raised against rabbit intestinal enzyme.[7] Thus, mouse intestinal fructose-1,6-bisphosphatase is a different isozyme from the liver type.

Mouse intestinal enzyme is also distinguishable from the muscle enzyme by immunotitration with the antiserum against the intestinal enzyme and by the measurement of the isoelectric point of the enzymes.[14] The intestinal enzyme in mouse is a new isozyme different from the liver or muscle enzyme. On the other hand, no evidence has been found that rabbit intestinal fructose-1,6-bisphosphatase is different from the liver enzyme.

The intestinal type of the enzyme also seems to be present in rat and golden hamster.[14]

Content in Intestinal Mucosa. In mouse, fructose-1,6-bisphosphatase activity per gram of tissue wet weight is 10–15% of that in liver, and ninefold as much as that in muscle. In rabbit the activity is approximately 60% of that in liver.

---

[14] H. Mizunuma and Y. Tashima, *Arch. Biochem. Biophys.* **217**, in press (1982).

[61] Fructose-1,6-bisphosphatase from Bumblebee Flight Muscle

By KENNETH B. STOREY

Fructose 1,6-bisphosphate + H_2O → D-fructose 6-phosphate + P_i

Although fructose-1,6-bisphosphatase (FBPase; EC 3.1.3.11) is generally considered to be a key, irreversible step in gluconeogenesis, the enzyme from bumblebee flight muscle appears to be geared for an alternate role. Flight muscle FBPase is present in high activities (up to 80 units/gram wet weight) in the absence of other gluconeogenic enzymes, such as phosphoenolpyruvate carboxykinase.[1] The ratio of FBPase to phosphofructokinase (PFK) activity is also surprisingly high at approximately 1 : 1, and bumblebee FBPase is unique in being unaffected by AMP as an inhibitor.[1,2] Taken together, these observations have suggested that FBPase and PFK are involved in the futile cycling of fructose 6-phosphate and fructose 1,6-bisphosphate in flight muscle, the net result of which is the hydrolysis of ATP.[1] This cycling has been investigated and is believed to be involved in metabolic heat production, allowing the bumblebee to warm up before flight when ambient temperature is low.[3] This high-activity FBPase from bumblebee flight muscle can be purified to homogeneity with a single purification step and with a high yield.

Assay Method

Principle. The production of fructose 6-phosphate is coupled to the reduction of NADP using phosphoglucoisomerase and glucose-6-phosphate dehydrogenase with the increase in absorbance at 340 nm measured spectrophotometrically.

Reagents. All assay reagents are made up in the appropriate buffer with the pH readjusted, if necessary, by addition of solid buffer or 1 M HCl or KOH.

Tris-HCl, 0.05 M, pH 7.4 or
Glycine-KOH, 0.2 M, pH 9.3
$MgCl_2$, 0.1 M

---

[1] E. A. Newsholme, B. Crabtree, S. Higgins, S. Thornton, and C. Start, *Biochem. J.* **128**, 89 (1972).

[2] K. B. Storey, *Biochim. Biophys. Acta* **523**, 443 (1978).

[3] M. Clark, D. Bloxham, P. Holland, and H. Lardy, *Biochem. J.* **134**, 589 (1973).

NADP, 0.01 M

Fructose 1,6-bisphosphate, 1 mM

Phosphoglucoisomerase, 70 units/ml, and glucose-6-phosphate dehy-drogenase, 70 units/ml, dialyzed against 1000 volumes of Tris-HCl, 0.05 M, pH 7.4 for 12 hr.

Procedure. The assay is performed at 24° using a spectrophotometer with thermostatted cell holder. To a cuvette add 0.06 ml of MgCl$_2$ (6 mM), 0.02 ml of NADP (0.2 mM), 0.04 ml of fructose 1,6-bisphosphate (0.04 mM), 0.01 ml of phosphoglucoisomerase–glucose-6-phosphate dehydro-genase mixture (0.7 unit of each), and 0.87 ml of buffer for a total volume of 1 ml. For standard assays use the Tris buffer. The reaction is started by the addition of the FBPase preparation. For unpurified FBPase prepara-tions, the presence of a nonspecific reduction of NADP should be checked in control assays omitting fructose 1,6-bisphosphate.

Definition of Unit and Specific Activity. One unit of FBPase activity is defined as that amount of enzyme producing the reduction of 1 μmol of NADP per minute at 24°. Specific activity is given in units per milligram of protein. Protein is measured by the method of Bradford[4] using the pre-pared reagent (Bio-Rad Laboratories, Richmond, California) with bovine γ-globulin as the protein standard. The standard assay procedure is used for the crude homogenate, and the microassay procedure is used for the purified enzyme.

Purification Procedure

Bumblebee Flight Muscle. Female worker *Bombus terrestris* are used. For preparations utilizing fresh muscle, bees should be anesthetized either by placing them in a refrigerator at 4° for 30 min or by flushing their container with CO$_2$ gas. Bees are then killed by decapitation; the abdo-men, legs, and wings are cut off, and the thorax is pinned out and opened ventrally. Gut and thoracic fat body are scraped away, and then the flight muscles are dissected out. About 500 mg of muscle can be obtained from 30 small to medium-sized bees. Equally active enzyme can be obtained from bees that have been stored deep-frozen at −80°. As an alternative to the careful dissection needed to obtain isolated muscle, whole thoraces can be utilized as the starting material. Legs, wings, and head are re-moved, and the gut is removed from the thorax by gentle pulling while removing the abdomen. The purification procedure described here is per-formed using isolated muscle from previously frozen bees.

Homogenization. Flight muscle (481 mg from 30 bees, FBPase activity = 69 units per gram wet weight) is homogenized in 10 volumes (4.81 ml) of

---

[4] M. Bradford, *Anal. Biochem.* **72**, 248 (1976).

ice-cold Tris-HCl buffer, 0.05 M, pH 7.4 containing 1 mM EDTA and 15 mM 2-mercaptoethanol using a glass–glass hand-held homogenizer. Muscle fibers are quite easily broken up, particularly in previously frozen material. The homogenate is centrifuged at 39,000 g for 40 min at 4°; the resulting supernatant contains >99% of FBPase activity.

Pretreatment of Cellulose Phosphate. Extensive precycling of cellulose phosphate (Sigma Chemical Co., St. Louis, Missouri; No. C-2383) is necessary to ensure success with this cation-exchange chromatography procedure. To precycle, wash the exchanger sequentially with the following solutions: 0.5 M HCl, distilled water, 0.5 M NaOH, distilled water. For each wash allow the exchanger to stir gently for 30 min, pour off the fines, and then collect the resin as a damp cake using a Büchner funnel. Resuspend in the next solution and take the exchanger through at least three cycles of this four-step procedure. Air-dry the final damp cake. This cake can then be resuspended in buffer and used as required.

Cellulose Phosphate Chromatography. All purification steps are now performed at room temperature. Precycled cellulose phosphate is resuspended in malonic acid–KOH buffer, 0.1 M, pH 6.0, containing 30 mM 2-mercaptoethanol, and the exchanger is equilibrated in this buffer by stirring gently with several changes of buffer until the pH of the wash returns to 6.0. The preequilibrated cellulose phosphate (5 ml of settled cellulose phosphate) is then poured into a column (1.5 × 5.0 cm) and washed with a further 10 bed volumes of the malonic acid buffer.

Malonic acid (0.1 volume of 1 M malonate, pH 6.0) is added to the high speed supernatant to readjust the pH to 6.0. The supernatant is then slowly applied to the column at a flow rate of 10 ml/hr. Essentially all FBPase activity is bound by the cellulose phosphate. The column is then washed extensively with malonic acid buffer until the eluent protein concentration drops below 1.0 μg/ml, the limit of detection of the Bio-Rad microassay for protein. Typically this requires at least 200 ml of buffer at a flow rate of 30–40 ml/hr.

FBPase is eluted from the column using a linear gradient of fructose 1,6-bisphosphate, from 0 to 100 μM fructose 1,6-bisphosphate in 100 ml of malonic acid buffer, with a flow rate of 30–40 ml/hr (Fig. 1). One-milliliter fractions are collected. FBPase is eluted at fructose 1,6-bisphosphate concentrations of between 10 and 20 μM in a peak of 4 to 15 ml. Tailing of the peak can sometimes occur. The gradient is distorted somewhat in the peak FBPase fractions presumably due to binding of fructose 1,6-bisphosphate by FBPase. Actual fructose 1,6-bisphosphate concentrations can be quantitated, if necessary, using the coupled enzyme assay of Lowry and Passonneau.[5] Peak tubes of FBPase activity are combined.

[5] O. H. Lowry and J. V. Passonneau, "A Flexible System of Enzymatic Analysis." Academic Press, New York, 1972.

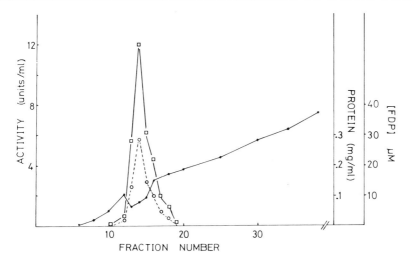

FIG. 1. Elution profile of bumblebee fructose-1,6-bisphosphatase from cellulose phosphate using a gradient of fructose 1,6-bisphosphate (0 to 100 μM) in malonic acid buffer, 0.1 M, pH 6.0 containing 30 mM 2-mercaptoethanol. Fractions are 1 ml. □——□, Fructose-1,6-bisphosphatase activity; ○---○, protein concentration; ●——●, fructose 1,6-bisphosphate concentration [FDP].

A typical purification is summarized in the table. This cellulose phosphate chromatography procedure is based on that described by Black *et al.*[6] as modified by Storey[2] for bumblebee flight muscle. In my experience, FBPase isolated by this procedure is homogeneous as judged by sodium dodecyl sulfate–gel electrophoresis with staining for protein using Coomassie Blue.[7] The mean final specific activity obtained is 42 units per milligram of protein; the highest specific activity that we have achieved is 47 units per milligram of protein.

Removal of Fructose 1,6-Phosphate from the Enzyme Preparation. To prepare the purified enzyme for kinetic studies, Tris base and MgCl₂ are added to the preparation to final concentrations of 50 mM and 20 mM, respectively, and the pH is adjusted to 7.5. The preparation is then incubated for 2 hr at 24° to hydrolyze fructose 1,6-bisphosphate followed by dialysis against 1000 volumes of 10 mM Tris-HCl, pH 8.0, containing 15 mM 2-mercaptoethanol for 12 hr at 4° to remove hexose phosphates. The purified enzyme is stable for several days at 4° or months at −80°.

Sephadex G-200 Gel Filtration Chromatography. A gel filtration step can be added to the purification for one of two reasons: (*a*) to achieve complete homogeneity of FBPase if this has not been achieved with the cel-

[6] W. Black, A. van Tol, J. Fernando and B. L. Horecker, *Arch. Biochem. Biophys.* **151**, 576 (1972).
[7] K. Weber and M. Osborn, *J. Biol. Chem.* **244**, 4406 (1969).

PURIFICATION OF FRUCTOSE-1,6-BISPHOSPHATASE FROM BUMBLEBEE
FLIGHT MUSCLE

Step	Volume (ml)	Total protein (mg)	Total activity (units)	Yield (%)	Specific activity (units/mg protein)
Crude homogenate	4.8	47.4	33.2	—	0.70
Cellulose phosphate	5.0	0.67	28.2	85	42.0
Sephadex G-100	5.0	0.60	25.2	76	42.0

lulose phosphate step alone, and in some hands this might be necessary; or (b) to replace dialysis as the means of removing hexose phosphates from the FBPase preparation. The FBPase preparation is concentrated prior to gel filtration by dialysis (in cellulose tubing, M_r cutoff 12,000, Fisher Chemical Co.) against dry, solid polyethylene glycol (M_r 20,000, Fisher Chemical Co.) for 1 hr or until the sample volume is reduced to approximately 0.5 ml. The sample is then loaded onto a Sephadex G-200 column (1.0 × 90 cm) equilibrated in Tris-HCl buffer, 0.05 M, pH 7.4 containing 15 mM 2-mercaptoethanol. The column is eluted with this buffer at 15 ml/hr, 1-ml fractions are collected and pooled, and the enzyme is used for kinetic study either directly or after further concentration using polyethylene glycol.

Comments. The preparation described here utilizes muscle from 30 bumblebees. The procedure can be scaled up for larger amounts of tissue and can also easily be scaled down. The high yield of enzyme from this one-step purification coupled with the high activities of FBPase in bumblebee flight muscle allows sufficient enzyme for several hundred assays to be isolated from a single bumblebee. When you must catch your bees individually from flower gardens, this is a true advantage.

Species identification and sexing of bumblebees can be difficult; when catching your own, an expert should be consulted. Bumblebee species other than B. terrestris may be equally good sources of FBPase. Honeybees, however, do not generate heat by substrate cycling and do not have the high activities of FBPase found in bumblebee flight muscle.

Properties

During purification of FBPases from other sources, proteolysis can result in a shifting of the pH optimum of the enzyme toward a more alkaline pH. There is no evidence that this occurs with the flight muscle enzyme, and the ratio of activity at pH 7.4 compared to pH 9.3 remains constant at 2 : 1 at all stages of the purification procedure.

The pH optimum of bumblebee FBPase is 8.0, and the molecular weight is 142,000 ± 10,000. The K_m for fructose 1,6-bisphosphate is 11 μM (with Mg^{2+} as the metal ion) and 8 μM (with Mn^{2+}) at 35°. The K_m for Mg^{2+} is 0.2 mM, and that for Mn^{2+} it is 0.03 mM. The enzyme is not affected by AMP; however, oleate is an effective activator of enzyme activity (K_a = 10 μM). The enzyme is strongly and specifically inhibited by Ca^{2+} (K_i = 30 μM) and Li^+ (K_i = 80 μM) with Li^+ potentiating Ca^{2+} inhibition by lowering the K_i for Ca^{2+}. These ions are perhaps the key effectors regulating flight muscle FBPase activity *in vivo* and thereby controlling the rate of substrate cycling in the muscle. Other details of flight muscle FBPase kinetics, including the effects of temperature on FBPase kinetics are given by Storey.[2]

Acknowledgments

Dr. R. C. Plowright, University of Toronto, is thanked for generously providing bumblebees for this study. Thanks are due also to Dr. E. A. Newsholme, University of Oxford, in whose laboratory I first studied bumblebee FBPase.

[62] Fructose-bisphosphatase from Spinach Leaf Chloroplast and Cytoplasm

By GRAHAME J. KELLY, GERHARD ZIMMERMANN, and ERWIN LATZKO

$$\text{D-Fructose 1,6-bisphosphate} + H_2O \xrightarrow{Mg^{2+}} \text{D-fructose 6-phosphate} + P_i$$

Photosynthetic cells require fructose-bisphosphatase (D-fructose-1,6-bisphosphate 1-phosphohydrolase, EC 3.1.3.11) both in the chloroplast and in the cytoplasm. In the chloroplast, it is one of the enzymes constituting the Calvin cycle of CO_2 fixation. In the cytoplasm, it is required for the synthesis of sucrose from triose phosphate, which is the major form of reduced carbon exported from chloroplasts during photosynthesis (Fig. 1). Both enzymes have potential regulatory roles, since the estimated *in vivo* activities are little more than sufficient to accommodate the observed rates of CO_2 fixation and sucrose formation.

METHODS IN ENZYMOLOGY, VOL. 90

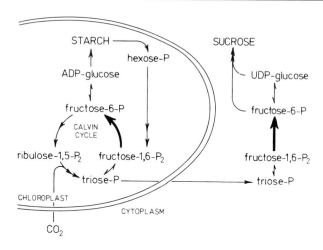

Fig. 1. Requirement for fructose-bisphosphatase (heavy arrows) in both chloroplasts and cytoplasm of leaf cells.

Assay Method

Principle. Fructose-bisphosphatase may be assayed either colorimetrically by estimating the released P_i,[1] or spectrophotometrically by coupling the production of fructose 6-phosphate to the reduction of $NADP^+$ using the enzymes phosphoglucose isomerase and glucose-6-phosphate dehydrogenase. The latter approach is the method of choice, since it permits the reaction to be followed continuously and can be used even with most crude extracts (see below). However, P_i release is useful in certain instances, e.g., for examining substrate specificity.

Reagents

Tris-HCl buffer, 1 M, pH 8.8 (for chloroplast enzyme)
Imidazole-HCl buffer, 1 M, pH 7.5 (for cytoplasmic enzyme)
$MgCl_2$, 0.1 M
$NADP^+$, 3 mM
EDTA, 10 mM
Fructose 1,6-bisphosphate, sodium salt, 12 mM and 1.2 mM
Phosphoglucoisomerase–glucose-6-phosphate dehydrogenase (PGI/ G6PDH): 40 μl of phosphoglucose isomerase (Boehringer yeast enzyme, 2 mg/ml, ca. 350 units/mg) and 20 μl of glucose-6-phosphate dehydrogenase (Boehringer yeast enzyme, 5 mg/ml, ca. 140 units/ mg) are diluted to 1 ml with 20 mM Tris-HCl or imidazole-HCl

[1] H. H. Taussky, E. Shorr, and G. Kurzmann, *J. Biol. Chem.* **202**, 675 (1953).

buffer (see above), and the solution is dialyzed against 500 ml of this same buffer for several hours to remove ammonium sulfate.

Procedure for Chloroplast Fructose-bisphosphatase. The following are added to a cuvette of 1.3 ml capacity and with an optical path of 1 cm: 0.1 ml of Tris-HCl buffer, 0.1 ml of $MgCl_2$, 0.05 ml of EDTA, 0.1 ml of $NADP^+$, 0.05 ml of PGI/G6PDH, ca. 0.02 unit of fructose-bisphosphatase, and water to a volume of 0.95 ml. The cuvette is equilibrated for 3–5 min at 25° in the spectrophotometer cell holder, then the reaction is initiated by addition of 0.05 ml of 12 mM fructose 1,6-bisphosphate. Enzyme activity is calculated from the recorded change in absorbance at 340 nm.

Procedure for Cytoplasmic Fructose-bisphosphatase. The procedure is basically as described above for the chloroplast enzyme. However, the composition of the reaction mixture is 0.1 ml of imidazole-HCl buffer, 0.05 ml of $MgCl_2$, 0.1 ml of $NADP^+$, 0.05 ml of PGI/G6PDH, ca. 0.02 unit of fructose-bisphosphatase, water to a volume of 0.95 ml, and (after 3–5 min) 0.05 ml of 1.2 mM fructose 1,6-bisphosphate.

Definition of Unit. The amount of enzyme required to hydrolyze 1 μmol of fructose 1,6-bisphosphate per minute is defined as 1 unit of activity.

Application of the Procedures to Crude Spinach Leaf Homogenates. The coupled-enzyme assay operates without interference when crude homogenates are used as the source of enzyme. If spinach leaves are held in darkness for at least 1 hr prior to extraction, a correct estimation of the cytoplasmic fructose-bisphosphatase activity may be obtained by assaying at pH 7.5 as described above; at this pH the chloroplast enzyme is completely inactive providing that reducing agents such as dithiothreitol are not included in the reaction mixture.[2] A reasonable estimate[3] of the activity of chloroplast fructose-bisphosphatase may then be obtained by assaying for this enzyme at pH 8.8 as described above and subtracting from the observed activity that activity previously measured at pH 7.5 and therefore attributable to the cytoplasmic enzyme.

Purification

The first four steps of the purification procedures are common to both enzymes. Chromatography on DEAE-Sephadex A-50 (step 4) separates the chloroplast enzyme from the cytoplasmic enzyme, and thereafter the procedures differ.

[2] G. J. Kelly, G. Zimmermann, and E. Latzko, *Biochem. Biophys. Res. Commun.* **70**, 193 (1976).
[3] The estimate obtained is not necessarily a reflection of the physiological value. The *in vivo* activity is best estimated using reaction mixtures of pH 8.0 and containing in addition 5 mM dithiothreitol, since the stroma of illuminated chloroplasts has a pH of about 8.0 and is a reductive environment.

TABLE I
PURIFICATION OF SPINACH CHLOROPLAST FRUCTOSE-BISPHOSPHATASE[a]

Step	Total protein (mg)	Total activity (units)	Specific activity (units/mg protein)	Recovery (%)
1. Extract	16,400	900	0.055	(100)
2. First ammonium sulfate	2,650	810	0.31	90
3. Second ammonium sulfate	440	760	1.72	85
4. DEAE-Sephadex A-50	16	740	46	82
5. Sephadex G-200	6.6	720	109	80

[a] Data from Zimmermann et al.[4]

Chloroplast Fructose-bisphosphatase

The following procedure has been used successfully a number of times in the authors' laboratory[4] and in at least three other laboratories.[5-7] Typical results are presented in Table I.

Step 1. Spinach (Spinacia oleracea L.) leaves (approximately 2 kg) obtained from a local market are washed, and the midribs are removed. The leaf tissue is cooled, and all subsequent operations, unless otherwise stated, are at 2–4°. The tissue is homogenized by pressing it through a kitchen juicer to yield up to 1 liter of spinach juice. To each 100 ml of this juice is added 10 ml of buffer A (0.1 M sodium phosphate buffer, pH 7.5, containing 2 mM EDTA). The pH of the extract is adjusted to 7.5, then it is centrifuged at 20,000 g for 20 min. The supernatant is collected and termed the crude extract.

Step 2. Solid ammonium sulfate is slowly (ca. 20 min) added to the crude extract at a rate of 24 g/100 ml. After stirring for a further 20 min, the preparation is centrifuged at 20,000 g for 20 min, and the supernatant is collected. Additional ammonium sulfate is similarly added to this supernatant at a rate of 17 g/100 ml. After centrifugation, the supernatant is discarded, tubes are dried of any adhering supernatant, and the precipitate is dissolved in 50 ml of buffer A. This preparation contains a low concentration of ammonium sulfate from that trapped with the precipitate.

[4] G. Zimmermann, G. J. Kelly, and E. Latzko, Eur. J. Biochem. 70, 361 (1976).
[5] T. Takabe, H. Ishikawa, M. Miyakawa, K. Takenaka, and S. Nikai, J. Biochem. (Tokyo) 85, 203 (1979).
[6] J. Buc, J. Pradel, J.-C. Meunier, J.-M. Soulie, and J. Ricard, FEBS Lett. 113, 285 (1980).
[7] S. A. Charles and B. Halliwell, Biochem. J. 185, 689 (1980).

Step 3. The solution obtained from step 2 is again fractionated by addition of solid ammonium sulfate. The procedure is identical to that in step 2, except that the first salt addition is at a rate of 18 g/100 ml and the second is at 14 g/100 ml. The precipitate obtained after the second salt addition is dissolved in 10 ml of buffer B (20 mM sodium phosphate buffer, pH 7.5, containing 50 mM NaCl and 0.4 mM EDTA).

Step 4. The preparation obtained from step 3 is applied to a DEAE-Sephadex A-50 column (40 cm × 8 cm^2) equilibrated with buffer B. After washing the column with 360 ml of the same buffer, protein is eluted with 1.2 liters of a linear gradient of NaCl between 0.05 M and 0.70 M. The flow rate is 20 ml/hr. Chloroplast fructose-bisphosphatase activity is eluted when the NaCl concentration reaches approximately 0.5 M and is contained in about 90 ml. Protein is precipitated from this preparation by adding ammonium sulfate (56 g of solid salt per 100 ml), then collected at the centrifuge (20,000 g, 20 min) and dissolved in 5 ml of buffer C (0.1 M sodium phosphate, pH 7.5, containing 2 mM EDTA and 0.1 M NaCl).

Step 5. The preparation from step 4 is filtered through a column of Sephadex G-200 (90 cm × 5.3 cm^2) previously equilibrated with buffer C. The flow rate is 10 ml/hr. Most of the activity appears in 30 ml of eluate that is concentrated to 3 ml over a Diaflo XM-50 ultrafiltration membrane using a pressure from N$_2$ of 360 kPa.

Comments. The preparation obtained appears homogeneous when examined by disc gel electrophoresis. It contains 80% of the original chloroplast fructose-bisphosphatase activity measured in the crude extract, and represents a 1700-fold purification. The activity is quite stable: almost no activity is lost during storage for 6 months at −20°.

Leaf Cytoplasmic Fructose-bisphosphatase

The following procedure is that described by Zimmermann *et al.*[8]; it is summarized in Table II. It has been used successfully on six occasions in the authors' laboratory.

Steps 1–3. These are identical to steps 1–3 described above for chloroplast fructose-bisphosphatase, except that 10 mM dithiothreitol is included in buffers A and B in steps 2 and 3.

Step 4. The preparation obtained from step 3 is applied to a DEAE-Sephadex A-50 column (40 cm × 8 cm^2) equilibrated with buffer B containing 10 mM dithiothreitol. The cytoplasmic fructose-bisphosphatase does not bind to the ion-exchanger, and so is eluted during subsequent washing of the column with 360 ml of the equilibration buffer. Fractions containing enzyme activity are combined, and protein in this preparation

[8] G. Zimmermann, G. J. Kelly, and E. Latzko, *J. Biol. Chem.* **253**, 5952 (1978).

TABLE II
PURIFICATION OF SPINACH LEAF CYTOPLASMIC FRUCTOSE-BISPHOSPHATASE[a]

Step	Total protein (mg)	Total activity (units)	Specific activity (units/mg protein)	Recovery (%)
1. Extract	16,400	613	0.037	(100)
2. First ammonium sulfate	2,550	482	0.20	79
3. Second ammonium sulfate	740	405	0.55	66
4. DEAE-Sephadex A-50	285	309	1.08	50
5. Acetone	56	202	3.60	33
6. Sephadex G-200	24	126	5.25	21
7. Electrofocusing[b]	0.8	51	62	8

[a] Data from Zimmermann et al.[8]

[b] Values were calculated on the basis that 24 mg of protein from step 6 were electrofocused. In practice, no more than 8 mg can be electrofocused at one time (see text).

is precipitated by ammonium sulfate and collected as described above (step 4, chloroplast enzyme procedure), then dissolved in 60 ml of buffer D (buffer C containing 10 mM dithiothreitol).

Step 5. The preparation from step 4 is placed in an ice–salt mixture, and acetone at $-20°$ is added dropwise to a concentration of 30% (v/v). Precipitated protein is removed by centrifugation at $-10°$ (20,000 g, 10 min) and discarded. The acetone concentration of the supernatant is increased to 48% (v/v), and the precipitated protein, collected by centrifugation (as above), is dissolved in 4 ml of buffer D. This preparation is finally clarified by centrifugation.

Step 6. The preparation from step 5 is filtered through a column of Sephadex G-200; details of this step are identical to those outlined above for the chloroplast enzyme, step 5.

Step 7. A portion of the preparation from step 6, containing not more than 8 mg of protein, is electrofocused using an LKB 8101 Ampholine column. A continuous gradient is prepared by adding 1.25 ml of Ampholine (pH 4–6) in 59 ml of H_2O to 3.75 ml of Ampholine in 38 ml of H_2O containing 28 g of sucrose, using an LKB 8121 gradient mixer at a delivery rate of 3 ml/min. After about one-third of the gradient has been formed in the column, the enzyme preparation is added to the mixing vessel for inclusion in the gradient. Electrofocusing is run at $2°$ with 500 V for no longer than 14 hr. The gradient contents are then pumped into 1.5-ml fractions (one fraction collected per minute) and those containing fructose-bisphosphatase activity are combined and filtered through a column of Sephadex G-50 (20 cm × 8 cm^2), previously equilibrated with

TABLE III

COMPARISON OF THE PROPERTIES OF THE FRUCTOSE-BISPHOSPHATASES
FROM SPINACH LEAF CHLOROPLASTS AND CYTOPLASM[a]

Property	Chloroplast enzyme	Cytoplasmic enzyme
Molecular weight	160,000	130,000
Number of subunits	4	4
Isoelectric point	4.5	5.2
Optimum pH	8.8[b]	8.0
$s_{0.5}$ (fructose 1,6-bisphosphate)[c,d]	80 μM	2.5 μM
$s_{0.5}$ (Mg^{2+})[d]	10 μM	0.13 mM
Effect of 1 mM AMP	No effect	Inhibition
Substrate specificity	Specific for fructose 1,6-bisphosphate[e]	Specific for fructose-1,6-bisphosphate

[a] Data from Zimmermann et al.[4,8] and Zimmermann.[9]
[b] In the absence of dithiothreitol. Enzyme preincubated with dithiothreitol displays a broad optimum between 7.5 and 8.5.
[c] Assayed at the optimum pH in the absence of dithiothreitol.
[d] The chloroplast enzyme displays positive cooperativity toward fructose 1,6-bisphosphate and Mg^{2+}; the cytoplasmic enzyme displays Michaelis–Menten kinetics.
[e] E. Racker, this series, Vol. 5 [29c].

buffer C, to remove the carrier ampholyte. Finally, active fractions from this column are combined and concentrated to 1 ml over a Diaflo XM-50 membrane as described above (step 5, chloroplast enzyme).

Comments. The preparation obtained from step 7 appears to be homogeneous when examined by disc gel electrophoresis. It represents a 1700-fold purification of the crude enzyme. The quantity of enzyme that can be prepared is limited, however, by the relatively high cost of electrofocusing. This last step is successful only if smaller samples (8 mg or less) are electrofocused for a minimum of time (no more than 14 hr). The preparation may be stored in buffer C at −20° for 2 weeks without loss of activity, and for a further 4 weeks with about 20% loss of activity.

Properties

A comparison of the chloroplast and cytoplasmic fructose-bisphosphatases with respect to several physical and kinetic parameters is presented in Table III. The properties of the leaf cytoplasmic enzyme are more or less comparable to those of mammalian fructose-bisphosphatases, but the chloroplast enzyme is somewhat unique in being

[9] G. Zimmermann, Dr. Agr. Thesis, Technische Universität, München (1976).

insensitive to AMP[10] and highly sensitive to the redox state of its environment. This latter property has been extensively investigated in view of the observed *in vivo* light-induced increase of activity at physiological pH[2] and the mimicking of this light effect *in vitro* by dithiothreitol, which simultaneously increases the number of sulfhydryl groups on the enzyme molecule and leads to the appearance of activity at physiological pH.[4] The activation is a consequence of a greatly increased affinity for the substrate fructose 1,6-bisphosphate.[11] The nature of the light-generated reductant that activates the enzyme *in vivo* is not firmly established at present; thioredoxin,[12] an iron–sulfur protein termed ferralterin,[13] and a thylakoid membrane-bound component[14] are all being considered.

Other properties of chloroplast fructose-bisphosphatase are: proclivity to dissociation into two equal halves when diluted at pH 8.8,[4,6] inhibition of activity by Ca^{2+},[9,15] and hysteretic activation by the substrate fructose 1,6-bisphosphate.[16]

[10] B. B. Buchanan, P. Schürmann, and P. P. Kalberer, *J. Biol. Chem.* **246**, 5952 (1971).
[11] R. von Garnier and E. Latzko, *Photosynth., Two Centuries Its Discovery Joseph Priestly, Proc. Int. Congr. Photosynth. Res. 2nd, 1971* p. 1839 (1972).
[12] R. A. Wolosiuk, P. Schürmann, and B. B. Buchanan, this series, Vol. 69 [36].
[13] C. Lara, A. de la Torre, and B. B. Buchanan, *Biochem. Biophys. Res. Commun.* **94**, 1337 (1980).
[14] L. E. Anderson, H.-M. Chin, and V. K. Gupta, *Plant Physiol.* **64**, 491 (1979).
[15] S. A. Charles and B. Halliwell, *Biochem. J.* **188**, 775 (1980).
[16] C. Chehebar and R. A. Wolosiuk, *Biochim. Biophys. Acta* **613**, 429 (1980).

[63] Fructose-bisphosphatase from *Rhodopseudomonas palustris*

By CLARK F. SPRINGGATE and CHESTER S. STACHOW

Fructose 1,6-bisphosphate + H_2O → fructose 6-phosphate + P_i

Fructose-1,6-bisphosphatase (Fru-P_2ase) (EC 3.1.3.11) was originally discovered in rabbit liver by Gomori.[1] Mammalian Fru-P_2ases are allosterically inhibited by AMP.[2] The physiological consequence of an AMP-regulated Fru-P_2ase in animals is that of control of gluconeogenesis.[3–6] In

[1] G. Gomori, *J. Biol. Chem.* **148**, 139 (1943).
[2] K. Taketa and B. M. Pogell, *J. Biol. Chem.* **240**, 651 (1965).
[3] R. Blanchetti and M. L. Sartirana, *Biochem. Biophys. Res. Commun.* **27**, 378 (1967).

photosynthetic organisms, however, Fru-P_2ase may be involved in the regulation of the Calvin–Bassham carbon reduction cycle and is AMP insensitive.[7-11] The Fru-P_2ase of the photosynthetic bacterium *Rhodopseudomonas palustris* is allosterically regulated by substrate (fructose 1,6-bisphosphate), metal cofactor (Mn^{2+}), and inhibitor (GTP).

Assay Method

Principle. Fructose-bisphosphatase catalyzes the hydrolysis of D-fructose 1,6-bisphosphate to D-fructose 6-phosphate and P_i. The Fru-P_2ase activity may be measured spectrophotometrically by following the reduction of $NADP^+$ in the presence of excess phosphohexoisomerase and glucose-6-phosphate dehydrogenase. Alternatively, Fru-P_2ase activity may be determined by measuring the release of inorganic phosphate employing the micro method of Chen *et al.*[12]

Procedure. Dithiothreitol is obtained from Calbiochem. All other reagents are obtained from Sigma. Reaction mixtures (1.0 ml) contain 0.5 mM fructose 1,6-bisphosphate, 100 mM Tris-HCl buffer, pH 8.5, 1.0 mM $NADP^+$, 0.1 mM $MnCl_2$, 1.0 mM dithiothreitol (DTT), 10 μg of phosphoglucoisomerase, and 2 μg of glucose-6-phosphate dehydrogenase. The reaction is initiated by the addition of Fru-P_2ase, and formation of NADPH is followed at 340 nm on a Gilford 2400 spectrophotometer. All assays are performed at 25°.

Definition of Enzyme Unit. A unit of enzyme is defined as the amount catalyzing the formation of 1 mol of fructose 6-phosphate per minute at pH 8.5.

Purification Procedure

The organism *R. palustris* (ATCC 17007, van Niel 2.1.37) was obtained from Dr. J. Orlando, Biology Department, Boston College. Stock cultures are maintained as a stab culture continually exposed to light in the chemically defined medium of Cohen-Bazire *et al.*[13] containing 0.5% casamino acids and 1.5% agar. The defined medium of Cohen-Bazire *et al.*[13] is

[4] J. Priess, M. L. Biggs, and D. Greenberg, *J. Biol. Chem.* **242**, 2292 (1967).
[5] D. G. Frankel, S. Pontremoli, and B. L. Horecker, *Arch. Biochem. Biophys.* **144**, 4 (1966).
[6] O. M. Rosen, S. M. Rosen, and B. L. Horecker, *Arch. Biochem. Biophys.* **112**, 411 (1965).
[7] C. Springgate and C. Stachow, *Arch. Biochem. Biophys.* **152**, 1 (1972).
[8] C. Springgate and C. Stachow, *Arch. Biochem. Biophys.* **152**, 12 (1972).
[9] C. Stachow and C. Springgate, *Biochem. Biophys. Res. Commun.* **39**, 637 (1970).
[10] C. Springgate and C. Stachow, *Biochem. Biophys. Res. Commun.* **49**, 522 (1972).
[11] B. Buchanan, P. Schiramann, and P. P. Kalberer, *J. Biol. Chem.* **246**, 5952 (1971).
[12] P. S. Chen, T. Y. Toribara, and H. Warner, *Anal. Chem.* **28**, 1756 (1956).
[13] G. Cohen-Bazire, W. R. Sistrom, and J. Stanier, *J. Cell. Comp. Physiol.* **25**, 49 (1957).

prepared from the three stock solutions. Stock solutions (A) $1.0\,M$ potassium phosphate, pH 6.8, and (B) $1.0\,M$ ammonium DL-malate, pH 6.8, are prepared fresh. A stock solution (C) of concentrated base is stored at 4°. The following components constitute 1 liter of concentrated base: nitrilotriacetic acid, 10 g; $MgSO_4$, 14.45 g; $CaCl_2 \cdot 2\ H_2O$, 3.335 g; $(NH_4)Mo_7O_4 \cdot 4\ H_2O$, 9.25 mg; $FeSO_4 \cdot 7\ H_2O$, 99 mg; nicotinic acid, 50 mg; thiamine $\cdot$ HCl, 25 mg; biotin, 0.5 mg; metals "44," 50 ml; and distilled water to 1000 ml. Metals "44" contained (per 100 ml distilled water): ethylenediaminetetraacetic acid, 250 mg; $ZnSO_4 \cdot 7\ H_2O$, 1095 mg; $FeSO_4 \cdot 7\ H_2O$, 500 mg; $MnSO_4 \cdot H_2O$, 154 mg; $CuSO_4 \cdot 5\ H_2O$, 39.2 mg; $Co(NO_3)_2 \cdot 6\ H_2O$, 24.8 mg; $Na_2B_4O_7 \cdot 10\ H_2O$, 17.7 mg. A few drops of sulfuric acid are added to prevent the formation of a precipitate. To prepare the concentrated base solution, the nitrilotriacetic acid is dissolved and neutralized with KOH before the rest of the components are added. Prior to making to volume, the solution is adjusted to pH 6.6–6.8. To prepare 1 liter of complete medium, 20 ml of each stock solution (A, B, and C) are diluted with distilled water. After the addition of sodium glutamate and sodium acetate to a final concentration of 0.1% and 200 μg of p-aminobenzoic acid, the solution is brought to volume with distilled water. The sodium glutamate and sodium acetate may be replaced by 0.5% casamino acids. *Rhodopseudomonas palustris* is mass cultured photosynthetically in defined medium[13] and harvested in late log phase (96 hr) by allowing the bacteria to settle to the bottom of the culture bottle, removing the clear medium with the aid of suction and finally low-speed centrifugation of settled cells. Cells are washed several times with 0.01 M Tris-HCL buffer, pH 7.4 and stored at $-20°$. The Fru-P_2ase of frozen cells is stable for at least 3 months when stored under these conditions.

Step 1. Crude Extract. All purification procedures are performed at 4°. Protein concentration is determined by the procedure of Lowry *et al.*[14] In a typical procedure 100 g wet weight of frozen or fresh cells are suspended (1/2 w/v) in 0.1 Tris-HCl buffer (pH 7.4) containing DTT (1.0 mM). Cells are ruptured by forcing the cell suspension twice through a precooled French pressure cell at 10,000 psi. The resulting thick extract is diluted twofold with extracting buffer. The extract is then incubated for 5 min with $MnCl_2$ (0.01 M), and the aggregated chromatophores are removed by centrifugation at 25,000 g for 15 min. The supernatant fraction contains all the Fru-P_2ase activity (fraction I, 610 ml).

All subsequent operations are in buffers containing 1 mM DTT and 0.1 mM $MnCl_2$.

[14] O. H. Lowry, N. J. Rosebrough, A. L. Farr, and R. J. Randall, *J. Biol. Chem.* **193**, 265 (1951).

Step 2. Ammonium Sulfate Fractionation. The supernatant fraction (610 ml) is treated with 142 g of solid ammonium sulfate (23 g/100 ml), and stirred for 15 min; the precipitate is discarded after centrifugation. To the resulting supernatant solution are added 102 g of ammonium sulfate (16 g/100 ml). After a 30-min incubation the precipitate collected after centrifugation is suspended in 0.2 volume of fraction I in 0.1 M Tris-HCl (pH 7.4) (fraction II, 122 ml).

Step 3. Protamine Sulfate Fractionation. Protamine sulfate (1.0% solution, pH 7.4) is added with stirring to fraction II to a final concentration of 0.2%. The precipitate that forms after a 30-min incubation is removed by centrifugation at 25,000 g for 15 min and discarded (fraction III, 151 ml).

Fraction III is made up to 3.6 times the volume of fraction II (final volume 440 ml) with 318 ml of 0.05 M Tris-HCl buffer (pH 7.4). Lowering the ionic strength of the Tris-HCl buffer from 0.10 M to 0.06 M greatly improves subsequent fractionations.

Step 4. Ammonium Sulfate Fractionation. Ammonium sulfate (27.5 g/100 ml) is added to fraction III with stirring and incubated for 30 min; the resulting precipitate is discarded after centrifugation. The supernatant fraction is treated with 29 g of ammonium sulfate (6.5 g/100 ml) and incubated for 3 hr before the precipitate is collected by centrifugation. The pellet is taken up in 80 ml of 5 mM Tris-HCl buffer, pH 7.4 (fraction IV, 84 ml).

Step 5. Alumina C_γ Adsorption Elution. The alumina C_γ (aged) used in this step is purchased from Sigma Chemical Co. Alumina C_γ (stock, 22 mg dry weight per milliliter) is added to fraction IV at a ratio of 1.5 mg per milligram of protein, and the mixture is gently stirred for 30 min. Protein is determined by the ratio of adsorbance at 280 nm and 260 nm. The centrifuged alumina C_γ pellet is washed once with 5 mM Tris-HCl buffer (pH 7.4). Enzyme is then eluted twice with 0.05 M potassium phosphate buffer (pH 7.4). The eluents are pooled (100 ml) and concentrated by the addition of 47 g of ammonium sulfate; the collected precipitate is dissolved in 25 ml of 5 mM Tris-HCl buffer (pH 7.4). The sample is prepared for DEAE column chromatography by passage through a Sephadex G-25 column (2.5 cm × 40 cm) that had been equilibrated with 5 mM KHCO$_3$ buffer (pH 7.8). Fractions of 5 ml are collected at a flow rate of 4 ml/min. Fractions containing Fru-P$_2$ase activity are pooled (fraction V, 33 ml).

Step 6. DEAE Column Chromatography. Fraction V was applied to a column of DEAE-cellulose (2.5 × 90 cm) equilibrated with 5 mM KHCO$_3$ buffer, pH 7.8, which contains 1 mM DTT and 0.2 mM MnCl$_2$. The column is then washed with 200 ml of the equilibrating buffer. A linear gradient of 0.07 to 0.45 M Tris-HCl buffer, pH 7.8, was applied for the

PURIFICATION OF *Rhodopseudomonas palustris* FRUCTOSE-BISPHOSPHATASE

Fraction and step	Volume (ml)	Total protein (mg)	Total activity (units)	Specific activity (units/mg protein)	Purification (fold)
1. Crude extract	610	11,100	121	0.011	—
2. Ammonium sulfate fractionation	122	6,810	143	0.021	1.9
3. Protamine sulfate fractionation	151	6,810	143	0.021	1.9
4. Ammonium sulfate fractionation	84	2,840	122	0.043	3.9
5. Alumina C_γ	33	864	95	0.11	10
6. DEAE-cellulose chromatography	20	30.2	58	1.92	171
7. Ammonium sulfate fractionation	2	10.4	37	3.86	353
8. Sephadex G-200 chromatography	15	3.04	29	8.81	797

elution of Fru-P_2ase activity. The total volume of the gradient is 1.6 liters. Fractions (5 ml) are collected at a flow rate of 1.5 ml/min. Tubes containing Fru-P_2ase activity (usually fractions 145–165) are pooled and concentrated by the addition of ammonium sulfate (48 g/100 ml). After gentle stirring for 2 hr, the precipitate is collected by centrifugation and suspended in 0.1 M Tris-HCl buffer, pH 7.4 (fraction VI, 20 ml).

Step 7. Ammonium Sulfate Fractionation. Fraction VI is adjusted to 45% saturation by addition of solid ammonium sulfate (27.5 g/100 ml) and stirred for 30 min. The precipitate is discarded after centrifugation. The supernatant solution is then adjusted to 55% saturation by adding 6.5 g of ammonium sulfate per 100 ml and stirred for 45 min. The precipitate is collected by centrifugation and dissolved in 2 ml of 0.05 M Tris-HCl buffer, pH 7.4 (fraction VII, 2 ml).

Step 8. Sephadex G-200 Column Chromatography. Fraction VII is applied to a Sephadex G-200 column (2.5 × 82 cm) previously equilibrated with 0.10 M Tris-HCl buffer, pH 7.4, which contains 1 mM DTT and 0.1 mM MnCl$_2$. Upward flow elution is employed at a rate of 9.6 ml/hr, and 2.5-ml fractions are collected. The peak fractions (usually tubes 98–103) are pooled and stored at $-20°$ (fraction VIII, 15 ml).

The overall purification procedure is summarized in the table. About 3 mg of 800-fold purified Fru-P_2ase are obtained from 100 g of cells. The overall recovery is about 24%.

Properties

Stability. The purified enzyme could be stored at $-20°$ in the presence of 0.1 mM MnCl$_2$ and 1 mM DTT in 0.01 M Tris-HCl buffer (pH 7.4) for at least 3 months without loss of catalytic activity. When stored in the absence of the metal ion, approximately 50% of catalytic activity is lost after 2 weeks and 90% after 3 weeks. Loss of activity could not be reversed by the addition of cofactor (Mn^{2+}), substrate (fructose bisphosphate), or sulfhydryl reagent (DTT).

Molecular Weight. Molecular weight determinations in sucrose density gradients[15] of freshly prepared enzyme (stored for less than 1 week) yielded a sedimentation coefficient (s_w^0) of 7.2 S and a molecular weight of $130,000 \pm 3000$. Fru-P$_2$ase stored in the presence of 1.0 mM MnCl$_2$, 1 mM DTT, and 0.10 M Tris-HCl buffer at $-20°$ for more than 3 weeks retained full catalytic activity but disassociated into components of $65,000 \pm 2000$ as determined by sucrose density gradients and gel filtration on Sephadex G-200. The sedimentation coefficient of disassociated enzyme was calculated to be $s_{20w}^0 = 4.3$ S.

pH Optimum, Chelating Agents, Sulfhydryl Reagents. The maximum enzymic activity for Fru-P$_2$ase occurred at pH 8.5. At pH 7.5 the rate of catalysis was approximately 65% of the hydrolytic rate observed at pH 8.5. EDTA (0.1 mM) completely inhibited Fru-P$_2$ase activity. Incubation of Fru-P$_2$ase with sulfhydryl reagents at concentrations of 1 mM resulted in a 3.8-fold stimulation of enzymic activity. Maximum stimulation was observed with glutathione followed by cysteine, dithiothreitol, and 2-mercaptoethanol. The —SH group reagent p-hydroxymercuribenzoate (PHMB) was shown to be a powerful inhibitor of Fru-P$_2$ase activity. After a 5-min incubation with 4 mM PHMB, the enzyme was completely inactivated.

Substrate Specificity. Of the following compounds—fructose 6-phosphate, glucose 6-phosphate, 6-phosphogluconate, ribose 1-phosphate, ribulose 1,5-diphosphate, glucose 1-phosphate, sedoheptulose 1,7-bisphosphate, and fructose 1,6-bisphosphate—only the latter two were cleaved to any extent. Sedoheptulose 1,7-bisphosphate was cleaved by Fru-P$_2$ase at 22.4% of the fructose 1,6-bisphosphate hydrolysis rate.

Kinetics. Rhodopseudomonas palustris Fru-P$_2$ase exhibited an absolute requirement for divalent cations that could be satisfied by Mn^{2+} (apparent $K_m = 7.8 \times 10^{-5} M$) or Mg^{2+} (apparent $K_m = 3.3 \times 10^{-3} M$). Substrate kinetic studies demonstrated that *R. palustris* Fru-P$_2$ase had a K_m for fructose bisphosphate and for sedoheptulose 1,7-bisphosphate of $4.5 \times 10^{-5} M$ and $30 \times 10^{-5} M$, respectively.

[15] R. G. Martin and B. N. Ames, *J. Biol. Chem.* **236**, 1372 (1961).

Regulation. Kinetic studies of *R. palustris* Fru-P$_2$ase revealed allosteric regulation by substrates (fructose bisphosphate and sedaheptulose 1,7-bisphosphate), cofactor (Mn^{2+}), and inhibitor (GTP). The K_i for fructose bisphosphate hydrolysis by GTP was 10 μM as compared to a K_i of 290 μM for sedoheptulose 1,7-bisphosphate hydrolysis. The extent of allosteric interactions exhibited by FDP, Mn^{2+}, or GTP was pH dependent. Cooperativity exhibited at neutral hydrogen ion concentration (pH 7.4) was abolished under alkaline conditions (pH 8.5). The physiological significance of the regulation of *R. palustris* Fru-P$_2$ase may be that of controlling the Calvin–Basshan carbon-reduction cycle.

[64] Fructose-1,6-bisphosphatase from *Bacillus licheniformis*

By DENNIS J. OPHEIM and ROBERT W. BERNLOHR

$$\text{Fructose 1,6-bisphosphate} \rightarrow \text{fructose 6-phosphate} + P_i$$

Fructose-1,6-bisphosphatase (D-fructose-1,6-bisphosphate 1-phosphohydrolase, EC 3.1.3.11) has long been considered to be a key regulatory enzyme in the control of the Embden–Meyerhof pathway, especially in systems where the flow of carbon through this pathway must be reversed at different periods of growth or differentiation.[1] Fructose-1,6-bisphosphatases have been isolated from a number of bacterial and mammalian systems,[2] but no satisfactory mechanism for the *in vivo* regulation of this enzyme has been established. Although the enzyme is usually strongly inhibited by AMP, it exhibits a K_i value for AMP inhibition that is usually 5- to 10-fold lower than the intracellular concentration of AMP.[3,4] The enzyme from *Bacillus licheniformis* has a number of unique and interesting properties that allow the activity of the enzyme to be modulated under physiological concentrations of effector molecules.[5]

---

[1] J. Ashmore, *in* "Fructose 1,6-Diphosphatase and Its Role in Gluconeogenesis" (R. W. McGilvery and B. M. Pogell, eds.), p. 43. Am. Inst. Biol. Sci., Washington, D.C., 1974.

[2] S. Pontremoli and B. Horecker, *in* "The Enzymes" (P. D. Boyer, ed.), 3rd ed., Vol. 4, p. 611. Academic Press, New York, 1972.

[3] D. G. Fraenkel, S. Pontremoli, and B. Horecker, *Arch. Biochem. Biophys.* **114**, 4 (1966).

[4] O. H. Lowry, J. Carter, J. B. Ward, and L. Glaser, *J. Biol. Chem.* **246**, 6511 (1971).

[5] D. J. Opheim and R. W. Bernlohr, *J. Biol. Chem.* **250**, 3024 (1975).

Assay Method

Principle. Fructose-1,6-bisphosphatase is assayed by measuring the hydrolysis of fructose 1,6-bisphosphate. The products of the reaction, fructose 6-phosphate and inorganic phosphate, can be measured by either a coupled spectrophotometric assay or by the Fiske and SubbaRow inorganic phosphate assay.[6]

Reagents

Sodium fructose 1,6-bisphosphate, 50 mM
NADP, 100 mM
Sodium phosphoenolpyruvate, 100 mM
Glycylglycine, 0.2 M, pH 8.0
MnCl$_2$, 2 mM
Phosphoglucose isomerase (EC 5.3.1.9), 5 mg/ml in 30% glycerol
Glucose-6-phosphate dehydrogenase (EC 1.1.1.49), 5 mg/ml in 30% glycerol

Procedure. The standard assay mixture (1 ml) for determining fructose-1,6-bisphosphatase activity contains 0.02 ml of fructose 1,6-bisphosphate, 0.01 ml of NADP, 0.01 ml of phosphoenolpyruvate, 0.01 ml of glycylglycine, 0.01 ml of MnCl$_2$, and 2 IU of yeast phosphoglucose isomerase (D-glucose-6-phosphate ketol-isomerase, EC 5.3.1.9) and glucose-6-phosphate dehydrogenase (D-glucose-6-phosphate : NADP$^+$ oxidoreductase, EC 1.1.1.49). The reaction was initiated by the addition of 0.006 to 0.03 IU of fructose-1,6-bisphosphatase (Fru-P$_2$ase). The production of NADPH was followed by measuring the increase in optical density at 340 nm. Alternatively, inorganic phosphate is measured by adding 0.5 ml of reaction mixture to 0.5 ml of Fiske and SubbaRow reagents and quantitating the colored complex that forms by its optical density at 660 nm.

Definition of Unit and Specific Activity. One unit of enzyme activity corresponds to the amount of enzyme that produces 1 μmol of fructose 6-phosphate or inorganic phosphate per minute at 30° under standard assay conditions. Specific activity is defined as units per milligram of protein.

Culture and Harvest of Cells

Bacillus licheniformis A5, type 1 was used through this study. The culture medium contained (per liter) 4 mmol of MgSO$_4$, 0.03 mmol of MnSO$_4$, 0.06 nmol of FeSO$_4$, 0.06 mmol of FeSO$_4$(NH$_2$)$_4$SO$_4$, 6.8 mmol of

[6] C. H. Fiske and Y. SubbaRow, *J. Biol. Chem.* **66**, 375 (1925).

NaCl, 5.4 mmol of KCl, 14.6 mmol of phosphate (K) buffer (pH 6.8), 1.6 mmol of potassium citrate, 1.0 mmol of $CaCl_2$, 10 mmol of NH_4Cl, and 15 mmol of glucose at a final pH of 6.9.

Cultures were started by adding 10^{10} spores to 1 liter of medium supplemented with 1.5 mM L-alanine. Cultures were aerated by shaking on a gyratory shaker (New Brunswick Scientific Company) at 37°. When cultures reached (2–3) × 10^8 cells per milliliter, the culture was transferred to a 10-liter fermentation vessel of a fermentor (New Brunswick Scientific Company, Model FS-307) containing 9 liters of prewarmed, aerated, sterile medium.

Volumes of culture were cooled by the addition of ice and harvested by centrifugation (17,000 rpm) in a Beckman J-27 centrifuge equipped with a JCF-2 continuous flow rotor. Cells were washed once with a 10 mM Tris-HCl (pH 7.5) buffer containing 1.0 mM $MnCl_2$, 10 mM $MgCl_2$, and 100 mM NaCl (TMMN buffer) supplemented with 1.0 mM phosphoenolpyruvate. The sedimented cells were used to prepare cell-free extracts immediately or were stored at −60°.

Purification Procedure

Step 1. Preparation of Cell-Free Extract. Sixty grams (wet weight) of late exponential-phase cells (grown on glucose) are suspended at 0–3° in 90 ml of TMMN buffer, pH 7.5, supplemented with 1 mM phosphoenolpyruvate. The $MgCl_2$ and NaCl are included in the buffer to prevent precipitation of nucleic acids and Fru-P_2ase by $MnCl_2$. The cells are ruptured by two passes through a French pressure cell at 11,000–12,000 psi. One microgram of bovine pancreas deoxyribonuclease (EC 3.1.4.5) is added to decrease the viscosity of the solution. The homogenate is centrifuged at 78,000 g for 1.5 hr in a Spinco Model L centrifuge, and the resultant supernatant fraction yields 100 ml of a solution that contains 3400 mg of protein and 415 IU of Fru-P_2ase activity. This preparation contains no 6-phosphogluconate dehydrogenase activity under the assay conditions used. The NADPH oxidase activity is so low under assay conditions that it does not interfere with the standard spectrophotometric assay for Fru-P_2ase activity.

Step 2. Ammonium Sulfate Precipitation. Powdered ammonium sulfate is added slowly to a total of 24.3 g per 100 ml of extract, and the solution is stirred for 15 min at 0°. The precipitated protein is sedimented by centrifugation at 40,000 g for 10 min, and the soluble fraction is treated with an additional 3.3 g of $(NH_4)_2SO_4$. The above procedure is repeated (at 3.3 g/100 ml increments) and the fractions containing most of the Fru-P_2ase (usually between 24.3 and 27.6 g/100 ml) are dissolved in 10 ml of TMMN buffer, to give a concentration of 30 mg of protein per milliliter.

Step 3. DEAE-Cellulose Chromatography. A 10-ml column (2×3.5 cm) of DEAE-cellulose (Whatman) is equilibrated with 10 mM glycylglycine buffer (pH 7.5) containing 1.0 mM MnCl$_2$ and 100 mM NaCl. The protein fraction, dialyzed against the same buffer, is applied to this column and eluted with the same buffer. The eluate from the column, which contains very little DNA (OD 280/OD 260 = 1.3), contains 240 mg of protein and 280 IU of Fru-P$_2$ase activity. The main purpose of this step is to remove the nucleic acids, which cause loss of activity during the MnCl$_2$ precipitation (below).

The eluate from the DEAE column is dialyzed (1 : 1000) twice against a Tris-HCl (pH 7.5) buffer containing 5 mM MnCl$_2$ for 1.5 hr. The precipitated protein is collected by centrifugation at 40,000 g for 10 min and then solubilized by dialysis against a Tris-HCl buffer (pH 7.5) containing 1 M NaCl, 2 mM MnCl$_2$, and 10 mM MgCl$_2$. This fraction (10 ml) contains 40 mg of protein and 250 IU of Fru-P$_2$ase activity.

Step 4. Absorption to α-Alumina Gel. One milliliter of α-alumina gel (Bio-Rad, Richmond, California) is mixed with the protein solution, and after 10 min the mixture is centrifuged at 5000 g for 5 min. The supernatant solution is discarded, and the sedimented gel is washed twice with TMMN buffer. The gel is then resuspended in 20 mM glycylglycine buffer (pH 8.7) containing 100 mM (NH$_4$)$_2$SO$_4$, and 1.0 M NaCl, mixed, and centrifuged; the supernatant solution (1 ml), which contains 0.8 mg of protein and 82 IU of Fru-P$_2$ase activity, is saved.

This protein solution is dialyzed against TMMN buffer and stored at $-20°$ in 30% glycerol. The activity in this preparation has a half-life of approximately 6 months. A typical purification scheme is presented in the table. The range of specific activities of different preparations was 50–170 IU of Fru-P$_2$ase activity per milligram of protein. These preparations contained less than 0.03 IU of alkaline phosphatase, NADPH oxidase, or

PURIFICATION OF FRUCTOSE-1,6-BISPHOSPHATASE

Step	Total protein[a] (mg)	Total activity (IU)	Specific activity (IU/mg protein)	Purification (fold)	Yield (%)
78,000 g centrifugation	3400	415	0.12	1	100
(NH$_4$)$_2$SO$_4$ fraction	360	280	0.75	6.8	69
DEAE chromatography	240	280	1.16	9.7	69
MnCl$_2$ precipitation	41	256	6.3	53	63
Alumina gel	0.82	82	100	835	20

[a] Protein was assayed by the procedure of O. H. Lowry, N. J. Rosebrough, A. L. Farr, and R. J. Randall, *J. Biol. Chem.* **193**, 265 (1951).

6-phosphogluconate dehydrogenase activity per milligram of protein. The specific activity of the most purified fraction appears to be about fivefold higher than that found for most other fructose-bisphosphatases. The glycerol was removed by dialysis against a 5.0 mM Tris-HCl buffer (pH 7.5) containing 100 mM NaCl and 0.2 mM MnCl$_2$ before any kinetic analysis was performed on the enzyme.

Properties

Physical Characteristics

Fructose-1,6-bisphosphatase from *B. licheniformis* has a molecular weight of approximately 500,000 by density gradient centrifugation and molecular sieve chromatography. There is no indication of multiple enzyme forms by either disc gel electrophoresis or immunological analysis.[5]

Stabilization

Mn^{2+} Ion Stabilization. Mn^{2+} ion is an absolute requirement for the stabilization of partially purified Fru-P$_2$ase activity; 0.5 mM MnCl$_2$ provides 100% stabilization of activity, whereas the enzyme activity has a half-life of 2 min in the absence of Mn^{2+} ion. Mg^{2+} or Co^{2+} could not replace the requirement for Mn^{2+}.

AMP Inactivation. AMP is able to cause the inactivation of purified Fru-P$_2$ase. The inactivation is first order with respect to time, and the rate of inactivation is directly proportional to the concentration of AMP (0.1–1.0 mM). Phosphoenolpyruvate is able to block the inactivation of Fru-P$_2$ase, but not reactivate inactivated enzyme.

Protein Stabilization. Even under optimal conditions there is a loss of enzyme activity that occurs when the protein concentration is less than 15 mg/ml. Addition of 10% glycerol prevents this inactivation in crude or purified enzyme preparations.

Kinetic Factors

Substrate Specificity. The purified enzyme is specific for fructose 1,6-bisphosphate (Fru-P$_2$). A variety of phosphorylated compounds, including fructose 1,7-diphosphate and p-nitrophenyl phosphate are attacked at a rate of less than 1% that of Fru-P$_2$, at pH 7.0, 8.0, and 9.0.

pH Optimum. The pH optimum of Fru-P$_2$ase is approximately 8.0–8.5 for both crude and purified enzyme. Reactions were usually buffered at pH 8.0 because the oxidation of Mn^{2+} to an insoluble precipitate was a problem at higher pH values.

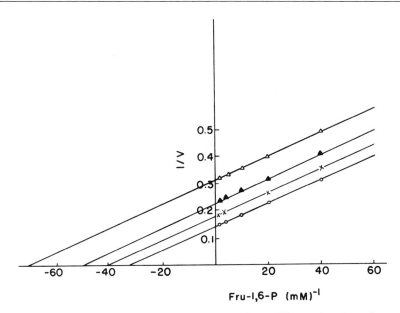

FIG. 1. Kinetics of fructose 1,6-bisphosphate saturation at different phosphoenolpyruvate concentrations. The reaction mixture (1.0 ml), containing 0.2 mM MnCl$_2$, 20 mM glycylglycine (pH 8.0), 2 mM NADP$^+$, 2 IU of glucose-6-phosphate dehydrogenase and glucose-6-phosphate isomerase, and varied amounts of fructose 1,6-bisphosphate and phosphoenolpyruvate ($\triangle$, 2 μM; $\blacktriangle$, 5 μM; $\times$, 10 μM; $\bigcirc$, 100 μM), was initiated by the addition of 3 μg of 57-fold purified fructose-1,6-bisphosphatase. The reciprocal plot of these data is presented. Reprinted from Opheim and Bernlohr[5] with permission of the publisher.

Cations and Sulfhydryl Reagents. Since Mn^{2+} is required for the stabilization of Fru-P$_2$ase, it is difficult to estimate the effect of Mn^{2+} on the rate of hydrolysis. No effect is found when 10 mM MgCl$_2$, CoCl$_2$, NH$_4$Cl, or sodium phosphate is added to the reaction mixture containing 0.1 mM MnCl$_2$. No effect on enzyme activity is found when 1 mM cysteine, dithiothreitol, mercaptoethanol, or reduced glutathione is added to the reaction mixture.

Fructose 1,6-Bisphosphate. The saturation kinetics of Fru-P$_2$ase by Fru-P$_2$ is presented in Fig. 1. Phosphoenolpyruvate is found to have a significant effect on these kinetics, causing an increase in both the V_{max} and K_m values for Fru-P$_2$.

Phosphoenolpyruvate. Saturation kinetics indicates that the enzyme has an apparent K_m value of 2 μM for phosphoenolpyruvate, and this value is unaffected by changes in Fru-P$_2$ concentration. The V_{max} values exhibit minimal changes over concentrations of Fru-P$_2$ (0.1–1.0 mM) that may be expected to occur under physiological conditions. The K_m value

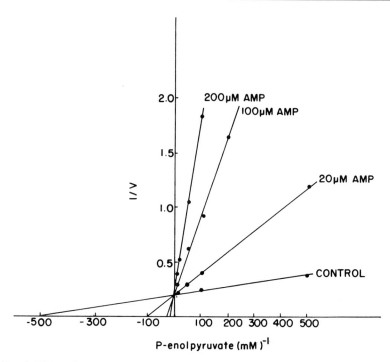

FIG. 2. Effect of adenosine 5'-monophosphate (AMP) on the relationship between initial velocity and phosphoenolpyruvate (PEP) concentration. The reaction mixture (1.0 ml), containing 0.2 mM MnCl$_2$, 0.2 mM Fru-P$_2$, 20 mM glycylglycine (pH 8.0), 2 mM NADP$^+$, 2 IU of glucose-6-phosphate dehydrogenase and glucose-6-phosphate isomerase, and varied amounts of PEP and AMP, was initiated by the addition of 3 μg of 53-fold-purified Fru-P$_2$ase. Reprinted from Opheim and Bernlohr[5] with permission of the publisher.

for phosphoenolpyruvate is significantly altered by AMP (Fig. 2). Further analysis indicates that AMP is a linear competitive inhibitor of phosphoenolpyruvate binding.

Adenosine 5'-Monophosphate. In order to evaluate the effect of AMP on enzyme activity, it was necessary to differentiate AMP inhibition from AMP inactivation. To do this the rate of reaction was estimated over the first 15 sec of reaction. To ensure that the decrease in activity was not due to inactivation, 10 mM phosphoenolpyruvate was then added to the reaction mixture, which blocks both AMP inhibition and inactivation. Using this procedure, the effect of AMP on enzyme activity was determined (Fig. 3). The concentration of AMP that causes 50% inhibition of Fru-P$_2$ase activity is approximately 5 μM in the absence of phosphoenolpyruvate and increases as the concentration of phosphoenolpyruvate in the reaction mixture increases.

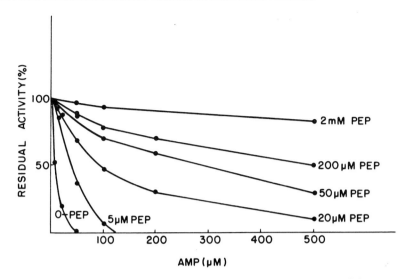

FIG. 3. The effect of phosphoenolpyruvate (PEP) on the inhibition of fructose 1,6-bisphosphatase by adenosine 5′-monophosphate (AMP). The reaction mixture (1.0 ml), containing 0.1 mM Fru-P$_2$, 10 mM glycylglycine (pH 8.0), 2 mM NADP$^+$, 2 IU of glucose-6-phosphate dehydrogenase and glucose-6-phosphate isomerase, 0.2 mM MnCl$_2$, and varied concentrations of AMP and PEP, was initiated by the addition of 3 μg of 50-fold-purified Fru-P$_2$ase. The initial velocity in the presence of AMP was compared with the initial velocity determined in the absence of AMP. The initial velocity was determined over the first 15 sec of the reaction. Reprinted from Opheim and Bernlohr[5] with permission of the publisher.

The low K_i of most Fru-P$_2$ases for AMP, coupled with the relatively high intracellular concentration of AMP, has always presented a problem in understanding the regulation of this enzyme. The net result of these factors is that the enzyme should be inhibited to a major degree under all physiological conditions. To our knowledge this is the only system where AMP inhibition of Fru-P$_2$ase can be completely reversed by an intracellular intermediate at physiological concentrations. Analysis of the intracellular concentrations of Fru-P$_2$ase, Fru-P$_2$, AMP, and phosphoenolpyruvate of *B. licheniformis* suggests that physiological changes in the concentrations of phosphoenolpyruvate may regulate the activity of this constitutive enzyme *in vivo*. The intracellular concentration of phosphoenolpyruvate increases two- to fourfold during the shift from glycolysis to gluconeogenesis.[4,7] This could result in a fivefold increase in Fru-P$_2$ase activity (under physiological concentrations of Fru-P$_2$ and AMP) when the cell requires this enzyme activity during gluconeogenic growth.

[7] T. J. Donohue and R. W. Bernlohr, *J. Bacteriol.* **135**, 363 (1978).

[65] Sedoheptulose-1,7-bisphosphatase from Wheat Chloroplasts

By IAN E. WOODROW

Sedoheptulose 1,7-bisphosphate + H_2O → sedoheptulose 7-phosphate + P_i

Wheat chloroplast sedoheptulose-1,7-bisphosphatase (SBPase) [EC 3.1.3.37], which catalyzes the hydrolytic cleavage of P_i from the C-1 position of sedoheptulose 1,7-bisphosphate, is one of several enzymes of the reductive pentose phosphate pathway that undergo a light-mediated activation.[1-3] The rate of activation *in vitro* is modulated by the concentration of sedoheptulose 1,7-bisphosphate, Mg^{2+}, and a reductant such as dithiothreitol.[4] Activation is rapid enough in the presence of these three effectors to observe the final steady-state reaction velocity in the course of a standard assay. Alternatively, the enzyme can be preactivated by incubating with a reducing agent alone for several hours.

Assay Methods for Sedoheptulose-1,7-bisphosphatase

NADH Oxidation Method

The SBPase activity is measured by coupling the sedoheptulose 7-phosphate formation to the oxidation of NADH in the presence of Mg^{2+} and dithiothreitol.

$$\text{Sedoheptulose 1,7-bisphosphate} + H_2O \rightarrow \text{sedoheptulose 7-phosphate} + P_i \quad (1)$$
$$\text{Sedoheptulose 7-phosphate} + ATP \rightarrow \text{sedoheptulose 1,7-bisphosphate} + ADP \quad (2)$$
$$\text{Phosphoenolpyruvate} + ADP \rightarrow \text{pyruvate} + ATP \quad (3)$$
$$\text{Pyruvate} + NADH + H^+ \rightarrow \text{lactate} + NAD^+ \quad (4)$$

The oxidation of NADH is followed spectrophotometrically at 340 nm. In the steady state (when intermediates of the sequence reach steady-state levels) the rate of NADH oxidation represents the velocity of the SBPase reaction. The assay is not suited to estimation of SBPase activity in crude tissue extracts. Enzyme preparations should be free from adenylate kinase, aldolase, ATPase, and nonspecific phosphatase activity. Reaction

[1] P. Schürmann and B. B. Buchanan, *Biochim. Biophys. Acta* **376**, 189 (1975).
[2] M.-L. Champigny and E. Bismuth, *Physiol. Plant.* **36**, 95 (1976).
[3] I. E. Woodrow and D. A. Walker, *Biochem. J.* **191**, 845 (1980).
[4] I. E. Woodrow and D. A. Walker, *Arch. Biochem. Biophys.* (in press).

(2) is catalyzed by fructose-6-phosphate kinase from rabbit muscle, which can also utilize sedoheptulose 7-phosphate as substrate.[5]

Reagents

Tricine-NaOH, 1 *M*, pH 8.2
$MgCl_2$, 0.25 *M*
Dithiothreitol, 1 *M*
Sedoheptulose 1,7-bisphosphate (sodium salt), 2 m*M*; obtained from Sigma Chemical Co. (Poole, Dorset, U.K.)
KCl, 1 *M*
ATP, 5 m*M*
NADH, 10 m*M*
Phosphoenolpyruvate, sodium salt, 0.05 *M*
Pyruvate kinase, 1 mg/ml
Lactate dehydrogenase, 5 mg/ml; both enzymes purchased from Boehringer, Mannheim, West Germany, as crystalline suspensions in ammonium sulfate
Fructose-6-phosphate kinase (rabbit muscle), 2.1 mg/ml purchased from Sigma
SBPase: dilute the enzyme solution to a concentration of 0.5 unit/ml.

Procedure. The reaction is carried out in a cell (1-cm light path) containing, in a final volume of 1 ml, 0.1 ml of buffer, 0.04 ml of $MgCl_2$, 0.02 ml of KCl, 0.02 ml of ATP, 0.02 ml of phosphoenolpyruvate, 0.05 ml of sedoheptulose 1,7-bisphosphate, 0.015 ml of NADH, 0.03 ml of dithiothreitol, 0.004 ml of pyruvate kinase, 0.002 ml of lactate dehydrogenase, 0.008 ml of fructose-6-phosphate kinase, and the amount of enzyme to be assayed. The coupling enzymes are normally dialyzed before use. In order to monitor enzyme activation, the reaction is initiated with SBPase. However, the enzyme can be preincubated in the reaction mixture for 2 hr and the reaction initiated with sedoheptulose 1,7-bisphosphate to eliminate the activation transient. Absorbance is followed at 340 nm. As the reaction substrate is recycled, the reaction can be monitored for long periods without a significant change in the substrate concentration. The enzyme is stable under the assay conditions for several hours.

Definition of Unit and Specific Activity. One unit of enzyme is defined as the amount that catalyzes the formation of 1 μmol of sedoheptulose 7-phosphate per minute at 20° under the above conditions. The specific activity is expressed as units of enzyme activity per milligram of protein. Protein is determined by the procedure of Lowry *et al.*[6]

[5] N. S. Karadsheh, G. A. Tejwani, and A. Ramaiah, *Biochim. Biophys. Acta* **327**, 66 (1973).
[6] O. H. Lowry, N. J. Rosebrough, A. L. Farr, and R. J. Randall, *J. Biol. Chem.* **193**, 265 (1951).

Phosphate Liberation Method

The hydrolysis of sedoheptulose 1,7-bisphosphate is followed by estimating the quantity of P_i cleaved during specific time periods. The technique of Taussky *et al.*[7] is used to determine P_i.

Reagents

Tricine-NaOH, 1 M, pH 8.2
$MgCl_2$, 0.25 M
Dithiothreitol, 1 M
Sedoheptulose 1,7-bisphosphate (sodium salt), 20 mM
SBPase: dilute enzyme to a concentration of 0.5 unit/ml.

Procedure. The reaction is carried out in centrifuge tubes containing, in a final volume of 1 ml, 0.1 ml of buffer, 0.04 ml of $MgCl_2$, 0.03 ml of dithiothreitol, and 0.01 ml of sedoheptulose 1,7-bisphosphate. The enzyme is activated by preincubation for 2 hr in the reaction mixture, after which the reaction is initiated by adding sedoheptulose 1,7-bisphosphate. After 30 min the reaction is terminated by adding 1 ml of 30% (w/v) trichloroacetic acid. The protein precipitate is removed by centrifugation, and the supernatant is analyzed for P_i. The assay is essentially linear throughout and is suitable for crude extracts.

Definition of Unit and Specific Activity. One unit of enzyme is defined as the amount that catalyzes the liberation of 1 μmol of P_i per minute under the above conditions. The specific activity is expressed as units per milligram of protein.

Purification Procedure

All operations are performed at 4°.

Step 1. Preparation of Leaf Extract. Twelve-day-old wheat leaves (100 g) are blended for 1 min in 150 ml of a medium containing 50 mM Tris-HCl (pH 7.8), 0.5 mM EDTA, and 10 mM 2-mercaptoethanol. The homogenate is filtered through four layers of muslin and centrifuged at 30,000 g for 15 min. The supernatant, containing all the SBPase activity, is collected.

Step 2. Ammonium Sulfate Fractionation. The supernatant is fractionated by adding solid ammonium sulfate. Protein precipitating between 2.2 M and 4.6 M ammonium sulfate is collected by centrifugation (30,000 g for 15 min) and dissolved in a minimum volume of a buffer containing 50 mM Tris-HCl (pH 8.2), 0.5 mM EDTA, and 10 mM NaCl. The protein solution is desalted on a Sephadex PD-10 column of G-25M equilibrated with the same buffer.

[7] H. H. Taussky, E. Shorr, and G. Kurzmann, *J. Biol. Chem.* **202**, 675 (1953).

PURIFICATION OF SEDOHEPTULOSE-1,7-BISPHOSPHATASE FROM WHEAT LEAVES

Fraction	Total protein (mg)	Total activity (units)	Yield (%)	Specific activity (units/mg protein)
Leaf extract	1881	64	100	0.034
$(NH_4)_2SO_4$ precipitate	945	63	99	0.067
DEAE-Sephadex A-50	423	55	86	0.130
Sephadex G-100	12.3	51	80	4.15

Step 3. DEAE-Sephadex A-50 Chromatography. The enzyme solution is applied to a column (3.6 cm² × 20 cm) of DEAE-Sephadex A-50 equilibrated with 50 mM Tris-HCl (pH 8.2), 0.5 mM EDTA, and 10 mM NaCl. The column is washed with the same buffer until the eluate is free from protein (determined by absorbance at 280 nm). The SBPase activity is eluted in a linear gradient of NaCl (10 to 300 mM) made up in the Tris buffer. The active fractions are pooled and concentrated to 3 ml in an Amicon ultrafiltration cell with a Diaflo PM-10 membrane.

Step 4. Sephadex G-100 Chromatography. The concentrated fraction is then applied to a column (2 cm² × 90 cm) of a Sephadex G-100 (superfine grade) equilibrated with 50 mM Tris-HCl (pH 8.2), 0.5 mM EDTA 10 mM 2-mercaptoethanol, and 0.2 M NaCl. SBPase activity is eluted (6 ml/hr) in a single peak behind the main protein peak. The active fractions are combined, concentrated in an Amicon ultrafiltration cell with a Diaflo PM-10 membrane, and desalted on a Sephadex PD-10 column equilibrated with 20 mM Tricine–NaOH (pH 8.2).

The enzyme solution can be stored at 4°, 50% activity being lost in 3–4 weeks.

A typical purification is summarized in the table.

Properties

pH Optimum. The optimum pH for SBPase activity, in the presence of 10 mM Mg^{2+}, is in the range 8.2–8.5. This range is also optimal for enzyme activation by dithiothreitol.

Catalytic Specificity. The SBPase preparation is specific for sedoheptulose-1,7-bisphosphatase. Using P$_i$ cleavage as a measure of activity, none of the following substrates appreciably support enzyme activity: fructose 1,6-bisphosphate, fructose 6-phosphate, sedoheptulose 7-phosphate, glucose 6-phosphate, ribulose 5-phosphate, dihydroxyacetone phosphate, or erythrose 4-phosphate at a concentration of 0.8 mM.

Effect of Substrate Concentration. No substrate inhibition is observed at sedoheptulose 1,7-bisphosphate concentrations up to 2 mM. The K_m value calculated from Lineweaver–Burk plots is found to be 14 μM at pH 8.2 in the presence of 10 mM Mg^{2+}.

Metal Ion Requirement. Wheat leaf SBPase requires Mg^{2+} for catalytic activity.

Inhibitors. Ca^{2+}, at a concentration of 0.7 mM, inhibits SBPase activity 50%, in the presence of 10 mM Mg^{2+} at pH 8.2. P_i competitively inhibits SBPase activity. The K_i calculated from Lineweaver–Burk plots is found to be 2.2 mM. The rate of activation of SBPase is also inhibited by P_i.

Activation. SBPase activation requires the presence of a reducing agent. Both dithiothreitol and 2-mercaptoethanol were effective, whereas cysteine and reduced glutathione were not. The rate of activation of SBPase is at least 20-fold faster in the presence of 10 mM Mg^{2+} and 0.1 mM SBP together with the reductant.

Homogeneity and Molecular Weight. The SBPase preparation, although not completely homogeneous, shows a single major band in sodium dodecyl sulfate–polyacrylamide gel electrophoresis. The molecular weight, as determined by Sephadex G-100 chromatography under the described conditions, is approximately 56,000.

[66] Glucose-6-phosphatase from Cerebrum

By MANFRED L. KARNOVSKY, JAMES MICHAEL ANCHORS, and MICHAEL A. ZOCCOLI

Glucose 6-phosphate + H$_2$O → glucose + P$_i$

The brain has usually been regarded as containing negligible amounts of glucose-6-phosphatase, if any.[1] In teleological terms this would not be unexpected, since this enzyme catalyzes the final step in the release of glucose from an intracellular pool of phosphorylated intermediates or from glycogen, for "export" to other locales. Most important, the reaction catalyzed by this enzyme is the final step in gluconeogenesis (reviewed in Nordlie[2]). The brain, by virtue of its own heavy dependence on glucose as its principal source of energy, its small stores of glycogen, and its modest content of protein and lipid available for provision of energy, would not be expected to require or exhibit much glucose-6-phosphatase activity.

---

[1] G. Weber and A. Cantero, *Endocrinology* **61**, 701 (1957).

[2] R. C. Nordlie, *in* "The Enzymes" (P. D. Boyer, ed.), 3rd Ed., Vol. 4, p. 543. Academic Press, New York, 1971.

ENZYME ACTIVITY OF RAT BRAIN AND LIVER[a,b]

Tissue	Enzyme activity (nmol min⁻¹ mg⁻¹ protein) of:		
	Glucose 6-phosphatase	PP_i/glucose phosphotransferase	Hexokinase
Brain	4.8	1.2	138.0
Liver	233	95.7	8.2

[a] Reprinted in modified form from Anchors et al.,[7] with permission.
[b] Each value represents the mean of four determinations. The standard deviations were less than 10% of the means cited.

However, in 1975, glucose-6-phosphatase (EC 3.1.3.9) was extracted from brain and purified to homogeneity, or near it. The brain enzyme is apparently more susceptible of purification than the enzyme of liver (and kidney). It was found to have a function in slow-wave sleep—i.e., to manifest changes of activity with state of consciousness.[3] The biochemical observation that rat brain contains significant amounts of glucose-6-phosphatase was buttressed by cytochemical findings in a different laboratory.[4] The cerebral enzyme was demonstrated at the cellular and subcellular level by a technique that involved the precipitation of inorganic phosphate released from the substrate as lead phosphate. The enzyme was found in all cell types of the cerebral cortex, the cerebellum, and the brainstem. Large neurons were more reactive than smaller cells. Thus, in the cerebellum the Purkinje cells were most active; and in the cerebral cortex, the pyramidal cells. Of importance was the fact that the activity was associated largely with the endoplasmic reticulum and with the nuclear envelope, but not the Golgi. This is consistent with the location of the enzyme in liver cells.[5] The authors of this cytochemical study were at pains to perform controls and to establish the specificity of their method to avoid confusion due to nonspecific phosphatases.[4]

Liver glucose-6-phosphatase is a multifunctional enzyme,[6] and the brain enzyme has similar properties.[3] In particular, both enzymes are known to act as phosphotransferases under appropriate conditions and to act also as an inorganic pyrophosphatase. Prior to describing extraction and purification of the enzyme, it would be useful to provide an idea of the levels in brain. A quantitative comparison of brain with liver is given in the table, and data for an independent activity, hexokinase, are also cited.[7]

[3] J. M. Anchors and M. L. Karnovsky, J. Biol. Chem. 250, 6408 (1975).
[4] H. R. Stephens and E. B. Sandborn, Brain Res. 113, 127 (1976).
[5] H. M. Gunderson and R. C. Nordlie, J. Biol. Chem. 250, 3552 (1975).
[6] R. C. Nordlie, Curr. Top. Cell Regul. 8, 33 (1974).
[7] J. M. Anchors, D. F. Haggerty, and M. L. Karnovsky, J. Biol. Chem. 252, 7053 (1977).

Since this enzyme is located in the endoplasmic reticulum, it is recovered predominantly in the microsomes after homogenization and centrifugation. *Full* activity of the enzyme of broken-cell preparations is realized only after disruption of membrane structures by extensive sonication or, better, by the use of deoxycholate, which may be included in the assay system.

Assays

The well-known method of Zakim and Vessey[8] for measuring glucose-6-phosphatase has been employed successfully in the case of brain.[7] Glucose-6-phosphatase activity has also been assayed[3] by the release of P_i from $0.08\,M$ glucose 6-phosphate (or other substrates tested) at $25°$ in 10 min in the presence of $0.1\,M$ Tris-maleate buffer, pH 6.8, and 0.2% (w/v) sodium deoxycholate, in which case P_i was conveniently determined by the method of Martin and Doty.[9]

In order to apply the assay procedure, a suitable homogenate of the tissue must be made. One such, appropriate for the assay of Zakim and Vessey,[8] was made in 20 mM Tris, 10 mM NaCl, 1 mM $CaCl_2$ and 0.05% (v/v) mercaptoethanol, pH 7.4. A glass homogenizer with a Teflon pestle was used, and the ratio of tissue to medium was 1 : 10 (w/v). Microsomes could be prepared if desired. Full expression of enzyme activity required the presence of deoxycholate (0.03% w/v).

An alternative method for assay of glucose-6-phosphatase follows. It is applicable on a microscale and has greater specificity than methods that measure release of P_i. [^{14}C]Glucose 6-phosphate is used as substrate; [^{14}C]glucose is the product determined, and glycerol 2-phosphate can be included in the reaction mixture to eliminate false high values of activity due to nonspecific phosphatases. The method is based on the procedures of Colby and Romano[10] for determining free and phosphorylated 2-deoxyglucose in cells. After the reaction is stopped, aliquots of the mixture are treated by ion-exchange chromatography to separate the uncharged product ([^{14}C]glucose) from the charged substrate.

A brain homogenate is made in $0.5\,M$ Tris-buffer, pH 8.0, $0.32\,M$ sucrose. The buffer volume is four times the fresh weight of brain. (For measurement of *total* hydrolase activity, the homogenate must include deoxycholate at a concentration of 0.2%.)

[8] D. Zakim and D. A. Vessey, *Methods Biochem. Analy.* **21**, 1 (1973).
[9] J. B. Martin and D. M. Doty, *Anal. Chem.* **21**, 965 (1949).
[10] C. Colby and A. H. Romano, *J. Cell. Physiol.* **85**, 15 (1975).

Enzyme Reaction System

Sodium acetate buffer, 1 M, pH 5.75, 0.1 ml
β-Glycerophosphate, 0.4 M, pH 5.75, 0.2 ml
[^{14}C]Glucose 6-phosphate, 0.4 M (0.1 μCi; New England Nuclear)
Distilled H_2O, 0.4 ml
0.1 ml homogenate above, or other similar homogenate.

A control is run with 0.1 ml of distilled H_2O instead of homogenate. Samples are incubated at 37° for up to 30 min (the linearity of the reaction should be checked). The reaction is stopped by immersing the tubes in boiling H_2O for 2 min. The samples are placed on ice and then spun in a refrigerated centrifuge for 20 min at 2500 rpm. Aliquots of the supernatant extract (0.2 ml) are pipetted into a centrifuge tube, and 0.8 ml distilled H_2O is added. The tube is mixed well. One milliliter is placed on a Dowex column (AG 1 × 2, 200–400 mesh, formate form, 30 × 5 mm. A Pasteur pipette makes a convenient column.) Total effluent is collected in a graduated tube (1 ml). The column is washed with 1.5 ml of distilled H_2O and the entire (1.5 ml) effluent is collected in the same tube. Aliquots of effluent (0.5 ml each from total of 2.5 ml) are counted in duplicate on a scintillation counter. A scintillation fluid is employed (e.g., Aquasol, New England Nuclear) that permits use of water in the counting mixture. Blank values are approximately 200 cpm, while enzyme measurements as described above are in the range of 2000 cpm.

Preparation of Crude Brain Microsomes

The major amount of enzyme, as expected from the cytochemical observations above, and by analogy with liver, is obtained in the microsomal (endoplasmic reticulum vesicle) fraction upon cellular fractionation of brain.

Rats are guillotined and the brains are immediately removed. Brain volume is measured at 0° by displacement of buffer. The tissue is homogenized in 9 parts of medium (0.25 M sucrose; 25 mM KCl; 4 mM MgCl$_2$; 50 mM Tris-HCl, pH 7.4), with five up-and-down strokes in a Teflon–glass homogenizer.[11] The homogenate is centrifuged in the cold at 15,000 g for 15 min. The supernatant fluid is decanted and recentrifuged at 105,000 g for 2 hr in a refrigerated centrifuge. The pellet is suspended at a final volume of 5 mg of protein per milliliter. As would be expected, the yield is small, but for many purposes this preparation is adequate. The pellets (above) can be recycled through the procedure to yield a cleaner

[11] V. P. Whittaker and L. A. Barker, *Methods Neurochem.* 2, 1 (1972).

preparation. A method for preparing brain microsomes of higher purity, based on an homogenization similar to that above, is described by Bloemendal et al.[12]

Extraction and Purification of Enzyme from Rat Brain[12a]

The strategy involves homogenization, delipidation by removing the myelin fraction by flotation, recovery of the membrane pellet, extraction of the latter with a buffered deoxycholate solution, known to "solubilize" the enzyme, and purification by gel filtration in the presence of deoxycholate. The description that follows is taken from Anchors and Karnovsky[3] with minor modifications.

The frozen brains are homogenized, 20 at a time, in 4 volumes of cold $0.5 M$ Tris buffer, pH 8.0, containing $0.32 M$ sucrose. The suspension is centrifuged at 59,000 g at 4° for 1 hr in the type 21 preparative rotor of a Spinco Model T centrifuge (Beckman). This procedure, which separates myelin from the insoluble proteins, is modified from a published procedure[13] originally intended for the isolation of pure myelin. In the present procedure, however, the supernatant fluid, including the band of myelin at the 0.32 to $0.85 M$ sucrose interface, is discarded. The pellet is washed twice with about 10 volumes of cold buffer containing, in addition, 0.1% Triton X-100 (v/v). The pellet is finally washed with cold distilled water. It is then dried under a stream of nitrogen and stored as a powder at $-30°$ in a desiccator over $CaCl_2$.

Ten grams of this powder are extracted twice with 100 ml of 10 mM sodium phosphate, pH 7.0, containing 1.0% sodium deoxycholate (w/v). The extraction is carried out at 25° for 1 hr. The extracts are combined and centrifuged at 100,000 g in the Ti40 head of a Beckman L-2 ultracentrifuge.

The extract is submitted to gel filtration on a 3-liter column of Sephadex G-100. The filtration is performed at 4° on a specially designed Lucite column that enables this step to be completed in 8 hr. This large column is built up of separate 700-ml sections with porous polyethylene disks at their ends. When the column is assembled, these disks are interspersed along its length. This feature permits the use of a peristaltic pump to drive the elution at the maximum flow rate recommended by the manufacturers of Sephadex. The column is eluted with 10 mM sodium phosphate buffer in which the detergent concentration is reduced to 0.1% (w/v). Fractions (25 ml each) are collected and assayed for activity.

---

[12] H. Bloemendal, W. S. Bont, M. de Vries, and E. L. Benedetti, *Biochem. J.* **103**, 177 (1967).
[12a] Reprinted in modified form from Anchors and Karnovsky,[3] with permission.
[13] M. L. Cuzner, A. N. Davison, and N. A. Gregson, *J. Neurochem.* **12**, 469 (1965).

The combined active fractions are loaded directly onto a 100-ml column of DEAE-Sephadex that has been equilibrated with 0.1% detergent buffer. The DEAE-Sephadex column is washed with 0.1% detergent buffer. The column is then developed with a linear gradient of ionic strength from 0.05 M sodium phosphate, pH 7.0, containing 0.1% detergent to 0.10 M sodium phosphate, pH 7.0, containing 0.1% detergent. It is also possible, and more convenient, to develop the DEAE-column in two steps of 0.05 M and 0.10 M phosphate buffer containing 0.1% detergent.

The 0.10 M phosphate eluate, which contains the activity, is lyophilized to dryness and reconstituted with 20 ml of distilled water. The solution is dialyzed at 4° overnight against 5 liters of distilled water. The dialyzate is concentrated to 2 ml and redialyzed. Toward the end of the second dialysis, the slightly turbid solution begins to clear, and flakes form and settle. The final dialyzate is lyophilized to dryness and stored at −30° in a desiccator over $CaCl_2$.

Note: In the original purification of the brain enzyme, active fractions were identified by the ratio of ^{32}P and ^{33}P in a phosphoprotein fraction from brains of sleeping and waking rats. This phosphoprotein was the enzyme intermediate of glucose-6-phosphatase (a histidine-phosphate, where the phosphate derives from the substrate). The amount of label in the enzyme protein of the sleeper is almost an order of magnitude greater than that of the waker. Brain powder from a rat in each state was prepared[3] (see above). These preparations were mixed, and a small portion of the mixture was added to the large preparation of unlabeled brain powder described above. The chromatographic fractions with the highest ratio of sleeper-isotope to waker-isotope were those that contained the glucose-6-phosphatase activity. This stratagem for following the enzyme purification is clearly impractical if one wishes simply to prepare the enzyme. The purification can easily be followed by measuring enzyme activity— preferably by the radioactive enzymic method described earlier employing [^{14}C]glucose 6-phosphate.

Comparison of Brain Enzyme with Liver Enzyme

In addition to points of similarity outlined at the beginning of this article, some points of difference should be detailed:

1. Rat brain microsomal glucose-6-phosphatase was found to have a latency with respect to saturating concentrations (20 mM) of glucose 6-phosphate *and* mannose 6-phosphate of about 60–70%. Latency was relieved by the use of deoxycholate (DOC) at 0.2%. The usual formula for percentage of latency was employed:

$$\frac{(\text{Activity in presence of DOC}) - (\text{activity without DOC})}{\text{activity in presence of DOC}} \times 100$$

Liver microsomes have a much smaller latency (i.e., ca. 30%) toward glucose 6-phosphate, but latency toward mannose 6-phosphate is virtually total.[14]

2. Brain microsome preparations did not exhibit sensitivity to diisothiocyanostilbene disulfonate (DIDS) under conditions[15] that inhibited the glucose-6-phosphatase of intact liver microsomes.

The sensitivity to DIDS of liver enzyme system has been used in support of the concept that the microsomes have a transport protein to convey glucose 6-phosphate from the cytoplasmic side of the endoplasmic reticulum to the lumenal side.[14] The action of DIDS is blocked by the simultaneous presence of glucose 6-phosphate, but not of mannose 6-phosphate.[15] The apparent lack of a transport protein for glucose 6-phosphate distinguishes the brain enzyme from the liver preparation.

[14] A. J. Lange, W. J. Arion, and A. L. Beaudet, *J. Biol. Chem.* **255**, 8381 (1980).
[15] M. A. Zoccoli and M. L. Karnovsky, *J. Biol. Chem.* **255**, 1113 (1980).

[67] Pyruvate Dehydrogenase Phosphatase from Bovine Heart

By Flora H. Pettit, W. Martin Teague, and Lester J. Reed

(Phospho)$_n$pyruvate dehydrogenase (E$_1$) + n H$_2$O → pyruvate dehydrogenase + n P$_i$

Pyruvate dehydrogenase phosphatase is a mitochondrial enzyme that catalyzes dephosphorylation and concomitant reactivation of the phosphorylated, inactive pyruvate dehydrogenase component (E$_1$) of the pyruvate dehydrogenase complex. It is a Mg^{2+}-dependent and Ca^{2+}-stimulated phosphatase.

Assay Method

Principle. Assay of pyruvate dehydrogenase phosphatase activity is based on measurement of the initial rate of reactivation of phosphorylated, inactive pyruvate dehydrogenase complex from bovine kidney or heart. The method is modified from that of Linn *et al.*[1] Phosphatase activity may also be determined by measuring the rate of release of ^{32}P-labeled phosphoryl groups from enzyme complex that has been phosphorylated with [γ-^{32}P]ATP.[2,3] These assay methods may not be applicable to whole tissue extracts in view of reports[4,5] of the existence of substantial amounts of extramitochondrial pyruvate dehydrogenase phosphatase-like activity. It seems likely that this activity is due to a cytosolic phosphoprotein phosphatase of broad substrate specificity.[6]

Reagents

Assay buffer: 0.01 M sodium 2-(N-morpholino)propane sulfonate (MOPS), pH 7.0, 0.1 mM CaCl$_2$, 0.5 mM MgCl$_2$, 0.5 mM dithiothreitol

MgCl$_2$, 0.1 M

ATP, 0.01 M

Glucose, 0.1 M

Hexokinase (Boehringer), 1 mg/ml

Preparation of Substrate. Phosphorylated, inactive pyruvate dehydrogenase complex is prepared by incubating a solution containing 1 mg of highly purified complex from bovine kidney,[7] essentially free of pyruvate dehydrogenase phosphatase, in 1 ml of assay buffer with 0.002 ml of 0.01 M ATP for 15 min at 30°. A 0.01-ml aliquot of the incubation mixture is assayed for pyruvate dehydrogenase complex activity,[7] which should be less than 5% of the activity of the untreated complex. One-hundredth milliliter of 0.1 M glucose and 0.001 ml of hexokinase solution are added to scavenge the remaining ATP. After 1 min at 30°, the preparation is placed in an ice bath.

Procedure. To 0.08 ml of phosphorylated, inactive pyruvate dehydrogenase complex (80 μg) is added 0.01 ml of phosphatase solution contain-

[1] T. C. Linn, J. W. Pelley, F. H. Pettit, F. Hucho, D. D. Randall, and L. J. Reed, *Arch. Biochem. Biophys.* **148**, 327 (1972).

[2] E. A. Siess and O. H. Wieland, *Eur. J. Biochem.* **26**, 96 (1972).

[3] R. M. Denton, P. J. Randle, and B. R. Martin, *Biochem. J.* **128**, 161 (1972).

[4] D. Stansbie, R. M. Denton, B. J. Bridges, H. T. Pask, and P. J. Randle, *Biochem. J.* **154**, 225 (1976).

[5] E. A. Siess and O. H. Wieland, *FEBS Lett.* **65**, 163 (1976).

[6] A highly purified preparation of a low-molecular-weight (32,000) phosphorylase phosphatase (designated protein phosphatase C-I), kindly furnished by S. R. Silberman and E. Y. C. Lee, exhibited activity toward phosphorylated pyruvate dehydrogenase complex comparable to that of the homogeneous pyruvate dehydrogenase phosphatase described in this article.

[7] F. H. Pettit and L. J. Reed, this series, Vol. 89 [65].

ing 0.1–0.4 unit.[8] The solution is incubated at 30° for 1 min, and the reaction is started by addition of 0.01 ml of 0.1 M MgCl$_2$. The control contains buffer instead of phosphatase. After 2 min, a 0.02-ml aliquot is withdrawn for assay of pyruvate dehydrogenase complex activity.[7]

Units. One unit of pyruvate dehydrogenase phosphatase activity is defined as the amount of enzyme that reactivates one unit of pyruvate dehydrogenase complex activity per minute at 30°. Specific activity is expressed as units per milligram of protein. Pyruvate dehydrogenase complex activity is expressed as micromoles of NADH produced per minute at 30°.[7]

Purification Procedure

The procedure is applicable to purification of pyruvate dehydrogenase phosphatase from both heart and kidney mitochondria. However, heart mitochondria contain at least 3 times as much phosphatase activity as kidney mitochondria and is the preferred source for isolation of the phosphatase. Forty to sixty percent of the phosphatase is associated with the membrane fraction after extraction of the pyruvate dehydrogenase complex from mitochondria. The remainder of the phosphatase copurifies with the complex. The phosphatase binds to the dihydrolipoyl transacetylase (E$_2$) component of the complex in the presence of Ca^{2+}.[9] The phosphatase is separated from the complex by ultracentrifugation in the presence of the Ca^{2+}-chelating agent ethylene glycol–bis(2-aminoethyl)-N,N'-tetraacetate (EGTA). This portion of the total phosphatase is combined with that extracted from the membrane fraction. A key step in the purification procedure is affinity chromatography on E$_2$ coupled to Sepharose 4B. In the presence of Ca^{2+} the phosphatase binds to the absorbent and is subsequently released in the presence of EGTA.

The purification procedure is summarized in the table. Unless specified otherwise, all operations are performed at 4°.

Preparation of E$_2$-Sepharose 4B. A mixture of 15 g of CNBr-activated Sepharose 4B (Sigma Chemical Co., St. Louis, Missouri) and 100 ml of 0.2 M NaHCO$_3$ is adjusted to pH 9.2 with NaOH. Highly purified dihydrolipoyl transacetylase (80 mg) from bovine heart[10] is added, and the mixture is stirred gently for 20 hr. The gel is washed on a sintered-glass

---

[8] When possible, phosphatase is added to the assay mixture from concentrated solution to minimize dilution of the enzyme. If dilution is necessary, assay buffer containing 0.3% bovine serum albumin is used.

[9] F. H. Pettit, T. E. Roche, and L. J. Reed, *Biochem. Biophys. Res. Commun.* **49,** 563 (1972).

[10] F. H. Pettit, S. J. Yeaman, and L. J. Reed, this volume [30]. Dihydrolipoyl transacetylase obtained from step 2 of the purification procedure is used.

PURIFICATION OF PYRUVATE DEHYDROGENASE PHOSPHATASE FROM BOVINE HEART[a]

Fraction	Volume (ml)	Protein (mg)	Specific activity[b]	Recovery (%)
Membrane extract	1830	6240	2.1	—
First protamine precipitation	283	1190	9.5	—
Ultracentrifuge supernatant[c]	260	923	8.5	—
Combined fractions 2 and 3	543	2114	9.0	100
Second protamine precipitation	292	742	22.6	78
Ultracentrifugation	280	403	36.3	69
Affinity chromatography concentrate	1.8	5.8	1655	45

[a] From about 80 pounds of heart.
[b] Units of pyruvate dehydrogenase complex reactivated per minute per milligram of protein.
[c] Phosphatase released from partially purified pyruvate dehydrogenase complex by ultracentrifugation in the presence of EGTA.

funnel with 0.2 M NaHCO$_3$ and then with deionized water. The gel is treated with a solution of 250 mg of bovine serum albumin in 100 ml of 0.2 M NaHCO$_3$ for 4 hr and then washed well with 0.2 M NaHCO$_3$ and deionized water.

Step 1. Extraction of Mitochondrial Membrane Fraction. The membrane fraction obtained after extraction of the pyruvate dehydrogenase complex from bovine heart mitochondria[7] is suspended in 2 liters of 0.1 M MOPS, pH 7.0, at 23°. The suspension is stirred for 1.5 hr and then centrifuged at 30,000 g for 30 min at 20°; the pellets are discarded.

Step 2. First Protamine Precipitation. The supernatant fluid from step 1 is cooled to 4°, and 0.011 volume of 2% protamine sulfate (Elanco Products[10a]) is added dropwise with stirring. After 15 min the precipitate is collected by centrifugation at 30,000 g for 15 min; the supernatant fluid is discarded. The protamine precipitate is resuspended, by means of a large glass homogenizer equipped with a motor-driven Teflon pestle, in 300 ml of buffer A (0.02 M MOPS, pH 7.0; 5 mM MgCl$_2$; 10% (v/v) glycerol; 0.5 mM dithiothreitol) containing 0.16% yeast sodium ribonucleate (ICN Pharmaceuticals[10a]). The suspension is stirred overnight, then 0.01 volume of 0.2 M EGTA is added. The mixture is warmed to 23° for 20 min and then centrifuged at 30,000 g for 15 min at 20°. The clear, yellow supernatant fluid is cooled to 4°; the pellets are discarded.

Step 3. Second Protamine Precipitation. The supernatant fluid from step 2 is combined with the supernatant fluid from the first ultracentrifuga-

10a Elanco Products, Indianapolis, Indiana; ICN Pharmaceuticals, Irvine, California; Bio-Molecular Dynamics, Beaverton, Oregon.

tion step in the purification of the pyruvate dehydrogenase complex.[7] To this solution is added dropwise, with stirring, 0.01 volume of 2% protamine sulfate. After 15 min, the precipitate is collected by centrifugation for 15 min. The protamine precipitate is extracted exactly as described in step 2.

Step 4. Ultracentrifugation. The supernatant fluid from step 3 is centrifuged at 105,000 g for 3.5 hr; the pellets are discarded. To the supernatant fluid is added 0.05 volume of 0.1 M CaCl$_2$.

Step 5. Affinity Chromatography. The phosphatase solution from step 4 is applied slowly to a column (1.5 × 26 cm) of dihydrolipoyl transacetylase (E$_2$)-Sepharose 4B that has been equilibrated with buffer A containing 2 mM CaCl$_2$. The column is washed with 150 ml of this equilibration buffer. The phosphatase is eluted with buffer A containing 2 mM EGTA and 0.02% sodium azide. The active fractions are combined and concentrated by vacuum dialysis against buffer A in a Bio-Molecular Dynamics[10a] concentrator.

Properties

Enzyme obtained from step 5, from both heart and kidney mitochondria, gives one band on polyacrylamide gel electrophoresis with Tris-(hydroxymethyl)aminomethane-glycine buffer in the absence of sodium dodecyl sulfate (SDS).[11] The phosphatase has a molecular weight by sedimentation equilibrium of about 150,000 and a sedimentation coefficient ($s_{20,w}$) of about 7.4 S.[12] It consists of two subunits with molecular weights of about 97,000 and 50,000 as estimated by SDS–gel electrophoresis.[13] Phosphatase activity appears to reside in the latter subunit, which is sensitive to proteolysis. Pyruvate dehydrogenase phosphatase preparations from bovine heart and kidney[1] and pig heart[2] were reported to have a molecular weight of 95,000–100,000 as estimated by SDS–polyacrylamide gel electrophoresis and by filtration through a calibrated column of Sephadex G-100. The activity of these phosphatase preparations was less than 10% of the activity of the homogeneous preparation described in this chapter. It appears that the earlier preparations were deficient in the protease-sensitive subunit of molecular weight 50,000.

Catalytic Properties. Pyruvate dehydrogenase phosphatase catalyzes dephosphorylation at all three phosphorylation sites on E$_1$. The phosphatase requires a divalent cation (Mg^{2+} or Mn^{2+}). The apparent K_m for

[11] B. J. Davis, *Ann. N.Y. Acad. Sci.* **121**, 404 (1964).
[12] F. H. Pettit, W. M. Teague, and L. J. Reed, unpublished results.
[13] K. Weber and M. Osborn, *J. Biol. Chem.* **244**, 4406 (1969).

Mg^{2+} is about 2 mM.[14] Ca^{2+} markedly stimulates phosphatase activity[3,9] in the presence, but not in the absence, of dihydrolipoyl transacetylase (E_2).[9] The apparent K_m for phosphorylated E_1 is about 58 μM. In the presence of Ca^{2+} and E_2 this value is decreased to about 2.9 μM. There is some uncertainty as to the apparent K_m for Ca^{2+}. Values in the range of 1 μM to 100 μM have been reported.[15,16] Highly purified pyruvate dehydrogenase phosphatase from bovine heart binds one Ca^{2+} with a dissociation constant (K_d) of about 8 μM.[17] In the presence of E_2, the phosphatase binds a second Ca^{2+} with a K_d of about 5 μM. It appears that Ca^{2+} may play a structural role. In the presence of Ca^{2+}, the phosphatase binds to E_2, thereby facilitating the Mg^{2+}-dependent dephosphorylation of phosphorylated E_1. Phosphatase activity is inhibited by NADH, and this inhibition is reversed by NAD^+.[18] Phosphatase activity is inhibited by fluoride ion, by inorganic orthophosphate, and by increasing the ionic strength.[2,14]

Pyruvate dehydrogenase phosphatase exhibits only slight activity toward phosphorylase a, about 10% of the activity observed with phosphorylated pyruvate dehydrogenase complex as substrate. The phosphatase is inactive toward p-nitrophenyl phosphate. It is not inhibited by protein phosphatase inhibitor-1 or inhibitor-2 (kindly furnished by Dr. Philip Cohen).

[14] F. Hucho, D. D. Randall, T. E. Roche, M. W. Burgett, J. W. Pelley, and L. J. Reed, *Arch. Biochem. Biophys.* **151**, 328 (1972).
[15] A. L. Kerbey and P. J. Randle, *FEBS Lett.* **108**, 485 (1979).
[16] R. M. Denton and J. G. McCormack, *FEBS Lett.* **119**, 1 (1980).
[17] T-L. Wu and L. J. Reed, unpublished results.
[18] F. H. Pettit, J. W. Pelley, and L. J. Reed, *Biochem. Biophys. Res. Commun.* **65**, 575 (1975).

[68] Phosphoprotein Phosphatase from Swine Kidney

By K. MUNIYAPPA, FREDRICH LEIBACH, and JOSEPH MENDICINO

Phosphoprotein phosphatase (EC 3.1.3.16) catalyzes the reaction

$$\text{Phosphoenzyme} + n \text{ H}_2\text{O} \xrightarrow{\substack{\text{phosphoprotein} \\ \text{phosphatase}}} \text{dephosphoenzyme} + n \text{ P}_i$$

The homogeneous phosphoprotein phosphatase isolated from swine kidney extracts catalyzes the dephosphorylation of the phosphorylated forms of glycogen synthase, phosphorylase, phosphofructokinase, pyruvate kinase, fructose-1,6-bisphosphatase and histone. The enzyme hydrolyzes phosphoryl groups attached to specific serine residue in polypeptide substrates,[1,2] and shows little or no activity with other low-molecular-weight monophosphate esters. High-molecular-weight complex forms of the enzyme are found in many tissues.[3,4] However, the active form of the purified enzyme contains only a single polypeptide chain with a molecular weight of about 34,000. The enzyme found in the cytosolic fraction of swine kidney extracts, however, is able to dephosphorylate proteins present in particulate fractions.[5] A homogeneous preparation of the enzyme has been purified from the soluble fraction of swine kidney.

Assay Methods

Activity can be measured by the rate of $^{32}P_i$ release from ^{32}P-labeled enzymes or by changes in the activity of the phosphorylated forms of phosphorylase or glycogen synthase. One enzymic assay is based on the ability of phosphoprotein phosphatase to catalyze the conversion of glycogen phosphorylase to a dephosphorylated form that is not dependent on the presence of AMP for activity.[6] The other assay is based on the ability of the phosphatase to convert glycogen synthase to a dephosphorylated form dependent on the presence of glucose 6-phosphate for activity.[7] Two incubations are required in both assays. The phosphoprotein phosphatase is incubated with the completely phosphorylated forms of glycogen phosphorylase or glycogen synthase. The incubation mixture is then added to a buffer that terminates the phosphatase reaction, and the activity of glycogen phosphorylase or glycogen synthase is assayed under conditions that measure the extent of conversion of the phosphorylated enzyme to the dephosphorylated form based on the dependence of the enzyme on its allosteric activator. The reaction mixtures are incubated under the same conditions used in the isotopic assays so that the release of $^{32}P_i$ can be directly compared with alterations in enzymic activity.

Reagents

Sucrose, 2.0 M
Imidazole-HCl, pH 7.5, 1 M
Tris-HCl, pH 7.5, 0.1 M

[1] W. D. Wosilait and E. W. Sutherland, *J. Biol. Chem.* **218**, 469 (1956).
[2] E. G. Krebs and E. H. Fisher, *Biochim. Biophys. Acta* **20**, 150 (1956).
[3] W. Merlevede and G. A. Riley, *J. Biol. Chem.* **241**, 3517 (1966).
[4] H. N. Torres and C. A. Chelala, *Biochim. Biophys. Acta* **198**, 495 (1970).
[5] J. Larner and C. Villar-Palasi, *Curr. Top. Cell. Regul.* **3**, 195 (1971).
[6] R. Medicus and J. Mendicino, *Eur. J. Biochem.* **40**, 63 (1973).
[7] H. Abou-Issa and J. Mendicino, *J. Biol. Chem.* **248**, 685 (1973).

2-(N-morpholino)ethanesulfonic acid, pH 6.5, 1 M
Dithiothreitol, 0.1 M
MgCl$_2$, 0.5 M
Theophylline, 0.1 M
Histone, 20 mg/ml
Protein kinase, 0.27 mg/ml
Adenosine [γ-^{32}P]triphosphate, 2 mM, specific activity, 100 cpm/pmol
^{32}P-labeled histone, 10 mg/ml, 160 cpm/nmol
Trichloroacetic acid, 25%
Perchloric acid, 2.8 N–phosphotungstic acid, 2%–NaH$_2$PO$_4$, 10 mM–sodium pyrophosphate, 3 mM

Procedures. The ^{32}P-labeled histone, glycogen phosphorylase, and glycogen synthase used as substrates in the isotopic assay are prepared by incubating the dephosphorylated forms of these proteins with [γ-^{32}P]ATP, Mg^{2+}, and homogeneous catalytic subunit of swine kidney protein kinase.[8] The [^{32}P]casein, [^{32}P]protamine, and [^{32}P]histone[9,10] used in the standard assay are prepared by incubating 0.15 ml of 2-(N-morpholino)ethanesulfonic acid, 0.45 ml of sucrose, 0.45 ml dithiothreitol, 0.12 ml MgCl$_2$, 1.5 ml of histone, 0.15 ml of [γ-^{32}P]ATP, 1.5 ml of protein kinase (0.6 μmol), and water in a final volume of 5 ml for 3 hr at 30°. Then 0.4 ml of trichloroacetic acid is added to precipitate protein kinase. After centrifugation the supernatant is adjusted to 30% trichloroacetic acid and the precipitated histone is collected by centrifugation at 3°. The pellet is dissolved in 5 ml of Tris-HCl, pH 7.5, and reprecipitated twice by the addition of trichloroacetic acid to 30%. The final pellet is washed twice with ethanol–ether (1 : 4) and dialyzed against 200 ml of Tris-HCl, pH 7.5.

Phosphofructokinase and phosphorylase isolated from swine kidney[7] are labeled with ^{32}P by the same procedure used for the preparation of [^{32}P]histone. Afterward the enzymes are precipitated by the addition ammonium sulfate (29.1 g/100 ml) at 3°. The suspension is centrifuged and dissolved in 1 ml of 0.1 M Tris-HCl, pH 7.2, containing 10% sucrose, 10 mM 2-mercaptoethanol, and 2 mM EDTA. The samples are applied to a Sephadex G-75 column (1.5 × 50 cm) that is eluted with the same buffer to remove the catalytic subunit of protein kinase and ammonium sulfate.

The isotopic assay can be carried out with [^{32}P]histone, [^{32}P]casein, [^{32}P]protamine, [^{32}P]phosphofructokinase, or [^{32}P]phosphorylase as the substrate. In a small tube, mix 5 μl of imidazole-HCl, pH 7.5, 5 μl of

[8] J. Mendicino, F. Leibach, and S. Reddy, *Biochemistry* **17**, 4662 (1978).
[9] M. H. Meisler and T. A. Langan, *J. Biol. Chem.* **244**, 4961 (1969).
[10] C. Nakai and J. A. Thomas, *J. Biol. Chem.* **249**, 6459 (1974).

dithiothreitol, 5 μl of theophylline, 2 μl of $MgCl_2$, 50 μl of sucrose, and 143 μl of phosphoprotein phosphatase plus water. This mixture is incubated for 3 min at 30°, and the reaction is initiated by the addition of 50 μl of [^{32}P]histone. Aliquots of 0.05 ml are removed at 5, 10, and 15 min, and they are added to small tubes containing 0.2 ml of perchloric acid–phosphotungstic acid mixture. Then 0.05 ml of histone is added as carrier protein, and the suspension is mixed and centrifuged. An aliquot, 0.1 ml, of the clear supernatant is added to a vial containing 5 ml of scintillation solvent and counted. The rate of dephosphorylation is calculated from the amount of ^{32}P$_i$ released during two time intervals. The rates of hydrolysis of ^{32}P-labeled substrates in this assay system were linear with time, and they were proportional to the amount of phosphoprotein phosphatase added up to 55–60% conversion of the phosphorylated substrate. A unit of activity is defined as the amount of phosphoprotein phosphatase required to produce 1 nmol of ^{32}P$_i$ from [^{32}P]histone per minute. Specific activity is expressed as units per milligram of protein.

Purification Procedure

Step 1. Homogenization. Swine kidneys, 1000 g, are cleaned and cut into pieces. They are homogenized in a Waring blender with 1.5 liters of 50 mM imidazole-HCl, pH 7.5, containing 5 mM EDTA, 0.5 mM dithiothreitol, and 0.1 mM phenylmethylsulfonyl fluoride. The homogenate is centrifuged at 27,000 g for 20 min, and the supernatant is passed through cheesecloth to remove insoluble lipid material. The pellet is reextracted with 500 ml of the same buffer. The two extracts are combined.

Step 2. Acid Precipitation. All the subsequent procedures are carried out at 3°. The crude extract is adjusted to pH 5.8 by the addition of 2 N acetic acid, and 3 g of Norit A are added with stirring. The homogenate is centrifuged at 27,000 g for 15 min, and the precipitate is reextracted with 500 ml of buffer at pH 5.8 containing 0.1 M NaCl. The combined supernatant and wash are adjusted to pH 7.2 with 1 M Tris base.

Step 3. Precipitation with Ammonium Sulfate. The solution is adjusted to 70% saturation at 3° by the addition of solid ammonium sulfate (43.5 g/100 ml), and the pH is adjusted to 7.5 with Tris base. The suspension is stirred for 15 min and centrifuged at 27,000 g for 10 min. The precipitate is dissolved in 800 ml of 0.1 M Tris-HCl, pH 8.0, containing 1 mM $MgCl_2$, and is then dialyzed against 12 liters of 50 mM Tris-HCl (pH 7.5)–10% sucrose–0.1 M NaCl–5 mM dithiothreitol overnight. The resulting turbid solution is centrifuged at 27,000 g for 10 min to remove insoluble material.

Step 4. Chromatography on DEAE-Sephadex A-50. The solution from step 3 is passed into a DEAE-Sephadex A-50 column (50 × 5 cm) previ-

PURIFICATION OF PHOSPHOPROTEIN PHOSPHATASE FROM SWINE KIDNEY

Fraction	Volume (ml)	Protein (mg/ml)	Total activity (units)	Specific activity (units/mg protein)	Recovery (%)
Homogenate	2200	49.4	13,700	0.12	100
Acid precipitation	2000	31.4	9,200	0.14	67
Precipitation with ammonium sulfate	950	25.2	8,600	0.35	62
Chromatography on DEAE-Sephadex A-50	120	5.2	6,358	10.1	46
Chromatography on Sephacryl S-200	450	0.2	4,260	47.3	31
Chromatography on Blue Dextran–Sepharose 4B and hexanediamine–Sepharose 4B	17	0.06	2,712	2658	20
Chromatography on polylysine–Sepharose 4B	10	0.05	1,400	2800	10
Gel filtration on Sephadex G-100	6	0.06	1,020	2833	7.5

ously equilibrated with the same buffer used for dialysis. The column is washed with about 3 bed volumes of the same buffer containing 0.18 M NaCl until the absorbance of the filtrate at 280 nm is the same as that of the buffer. The enzyme is then eluted from the column with buffer containing 0.3 M NaCl. Fractions of 20 ml are collected at a flow rate of 40 ml/hr. The phosphoprotein phosphatase elutes as a broad peak after 600 ml. Active fractions are collected, and 43.5 g per 100 ml of solid ammonium sulfate are added to the solution at 3°. The suspension is centrifuged, and the precipitate is dissolved in 25 ml of 50 mM Tris-HCl (pH 7.5)–10% sucrose–0.1 M NaCl–5 mM dithiothreitol.

Step 5. Gel Filtration on Sephacryl S-200. The enzyme solution from step 4 is applied to a Sephacryl S-200 column (5 × 45 cm) previously equilibrated with 0.1 M Tris-HCl, pH 7.5, containing 5% sucrose and 10 mM 2-mercaptoethanol. Fractions of 15 ml are collected at a flow rate of 28 ml/hr. The enzyme is eluted in a broad peak after about 500 ml of buffer have passed through the column. The fractions containing phosphatase activity are pooled.

Step 6. Chromatography on Blue Dextran–Sepharose 4B and Hexanediamine–Sepharose 4B. The solution from step 5 is passed into a Blue Dextran–Sepharose 4B column (2.8 × 18 cm), which was equilibrated with 0.1 M Tris-HCl, pH 7.5, containing 5% sucrose and 10 mM 2-mercaptoethanol. The column is washed with 60 ml of buffer. The filtrate and wash, which contain nearly all of the phosphatase activity, are combined and immediately applied to a hexanediamine–Sepharose 4B column (2.2 × 20 cm) previously equilibrated with the same buffer described above. The column is then washed with buffer until the absorption of the filtrate at 280 nm decreases to 0.02. The phosphatase is eluted from this column with a linear gradient made up to 300 ml of buffer containing 0.1 M NaCl in the mixing chamber and 300 ml of buffer containing 0.6 M NaCl in the reservoir. Fractions of 5 ml are collected at a flow rate of 20 ml/hr. The phosphatase activity is found in a single protein peak that emerges from the column after about 50 ml of the eluting solution have been collected. The desired fractions are pooled and dialyzed against 0.1 Tris-HCl, pH 7.5, 10 mM 2-mercaptoethanol containing 10% sucrose overnight (fraction VI).

Step 7. Chromatography on Polylysine–Sepharose 4B. The dialyzed fraction is passed into a polylysine–Sepharose 4B column (2.2 × 12 cm) that was previously equilibrated with 0.1 M Tris-HCl, pH 7.5, 10 mM 2-mercaptoethanol, and 0.1 mM MgCl$_2$. The column is washed with 10 bed volumes of the above buffer containing 0.1 M NaCl. The enzyme is then eluted with a linear salt gradient containing 100 ml of 0.1 M NaCl in 0.1 M Tris-HCL, pH 7.5, 10 mM 2-mercaptoethanol, 10% sucrose, and 0.1 mM MgCl$_2$ in the mixing chamber, and 100 ml of 0.3 M NaCl in the same buffer in the reservoir. The enzyme emerges from the column when the salt concentration is about 0.2 M. Fractions containing phosphoprotein phosphatase activity are pooled and concentrated by ultrafiltration using a Amicon Diaflo PM-10 membrane (fraction VII).

Step 8. Gel Filtration on Sephadex G-100. The concentrated enzyme solution (fraction VII) is applied to a Sephadex G-100 column (2.6 × 90 cm) that was equilibrated with 50 mM Tris-HCl, pH 7.5, 10 mM 2-mercaptoethanol, 10% sucrose, and 1 mM MgCl$_2$ containing 0.1 M NaCl; the enzyme is eluted with the same buffer. Phosphoprotein phosphatase activity is eluted as a single symmetrical protein peak with enzymic activity. The desired fractions are pooled and dialyzed against 60% glycerol containing 0.05 M Tris-HCl, pH 7.5, 10 mM 2-mercaptoethanol, and 1 mM MgCl$_2$. Afterward, the solution is further concentrated by ultrafiltration as described above. The purified enzyme can be stored at $-70°$ in 60% glycerol for at least 2 months with no loss of activity.

Properties

The homogeneous enzyme showed a single diffused band on polyacrylamide gel electrophoresis, and only one component was observed by sucrose density centrifugation. The molecular weight estimated by gel filtration on Sephadex G-100 was 70,000. A single protein band with a molecular weight of 70,000 was observed by polyacrylamide gel electrophoresis in the presence of sodium dodecyl sulfate, and a similar value, 70,000, was calculated from data obtained by sucrose density centrifugation. The specific activity of the purified enzyme from ten different preparations varied from 2.8 to 5.8 μmol/min per milligram of protein with histone as substrate (see table).

Substrate Specificity. The apparent K_m of phosphoprotein phosphatase for histone, casein, protamine, phosphorylase, and phosphofructokinase is 66.0, 6.6, 100.0, 2.8, and 3.3 μM, respectively. The pH optimum of the purified enzyme is 7.0, and 10 mM MgCl$_2$ is required for maximum activity.

Section VI

Phosphoenolpyruvate : Glycose Phosphotransferase System

[69] General Description and Assay Principles[1]

By Saul Roseman, Norman D. Meadow, and
Maria A. Kukuruzinska

General Description

The bacterial phosphoenolpyruvate : glycose phosphotransferase system (PTS) was discovered[2] in studies concerned with the metabolism of complex carbohydrates. The PTS has since been shown to be involved in a number of important cellular processes, and for this reason an extensive literature has become available on this subject. Various aspects of the PTS have been reviewed in recent years.[3,4]

Perhaps the best defined function of the PTS, i.e., translocation of PTS sugars (substrates) across the bacterial cytoplasmic membrane concomitant with their phosphorylation, is illustrated in Fig. 1. Thus the PTS catalyzes a solute transport process termed group translocation: the product of transport is a derivative of the solute, in this case a phosphate ester. Phosphorylation of the sugars during transport serves two important functions. First, the sugar is trapped in the cytoplasm, since bacterial membranes are impermeable to the sugar phosphate (except for the special case of cells induced for a hexose phosphate transport system).[5] By contrast, non-PTS sugars, such as lactose in *Escherichia coli*, efflux from the cell, particularly when the energy supply in the cell is impaired. The second important function is that the product of transport, the sugar phosphate, is the first substrate in the catabolic pathway of the sugar. Thus, the energy used for translocation of the sugar is conserved, and this has important consequences in the case of obligate anaerobes.[6]

The bacterial PTS also participates in other physiological processes. For example, it is involved in chemotaxis toward its sugar substrates.[7] It also regulates both adenylate cyclase[8] and certain non-PTS sugar and

[1] This work was supported by Grant CA 21901 from the National Institutes of Health. NDM was supported in part by a National Institutes of Health Special Fellowship HD57023 and AG05029.

[2] W. Kundig, S. Ghosh, and S. Roseman, *Proc. Natl. Acad. Sci. U.S.A.* **52**, 1067 (1964).
[3] P. W. Postma and S. Roseman, *Biochim. Biophys. Acta* **457**, 213 (1976).
[4] S. S. Dills, A. Apperson, M. R. Schmidt, and M. H. Saier, Jr., *Microbiol. Rev.* **44**, 385 (1980).
[5] G. W. Dietz and L. A. Heppel, *J. Biol. Chem.* **246**, 2881 (1971).
[6] S. Roseman, *J. Gen. Physiol.* **54**, 138 (1969).
[7] J. Adler and W. Epstein, *Proc. Natl. Acad. Sci. U.S.A.* **71**, 2895 (1974).
[8] A. Peterkofsky and C. Gazdar, *Proc. Natl. Acad. Sci. U.S.A.* **72**, 2920 (1975).

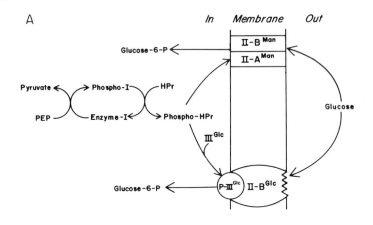

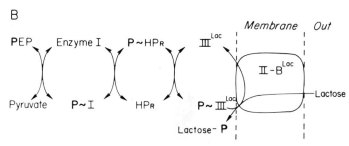

FIG. 1. Schematic representation of the phosphotransferase (PTS) system. (A) The glucose phosphotransferase systems in *Salmonella typhimurium*. The phosphoryl group is sequentially transferred from phosphoenolpyruvate (PEP), to Enzyme I, HPr, and then to one of the sugar-specific proteins, II-A^{Man} (an integral membrane protein) or III^{Glc} (a soluble and/or peripheral membrane protein). There is no evidence that the integral membrane proteins, II-B^{Man} and II-B^{Glc}, are phosphorylated. II-B^{Man} and II-B^{Glc} catalyze the transfer of the phosphoryl group from II-A^{Man} and III^{Glc} to the sugar (glucose) concomitant with the translocation of the sugar across the membrane. The II-A^{Man}/II-B^{Man} complex is designated II^{Man}, whereas II^{Glc} denotes the III^{Glc}/II-B^{Glc} complex. Methyl α-D-glucopyranoside is phosphorylated by II^{Glc}, and mannose and 2-deoxyglucose are phosphorylated by II^{Man}. (B) The lactose phosphotransferase system in *Staphylococcus aureus*. Almost all sugars are PTS sugars in *S. aureus*. The lactose PTS shown above can also phosphorylate analogs of lactose (e.g., methyl β-D-thiogalactopyranoside, TMG). From R. D. Simoni *et al.*, *J. Biol. Chem.*, **248**, 932 (1973).

amino acid transport systems,[9,10] thereby controlling the synthesis of the corresponding inducible proteins required for uptake and catabolism of these substances. The PTS therefore mediates the "glucose effect" or diauxie, where bacteria will utilize glucose (or another PTS sugar) in the

[9] M. H. Saier, Jr. and S. Roseman, *J. Biol. Chem.* **251**, 6606 (1976).
[10] M. Berman, N. Zwaig, and E. C. C. Lin, *Biochem. Biophys. Res. Commun.* **38**, 272 (1970).

growth medium before utilizing a non-PTS sugar, such as lactose, present in the same medium.

The PTS is widely distributed in obligate and facultative anaerobic bacteria, and has occasionally been found in strict aerobes. No sugar phosphorylation system dependent on phosphoenolpyruvate has been found in extracts of fungi, and while there have been occasional reports on the presence of the system in extracts from yeast and invertebrate intestine, no confirming studies have appeared. It should perhaps be emphasized that work reporting the absence of the PTS in higher organisms has always been based on PEP-dependent sugar phosphorylation. In these experiments, the underlying assumption is that the PTS has remained intact through evolution and is essentially the system shown in Fig. 1. However, the very complexity of the system would permit modification in many ways. For example, one modification would be that some other high-energy phosphate compound such as ATP would be the phosphate donor for a modified phosphotransferase system. This kind of modification would not be recognized as a derivative of the PTS under the usual assay conditions.

Whether or not a particular sugar is a substrate of the PTS depends on the organism. In *Staphylococcus aureus,* for example, essentially all sugars tested are PTS sugars,[3] including the disaccharides lactose, maltose, and melibiose. The same disaccharides are not PTS sugars in the enteric bacteria *E. coli* and *Salmonella typhimurium.* In addition, the following are not substrates of the PTS in the enteric bacteria (all D-configuration): glycerol, α-glycerophosphate, glucose-6-P, mannose-6-P, fructose-6-P, galactose, pentoses, and gluconate. The following are PTS sugars: glucose, mannose, fructose, glucosamine, N-acetylglucosamine, N-acetylmannosamine, hexitols, and β-glucosides. A number of analogs of the sugars, such as methyl α- and β-glucosides, and 2-deoxyglucose (but not 3-O-methylglucose), are also taken up and phosphorylated via the PTS and have been used extensively for studies with whole cells and bacterial membrane vesicles, and also in sugar phosphorylation assays *in vitro*.

The overall reaction catalyzed by the PTS is the transfer of the phosphoryl group from phosphoenolpyruvate to the sugar substrate, yielding the corresponding sugar-6-P, with the exception of fructose, which is phosphorylated by an inducible fructose-specific PTS at the C-1 position or, less effectively, by a constitutive PTS at C-6.[3] This overall phosphoryl transfer reaction requires two sets of proteins, one set designated general PTS proteins (required for all sugar substrates) and the other sugar-specific proteins. The general proteins are Enzyme I and HPr as shown in Fig. 1. The phosphoryl group is transferred from phosphoenolpyruvate to Enzyme I in the presence of a divalent cation, generally Mg^{2+}.[3]

Phospho-Enzyme I then interacts with the next phosphocarrier protein, HPr, yielding phospho-HPr.

The next steps in the sequence involve the sugar-specific proteins. Many or perhaps most of these are inducible (by growth of cells on the corresponding sugars). A few, such as those responsible for phosphorylation of sugars of the D-gluco and D-manno configurations, are constitutive, although the levels of these proteins can be increased severalfold by growth of the cells on the corresponding sugar, as is true also of HPr and Enzyme I.[3]

Transfer of the phosphoryl group from phospho-HPr to a given sugar requires the corresponding sugar-specific protein or proteins.[11] With two possible exceptions discussed below, each of the sugar-specific Enzyme II complexes have thus far been found to consist either of a pair of membrane components, called II-A and II-B, or a membrane component called II-B and a soluble component designated III. Since both the II-A and II-B proteins are membrane components, their separation is technically very difficult. While II-A and II-B have been separated from each other in one case[12] there is only preliminary evidence about the sequence of phosphoryl transfer, i.e., that the phosphoryl group is transferred from phospho-HPr to the II-A protein, followed by direct transfer to the sugar catalyzed by II-B. On the other hand, the sequence shown on Fig. 1 involving the transfer of the phosphoryl group via the soluble sugar-specific proteins has been unequivocally demonstrated with the lactose-specific protein (III^{Lac}) from S. aureus[13] and the glucose-(and methyl α-glucoside-)specific protein (III^{Glc}) from S. typhimurium.[14] Efforts to show that the phosphoryl group is transferred to the II-B protein have given negative results, and from these we conclude that II-B acts catalytically, transferring the phosphoryl group from phospho-III (or phospho-II-A) to the sugar concomitant with translocation of the sugar across the cell membrane.

---

[11] Prior to elucidation of the functions of the various proteins, Roman numerals were assigned to each of the fractions as they were isolated (except for HPr, which was named on the basis of its heat stability). This cumbersome and frequently confusing nomenclature is still used. Enzyme I and HPr have been defined above, and the term Enzyme II refers to the sugar-specific complex that catalyzes the transfer of the phosphoryl group from phospho-HPr to a given sugar. Abbreviated superscripts are used to define sugar specificity, so that II^{Glc} refers to the complex that phosphorylates glucose. Abbreviated subscripts are used only where necessary for clarity to define the organism from which the protein was isolated, and if necessary the growth conditions. For example, II$^{Lac}_{Sa,Gal}$ means the Enzyme II complex isolated from Staphylococcus aureus grown on galactose, which phosphorylates lactose. The subscripts are also used, only when necessary, for Enzyme I and HPr.

[12] W. Kundig and S. Roseman, J. Biol. Chem. 246, 1407 (1971).
[13] R. D. Simoni, J. B. Hays, T. Nakazawa, and S. Roseman, J. Biol. Chem. 248, 957 (1973).
[14] N. D. Meadow and S. Roseman, J. Biol. Chem. in press (1982).

Two reports have appeared that suggest that the Enzyme II complexes from *E. coli* responsible for the phosphorylation of β-glucosides[15] and of mannitol[16] contain only one sugar-specific protein and therefore catalyze the transfer of the phosphoryl group directly from phospho-HPr to the corresponding sugars. If the *E. coli* mannitol Enzyme II consists of one protein, it differs from the *S. aureus* mannitol phosphorylating system, which comprises two proteins.[17]

All the phosphoproteins of the PTS thus far characterized contain the phosphoryl group linked to a histidine moiety. In phospho-HPr, the linkage is to the N-1 position of the imidazole ring[3] whereas in Enzyme I_{St},[18] III_{St}^{Glc},[14] and III_{Sa}^{Lac},[3] the phosphoryl group is linked to the N-3 position. The phosphohistidines are extremely sensitive to acid hydrolysis, the N-1 phosphohistidine being even more sensitive than the N-3 derivative.[19] Both phosphohistidines are stable in alkali. The $t_{1/2}$ for the hydrolysis of phospho-HPr at 37°, pH 8.0, for example, is 13 min.

An earlier report in this series[20] described a paper electrophoretic assay for the PTS from *E. coli* when only one of the proteins, HPr, had been isolated in homogeneous form. The present report describes a simpler routine assay involving ion-exchange columns for measuring the rate of sugar phosphorylation and a spectrophotometric assay for directly measuring the phosphorylation of HPr and other purified proteins, such as III^{Glc}. Using these assay methods, procedures are described for isolating the following PTS proteins in homogeneous form: HPr and III^{Lac} from *S. aureus*; Enzyme I, HPr, and III^{Glc} from *S. typhimurium*.

Assay Principles

Two types of assays are used to measure the components of the phosphotransferase system, the spectrophotometric assay and the sugar phosphorylation assay.

The spectrophotometric assay is based on a simple principle. Knowing the stoichiometry of phosphate incorporation into certain PTS proteins, one can determine the quantity of the given PTS protein in the sample by measuring pyruvate formation resulting from phosphate transfer from phosphoenolpyruvate to the unknown. The following stoichiometries have been established: 1 mol of phosphoryl group per mole of HPr,[3] or

[15] C. Fox and G. Wilson, *Proc. Natl. Acad. Sci. U.S.A.* **59**, 988 (1968).

[16] G. R. Jacobson, C. A. Lee, and M. H. Saier, Jr., *J. Biol. Chem.* **254**, 249 (1979).

[17] R. D. Simoni, M. F. Smith, and S. Roseman, *Biochem. Biophys. Res. Commun.* **31**, 804 (1968).

[18] N. Weigel, M. A. Kukuruzinska, A. Nakazawa, and S. Roseman, *J. Biol. Chem.* in press (1982).

[19] D. E. Hultquist, *Biochim. Biophys. Acta* **153**, 329 (1968).

[20] W. Kundig and S. Roseman, this series, Vol. 9, p. 396.

III_{St}^{Glc},[14] and 3 mol per mole of III_{Sa}^{Lac} [3] (1 mol per monomeric subunit). The assay depends on continuous determination of pyruvate by reduction with NADH and lactate dehydrogenase. Thus, contaminating enzymes, such as those present in crude extracts, that lead either to hydrolysis of PEP or catalyze NADH oxidation preclude use of the method. The method is especially useful with purified preparations of HPr and the III proteins and generally can be used with relatively impure preparations of Enzyme I (after separation from HPr and III^{Glc}).

The sugar phosphorylation assay is based on the separation of radioactively labeled sugar phosphate (formed by the complete phosphotransferase system) from unreacted radioactive sugar. The II-B proteins of the PTS are integral membrane proteins that catalyze the transfer of the phosphoryl group from the III (or II-A) proteins to the sugar substrates. Thus far we have been unable to show that these are phosphorylated. The other proteins in the PTS act as phosphocarriers, and it is unclear whether to consider each of them separately as a substrate of other proteins in the PTS or as an "enzyme."

Enzyme I is generally quantitated by its ability to catalyze the transfer of the phosphoryl group from PEP to sugar. Depending on the activity to be measured, either Enzyme I or Enzyme II is used in rate-limiting amounts. With given quantities of Enzyme I and Enzyme II, any other component of the system (including Mg^{2+}, sugar, HPr, III) can be used at less than its optimum concentration, and the rate of phosphorylation will decrease accordingly. Assay conditions can thus be selected so that a linear response to the amount of any of the proteins is obtained. However, for measuring Enzyme I, all other components are best used at saturating concentrations. For measuring HPr, usually Enzyme II is made rate-limiting, and HPr concentrations are used that fall within the pseudolinear portion of the typical hyperbolic Michaelis–Menten curve for this protein. Similarly, for measuring III^{Lac} or III^{Glc}, the corresponding II-B proteins are made rate-limiting, and the quantities of the III proteins used for assay are in the "linear" range of the respective curves. Standard curves for HPr and the III proteins are determined by using either the homogeneous protein or at least partially purified proteins where the concentrations have been determined by the spectrophotometric assay. We emphasize that absolute catalytic activities cannot be assigned to partially purified preparations of Enzymes I or II, and therefore it is necessary to compare relative concentrations of HPr or III in different fractions or at different stages during purification by assaying these simultaneously with known quantities of Enzymes I and II. Further, in order to determine the exact concentrations of HPr and III; standard curves must be obtained at the same time.

Convenient, albeit semiquantitative, complementation assays can be used. These are performed with extracts from mutants defective in one of the PTS proteins. Such extracts contain all the other PTS proteins and need only be supplemented with the fractions being assayed to reconstitute the complete PTS, capable of phosphorylating sugar. This assay is particularly useful as a "scanning" assay, for example for the detection of activity in fractions obtained by column chromatography. This complementation method is semiquantitative because in a given extract of the mutant cells it is not known whether Enzyme I or Enzyme II-B is in excess (although usually it is Enzyme I), and whether the other PTS proteins in the extract are present at optimal levels for assay of the missing protein. The method is especially subject to error when it is used to assay crude extracts or fractions for a particular PTS protein, since in this case, more than one PTS protein is being added.

[70] Assays for the Phosphotransferase System from *Salmonella typhimurium* [1-3]

By E. BRUCE WAYGOOD and NORMAN D. MEADOW

Included here are procedures for the preparation of membranes and for the partial purification of Enzyme II-B^{Glc}, which, as sources of Enzyme II, are required for the sugar phosphorylation assays described below. Preparations of the other PTS proteins required for these assays (Enzyme I, HPr, and III^{Glc}) are described in subsequent articles.

Sugar Phosphorylation Assay

Principle. This assay is based on the separation, by ion-exchange chromatography, of radioactively labeled unreacted sugar from radioactive sugar phosphate formed by the complete phosphotransferase system. The lower limit of the range of the assays depends on the background activity from the particular batches of the PTS proteins being used. The

[1] This work was supported by Grant CA 21901 from the National Institutes of Health. N. D. M. was supported in part by a National Institutes of Health Special Fellowship HD 57023 and AG 05029.

[2] E. B. Waygood, N. D. Meadow, and S. Roseman, *Anal. Biochem.* **95**, 293 (1979).

[3] J. B. Stock, E. B. Waygood, N. D. Meadow, P. W. Postma, and S. Roseman, *J. Biol. Chem.* in press (1982).

upper limits are designed so that no more than 25% of the sugar substrate is phosphorylated during the assay period. In the following sections typical values are given.

Reagents

Mixture A: 2 ml of 1 M potassium phosphate buffer, pH 7.5; 1 ml of 0.5 M KF; 0.5 ml of 0.2 M dithioerythritol (DTE) or dithiothreitol (DTT); 0.5 ml of water

Mixture B: 1 ml of 0.2 M PEP, potassium or cyclohexylammonium salt; 1 ml of 0.1 M MgCl$_2$

Mixture C: Same as mixture A except that Tris-HCl buffer, pH 8.0, is used in place of the phosphate buffer

Mixture D: Same as mixture A except that pH 6.5 phosphate buffer is used

[^{14}C]Methyl α-D-glucopyranoside or 2-deoxyglucose, 0.1 M, specific activity $(1-2) \times 10^5$ cpm/μmol

Assay proteins: These depend on the requirements of the assay. For assaying the soluble PTS proteins (Enzyme I, HPr, III^{Glc}), either partially purified II-B^{Glc} (described below) or washed membranes from *S. typhimurium* strain SB2950 (deleted for Enzyme I, HPr, III^{Glc}) or SB1687 (II-B^{Man-}, for the assay of III^{Glc}) are used and are supplemented with the appropriate soluble proteins; it is essential that each of the assay proteins be free of other PTS proteins.

Crude extracts: (approximately 20–30 mg protein/ml) from mutant strains of *S. typhimurium* defective in Enzyme I (SB1690, *ptsI139 trpB223*)[4] or HPr (SB2226, *ptsH38 trpB223*)[4] are used for semiquantitative assays of the respective proteins in fractions obtained from wild-type strains. The preparation of crude extracts is described in the last section of this article.

Enzyme I. All incubation mixtures have 0.1 ml final volumes and contain the following components: 10 μl of mixture D; 10 μl of mixture B; 25 μg of HPr purified at least through step 4 (see this volume [72]); 5–10 units (0.3–0.6 mg) of washed SB2950 membranes; 10 μl of [^{14}C]methyl α-glucoside solution; and 0.03–0.1 unit of Enzyme I to be assayed. Prior to assay, it is essential that samples of Enzyme I be dialyzed in the cold against 0.05 M phosphate buffer, pH 6.5, 1 mM DTT, 1 mM EDTA, and that the dialyzate be warmed to 37° before being diluted for assay in the same buffer at 37°. It is then incubated for 5–15 min at 37° before being added to the components of the assay mixture.[5] The assays are conducted

[4] J. C. Cordaro and S. Roseman, *J. Bacteriol.* **112**, 17 (1972).
[5] N. Weigel, E. B. Waygood, M. A. Kukuruzinska, A. Nakazawa, and S. Roseman, *J. Biol. Chem.* in press (1982).

at 37° for 30 min. Reactions can be terminated by chilling to 0° or by adding 2 μmol of EDTA and immediately transferring to the ion-exchange columns described below. Under these conditions, the rate of sugar phosphorylation remains constant for at least 90 min provided that the quantity of Enzyme I added for assay is within the range specified.

HPr. Incubation mixtures contain the following components in volumes of 0.1 ml: 10 μl of Mixture A; 10 μl of mixture B; 10 units of purified or partially purified Enzyme I (at least through the AcA-44 column, step 4, described below); 10 μl of the labeled methyl α-glucoside solution; 0.35 to 1 unit of Enzyme II^{Man} (about 50 μg of washed SB2950 membranes); the HPr fraction to be assayed containing from 0.2 to 2 μg of HPr. For determining the quantity of HPr, a standard curve is constructed using the homogeneous protein. Other reaction conditions are as described above except that preincubation is not required for PTS proteins other than Enzyme I.

III^{Glc}. Incubation mixtures contain the following in 0.1-ml volumes: 10 μl of mixture C; 10 μl of mixture B; 5 units of Enzyme I; 3 μg of HPr; 0.2 unit of II-B$^{Glc^-}$, either partially purified or membranes from SB1687; the fraction to be assayed containing III^{Glc} equivalent to 1–10 μg of the homogeneous protein. Standard curves are obtained with the homogeneous protein. Control incubation mixtures are conducted in the absence of III^{Glc}.

Semiquantitative Complementation Assay for Enzyme I or HPr. Incubation mixtures contain the following components in 0.1 ml: 10 μl of mixture A (for HPr) or D (for Enzyme I); 10 μl of mixture B; 10 μl of [^{14}C]methyl α-glucoside; 50 μl of crude extract from SB1690 (for Enzyme I) or SB2226 (for HPr). It should be emphasized that complementation assays are, at best, semiquantitative for the reasons described above.

Enzyme II. These are assayed under conditions similar to those described for the phosphocarrier proteins. Incubation mixtures contain the following components in volumes of 0.1 ml: for II-B^{Man}, 10 μl of mixture A, 10 μl of mixture B, 10–20 units of Enzyme I (purified at least through step 5), 25 μg of HPr (purified at least through the CM-23 column), and 10 μl of 2-deoxyglucose; for II-B^{Glc}, 10 μl of mixture C, 10 μl of mixture B, 10–20 units of Enzyme I (purified at least through step 5), 2 μg of HPr (purified at least through the CM-23 column), 20 μg of III^{Glc} (purified at least through step 4 in procedure A or step 2 in procedure B, this volume [73]), 1 μl of [^{14}C]methyl α-glucoside. These conditions are suitable for the assay of up to 0.25 unit of II-B^{Man}, but of only 0.025 unit of II-B^{GLC}.[2,3]

Definition of Unit and Specific Activity. HPr and III^{Glc} are defined in absolute terms: quantity in milligrams or micromoles of pure protein; or they can be defined in terms of relative activities when comparing fractions. They can also be defined as phosphorylation equivalents based on the spectrophotometric assay. This is particularly useful in other bacteria where pure proteins are not available. Enzyme I and the Enzyme II are defined in terms of activity; 1 unit converts 1 μmol of sugar to sugar phosphate under the conditions described above (37°, 30 min) where the enzyme is rate-limiting. Specific activity is expressed as units per milligram of protein.

   Ion-Exchange Separation of Sugar-Phosphates and Sugars. An ion-exchange method for routine assay of incubation mixtures[6] has been modified to conserve expensive materials, such as liquid scintillation fluids, and to permit the use of small counting vials (7 versus 21 ml). In this procedure, the incubation mixture containing labeled sugar substrate and the product, labeled sugar phosphate, is passed over an ion-exchange column, washed with water to remove excess substrate, and then eluted with LiCl to release the sugar phosphate. The latter is then counted in a liquid scintillation counter.

   The columns consist of precision-bore glass tubing, 4 mm i.d. $\times$ 10 cm, drawn to a coarse point at one end (containing glass wool as a support for the resin), and sealed to a reservoir (1.5 $\times$ 3.5 cm) at the other. The columns are supported by Lucite blocks (19 $\times$ 19 $\times$ 2 cm) drilled to accommodate 25 columns, each of which is held in the block by its reservoir. Analytical grade anion-exchange resin is used (Bio-Rad AG 1-X2, 50–100 mesh) and is converted batchwise to the chloride form after washing extensively with HCl, NaOH, etc., in the usual manner. The resin, in 1 M NaCl, is placed in the column to a height of 3.5–4 cm and washed with water before use. All solutions used to wash and to elute the columns contain 0.01% Triton X-100. The incubation samples described above are quantitatively transferred to the columns, washed with at least two 7-ml portions of water, and eluted with three 0.75-ml portions of 1 M LiCl. During elution, the columns are placed over Lucite blocks containing 25 scintillation vials (7 ml capacity each) so that each eluate is collected in a vial. Liquid scintillation fluor (3 ml, National Diagnostic Hydrofluor) is added to each vial, and the samples of sugar phosphate are then counted.

   The columns are regenerated for use again with a minimum of 2 $\times$ 7 ml of 1 M NaCl followed by washing with water.

Spectrophotometric Assay[2]

   Principle. The extent (and with limiting Enzyme I, the rate) of phosphorylation of purified PTS proteins, such as HPr, III^{Glc}, III^{Lac}, can be

---

[6] W. Kundig and S. Roseman, *J. Biol. Chem.* **246**, 1407 (1971).

followed spectrophotometrically in a coupled assay system. In this assay, the pyruvate formed during the phosphorylation of the protein is measured by following the oxidation of NADH by lactate dehydrogenase as a function of time, using a recording spectrophotometer at 340 nm with a temperature-regulated water jacket (30°) surrounding the cuvette holder.

Reagents

Potassium phosphate buffer, 1 *M*, pH 7.5 (for Enzyme I)
Tris-HCl buffer, 1 *M*, pH 8.0 (for HPr or III^{Glc})
PEP, 0.2 *M*, potassium salt
MgCl$_2$, 0.1 *M*
NADH, 0.006 *M*
Lactate dehydrogenase (LDH) rabbit muscle, 10,000 units/ml (units as defined by supplier), Sigma type II
PTS proteins: Enzyme I, protein obtained from step 4 and beyond may be used[5]; HPr, protein obtained from step 4 and beyond may be used[7]; III^{Glc}, protein obtained from step 5 of Procedure A or step 2 of Procedure B (see this volume [73]).

The volume of the incubation mixture should be the smallest that can be accommodated in the spectrophotometer used, thus minimizing consumption of the PTS proteins. In the descriptions that follow, the volume of additions are given for 1-ml incubation mixtures.

HPr. The incubation mixture contains 50 μl of potassium phosphate buffer, 5 μl of PEP, 50 μl of MgCl$_2$, 20 μl of NADH, 4 μl of LDH, and 20 units of Enzyme I. Quantities of HPr greater than 10 nmol are used in order to obtain accurately measurable absorbance changes. The reaction is usually started by the addition of Enzyme I, which is contained in as small a volume as is practical. This quantity of Enzyme I should be sufficient to complete the reaction within a few minutes. Phospho-HPr spontaneously hydrolyzes, and thus there is a continual oxidation of NADH following the initial rapid rate, but corrections can be made for this hydrolysis reaction.[2] Controls contain either Enzyme I or the PTS protein fraction being assayed, but in the absence of LDH to ascertain that there is no NADH oxidation by contaminating enzymes.

III^{Glc}. The incubation mixtures contain 50 μl of Tris-HCl buffer, 5 μl of PEP, 50 μl of MgCl$_2$, 20 μl of NADH, 5 μl of LDH, 3–6 units of Enzyme I, HPr, 5% (or less) of the molar quantity of III^{Glc}. Quantities of III^{Glc} greater than 10 nmol are used in order to obtain accurately measurable absorbance changes. The reaction is usually started by the addition of either the Enzyme I or HPr, which is contained in as small a volume as practical. Phospho-III^{Glc} is more stable than phospho-HPr, making correction for

[7] D. A. Beneski, A. Nakazawa, N. Weigel, P. E. Hartman, and S. Roseman, *J. Biol. Chem.* in press (1982).

hydrolysis less important than it is during the measurement of HPr. Controls employed are the same as those for HPr described above.

Preparation of the Membrane Components of the PTS for Assays

Preparation of Membranes. Growth of cells and preparation of crude extracts are described in the last section of this article. A mutant strain, SB2950 (*trpB223 trzA-ptsHI crrΔ49*) is used to prepare membranes for assays since this strain is deleted in the structural genes for the three soluble proteins, Enzyme I, HPr, and III^{Glc}. The membranes contain two important Enzyme II components: (*a*) II-B^{Glc}, which phosphorylates methyl α-glucoside (K_m, 6 μM) and glucose (K_m, 10 μM) when supplemented with the three soluble proteins; (*b*) an Enzyme II complex designated II^{Man}, comprising a II-A and II-B component, which phosphorylates 2-deoxyglucose (K_m, 0.2 mM), glucose (K_m, 0.015 mM), and mannose (K_m, 0.045 mM). In assays with II^{Man}, the required supplements are Enzyme I and HPr, but not III^{Glc}. Since SB2950 contains relatively low levels of Enzyme II-B^{Glc}, membranes from another mutant strain, SB1687 (II-B^{Man-})[8] are used as a source of II-B^{Glc}.

Membranes are prepared from crude extract by centrifugation for 2 hr at 200,000 g. The pellet is washed twice with 0.025 M Tris-HCl buffer, pH 7.5, containing 1 mM EDTA and 0.2 mM DTE, and finally suspended in this buffer to a concentration of 20–40 mg of protein per milliliter. Since such membrane preparations cannot be frozen, they are stored at 0°; their Enzyme II activity is stable for about 10 days. Although the activity of Enzyme II-B^{Glc} in membranes from SB1687 is relatively high compared to the activity of Enzyme II-B^{Man}, the latter enzyme shows some background activity in the sugar phosphorylation assay. The following section describes preparation of partially purified Enzyme II-B^{Glc} containing very low activity of Enzyme II^{Man}.

Preparation of II-B^{Glc}.[2] This protein is required for assay of III^{Glc}. Most membrane preparations contain both this activity and II^{Man}; the aim of the purification procedure is to remove the latter as well as the soluble PTS proteins in order to reduce blank values (sugar phosphorylation in the absence of III^{Glc}).

To purify Enzyme II-B^{Glc}, 140 g (wet weight) of frozen *S. typhimurium* LT-2 cells are resuspended in 1 liter of 25 mM Tris-HCl, pH 7.5, containing 1 mM EDTA, and 0.2 mM dithioerythritol (buffer A). Crude extracts are prepared as described in the last section. The cell suspension is homogenized by two passages through a Manton–Gaulin (Gaulin Co., Everett, Massachusetts) press operated at 8000 psi. The membranes are

---

[8] M. H. Saier, Jr., R. D. Simoni, and S. Roseman, *J. Biol. Chem.* **251**, 6584 (1976).

separated from the soluble components by centrifugation of 410 ml of crude extract at 100,000 *g* for 3 hr in a type 42 rotor in a Beckman L2-65B centrifuge. Centrifugation and resuspension (to the original volume with buffer A) are repeated three times. The final suspension is adjusted to a protein concentration of 25 mg/ml. The washed membranes (50 ml) are treated with 5.5 ml of a solution of 10% sodium lauroyl sarcosinate (Sarkosyl NL-97, K and K Fine Chemicals, filtered through a glass-fiber filter, Reeve Angel, Clifton, New Jersey, 985H) and stirred for 1 hr at 0°. The 1% detergent solution is centrifuged at 160,000 *g* in a Ti-50 rotor (Beckman L2-65B centrifuge), and the pellets are discarded. The supernatant fraction (49 ml) is adjusted to 30% of saturation with ammonium sulfate by the addition of 26.3 ml of cold saturated ammonium sulfate previously adjusted to pH 7.0 with ammonium hydroxide. The solution is stirred for 1.5 hr at 0° and then centrifuged at 78,000 g (type 30 rotor) for 30 min. The supernatant is adjusted to 60% of saturation with ammonium sulfate by the addition of 47.2 ml of the saturated ammonium sulfate solution, stirred for 1 hr at 0°, and centrifuged as above. The protein is recovered as a pellicle floating on the clear ammonium sulfate solution.

Detergent extracts vary in their response to the first treatment with ammonium sulfate. For this reason, small samples of each preparation are tested at several concentrations of ammonium sulfate varying from 25 to 35%, and the concentration that results in maximum removal of extraneous proteins and minimum loss of Enzyme II-B^{Glc} activity is used to treat the remaining detergent supernatant.

The protein is dissolved in 12 ml of buffer A and dialyzed for 18 hr at 4° with three 500-ml changes of the same buffer. The preparation is diluted with buffer A to a protein concentration of 4 mg/ml, chilled to 0° in an ethylene glycol–Dry Ice bath, and stirred; acetone at −77° is added slowly to a final concentration of 60% (v/v) while the temperature of the protein solution is slowly reduced to −10°. The mixture is stirred for 20 min and centrifuged at 10,000 *g* at −10° for 15 min (Sorvall RC2B centrifuge, SS-34 rotor). The supernatant solutions are decanted, the tubes are inverted and allowed to drain in a freezer, and the pellets are resuspended in buffer A (10 ml) and dialyzed against two 500-ml changes of buffer over 18 hr at 4°. The preparation is frozen at −20° in small aliquots. Under these conditions, the preparation is stable for at least 6 months. Upon thawing, each aliquot is sonically dispersed by immersing the tube intermittently in a 50-W bath-type sonicator for a total of about 3–5 min of sonication.

The specific activity of Enzyme II-B^{Glc} from the final stage of this scheme is from 5- to 10-fold higher than that of the washed membranes. The recovery of Enzyme II-B^{Glc} activity is approximately 50%. Of greatest importance, however, is the fact that the activity of Enzyme II^{Man}

is almost totally lost during the preparation. In washed membranes from *S. typhimurium* LT-2, the ratio of activity of Enzyme II-B^{Glc} : Enzyme II^{Man} is approximately 1, whereas in the final stage the ratio is about 200 because much less than 1% of the activity of Enzyme II^{Man} remains.

Detergent concentration is followed by using ^{14}C-labeled sodium lauroyl sarcosinate prepared by the method of Jungermann *et al.*[9] from [^{14}C]sarcosine (California Bionuclear Corporation, Sun Valley, California). The concentration of Sarkosyl decreases from 1% in the original extract to 0.6% after centrifugation, 0.2% in the ammonium sulfate pellicle, and is not detectable in the acetone pellet. Activity in this pellet is stimulated 3- to 10-fold (depending on the preparation) by phosphatidylglycerol.[6] All the activities are determined with the optimal lipid concentration for each particular step of purification. The washed membranes are not stimulated by phospholipid.

Growth of Cells and Preparation of Crude Extracts

The *S. typhimurium* proteins are purified from the wild-type strain, LT-2, or from a strain auxotrophic for tryptophan, SB3507 (*trpB223*). These cells are grown in minimal medium containing glucose to induce high levels of Enzyme I, HPr, and III^{Glc}. The minimal salts medium used for these studies is a modified medium A[4] that contains 93.4 mM potassium phosphate buffer (pH 7.2), 7.5 mM (NH$_4$)$_2$SO$_4$, and 0.83 mM MgSO$_4$. The medium is supplemented with L-tryptophan, 20 mg/liter, for strains requiring this amino acid. A modified medium 63[8] lacking iron has also been used and contains the following components: 50 mM KH$_2$PO$_4$, 15.1 mM (NH$_4$)$_2$SO$_4$, and 0.81 mM MgSO$_4$; the pH is adjusted to 7.3 with KOH. The liquid medium is autoclaved, and sterilized glucose solution is added to the hot medium, immediately after autoclaving, to a final concentration of 0.2%. The deletion strain, SB2950, is grown in the same medium containing either 0.2% galactose or 0.4% DL-lactate (adjusted to pH 7.0). Cells are grown in the media in a New Brunswick incubator shaker at 37° with the gyratory shaker speed set at 200–250 rpm and using 5–10% inocula. When the cultures reach the late-exponential phase of growth (A_{500} = 1.5–1.8), the suspensions are chilled, the cells are harvested by centrifugation at 4° in a Sorvall RC2B centrifuge at 10,000 g for 15 min, washed with cold 0.9% KCl solution, and centrifuged. The pellet is suspended in cold (4°) 0.01 M Tris or 25 mM potassium phosphate buffer, pH 7.5, containing 1 mM EDTA and 0.2 mM dithioerythritol (4 ml of the solution for each gram of cell paste). Homogenates of the cells are prepared by passing the cold suspension twice

[9] E. Jungerman, J. F. Gerecht, and I. J. Krems, *J. Am. Chem. Soc.* **78**, 172 (1956).

through a French pressure cell (Amicon Corp., Lexington, Massachusetts) using an hydraulic pressure of 13 tons per square inch.

Cells are also grown in a 200-liter working capacity fermentor (Chemap AG, Switzerland) using the same media. Growth is continued until the stationary phase, the cells are harvested in a continuous-flow centrifuge (Carl Padberg GMBH, West Germany) at 50,000 g, and the cell paste is frozen at $-20°$. The usual yield from the fermentor is about 1000 g of cell paste. For large-scale preparations, 500 g of cell paste are suspended in the above buffer to a final volume of 2500 ml, and the cells are disrupted in a Manton-Gaulin press at 8000 psi. Crude extract is prepared from the homogenate by centrifugation at 16,300 g for 15 min at 4° to remove cell debris. Crude extracts from mutant strains SB1690 or SB2226 are stored frozen in 1-ml aliquots until they are used in the sugar phosphorylation assay described above.

[71] Enzyme I from *Salmonella typhimurium*[1]

By MARIA A. KUKURUZINSKA, NANCY WEIGEL,
and E. BRUCE WAYGOOD

Purification[2]

The purification of Enzyme I is started with 500 g of cell paste. Unless otherwise indicated, all steps are conducted at 0–4°, and all columns are packed by gravity.

The procedure takes advantage of two properties of the enzyme: first, that Enzyme I undergoes a temperature-dependent change in molecular weight; and second, that Enzyme I can be phosphorylated by incubation with phosphoenolpyruvate (PEP) and $MgCl_2$, which changes its charge and chromatographic properties.

Step 1. Precipitation with Protamine Sulfate. A 2% (w/v) solution of protamine sulfate (Sigma, Grade II) is gradually added with stirring to the crude extract (see this volume [70]) to a final concentration of 0.33%. The mixture is stirred for an additional 30 min, and the precipitate is removed by centrifugation at 16,300 g for 30 min.

Step 2. DEAE-Cellulose Chromatography. The supernatant solution is diluted 1 : 1 with 0.01 M Tris buffer, pH 7.5, containing 1 mM EDTA and

[1] This work was supported by Grant CA 21901 from the National Institutes of Health.
[2] N. Weigel, E. B. Waygood, M. A. Kukuruzinska, A. Nakazawa, and S. Roseman, *J. Biol. Chem.* in press (1982).

0.2 mM dithioerythritol (buffer B), and is applied to a 1500-ml Whatman DE-23 DEAE-cellulose column equilibrated with buffer B. The column is washed with 6 liters of buffer B and then successively with 6 liters each of 0.05 M KCl, 0.1 M KCl, 0.2 M KCl, and 0.25 M KCl, in buffer B to elute the other soluble proteins of the phosphotransferase system from *S. typhimurium* (PTS). HPr was eluted in the 0.05 M KCl fraction and III^{Glc} in the 0.2 M KCl fraction. Enzyme I is then eluted with 6 liters of 0.5 M KCl in buffer B.

Step 3. Precipitation with Ammonium Sulfate. The 0.5 M KCl eluate is concentrated 30-fold in a Pellicon concentrator with a 10,000 molecular weight (M_r) cutoff membrane (PTGC cassette, Millipore Corp.). The retained material (150 ml) is dialyzed against two 6-liter volumes of 10 mM potassium phosphate buffer, pH 7.5, containing 1 mM EDTA and 0.2 mM dithioerythritol (buffer C), and subsequently precipitated with ammonium sulfate (Enzyme Grade, Schwarz-Mann) at 90% of saturation. The mixture is stirred overnight and the precipitate is collected by centrifugation at 40,000 g for 20 min, resuspended in buffer C, and dialyzed for 24 hr against three 6-liter portions of the same buffer.

Step 4. Gel Filtration (AcA-44 Ultrogel). The dialyzed material (40 ml) is chromatographed on a 2-liter Ultrogel AcA-44 (LKB Produktur) column (5 × 100 cm) with buffer C containing 0.1 M NaCl, and eluted with the same solution at a flow rate of 0.5 ml/min. Enzyme I activity comigrates with the trailing portion of the major protein peak. The central portion of the Enzyme I peak, containing about 75% of the activity eluted from the column, is pooled (375 ml).

Step 5. The pooled fractions from step 4 (375 ml) are warmed to 37° and incubated with 2 mM PEP and 5 mM MgCl$_2$ for 15 min at 37° in order to phosphorylate the protein. The mixture is chilled in ice and applied to a 50-ml DEAE-Sephadex column equilibrated with 100 mM potassium phosphate buffer, pH 7.5, containing 1 mM EDTA and 0.2 mM dithioerythritol (buffer C). The column is washed with 150 ml of buffer and subjected to a 4-liter linear gradient of 0.1 to 0.3 M potassium phosphate buffer, pH 7.5, containing 1 mM EDTA and 0.2 mM dithioerythritol. During this treatment most of the contaminating proteins are washed from the column, while Enzyme I remains bound. Enzyme I is eluted with a second 4-liter linear gradient of 0.3 to 0.7 M potassium phosphate buffer, pH 7.5, containing 1 mM EDTA and 0.2 mM dithioerythritol. The enzyme migrates as a broad peak, a small portion of which overlaps with the major protein peak. Fractions containing Enzyme I, with the exception of the leading edge of the peak, are combined (1225 ml), concentrated 100-fold in an Amicon concentrator with a Millipore membrane type PTGC (10,000 M_r cutoff), and dialyzed overnight against 6 liters of 10 mM potassium

PURIFICATION OF ENZYME I FROM *Salmonella typhimurium*[a]

Step	Activity[b] (units/ml)	Concen- tration (mg/ml)	Specific activity (units/mg protein)	Purifi- cation (fold)	Total activity (units)	Yield (%)
1. Crude extract	1,000	45	22.2	1.0	1,200,000	100
protamine sulfate	300	6.7	44.8	2.0	765,000	64
2. DEAE-cellulose	3,772	24.5	154	6.9	562,028	47
3. Ammonium sulfate	8,688	55.5	157	7.1	564,720	47
4. AcA-44	1,134	3.44	330	14.9	425,250	35
5. DEAE-Sephadex	16,571	2.45	6764	305	290,000	24
6. AcA-34	35,714	4.0	8929	402	250,000	21

[a] Details of the purification procedure and assay conditions are described in the text. The crude extract and the protamine sulfate fraction are centrifuged at 160,000 g in the Ti 50 rotor for 2 hr to remove membranes before analysis.

[b] A unit of Enzyme I activity is defined as the amount of enzyme catalyzing the phosphorylation of 1 μmol of sugar in 30 min at 37° under the conditions described in this volume [70].

phosphate buffer, pH 6.5, containing 1 mM EDTA and 0.2 mM dithioerythritol.

Step 6. The dialyzed material is then warmed to room temperature and applied to a 1-liter AcA-34 Ultrogel column (4 × 80 cm), first equilibrated at room temperature with 0.1 M potassium phosphate buffer, pH 6.5, containing 1 mM EDTA, 0.5 mM dithioerythritol, and 5 mM MgCl$_2$. The proteins are eluted with same solution. The bulk of Enzyme I activity is eluted and is coincident with the major protein peak. The active Enzyme I fractions are combined into three pools, and the central portion of the peak (pool II, 60 ml) appears pure. The fractions are concentrated about 10-fold with an Amicon concentrator as described in step 5. When analyzed for purity in the sodium dodecyl sulfate–polyacrylamide gel electrophoresis system, pool II shows a single protein band. Pools I and III show trace quantities of contaminants. At this stage, the yield of Enzyme I (pool II alone) is 30 mg/500 g of cell paste.

The purification procedure is summarized in the table. The homogeneous material is obtained in about 20% yield, about 30 mg per 500 g of a wet cell paste.

Properties

Purity. The purified Enzyme I preparation is found to be homogeneous when examined by polyacrylamide gel electrophoresis under "native" conditions[3] and under denaturing conditions in sodium dodecyl sulfate.[4] Similarly, a single band is detected on urea–acetic acid denaturing gel electrophoresis.[5] Additional evidence of purity is obtained by sedimentation velocity studies in which the preparation sediments as a single symmetrical boundary.

Molecular Weight. The monomeric subunit of Enzyme I has a molecular weight of about 58,000. This value is obtained from the sedimentation equilibrium studies and from the gel filtration measurements under denaturing conditions (6 M guanidinium chloride).[6] The sodium dodecyl sulfate–gel electrophoresis according to the method of Laemmli[4] gives a molecular weight value of 60,000–62,000, while the Weber–Osborn procedure[7] shows a molecular weight of 68,000.

Association.[8] Enzyme I monomers undergo a reversible, temperature- and concentration-dependent association. Gel filtration chromatography under native conditions shows that at 4–6° the enzyme migrates as a 70,000 molecular weight protein (apparent), and at room temperature its apparent molecular weight increases to a value in the range of 130,000 to 150,000. Sedimentation equilibrium studies at 8° and at an initial concentration of 0.5 mg/ml in the absence of substrates (PEP, pyruvate) and cofactor (Mg^{2+}) show that the association is too weak to distinguish between monomer–dimer and isodesmic (indefinite) reaction mechanisms. At 21° preliminary scanning gel chromatography studies indicate that Mg^{2+} and PEP do not significantly affect the self-association of Enzyme I used at an initial concentration of 2 mg/ml.

Physical Properties.[2,8] The Enzyme I monomer is a globular molecule, with a frictional ratio very close to 1. The sedimentation coefficient of Enzyme I in buffer at 4° is 3.63 S ($s_{20,w} = 5.59$ S). The partial specific volume determined from the density of the protein solution and dialyzate is 0.72 ml/g.

Isoelectric Point. When Enzyme I is subjected to preparative isoelectric focusing by the method of Radola[9] its isoelectric point is about 4.5.

Amino Acid Composition.[2] The tryptophan content of Enzyme I is low as evidenced by the extinction coefficient value of 4.0–4.4 at 280 nm for a 10 mg/ml solution. The protein contains 8 histidine residues per subunit

[3] B. J. Davis, *Ann. N.Y. Acad. Sci.* **121**, 404 (1964).
[4] U. K. Laemmli, *Nature (London)* **227**, 680 (1970).
[5] K. Takayama, *Arch. Biochem. Biophys.* **114**, 223 (1966).
[6] K. G. Mann and W. W. Fish, this series, Vol. 26, p. 28.
[7] K. Weber and M. Osborn, *J. Biol. Chem.* **244**, 4406 (1969).
[8] M. A. Kukuruzinska, W. F. Harrington, and S. Roseman, *J. Biol. Chem.* in press (1982).
[9] B. J. Radola, *Biochim. Biophys. Acta* **295**, 412 (1973).

molecular weight, one of which becomes phosphorylated during the phosphotransfer reaction.

pH Optimum.[2] The optimum pH for Enzyme I activity is 7.2 in various buffer systems tested (Tris, phosphate, Tricine, HEPES, and MES).

Stability. Enzyme I is very sensitive to inhibition by sulfhydryl reagents and is relatively labile in the purified state.[10] The protein is most stable in 0.05 M potassium phosphate buffer, pH 6.5, supplemented with 0.2 mM dithioerythritol and 1 mM EDTA. After quick-freezing it can be stored at $-18°$ for several months without significant loss of activity.

Kinetic Properties.[2,11] Enzyme I interacts with HPr and PEP with K_m values of 5.4 μM and 0.2 mM, respectively. The kinetics for the transfer of the phosphoryl group from PEP to HPr are consistent with a Bi-Bi Ping-Pong reaction mechanism. The enzyme requires Mg^{2+} for activity ($K_m = 0.53$ mM), but divalent ions such as Mn^{2+} and Co^{2+} ($K_m = 0.05$ mM in each case) can, in an appropriate concentration range, substitute for Mg^{2+}. At higher concentrations Co^{2+} inhibits Enzyme I activity with a K_i of 0.7 mM. In the presence of Zn^{2+} no enzyme activity is detected.[2]

During the phosphotransfer reaction, close to 1 mol of phosphoryl group is incorporated per mole of Enzyme I monomer. Current evidence suggests that the monomer is not catalytically active. The phosphorylation of Enzyme I requires Mg^{2+}; on the other hand, the transfer of the phosphoryl group from phospho-Enzyme I to HPr proceeds in the presence of 20 mM EDTA. The rate of Enzyme I phosphorylation is biphasic, suggesting that the enzyme is phosphorylated in the associated state. The apparent equilibrium constant for Enzyme I phosphorylation at low substrate (PEP, pyruvate) concentrations is about 1.5 and the reaction appears independent of pH in the range of 6.5 to 8.0. At high substrate concentrations only about 50% of Enzyme I is phosphorylated and the reaction is much more complex. The apparent equilibrium constant for the transfer of the phosphoryl group from PEP to HPr catalyzed by Enzyme I is about 11.

Nature of the Phosphoryl Linkage in Phospho-Enzyme I.[11] In phospho-Enzyme I the phosphoryl group is linked to position N-3 in the imidazole ring of a histidine residue. This conclusion is based on the pH stability of phospho-Enzyme I and its sensitivity to hydrolysis in the presence of pyridine and hydroxylamine. In addition, 3-phosphohistidine has been isolated after alkaline hydrolysis of ^{32}P-labeled phospho-Enzyme I.

Properties of Phospho-Enzyme I.[11] The half-life of phospho-Enzyme I is

[10] P. W. Postma and S. Roseman, *Biochim. Biophys. Acta* **457,** 213 (1976).

[11] N. Weigel, M. A. Kukuruzinska, A. Nakazawa, and S. Roseman, *J. Biol. Chem.* in press (1982).

2.5–3 hr at pH 6.5 and at room temperature. Phospho-Enzyme I can transfer its phosphoryl group to pyruvate (to form PEP) and to HPr (to form phospho-HPr). In addition, in the presence of HPr and appropriate sugar-specific proteins the phosphoryl group can be transferred from phospho-Enzyme I to methyl α-glucoside (to form sugar phosphate). The standard free energy of hydrolysis of phospho-Enzyme I, -14.3 to -14.7 kcal/mol, is very close to that of PEP, making the phosphate transfer potential of phospho-Enzyme I among the highest of known biological phosphate derivatives.

[72] HPr from *Salmonella typhimurium* [1,2]

By ATSUSHI NAKAZAWA and NANCY WEIGEL

Purification

The preparation of HPr starts with 500 g of cell paste. Unless otherwise indicated, all steps are conducted at 0–4° and centrifugations are at 16,300 g for 15 min. All columns are packed by gravity.

Steps 1 and 2. Protamine Sulfate Precipitation and DEAE-Cellulose Chromatography. The first two steps in the purification of HPr are the same as those used for the purification of Enzyme I and III[Glc] (this volume [71] and [73]). HPr is separated from the latter two proteins on the DEAE-cellulose column; HPr is eluted from this column with 0.05 M KCl, III[Glc] with 0.2 M KCl, and Enzyme I with 0.5 M KCl.

The 0.05 M KCl fraction from the DEAE-cellulose column (step 2) is concentrated 20-fold at room temperature in a Pellicon concentrator with a 10,000 M_r cutoff membrane (PTGC cassette, Millipore Corp.). The retained material is lyophilized, suspended in less than 20 ml of water, and dialyzed against 6 liters of water for 2 hr. In this and subsequent dialyses, 3500 M_r cutoff dialysis tubing (A. H. Thomas Co., Philadelphia, Pennsylvania) is used to prevent loss of HPr. The retained material is centrifuged at 37,000 g for 15 min to remove insoluble material.

Step 3. Gel Filtration. The concentrated, dialyzed sample from step 2 is applied to a 6- × 130-cm column of AcA-54 Ultrogel. The column is first equilibrated with 0.01 M NaCl, and, after application of the sample, the

[1] This work was supported by Grant CA 21901 from the National Institutes of Health.
[2] D. A. Beneski, A. Nakazawa, N. Weigel, P. E. Hartman, and S. Roseman, *J. Biol. Chem.* in press (1982).

column is eluted with 4 liters of 0.01 M NaCl. HPr is eluted after most of the other protein.

Step 4. Carboxymethylcellulose Chromatography. The active fractions from step 3 are combined, diluted with 2 volumes of 0.01 M sodium acetate buffer, pH 5.5, and applied to a 200-ml column of carboxymethyl cellulose (CM-23, Whatman) previously equilibrated with the acetate buffer. After application of the samples, the column is washed with 600 ml of the buffer and eluted with a linear gradient, total volume 4 liters, containing 0 to 0.2 M NaCl in the acetate buffer. HPr is eluted at about the middle of the gradient, followed immediately by (and partially overlapping with) a yellow substance. Fractions containing HPr are combined, with the exception of the trailing edge, which is yellow-brown. The pooled material is lyophilized and dialyzed overnight against 4 liters of 0.01 M Tris buffer, pH 7.5.

Step 5. DEAE-Cellulose Chromatography. The dialyzed material from step 4 is diluted with 2 volumes of 0.01 M Tris buffer, pH 7.5, and applied to a 50-ml column of DEAE-cellulose (DE-23, Whatman) previously equilibrated with the same buffer. The column is then washed with 150 ml of the same buffer and eluted with a 4-liter (total volume) linear gradient containing 0 to 0.1 M KCl in the 0.01 M Tris buffer, pH 7.5. Native HPr is eluted at about 0.02 M KCl, while HPr-1,[2] when present, is eluted at about 0.05 M KCl.

Each fraction is carefully assayed by disc gel electrophoresis in the system described by Davis[3]; the analytical gels are heavily loaded with the protein samples in order to detect trace contaminants. Generally, the first few fractions, containing such contaminants, are discarded. Occasional preparations at this stage may require additional DE-23 column chromatography in order to remove the last traces of HPr-1.

The relevant fractions are combined, dialyzed against water, lyophilized, and stored frozen at 0.1–2 mg of protein per milliliter. Under these conditions, HPr appears to be stable, although over a period of many months or years a small percentage of HPr-1 is gradually formed in the samples.

The purification procedure is summarized in the table.

Properties

Purity. The isolated HPr is shown to be homogeneous when checked for purity by disc gel electrophoresis under native conditions in the system described by Davis.[3] The preparations were free of carbohydrate.

[3] B. J. Davis, *Ann. N.Y. Acad. Sci.* **121**, 404 (1964).

PURIFICATION OF HPr FROM *Salmonella typhimurium*[a]

Step	HPr[b] (nmol)	Protein (mg)	Specific activity (nmol HPr/mg)	Purification (fold)	Yield (%)
Crude extract	23,000	27,300	0.85[c]	1.0	100
1. Protamine sulfate	17,000	17,200	1.0[c]	1.2	74
2. 0.05 M KCl pool	15,600	3,800	4.2	4.9	68
3. AcA-54 pool	9,400	640	15	18	41
4. CM-23 pool	8,400	103	81	96	36
5. DE-23 pool	6,900	62	110	130	30

[a] Details of the methods are given in the text. Protein is determined by the method of S. Zamenhof (this series, Vol. 3, p. 696). All assays are performed simultaneously with the same preparations of Enzyme I and membranes.
[b] No carbohydrate is detectable in the purified protein fractions.
[c] These fractions are centrifuged at 226,000 *g* for 2 hr to remove membranes before assays or protein determinations are performed.

Molecular Weight. Salmonella typhimurium HPr is a small, monomeric protein, with a molecular weight of 9017. The value for the molecular weight is obtained from the complete amino acid sequence of the protein.[4,5]

Amino Acid Composition. The HPr has an unusual amino acid composition in that it does not contain tryptophan, tyrosine, or cysteine. The protein has two histidine residues, one of which is phosphorylated in the phosphotransfer reaction.

Heat Stability. Heating at 100° modifies homogeneous HPr, i.e., causes simultaneous loss of amide nitrogen and PTS activity.[2] From heat-treated extracts two species of HPr can be obtained, HPr-1 and HPr-2, which contain one and two or three fewer amide nitrogens, respectively, and have greatly reduced (20–80%) specific activity.

Stability. Homogeneous HPr is stable. It is stored frozen at concentrations of 0.1–2 mg/ml in water. When stored for a long period of time there is a gradual increase in the quantity of HPr-1, but the latter never exceeds a few percent of the total.

Nature of the Phosphoryl Linkage in Phospho-HPr. In phosphorylated HPr the phosphoryl group is attached to position N-1 of the imidazole ring of the histidine moiety, residue 15 of the protein. This conclusion is based

[4] Some of the properties described in this section were determined with homogeneous protein isolated from *Escherichia coli* [B. Anderson, N. Weigel, W. Kundig, and S. Roseman, *J. Biol. Chem.* **246**, 7023 (1971)]. The HPr proteins from *E. coli* and *S. typhimurium* appear to be identical.[5]
[5] N. Weigel, D. A. Powers, and S. Roseman, *J. Biol. Chem.* in press (1982).

on the pH stability of phospho-Hpr, its sensitivity to hydrolysis in the presence of pyridine and hydroxylamine, on the identification of the phosphorylated amino acid after alkaline hydrolysis of ^{32}P-labeled phospho-Hpr, and on the isolation of ^{32}P-labeled peptide.

Properties of Phospho-Hpr. In phospho-Hpr 1 mol of phosphate is incorporated per mole of the protein. The phosphorylated Hpr derivative is unstable, with a half-life of 13 min at pH 8.0 and 37°. The stability of phospho-Hpr increases with increasing pH values, and at pH 9.2 the half-life is about 1 hr. The free energy of hydrolysis of phospho-Hpr is − 13 kcal/mole, making its phosphate transfer potential one of the highest, next to phosphoenolpyruvate and phospho-Enzyme I, of the known biological phosphate derivatives.

[73] III^{Glc} from *Salmonella typhimurium*[1,2]

By NORMAN D. MEADOW

Purification

III^{Glc} can be purified by two procedures described below. Pure protein is first produced by procedure A, and this protein can then be used to obtain antibodies. Once the antibodies are available, the simpler and much more rapid procedure B is used. The purification of III^{Glc} is monitored with anti-III^{Glc} antiserum using rocket immunoelectrophoresis[3] and related techniques.[4,5] The sugar phosphorylation assay (see this volume [70]) is used to confirm the results obtained by immunoelectrophoresis methods. The sugar phosphorylation assay is used with caution to measure III^{Glc} activity in crude preparations, since its concentration in these preparations is low and the preparations are contaminated with small amounts of Enzyme II-B^{Glc}.

Procedure A

Steps 1 through 3. Preparation of Crude Extract, Precipitation with Protamine Sulfate, and DEAE-Cellulose Chromatography. Preparation of the

[1] This work was supported by Grant CA 21901 from the National Institutes of Health. N.D. M. was supported in part by a National Institutes of Health Special Fellowship HD 57023 and AG 05029.
[2] N. D. Meadow and S. Roseman, *J. Biol. Chem.* in press (1982).
[3] C.-B. Laurell, *Scand. J. Clin. Lab. Invest.* **29**, Suppl. 124, 21 (1972).
[4] B. Weeke, *Scand. J. Immunol.* **2**, Suppl. 1, 69 (1973).
[5] B. Weeke, *Scand. J. Immunol.* **2**, Suppl. 1, 47 (1973).

crude extract and the first two steps in the purification procedure are the same as those used for purification of Enzyme I (this volume [71]) and HPr (this volume [72]). III^{Glc} was separated from the latter two proteins on the DEAE-cellulose column; III^{Glc} was eluted with 0.2 M KCl, HPr with 0.05 M KCl, and Enzyme I with 0.50 M KCl.

The recovery of III^{Glc} activity from the DEAE-cellulose column is frequently very low unless the fine particles are very thoroughly removed from the DEAE-cellulose in order to obtain the highest possible flow rate.

The volume of the 0.2 M KCl eluate is 6 liters. This fraction is rapidly concentrated to 300 ml with the same apparatus used for the concentration of Enzyme I (this volume [71]) and either frozen at $-20°$ or immediately used for the next purification step. With the Pellicon PTGC membrane (10,000 M_r cutoff; Millipore), III^{Glc} is recovered in greater than 90% yield.

Step 4. Sephadex G-75 Column. The entire fraction obtained from the DEAE-cellulose column is applied to a Sephadex G-75 column (10 × 120 cm) equilibrated with a solution containing 25 mM Tris-HCl, pH 7.5, at 4°, 1 mM EDTA, 0.2 mM dithioerythritol, and 50 mM KCl. The column is developed with the same solvent at a rate of 5 ml/min; 20-ml fractions are collected.

III^{Glc} is eluted from this column as a main peak, corresponding to a molecular weight of about 20,000, preceded by an elongated leading edge. When material from the leading edge is rechromatographed on a smaller Sephadex G-75 column, the III^{Glc} is now eluted at a position corresponding to that of III^{Glc} from the main pool of the first column. The pool of the main peak is reduced in volume from 750 ml to 7 ml by ultrafiltration as described above, followed by lyophilization and dialysis against 2 mM potassium phosphate buffer, pH 7.5, containing 5% glycerol (v/v).

Step 5. Preparative Isoelectric Focusing. Preparative isoelectric focusing is performed by the method of Radola[6] with modifications suggested by LKB Produktur, Rockville, Maryland (LKB application note No. 198 and Instructions for the use of the Ultrodex), using Ultrodex gel cast in a trough 9 × 18 × 1 cm deep. The volume of the bed after drying is 100 ml. The ampholytes (Ampholine-LKB) are a mixture of equal parts of pH ranges 3.5–5.0 and 4.0–6.0, and their final total concentration is 4% (w/v) in the dried bed, which also contains 5% glycerol (v/v). The electrolytes are 1 M H$_3$PO$_4$ and 1 M NaOH and are contained in electrofocusing wicks (LKB). The isoelectric focusing is performed with the trough on an aluminum plate cooled with cold tap water (about 10°). The bed is prefocused until its electrical resistance is constant (in the bed described above, 56,000 ohms). The gel is then removed from an area 1.5 × 9 cm

[6] B. J. Radola, *Biochim. Biophys. Acta* **295**, 412 (1973).

centered on pH 4.4 and mixed with the sample; the mixture is returned to the slot in the bed and focused for 10–12 hr at approximately 400 V.

The 7-ml solution from step 4 is divided into two equal parts, each of which is focused separately. III^{Glc} is found primarily in the area corresponding to pH 4.2–4.6. Small amounts of III^{Glc} (less than 10% of the total) are found outside this pH range, but this material contains contaminating proteins difficult to remove in the subsequent step and is discarded.

At the end of the focusing period, the gel is removed from the trough in bands 1-cm wide and placed in columns. The proteins are eluted with 3 bed-volumes of 0.5 M Tris-HCl, pH 7.5, containing 2 M KCl, 1 mM sodium EDTA, and 5% glycerol, and the eluates are dialyzed against 10 times their volume of the same buffer for 3 days at 4°, with a change of buffer each day. They are finally dialyzed against 25 mM Tris-HCl, pH 7.5, containing 1 mM EDTA for 1 day at 4°. The pooled III^{Glc} fractions from both focusing experiments are combined after elution of the gels; the total volume is approximately 50 ml.

Step 6. Preparative Polyacrylamide Gel Electrophoresis. Preparative polyacrylamide gel electrophoresis is performed in an Ultraphor apparatus (Colora Messtechnik, Lorch, West Germany) following the manufacturer's directions. Of the buffer systems for polyacrylamide gel electrophoresis published by Jovin[7] and Jovin et al.,[8] system No. 3495 is optimal for the purification of III^{Glc}. This system is used with an alternative stacking gel buffer, so that both stacking and resolution occur at pH 7.5; it is designated 3495.8.[7] The buffers are as follows: upper buffer, 40 mM HEPES–16.5 mM Tris, pH 7.23; upper gel buffer, 68.6 mM acetic acid–45.1 mM Tris, pH 5.0; lower gel buffer, 50.4 mM acetic acid–40.0 mM Tris, pH 5.3; lower buffer, 62.5 mM Tris-HCl, pH 7.5. Electrophoresis is conducted at room temperature with the chamber cooled by tap water, and the current at 90 mA before elution of III^{Glc} and at 30 mA during elution of III^{Glc}. Before elution of the tracking dye (bromophenol blue), the elution and counterelution chamber are filled with lower gel buffer containing 5% (v/v) glycerol; after elution of the tracking dye, the chambers are filled with theoretical phase π (Jovin et al.[8]) [elution chamber, 23 mM Tris-HCl, pH 7.5 (0.9 π) containing 4.5% (v/v) glycerol; counterelution chamber, 104 mM Tris-HCl, pH 7.5 (4 π)]. The sample is added in upper buffer. In addition to the buffers given above, the gels contain the following components. Stacking gel: acrylamide–methylene

[7] T. M. Jovin, *Ann. N.Y. Acad. Sci.* **209**, 477 (1973).
[8] T. M. Jovin, M. L. Dante, and A. Chrambach, "Multiphasic Buffer Systems Output," PB Nos. 196089–196091, 259309–259312, National Technical Information Service, Springfield, Virginia, 1970.

bisacrylamide (3% acrylamide, 0.75% methylene bisacrylamide), 0.4 mM potassium persulfate, 2.2 mM TEMED, 6.5 μM riboflavin 5'-phosphate, and 5% (v/v) glycerol; polymerization of this gel occurs after 10–15 min of illumination. Resolving gel: acrylamide–methylene bisacrylamide (11.9% acrylamide, 0.63% methylene bisacrylamide), 0.31 mM potassium persulfate, 2.1 mM TEMED, 8.1 μM riboflavin 5'-phosphate, and 5% (v/v) glycerol; polymerization of this gel occurs after 3–4 min of illumination with fluorescent light.

The fractions containing the III^{Glc} from both preparative isoelectric focusing experiments are concentrated by ultrafiltration (Pellicon PTGC membrane) to approximately 20 ml and dialyzed against upper buffer containing 5% glycerol. Owing to the quantity of protein, the III^{Glc} obtained from the G-75 column is divided into four aliquots, each of which is separately fractionated by preparative polyacrylamide gel electrophoresis.

Two peaks are obtained, both of which contain protein that reacts with anti-III^{Glc} serum. The protein that migrates more rapidly is designated III$^{\text{Glc}}_{\text{Fast}}$, and the second protein is called III$^{\text{Glc}}_{\text{Slow}}$. Evidence will be given below that III$^{\text{Glc}}_{\text{Fast}}$ is derived from III$^{\text{Glc}}_{\text{Slow}}$ by the cleavage of the amino-terminal heptapeptide from III$^{\text{Glc}}_{\text{Slow}}$. Preparative polyacrylamide gel electrophoresis is the only technique available that separates these two proteins.

Procedure A is summarized in Table I.

Procedure B

After purification of III^{Glc} by procedure A, as described above, antibodies against III$^{\text{Glc}}_{\text{Fast}}$ are prepared in a goat, and the IgG fractions obtained are coupled to Sepharose 4B.[2,9,10]

Step 1. Adsorption and Elution from Antibody Column. Either crude extract or the protamine sulfate supernatant fraction made from crude extract is purified on this column. The capacity of the column (150 ml packed in a column 6 cm in diameter) is 20 mg of III^{Glc}, a quantity found in 300 ml of crude extract; generally, 1500 ml of crude extract obtained from 300 g of cell paste are processed.

The crude extract is loaded onto the column over a period of approximately 30 min at 4°. The quantity of III^{Glc} in 300 ml of crude extract slightly exceeds the capacity of the column. The column is then washed with 1 liter of 25 mM Tris-HCl, pH 7.5, 1 mM EDTA, 0.1 M NaCl. III^{Glc} is eluted with 700 ml of 0.5 N acetic acid, and the eluate is immediately

[9] J.-C. Jaton, D. C. Brandt, and P. Vassali, *in* "Immunological Methods" (I. Lefkovits and B. Pernis, eds.), pp. 44–67. Academic Press, New York, 1979.
[10] S. C. March, I. Parikh, and P. Cuatrecasas, *Anal. Biochem.* **60**, 149 (1974).

TABLE I
PURIFICATION OF III^Glc FROM *Salmonella typhimurium* BY PROCEDURE A[a]

Step	Total III^Glc (mg)	III^Glc activity (μmol sugar-P per 30 min)	Total protein (mg)	Purification (fold)
1. Crude extract	89.3 (100)[b]	ND[c]	54,500 (100)	1
2. Protamine sulfate supernatant fraction	98.4 (110)	ND	42,500 (79)	1.4
3. DEAE-cellulose column	68.3 (78)	0.27 (78)[d]	10,600 (19)	4.1
4. Sephadex G-75 column	51.0 (57)	0.22 (64)	260 (0.47)	121
5. Preparative isoelectric focusing	19.4 (22)	0.057 (16)	52 (0.1)	220
6. Preparative PAGE[e]				
III^Glc_Fast	5.6 (6.3)	—	5.6 (0.01)	630[f]
III^Glc_Slow	3.6 (4.0)	0.056 (16)	3.6 (0.007)	570[f]

[a] The purification procedure is described in the text. The total amount of III^Glc is determined by rocket immunoelectrophoresis by comparison with samples containing a known amount of III^Glc. Total protein is measured by the method of M. M. Bradford [*Anal. Biochem.* **72**, 248 (1976)] corrected for the final pure III^Glc fraction as described by Meadow and Roseman.[2]

[b] Numbers in parentheses indicate percentage of the starting values.

[c] ND, not determined. The phosphorylating activity of III^Glc cannot be determined accurately in crude preparations because the activity is low and the preparations are contaminated with small amounts of Enzyme II-B^Glc activity.

[d] The percentage of recovery is arbitrarily made equal to the percentage of recovery for this step as determined by rocket immunoelectrophoresis (column 2). This is done to show that both methods for the determination of III^Glc result in approximately the same proportional values.

[e] The yields and specific activities shown are those obtained with the pooled fractions containing either III^Glc_Fast or III^Glc_Slow. Intermediate fractions contain mixtures of both species and are not shown.

[f] Overall purification factors vary from one preparation to another, the lowest value being about 350-fold.

neutralized with 2 M Tris base. Less than 20 min is required for the elution and neutralization of the III^Glc solution.

The column is regenerated by washing with 1–2 liters of 0.1 M Tris-HCl, pH 7.5, containing 0.1 M NaCl followed by 1–2 liters of 25 mM Tris-HCl, pH 7.5, 1 mM EDTA. Each additional portion of crude extract is then processed. The neutralized acetic acid eluates from 1500 ml of crude extract are combined, and the pool (4.5 liters) is concentrated to approximately 20 ml by ultrafiltration as described under Procedure A, step 3.

The column is stored in 25 mM Tris-HCl, pH 7.5, 1 mM EDTA, 0.02% sodium azide. It can be used for 18 months without reduction of its capacity.

Step 2. Sephadex G-75 Column. The concentrated fraction from step 1 is transferred to a Sephadex G-75 column (4 × 61 cm) first equilibrated with a solution containing 25 mM Tris-HCl buffer, pH 7.5, containing 1 mM EDTA and 50 mM KCl at 4°. Elution (3-ml fractions collected) is accomplished with the same buffer mixture at a rate of 1 ml/min under hydrostatic pressure.

III^{Glc} is not always eluted as a single symmetrical peak, but sometimes exhibits an elongated leading edge, similar to the results obtained with the Sephadex G-75 column used for Procedure A.

Step 3. Preparative Polyacrylamide Gel Electrophoresis. The major peak of III^{Glc} from step 2 (fractions 45 to 74, approximately 90 ml) is concentrated to approximately 10 ml by ultrafiltration, and dialyzed against upper buffer containing 5% glycerol, before polyacrylamide gel electrophoresis. This step is conducted precisely as described for Procedure A, step 6. When 1500 ml of crude extract is processed by Procedure B, the protein from step 2 is divided into five or six aliquots, each of which is separately purified by the preparative polyacrylamide gel electrophoresis technique. Proportionally much less III$^{Glc}_{Fast}$ than III$^{Glc}_{Slow}$ is obtained from the preparative polyacrylamide gel electrophoresis in Procedure B than is obtained from the same step in Procedure A.

Procedure B is summarized in Table II.

Properties

Relationship between the Two Electrophoretic Forms of III^{Glc}. III$^{Glc}_{Fast}$ is derived from III$^{Glc}_{Slow}$ by cleavage of the latter protein between residues 7 and 8. If residue 1 of III$^{Glc}_{Fast}$ is aligned with residue 8 of III$^{Glc}_{Slow}$, the amino-terminal sequences of the two proteins are identical. Maps of the tryptic peptides of the two proteins suggest that their primary sequences are identical except for the two tryptic peptides associated with the first eight residues of III$^{Glc}_{Slow}$. The quantity of III$^{Glc}_{Fast}$ relative to III$^{Glc}_{Slow}$ increases as the time required to purify the protein increases; i.e., much more III$^{Glc}_{Fast}$ is obtained from Procedure A (which requires several weeks) than after step 2 of Procedure B (where the proteins are virtually free of contamination after 3 days). Even actively growing cells, however, are found to contain traces of III$^{Glc}_{Fast}$.

Purity. The purity of III^{Glc} is tested by using several different polyacrylamide gel electrophoresis systems. These are the 3495.8 system described above, which resolves at pH 7.5, and the 4051.V system[8] which

TABLE II
PURIFICATION OF III^{Glc} FROM *Salmonella typhimurium* BY PROCEDURE B[a]

Step	Total III^{Glc} (mg)	Total protein (mg)	Purification (fold)
Crude extract	90	35,000	1
1. Acetic acid elution from anti/III^{Glc} column	58.4 (88)[b]	160	140
2. Sephadex G-75 column	35 (53)	34	400
3. Preparative PAGE			
III$^{Glc}_{Fast}$	7.8 (12)	ND[c]	ND[d]
III$^{Glc}_{Slow}$	20.2 (31)	ND[c]	ND[d]

[a] The purification procedure is described in the text. Total III^{Glc} is determined by rocket immunoelectrophoresis by comparison with samples containing a known amount of III^{Glc}. Total protein is measured by the method M. M. Bradford [*Anal. Biochem.* **27**, 248 (1976)], corrected for the final pure III^{Glc} fractions as described by Meadow and Roseman.[2]

[b] Numbers in parentheses are the percentage of the III^{Glc} that bound to the column.

[c] ND, not determined.

[d] The purification factor increases very little at this stage of purification.

resolves at pH 9.5; four concentrations of acrylamide are used with each system. In addition, sodium dodecyl sulfate–polyacrylamide gel electrophoresis is used. When tested under all these conditions, preparations of III$^{Glc}_{Fast}$ or III$^{Glc}_{Slow}$ made by either Procedure A or B show much less than 1% contamination by extraneous proteins; the III$^{Glc}_{Slow}$ always contains traces of III$^{Glc}_{Fast}$, however. Preparations from step 2 of Procedure B contain only traces of contamination and may be useful for many purposes if a mixture of III$^{Glc}_{Fast}$ and III$^{Glc}_{Slow}$ can be used.

Molecular Weight. Both III$^{Glc}_{Fast}$ and III$^{Glc}_{Slow}$ are monomeric proteins with a molecular weight of about 20,000. The measurement was made by gel filtration under native conditions, gel filtration under denaturing conditions (6 *M* guanidinium chloride), and sodium dodecyl sulfate–polyacrylamide gel electrophoresis. III$^{Glc}_{Fast}$ has a slightly lower relative mobility than III$^{Glc}_{Slow}$ in sodium dodecyl sulfate–polyacrylamide gel electrophoresis even though it is a slightly smaller molecule. This anomaly probably is the result of differential proportions of sodium dodecyl sulfate binding between the two proteins.

Amino Acid Composition. Neither III$^{Glc}_{Fast}$ nor III$^{Glc}_{Slow}$ contains cysteine, tyrosine, or tryptophan.

Isoelectric Point. Both III^{Glc}_{Fast} and III^{Glc}_{Slow} are very insoluble near their isoelectric point and, therefore, produce smears rather than sharp bands of protein in analytical isoelectric focusing gels. From the results of preparative isoelectric focusing, however, it can be concluded that both proteins have their isoelectric points in the vicinity of pH 4.5.

Heat Stability. The activity of III^{Glc}_{Slow} in the sugar phosphorylation system is very stable to heat; 50% of its activity is retained after 1 hr at 100°.

Kinetic Properties. Both III^{Glc}_{Fast} and III^{Glc}_{Slow} are active in the sugar phosphorylation system, but the specific activity of III^{Glc}_{Fast} is only 2–3% of that of III^{Glc}_{Slow}. When the sugar phosphorylation assay is conducted with a large excess of Enzyme I (so that all of the III^{Glc} is phosphorylated), the apparent K_m of Enzyme II-B^{Glc} for phospho-III^{Glc}_{Slow} is approximately 3.5 μM. It is a substrate of Enzyme II-B^{Glc}. The activity of III^{Glc}_{Fast} is too low to measure a K_m or V_{max}. III^{Glc}_{Fast} does not inhibit the activity of III^{Glc}_{Slow}.

The extent of phosphorylation of III^{Glc}_{Fast} and III^{Glc}_{Slow} was studied by the spectrophotometric assay already described (this volume [69]) and by the direct isolation of ^{32}P-labeled protein. Both forms of this protein were found by these two methods to accept 1 mol of phosphate per mole of protein.

Properties of Phospho-III^{Glc}_{Slow}. Pure III^{Glc}_{Slow} labeled with ^{32}P can transfer its phosphoryl group to HPr. This reaction is reversible, and preliminary data show that the equilibrium constant for the reaction is approximately 0.1:

$$\text{Phospho-HPr} + III^{Glc} \rightleftharpoons \text{HPr} + \text{phospho-}III^{Glc}$$

The labeled phosphate from III^{Glc}_{Slow} is transferred to glucose or methyl α-glucoside in the presence of Enzyme II-B^{Glc}.

Nature of the Phosphoryl Linkage in Phospho-III^{Glc}_{Slow}. Studies of the pH stability of the phospho-III^{Glc}_{Slow} and of its stability in the presence of hydroxylamine or pyridine suggest that the phosphoryl group is linked to a histidinyl residue.

[74] Phosphotransferase System from *Staphylococcus aureus* [1]

By JOHN B. HAYS and ROBERT D. SIMONI

Assays

Reagents

[14]C-Labeled 0.1 M thiomethyl β-D-galactopyranoside or methyl α-D-glucopyranoside; specific activity 0.5 to 1×10^6 cpm/μmol
Potassium phosphate buffer, 0.1 M, pH 7.5
Phosphoenolpyruvate, 0.1 M, potassium salt (PEP)
Dithiothreitol (DTT), 0.1 M
MgCl$_2$, 0.1 M
KF, 0.1 M

Assay Method. For the principle of this sugar phosphorylation assay, see this volume [69]. Incubation mixtures contain the following components in final volumes of 250 μl: 0.1 ml of phosphate buffer; 10 μl each of the labeled sugar substrate, MgCl$_2$, and KF; 20 μl of PEP; 5 μl of DTT. The following proteins are used. HPr assay: 50 μg of Enzyme I (DEAE-cellulose fraction), 35 μg of III^{Lac} (Sephadex G-75 fraction), 240 μg of Enzyme II^{Lac} (deoxycholate-alkali pellet); HPr equivalent to 0.2–2.0 μg of homogeneous protein. III^{Lac} assay: Assay mixtures similar to those above can be used to determine III^{Lac} by making this protein rate-limiting and by adding excess HPr (20 μg). Complementation assays for routine screening of fractions can also be used. In this case, for III^{Lac}, 1 mg of a crude extract of strain F9 (III^{Lac-}) is used in place of the other PTS proteins, and fractions containing the equivalent of 0.1–5 μg of pure III^{Lac} are added for assay. Similarly, 0.5–1.2 mg of FM9 extract is used for assay of Enzyme I, and 0.8–1.6 mg of F15 extract for assay of II-B^{Lac}. After incubation for 30 min at 37°, 3 μmol of EDTA are used to terminate the reactions, the mixtures are chilled to 0°, and the labeled sugar phosphate is measured by the paper electrophoresis method[2] or by the ion-exchange column procedure described above. The rates of sugar phosphorylation are proportional to the quantity of the rate-limiting protein and constant with time of incubation for at least 90 min.

Preparation of II-B^{Lac} for Assay. A 10-fold purification of the membrane activity, suitable for the assays described, is obtained as follows. The washed membranes (described below), 16 ml containing 15 mg of protein per milliliter, are treated with 0.4 ml of 1 M Tris buffer, pH 9.0, 0.4 ml of

[1] R. D. Simoni, T. Nakazawa, J. B. Hays, and S. Roseman, *J. Biol. Chem.* **248**, 932 (1973).
[2] W. Kundig and S. Roseman, *J. Biol. Chem.* **246**, 1393 (1971).

0.2 M DTT, and 0.16 ml of 0.1 M EDTA. After incubation for 15 min at 37°, 1.8 ml of a 10% solution of sodium deoxycholate are added, and the mixture is maintained at 0° for 20 min. The suspension is centrifuged at 160,000 g for 60 min, the supernatant fluid is discarded, and the pellet is resuspended in 16 ml of a solution containing 1 mM Tris-HCl buffer, pH 8.3, 1 mM DTT, and 1 mM EDTA. The suspension is then incubated with 0.6 ml of 1 N NaOH for 10 min at 25°, centrifuged at 167,000 g (4°) for 60 min, and washed and resuspended in the 1 mM Tris, DTT, EDTA solution (16 ml).

Some of the properties of this partially purified preparation are described by Simoni et al.[1,3] In addition, studies with substrate analogs provide further information on phosphoryl transfer in this system.[4,5]

Preparation of Partially Purified Enzyme I for Assay. The first two steps are those used for purification of HPr (given below). The acid precipitate is dissolved and treated with ammonium sulfate as described for HPr, except that the fraction precipitating between 40 and 80% of ammonium sulfate saturation is collected. This fraction contains both Enzyme I and III^{Lac}. These are completely separated from each other (and from residual HPr) by Sephadex G-75 gel filtration as described in the purification of III^{Lac} below. The Enzyme I is eluted from the column and is further purified as follows.

The fractions containing Enzyme I from the preceding step are pooled and transferred to a DEAE-cellulose column, 2.5 × 25 cm, previously equilibrated with a solution containing 0.01 M potassium phosphate buffer, pH 7.5, 0.1 mM DTT, and 0.1 mM EDTA. The column is eluted with a linear gradient of 0.2 to 0.4 M KCl in the same buffer, total volume 2000 ml. Enzyme I is found in the fractions containing approximately 0.3 M KCl. The pooled Enzyme I fractions are dialyzed against the same buffer solution and concentrated by adsorption to a small DEAE-cellulose column, followed by elution with 0.5 M KCl in the buffer solution. After dialysis against the phosphate–DTT–EDTA buffer solution to remove the KCl, the preparation is stored in small aliquots at −18°. It is stable for at least several months under these conditions. While it is only about 20-fold purified at this stage, the preparation is free of HPr, III^{Lac}, II-B^{Lac}, etc., and is satisfactory for assay purposes.

Growth of Cells: Preparation of Crude Extracts

All $S.$ *aureus* strains were derived from $S.$ *aureus* 5601, kindly supplied by Dr. M. L. Morse, Department of Biophysics, University of Colorado Medical Center, Denver. This parental strain is inducible for lactose fer-

---

[3] R. D. Simoni, J. B. Hays, T. Nakazawa, and S. Roseman, *J. Biol. Chem.* **248**, 957 (1973).
[4] J. B. Hays, M. L. Sussman, and T. W. Glass, *J. Biol. Chem.* **250**, 8834 (1975).
[5] J. B. Hays and M. L. Sussman, *Biochim. Biophys. Acta* **443**, 267 (1976).

mentation, and an effective inducer in the growth medium described below is 0.01 M galactose 6-phosphate. It may be noted that *S. aureus* cleaves the transport product of lactose, lactose-phosphate [6-phosphogalactopyranosyl-(β,1 → 4)-glucose], via a phospho-β-galactosidase. A variety of mutant strains were derived from the parental strain 5601.[6] These include strain C22, constitutive for the lactose operon, and strains lacking one of the following: phospho-β-galactosidase, Enzyme I, III^{Lac}, or II-B^{Lac}. In addition, several strains simultaneously defective in three proteins of the lactose operon (II-B^{Lac}, III^{Lac}, and phospho-β-galactosidase) were obtained. No mutants defective in HPr were isolated. The original reference[6] should be consulted for methods of characterizing and storing the mutants.

Complementation assays with mutant extracts suffer from the problems described in this volume [69], but are satisfactory for use prior to separation of purified PTS proteins or for routine screening of column fractions. For assay of Enzyme I, for example, 0.5–1.2 mg of a crude extract of a mutant defective in this protein, strain FM9, is used in place of the purified proteins in the assay described above. Similarly, 0.8–1.6 mg of F15 extract is used to assay for II-B^{Lac}, and 0.5 to 1.5 mg of F9 to assay for III^{Lac}.

The wild-type strain, *S. aureus* 5601, is grown in liquid medium containing 2% polypeptone, 0.5% yeast extract (Baltimore Biological Laboratories, Baltimore), and 1% galactose, pH 7.0–7.5. The sugar is added to the autoclaved medium, as indicated in this volume [70]. For small quantities of cells, the cultures are grown in Erlenmeyer flasks half full of medium, using 5–10% inocula grown on the same medium; the flasks are rotated at 200 rpm in a New Brunswick incubator shaker at 37°. For larger quantities of cells, 10-liter cultures are grown in 14-liter glass fermentor jars (New Brunswick Co.) with stirring at 200 rpm, with a slow rate of air flow above, but not into, the liquid medium. The pH is maintained between 7.0 and 7.5 by adding NaOH as required. Since the specific activities of III^{Lac} and II-B^{Lac} decline as the cells reach the stationary phase of growth, cultures are grown to the middle of the exponential phase, chilled rapidly, and harvested by centrifugation. The cells are washed twice in the cold with a solution containing 0.05 M Tris-HCl buffer, pH 7.5, and 0.15 M NaCl. The cell paste can be stored at $-18°$ or used immediately for fractionation. From 1 to 2 g, wet weight, of cells are obtained per liter of culture.

Because *S. aureus* is very resistant to homogenization, the following two methods are suggested for cell rupture.

Lysostaphin Procedure. For quantities of cells up to 25 g, crude extracts are prepared by treating with lysostaphin, obtained from Schwarz-

[6] R. D. Simoni and S. Roseman, *J. Biol. Chem.* **248**, 966 (1973).

Mann, Inc. The following mixture is used per gram (wet weight) of cells: 4 ml of 0.05 M Tris-HCl buffer, pH 7.5, containing 0.15 M NaCl, 0.4 ml of 0.01 M MgCl$_2$, 0.5 mg of lysostaphin, and 0.1 mg of DNase (crystalline, pancreatic, Worthington Co.). After incubation for 30–60 min at 37°, the suspension is cooled in ice and subjected to five 1-min treatments of sonic oscillation in a 10-kc Raytheon sonic oscillator (with rapid cooling after each treatment).

Mechanical Rupture. For rupture of cells on a large scale, a heavy-duty mechanical mill is used with micro glass beads. The glass beads, 0.10–0.11 mm in diameter (Cat. No. 514140 (2883), B. Braun Melsungen Apparatebau, purchased from Will Scientific Co., Rochester, New York), are first washed extensively with the following solutions in the indicated sequence: 6 N HCl, water, 0.05 M Tris-HCl buffer, pH 7.5, containing 0.15 M NaCl, 10% horse serum in the same buffer, and finally with the buffered sodium chloride solution once again. In a typical preparation, 500 g of cell paste, 500 ml of glass beads (packed by gravity), and 250 ml of the buffered sodium chloride solution are chilled to 0° and homogenized in an Eppenbach Micromill (Gifford-Wood Co., Hudson, New York) for 20 min at a rheostat setting of 75, a stator gap of 36, and with coolant circulating through the bath at about −10°. The temperature of the mixture passing through the stator is maintained below 15°. The viscous suspension is then treated with 50 mg of DNase and adjusted to 1 mM MgCl$_2$, incubated at 37° for 15 min, the extract is separated from the glass beads by decantation, the beads are washed three times with 150-ml portions of the buffered sodium chloride solution, and the washings and extract are combined. The final volume of the extract is 1 liter. Unbroken cells and cell debris are removed from the extract by centrifugation at 10,000 g for 10 min in the cold.

Membranes containing II-B^{Lac} activity are separated from the soluble proteins by centrifugation for at least 2 hr at 200,000 g at 0–4°. The membranes are layered over a compact opaque pellet of cell wall material, and the translucent, amber membrane layer is carefully separated with a spatula from the dense pellet. The membranes are then washed twice by resuspension (followed by centrifugation) in a solution containing 1 mM Tris-HCl, pH 8.3, 1 mM DTT, and 1 mM EDTA. The membrane preparation is finally resuspended in this buffer and stored in small aliquots at −18°.

HPr from *Staphylococcus aureus*

Purification

All fractionation steps are conducted at 0–4°, and centrifugations at 25,000 g unless otherwise specified.

Step 1. Streptomycin Treatment. The supernatant fraction of the crude extract (1000 ml) is diluted to 20 mg of protein per milliliter and treated with streptomycin sulfate to a final concentration of 4% at pH 7.5. The mixture is stirred for 12 hr, and the precipitate is removed by centrifugation at 20,000 g for 30 min and discarded. The supernatant solution is dialyzed against 0.01 M potassium phosphate, pH 7.5, containing 0.1 mM DTT and 0.1 mM EDTA (buffer E) for 12 hr, during which time a copious precipitate forms; this is removed by centrifugation, and the solution is dialyzed against fresh buffer E for an additional 24 hr. The small amount of precipitate that forms during the second dialysis is also removed.

Step 2. Acid Precipitation of Enzyme I and Factor III^{Lac}. The solution from step 1 (about 1500 ml) is adjusted to pH 4.0 with 1 M sodium acetate buffer, pH 3.6, continuously stirred over a period of 30 min, and centrifuged. The precipitate, containing Enzyme I and Factor III^{Lac}, is dissolved in 100 ml of buffer E, adjusted to pH 7.5 with KOH, and used for the isolation of these two proteins. The supernatant solution contains HPr and is adjusted to pH 7.5 with KOH.

Step 3. Ammonium Sulfate Fractionation. The supernatant solution from step 2 is adjusted to 65% of saturation by adding powdered ammonium sulfate over a period of 1–2 hr with continuous stirring; during this step the apparent pH (glass electrode) is maintained at 7.5 with NH$_4$OH. After 30 min, the precipitate is sedimented by centrifugation and discarded. Additional ammonium sulfate is added as described above until saturation is achieved, and the solution is stirred for 10 hr. The precipitate is collected and dissolved in a minimal volume of buffer E (usually about 30 ml).

Step 4. Sephadex Chromatography. The ammonium sulfate fraction is applied to a Sephadex G-75 column (5 × 80 cm) that has been equilibrated with buffer E containing 0.05 M NaCl. The column is eluted with the same buffer at a flow rate of 0.5 ml/min. The fractions, 15 ml, are first tested for SO$_4^{2-}$; since HPr is eluted slightly before the ammonium sulfate, this provides a simple method for monitoring the elution. The Sephadex chromatography removes traces of Enzyme I and Factor III^{Lac} from the HPr.

Step 5. DEAE-Cellulose Chromatography. The combined fractions from step 4 are diluted twofold with buffer E, and applied to a DEAE-cellulose (DE-23, Whatman) column (2.5 × 25 cm) previously equilibrated with the buffer. The column is eluted with a linear gradient of 0 to 0.2 M KCl in buffer E (1500 ml total). HPr is eluted at about 0.1 M KCl.

Step 6. Hydroxyapatite Chromatography. The active fractions from the DEAE-cellulose column are pooled, diluted 10-fold with water, and applied to a column of hydroxyapatite (Bio-Rad) (3 × 10 cm) that had been previously equilibrated with 0.001 M potassium phosphate buffer, pH 7.0.

TABLE I
PURIFICATION OF HPr FROM *Staphylococcus aureus*

Step	Total protein (mg)	Apparent total activity (units)[a]	Apparent specific activity (units/mg protein)	Yield (%)
1. 4% Streptomycin sulfate supernatant	ND[b]	ND	ND	ND
2. Acid (pH 3.6) supernatant	8200	1480	0.18	(100)
3. Ammonium sulfate (65–100% fraction)	1600	1100	0.69	74
4. Sephadex G-75 eluate	140	1200	8.6	81
5. DEAE-cellulose eluate	56	1000	18.0	70
6. Hydroxyapatite eluate	30	900	30.0	60

[a] One unit of apparent activity equals the amount of protein required to catalyze the conversion of 1 μmol of thiomethyl β-D-galactoside (TMG) to TMG-phosphate, under the reaction conditions described. Aliquots of all fractions are assayed simultaneously.
[b] ND, not determined. Modified from reference 1, with permission.

The column is eluted with a linear gradient of potassium phosphate buffer, pH 7, 0.001 to 0.02 M (2000 ml total). The HPr activity is eluted in a single peak, at approximately 0.008 M potassium phosphate, that coincides with a single protein peak. The specific activity is constant throughout the peak.

Representative samples from the active fractions are checked for purity by disc gel electrophoresis.[7] A single protein band is observed with each fraction except those at the leading edge of the peak.

Step 7. Concentration Step. The homogeneous fractions from step 6 are combined and concentrated by adsorption of HPr to a small DEAE-cellulose column, which is eluted with 0.3 M KCl in buffer E.

A summary of the purification procedure is given in Table I.

Properties

The HPr protein of *Staphylococcus aureus* is a small single polypeptide with molecular weight about 9000. It contains no cysteine or tryptophan and has two tyrosines and a single histidine residue. When permitted to react with Enzyme I and an excess of [^{32}P]PEP, 1 mol of HPr accepts a maximum of 1 (0.93) mol of phosphoryl group; the equilibrium constant for this reaction is about 10. The phosphoryl group is probably linked to

[7] B. Anderson, N. Weigel, W. Kundig, and S. Roseman, *J. Biol. Chem.* **246**, 7023 (1971).

N-1 of the imidazole moiety of the histidine residue, since the linkage is acid-labile and alkali-stable, its hydrolysis is readily catalyzed by pyridine, and it is hydrolyzed (at pH 3.5, 46°) at the same rate as the *Escherichia coli* P-HPr.

III^{Lac} from Staphylococcus aureus [8]

Purification

Buffer F, used most frequently during the purification procedure, contains 10 mM potassium phosphate, pH 7.4, 0.1 mM EDTA, and 0.5 mM dithiothreitol. The composition of the other buffers is given where they are used. Unless otherwise specified, all purification steps are performed at temperatures between 0 and 4°, and all centrifugations at 25,000 g for 15 min.

Preliminary Steps. Growth and breakage of cells, removal of membranes from the soluble fraction, and streptomycin treatment of the soluble fraction are performed as described above.

Step 1. Ammonium Sulfate Precipitation. The dialyzed streptomycin supernatant (see step 1 of the purification of HPr, above) (volume about 1 liter, protein concentration about 5 mg/ml) is treated with sufficient ammonium sulfate to saturate the solution completely. The resulting suspension is stirred overnight at 4°, the precipitate is collected by centrifugation, the pellet is resuspended in a minimal volume of buffer B containing 50 mM KCl (about 70 ml), and the resulting suspension is dialyzed against the same buffer until all the protein dissolves (2–3 hr).

Step 2. Sephadex Chromatography. The ammonium sulfate fraction from step 1 is applied to a Sephadex G-75 column (10 × 100 cm) first equilibrated with buffer F containing 50 mM KCl. The column is eluted with the same buffer at 1.2 ml/min, thus separating Enzyme I, Factor III^{Lac}, and HPr from one another. The Factor III^{Lac} peak also contains phospho-β-galactosidase.

Step 3. Acid Precipitation. One volume of 2 M sodium acetate buffer, pH 4.65, is added over a 5-min period at 0° to 19 volumes (about 700 ml) of Factor III^{Lac} from step 2. The resulting solution, pH 4.7, is stirred for 20 min at 0°, and the precipitate is collected by centrifugation (20,000 g for 20 min). The pellet is redissolved in 2 volumes of 50 mM potassium phosphate buffer containing 0.1 mM EDTA and 1.0 mM dithiothreitol.

Step 4. Second Ammonium Sulfate Fractionation. An ammonium sulfate solution saturated at 4° is used for this step; the solution is first adjusted with ammonia so that a fourfold dilution is at pH 7.0 when mea-

[8] J. B. Hays, R. D. Simoni, and S. Roseman, *J. Biol. Chem.* **248**, 941 (1973).

sured with a glass electrode at room temperature. The ammonium sulfate concentrations described below are given as percentage of saturation at 23° although the actual fractionation steps are conducted between 0 and 4°. The suspension from step 3 is adjusted to a protein concentration of 5 mg/ml. The solution (usually about 120 ml) is maintained at 0° while the saturated ammonium sulfate solution is added over 1 hr with continuous stirring to 53% of saturation. The suspension is stirred for a further 30 min at 0°, and the precipitate is removed by centrifugation and discarded. Saturated ammonium sulfate solution is slowly added to the supernatant fraction to 75% of saturation. The suspension is stirred for 30 min at 0°, and the precipitate is collected by centrifugation and redissolved in a minimal amount (usually 6–10 ml) of 10 mM potassium phosphate buffer, pH 7.4, containing 0.1 mM EDTA, 1.0 mM dithiothreitol, and 50 mM KCl.

Step 5. Sephadex G-75 Chromatography. The fraction from step 4 is applied to a column (2.5 × 80 cm) of Sephadex G-75, previously equilibrated with buffer E containing 50 mM KCl and eluted with the same buffer at 0.25 ml/min. The Factor III^{Lac} in the pooled fractions from this column is about 95% pure, but still contains phospho-β-galactosidase activity.

Step 6. Hydroxyapatite Chromatography. For the procedures used in this step, increasing concentrations of potassium phosphate buffer, pH 6.8, are used, each containing 0.5 mM dithiothreitol. The pooled fractions from step 5 are dialyzed overnight against 1 mM phosphate buffer. The solution is applied to a column (2 × 7 cm; about 20 ml) of hydroxyapatite previously equilibrated with the same buffer. The column is washed with 360 ml of this buffer and then washed with 200 ml of 10 mM phosphate buffer. Little, if any, Factor III^{Lac} is eluted at this concentration of phosphate. The column is then eluted with 400 ml of 16 mM phosphate buffer at a flow rate of 0.5 ml/min. The Factor III^{Lac} is eluted as a single broad peak free from phospho-β-galactosidase. The precise phosphate concentrations required at each step of the hydroxyapatite chromatography vary from one preparation to another, and preliminary small-scale experiments are required to determine the optimal buffer concentration for each preparation. In a more recent experiment, elution of a small (8 ml) hydroxyapatite column by a linear phosphate gradient (400 ml, 2 to 25 mM) gave excellent results.

Step 7. DEAE-Cellulose Chromatography. The combined fractions from step 6 are adjusted to pH 7.4, and applied to a 40-ml DEAE-cellulose (DE-52, Whatman) (chloride form) column (1.5 × 20 cm) previously equilibrated with buffer E. The column is washed with 100 ml of buffer E containing 0.1 M KCl and eluted with a linear gradient of 0.1 to 0.2 M KCl in buffer E containing 1.0 mM dithiothreitol (2 liters, total volume) at a flow rate of 0.3 ml/min. Factor III^{Lac} activity is eluted by this gradient as a

TABLE II

PURIFICATION OF FACTOR III^{Lac} FROM *Staphylococcus aureus*[a]

Step	Volume (ml)	Total protein (mg)	Total activity (units)[b]	Relative specific activity (units/mg × 10²)
4% Streptomycin sulfate supernatant	1060	5610	307	5.5
1. Saturated ammonium sulfate pellet	115	4840	278	5.7
2. Sephadex G-75 eluate	700	560	193	35
3. Acid (pH 4.7) pellet	77	129	168	130
4. Ammonium sulfate, 53–75% fraction	6	45	79	175
5. Sephadex G-75 eluate	100	16.5	65	394
6. Hydroxyapatite eluate	318	12.0	39	325
7. DE-52 eluate	46	5.7	21	370

[a] Details of the purification procedure are described in the text. Aliquots of the III^{Lac} fractions are collected at each step and are stored at $-18°$. They are simultaneously thawed, dialyzed overnight against buffer E, and assayed as described.

[b] One unit is the quantity of Factor III^{Lac} that, when added to a standard assay mixture, results in the formation of 1 nmol of ^{14}C-labeled TMG-phosphate in 30 min. The units do not represent a true absolute activity, since TMG-P formation is proportional to the concentration of Enzyme II^{Lac} as well as Factor III^{Lac}. Since Factor III^{Lac} is a substrate, rather than a true enzyme, it is in fact not possible to define an absolute unit of catalytic activity. Modified from reference 8, with permission.

single broad peak. Disc gel electrophoresis of aliquots from the active fractions shows that almost all of them contain a single protein. Contaminating bands appear only at the leading and trailing edges of the entire III^{Lac} peak. The III^{Lac} fractions free of contaminating bands are pooled and dialyzed against buffer E with 1.0 mM dithiothreitol and then concentrated either by lyophilization or by adsorption on a 1-ml DEAE-cellulose column (DE-52, Whatman) followed by elution with 10 ml of buffer E containing 2.0 mM dithiothreitol and 0.3 M KCl.

A summary of the purification procedure is given in Table II. Steps 6 and 7 do not result in an increase in specific activity but are necessary for the removal of small (less than 5%) amounts of contaminants revealed as minor bands on disc gel electrophoresis.

Properties

The lactose-specific phosphocarrier protein of *Staphylococcus aureus*, Factor III^{Lac}, has a native molecular weight of about 35,000. Based on an estimated molecular weight of 12,000 under denaturing conditions (equilibrium sedimentation in 4.3 M guanidinium chloride, analytical gel filtration chromatography in 8 M guanidinium chloride), Factor III[Lac] appears to have the unusual number of three identical subunits. Electrophoresis in SDS–polyacrylamide yields an erroneous molecular weight estimate of 9200, presumably because complexes of small proteins with SDS resemble spheres rather than rods, as is the case with larger proteins. Sulfhydryl labeling experiments show one cysteine per monomer (2.6 per trimer); amino acid analysis reveals no tryptophan, but two tyrosine and four histidine residues per monomer. Partial amino acid sequencing confirms the hypothesis of identical subunits.

When permitted to react with Enzyme I, HPr, and excess [^{32}P]PEP, Factor III[Lac] accepts nearly three (2.5) phosphoryl groups. In contrast to HPr, the linkage appears to be at the N-3 position in the imidazole ring of one of the histidine residues. This assignment is based on hydrolysis studies (acid lability, alkali stability, and catalysis by hydroxylamine and pyridine with approximately equal efficiencies), and on chromatography of an alkaline hydrolyzate of ^{32}P-labeled Factor III[Lac] on Dowex 1 (HCO$_3$$^{-}$), in the presence of synthetic 1- and 3-phosphohistidine.

The circular dichroic spectrum of Factor III[Lac] and its optical rotatory dispersion reveal an unusual secondary structure. The estimated α-helical content, 60%, exceeds the values for most globular proteins.

Section VII

Mono- and Disaccharide-Binding Proteins

[75] Maltose-Binding Protein from *Escherichia coli*

By Odile K. Kellermann and Thomas Ferenci

The maltose-binding protein is a periplasmic protein of *Escherichia coli* involved in the transport of maltose and maltodextrins across the bacterial envelope[1] as well as in the chemotaxis[2,3] toward these sugars.

Binding Assay

Principle. Maltose binding is measured by equilibrium dialysis using [14C]maltose.

Reagents

[14C]Maltose
Tris-HCl, pH 7,3
$MgCl_2$
NaN_3
EDTA
Na_2CO_3
Bray's solution[4]: naphthalene, 60 g; 2,5-diphenyloxazole (PPO), 4 g; 1,4-di-4-methyl-5-phenyloxazole benzene, 0.2 g; absolute methanol, 100 ml; ethylene glycol, 20 ml; *p*-dioxane to 1 liter.

Procedure. Small dialysis bags are prepared from dialysis tubing boiled in 5% Na_2CO_3, thoroughly washed with distilled water, and stored in 1 mM EDTA. Two identical dialysis bags, from which liquid is removed as much as possible using Kleenex paper, are filled with 100 μl of binding protein solution and put into flasks containing 10 ml of 10 mM Tris-HCl pH 7, 3, 2 mM $MgCl_2$, 3 mM NaN_3, and [14C]maltose (1–10 μM). Dialysis is allowed to occur for 15–18 hr at 4° with gentle shaking. Equilibrium is attained under these conditions if the concentration of binding protein is lower than 40 μM. Duplicate 25-μl samples are removed from the outside medium for determination of free maltose and from each of the two bags for the determination of bound maltose plus free maltose. The samples are transferred into 10 ml of Bray's solution and counted in a scintillation counter. If different samples are dialyzed against the same concentration of [14C]maltose, the concentration of bound maltose in the

---

[1] O. Kellermann and S. Szmelcman, *Eur. J. Biochem.* **47**, 139 (1974).

[2] G. L. Hazelbauer and J. Adler, *Nature (London), New Biol.* **230**, 101 (1971).

[3] G. L. Hazelbauer, *J. Bacteriol.* **122**, 206 (1975).

[4] G. A. Bray, *Anal. Biochem.* **1**, 279 (1960).

bags is proportional to the concentration of active binding protein. To determine the actual concentration of active binding protein in any given sample, it is necessary to perform dialysis against several concentrations of [^{14}C]maltose, from about 0.25 μM to 75 μM, and to perform a Scatchard analysis.[5]

Purification Procedures

The maltose-binding protein is a periplasmic protein released from bacteria by a cold osmotic shock treatment. From osmotic shock fluids, extensive purification of the maltose-binding protein was first achieved by ion-exchange chromatography[1] and later by affinity chromatography.[6] The two procedures are presented.

Growth of Cells and Osmotic Shock Treatment. Any maltose-positive strain of *Escherichia coli* K12 can be used. It is grown in M63 maltose medium (13.6 g of KH$_2$PO$_4$, 2 g of (NH$_4$)$_2$SO$_4$, 0.2 g of MgSO$_4 \cdot$ 7 H$_2$O, 0.005 g of FeSO$_4 \cdot$ 7 H$_2$O, 4 g of maltose per liter) supplemented with the necessary nutritional requirements. Shock fluids are prepared from an exponentially growing culture harvested at 10^9 bacteria per milliliter. The cells are washed twice with one-tenth the culture volume of 10 mM Tris-HCl, pH 7.1, 30 mM NaCl, and then submitted to the osmotic shock procedure.[7] The cell pellet is suspended in one-twentieth the culture volume of 30 mM Tris-HCl, pH 7.1, containing 20% sucrose and 0.1 mM EDTA and mixed with a magnetic stirrer at room temperature for 15 min. The cells are centrifuged at 12,000 rpm for 10 min, and the pellet is spread on the sides of the beaker. An ice-cold hypotonic solution of 0.1 mM MgCl$_2$ (100 ml per liter of growth medium) is poured on the bacteria. The mixture is stirred for 10 min at 4° and centrifuged. The pH of the shock fluid supernatant is adjusted with 10 mM Tris-HCl, pH 7.1. The shock fluid is then filtered through a Millipore filter (type HA 0.45 μM) to remove whole bacteria and concentrated four- to fivefold by lyophilization or by ultrafiltration over a Diaflo PM-10 membrane (Amicon).

QAE-Sephadex Chromatography (Original Procedure). The shock fluid is chromatographed on QAE-Sephadex in a two-step procedure. Sixty to seventy milligrams of concentrated shock fluid protein, dialyzed against 2 mM MgCl$_2$, 10 mM Tris-HCl, pH 8, are applied onto a QAE-Sephadex column (2.5 × 45 cm) equilibrated with the same buffer. The column is extensively washed with the starting buffer and eluted stepwise with 300

[5] M. Schwartz, O. Kellermann, S. Szmelcman, and G. L. Hazelbauer, *Eur. J. Biochem.* **71**, 167 (1976).

[6] T. Ferenci and U. Klotz, *FEBS Lett.* **94**, 213 (1978).

[7] N. G. Nossal and L. A. Heppel, *J. Biol. Chem.* **241**, 3055 (1966).

PURIFICATION OF THE MALTOSE-BINDING PROTEIN[a]

Step	Total protein[b] (mg)	Maltose bound[c] (pmol/mg protein)	Yield (%)	Purification (fold)
Osmotic shock fluid (concentrated)	56	440	100	1
QAE-Sephadex, 1st column	19.5	1400	96	3
QAE-Sephadex, 2nd column	4.1	8100	83	15

[a] This purification was from 12 g (wet weight) of bacteria.
[b] The concentration of protein was determined by the method of Lowry et al.[8]
[c] In these assays the concentration of [14C]maltose was 5 μM.

ml of 0.25 M KCl in the same buffer, at a flow rate of 10 ml/hr. Fractions of 5 ml are collected and assayed for binding activity. The peak fractions are pooled, dialyzed against 2 mM MgCl$_2$, 10 mM Tris-HCl, pH 8, and poured over another identical QAE-Sephadex column. Elution was then achieved with a linear salt gradient from 50 mM to 0.5 M KCl in the same buffer. The maltose-binding protein eluted as a single sharp peak between 0.2 and 0.25 M KCl. The fractions containing the highest specific activity were pooled, dialyzed against 10 mM Tris-HCl, pH 8, 2 mM MgCl$_2$ and stored at $-30°$.

The maltose-binding protein is purified 15 times relative to the initial shock fluid. However, since the shock fluid contains only 3% of the total bacterial proteins, the binding protein is really purified 300 times.

A typical purification is summarized in the table.

Affinity Chromatography (Alternative Procedure). Chromatography is performed on cross-linked amylose[9] prepared as follows. One gram of amylose (Sigma grade III, from potato) is evenly suspended in 4 ml of water and warmed to 50° in a water bath; 6 ml of 5 N NaOH followed by 3 ml of epichlorohydrin are added with rapid stirring. The suspension, solidified into a gel within about 10 min, is allowed to cool to room temperature and left for 45 min. The gel is washed with water, transferred to a Waring blender, and fragmented by brief blending as a suspension in water. The broken gel is transferred to a measuring cylinder and suspended in three separate 100-ml washes of buffer containing 50 mM glycine-HCl, 0.5 M NaCl (pH 2.0). Between each wash, the gel is allowed to settle to separate fines remaining in the supernatant. Two further fining cycles in water and two in 10 mM Tris-HCl buffer (pH 7.2) are performed. Finally,

[8] O. M. Lowry, N. J. Rosebrough, A. L. Farr, and R. J. Randall, *J. Biol. Chem.* **193**, 265 (1951).
[9] I. Matsumoto and T. Osawa, this series, Vol. 34, p. 329.

the gel is stored by suspension in 10 mM Tris-HCl (pH 7.2) containing 0.02% (w/v) sodium azide.

The concentrated shock fluid (60 mg protein) is extensively dialyzed against 10 mM Tris-HCl pH 7.2 and applied on a column (2.6 × 9.5 cm) of cross-linked amylose equilibrated with the same buffer. Ascending elution is recommended to avoid compression of the soft gel. After extensive washing, with 10 mM Tris-HCl, 1 M NaCl buffer, pH 7.2, the binding protein can be eluted using 10 mM Tris-HCl, 10 mM maltose pH 7.2. The recovery of active protein from shock fluids is near-quantitative.

Properties

Both methods yield a protein that is more than 90% pure as estimated by electrophoresis on polyacrylamide gels in different conditions. Immunoelectrophoresis of the purified protein using antiserum prepared against fully induced sonicated E. coli K12 shows only one immunoprecipitation band. One milligram of pure protein has an absorbance $A_{280}^{1\,mg/ml} = 1.47$.

A molecular weight of 40,000 to 44,000 was determined by polyacrylamide gel electrophoresis in presence of sodium dodecyl sulfate. The weight-average molecular weight obtained by sedimentation equilibrium was 37,000. In electrofocusing, the protein has an isoelectric point at pH 5.0.

The affinity of the protein for various ligands was determined by one of two methods. The first is the technique of equilibrium dialysis described under Binding Assay, either by measuring the binding directly or by determining the competition for the binding of maltose. The second method is based on the fact that the binding of substrate induces a specific fluorescence change in the protein.[10] The protein exhibits binding affinities with dissociation constants in the micromolar range both for maltose and for linear maltodextrins up to 10 glucose residues.[11] Cyclohexaamylose and macromolecular $\alpha(1 \rightarrow 4)$-linked glucans are also bound with high affinity, as are the maltose analogs[12] methyl α-maltoside and 5-thiomaltose. Glucose, methyl α-glucoside, isomaltose, lactose, and trehalose are not bound.

Experiments performed by fluorescence spectroscopy demonstrated that the rate of dissociation of the complex between maltose and the protein is high. The normally observed low rate of exit of maltose from a dialysis bag containing maltose-binding protein results from the fact that a

[10] S. Szmelcman, M. Schwartz, T. J. Silhavy, and W. Boos, Eur. J. Biochem. 65, 13 (1976).
[11] C. Wandersman, M. Schwartz, and T. Ferenci, J. Bacteriol. 140, 1 (1979).
[12] T. Ferenci, J. Bacteriol. 144, 7 (1980).

molecule of free ligand has a much higher probability of reacting with a free binding site than of leaving the dialysis bag (retention effect[13]).

The maltose binding protein is coded by gene *malE,* one of the five cistrons[14] of the *malB* region involved in the transport of maltose and maltodextrins across the envelope of *E. coli.* The functioning of maltose binding protein in maltose transport seems to involve an interaction both with an outer membrane protein, the product of gene *lamB,* and at least one protein of the cytoplasmic membrane, the product of gene *malF, malG,* or *malK.*[15] The interesting multifunctional protein is also the subject of preliminary X-ray crystallographic analysis.[16]

[13] T. J. Silhavy, S. Szmelcman, W. Boos, and M. Schwartz, *Proc. Natl. Acad. Sci. U.S.A.* **72,** 2120 (1975).

[14] T. J. Silhavy, E. Brickman, P. J. Bassford, M. J. Casadaban, H. A. Shuman, V. Schwartz, L. Guarente, M. Schwartz, and J. R. Beckwith, *Mol. Gen. Genet.* **174,** 249 (1979).

[15] T. Ferenci and W. Boos, *J. Supramol. Struct.* **13,** 101 (1980).

[16] F. A. Quiocho, W. E. Meador, and J. W. Pflugrath, *J. Mol. Biol.* **133,** 181 (1979).

[76] L-Arabinose- and D-Galactose-Binding Proteins from *Escherichia coli*

By Robert W. Hogg

Periplasmic binding proteins are functional components of many nutrient uptake systems observed in gram-negative bacteria. They apparently mediate a recognition process in the periplasmic space that facilitates specific nutrient passage to the cytoplasmic membrane and subsequently into the cell. The L-arabinose-binding protein (ABP) has been identified and isolated from *Escherichia coli* B/r[1] and from *E. coli* K12.[2] The D-galactose-binding protein (GBP) was identified, and extensive studies have established the role of this protein in galactose uptake[3,4] and as a chemotaxis receptor.[5]

The strains of *E. coli* B/r and *E. coli* K12 used in studies of the arabinose binding protein, produce this protein when induced by L-arabinose. However, in addition to inducing ABP, L-arabinose induction also increases the normally low-level constitutive synthesis of the

[1] R. W. Hogg and E. Englesberg, *J. Bacteriol.* **100,** 429 (1969).

[2] R. Schleif, *J. Mol. Biol.* **46,** 185 (1969).

[3] Y. Anraku, *J. Biol. Chem.* **242,** 793 (1967).

[4] W. Boos, *Eur. J. Biochem.* **10,** 66 (1969).

[5] G. L. Hazelbauer and J. Adler, *Nature (London), New Biol.* **230,** 101 (1971).

galactose-binding protein, and thus the two proteins can be purified simultaneously from the same cells and separated by column chromatography as described.

Reagents

Potassium phosphate buffer, 0.1 M, pH 7.8
2-Mercaptoethanol

Assay

The simplest assay for the detection of binding proteins is by equilibrium dialysis of small samples of protein against a buffered solution containing radiolabeled ligand. The arabinose binding protein binds both L-arabinose and D-galactose with an affinity of $10^{-7} M$. The galactose-binding protein binds D-galactose and D-glucose with a similar affinity.[6] Therefore, to a 100-ml volume of 10 mM potassium phosphate buffer, pH 7.8, containing 15 mM 2-mercaptoethanol or 0.1 M dithiothreitol, add 0.1 μM radiolabelled L-arabinose or D-galactose containing 2.5 μCi. This level of radioactivity is sufficient to allow accurate determinations when 50 μl of sample and/or dialyzate are prepared for counting.[6] Dialysis samples (100–250-μl samples) equilibrate in a matter of hours, and appropriate samples can be removed for protein determination and counting.

After isolation of pure proteins, specific antibodies can be prepared and immunoassays used for detection of microgram quantities.[6]

Growth and Induction of Cells

Strains suitable for the preparation of ABP and/or GBP include all strains of *Escherichia coli* B/r or K 12 that lack isomerase (*araA*) or kinase (*araB*). *Escherichia coli* B/r *araA39* when grown as described will yield 1–2 mg of ABP and 0.5–1 mg of GBP per gram of wet cell paste.[6] Wild-type cells are less suitable, as yields of purified protein are considerably lower as a result of catabolite repression. Strains are available from the *E. coli* Genetic Stock Center (Department of Microbiology, Yale University School of Medicine, 310 Cedar Street, New Haven, Connecticut).

Cells are grown in a medium containing, per liter of water, 3 g of KH_2PO_4, 7 g of K_2HPO_4, 1 g of $(NH_4)_2SO_4$, 0.1 g of $MgSO_4 \cdot 7 H_2O$, 0.03 g of $MnCl_2 \cdot 4 H_2O$, and 10 g of casamino acids (Difco Laboratories, Detroit, Michigan). Three hours after addition of the inoculum at a 1:50 dilution, the cells are induced by addition of L-arabinose to 0.2%, and growth is continued at 37° for 15–18 hr with aeration. For example, 100

---

[6] R. G. Parsons and R. W. Hogg, *J. Biol. Chem.* **249**, 3602 (1974).

liters of medium (aerated at a rate of 5 CFM and stirred at 200 rpm), inoculated with 2 liters of log-phase cells at 5 PM, and induced by addition of 200 g of L-arabinose at 9 PM, will yield 750 g of packed cells when harvested at 8 AM the following day. It is not necessary to wash the cells before use, and they may be frozen rapidly for storage by forming a pancake of cell paste in a plastic bag and covering with crushed Dry Ice.

Purification Procedure

Preparation of Cell-Free Extract. Seven hundred grams of frozen or fresh cell paste are suspended in 700 ml of cold 10 mM potassium phosphate buffer, pH 7.8, 15 mM 2-mercaptoethanol. The suspended cells are disrupted in a French pressure cell at 0°. Approximately 5 mg each of crude DNase and RNase are added to the extract and allowed to stir for 5 min. The extract is diluted to 3 liters with 10 mM phosphate buffer, pH 7.8, and centrifuged (12,000 rpm for 90 min at 0°). The supernatant is decanted and brought to 3 liters with buffer.[6,7]

Ammonium Sulfate Precipitation. Solid $(NH_4)_2SO_4$ (35 g/100 ml) is slowly added, over a period of 45 min. During this process the pH of the undiluted supernatant is maintained at 7.8 by addition of 10% ammonium hydroxide. The mixture is allowed to stir for 1 hr at 0° prior to centrifugation (12,000 rpm for 30 min).

The supernatant is recovered, and an additional 34.4 g of solid ammonium sulfate per 100 ml of original solution are added. The solution is allowed to stir at 0° overnight. The 100% ammonium sulfate saturation pellet is recovered by centrifugation, resuspended in a minimal volume (100–200 ml) of 10 mM phosphate buffer, pH 7.8, and dialyzed against three or four changes of 10 volumes of 10 mM potassium phosphate buffer, pH 7.8, containing 15 mM 2-mercaptoethanol at 0°.[6,7]

DEAE Ion-Exchange Chromatography. A slurry of DEAE-cellulose, equilibrated with 10 mM KHPO$_4$ buffer at pH 7.8 is collected on a Büchner funnel to form a dried packed layer approximately 5 inches in diameter by 1.5–2 inches deep. The dialyzed resuspended precipitate of the 100% ammonium sulfate fractionation step is allowed to pass through the DEAE-cellulose layer followed by two rinses of 150–200 ml each of 10 mM buffer. Liquid trapped in the DEAE-cellulose is drawn out under vacuum and the filtrate is retained. If crystals of the purified protein are desired, the filtrate should be concentrated by ultrafiltration. If one simply wants pure protein, the filtrate can be lyophilized and resuspended in 20–50 ml of H_2O before proceeding.

The concentrated filtrate from the DEAE-cellulose passage is dialyzed

---

[7] R. W. Hogg and M. A. Hermodson, *J. Biol. Chem.* **252**, 5135 (1977).

extensively against 1 mM potassium phosphate buffer, pH 7.8, containing 15 mM 2-mercaptoethanol and then applied to a column of DEAE-cellulose (2.5 × 100 cm), packed under 3 psi of N_2 and equilibrated with the same buffer. As ABP and GBP have a weak affinity for DEAE-cellulose, it is advantageous to rinse the column with several hundred milliliters of water before applying the sample. The sample is washed in to the column with 250 ml of water (or 1 mM phosphate buffer) and the column attached to a 3-liter gradient consisting of 1.5 liters of 1 mM phosphate buffer, pH 7.8, and 1.5 liters of 50 mM KCl in the same buffer. Both reservoirs contain 15 mM 2-mercaptoethanol. The gradient is pumped through the column at a flow rate of 100 ml/hr, and 10-ml fractions are collected. Two somewhat overlapping protein peaks elute between 3 and 10 mM KCl. ABP elutes before GBP. The first third will contain ABP free of GBP, the middle third will contain both ABP and GBP, and the last third will contain GBP free of ABP. The separated binding proteins are obtained by pooling samples on the basis of arabinose- and glucose-binding capacities.

Sizing Column Chromatography. The pooled material from the DEAE-cellulose column is concentrated (by vacuum dialysis or lyophilization—see earlier note), suspended in a minimal volume of buffer (2–5 ml), and weighted with sucrose if necessary. This fraction is then applied to a 2.5- × 100-cm column of Sephacryl 200 (or Sephadex G-150) equilibrated with 0.1 M potassium phosphate buffer, pH 7.8, containing 15 mg of dithiothreitol per liter and eluted with the same buffer mixture; 5-ml fractions are collected. Material from the gel filtration column is usually pure as judged by standard Davis[8] or SDS[9]–polyacrylamide gel electrophoresis. Protein in the overlap regions and the sides of peaks can be collected and purified by repeating the ion-exchange and gel-filtration steps.

Alternative Methods. Binding proteins in general can also be purified by a method that involves subjecting exponentially growing cells to osmotic shock as described by Nossel and Heppel[10] and modified by Willis *et al.*[11] The shockate obtained by this procedure is concentrated, dialyzed, and subjected to appropriate column chromatography. This procedure provides a starting material that contains fewer protein species; however, we have found that from some strains excessive carbohydrate polymers are present that clog initial column steps, and the volumes of the shockate

[8] B. J. Davis, *Ann. N.Y. Acad. Sci.* **121**, 404 (1964).
[9] U. K. Laemmli, *Nature (London)* **227**, 680 (1970).
[10] N. Nossal and L. A. Heppel, *J. Biol. Chem.* **241**, 3055 (1966).
[11] R. C. Willis, R. G. Morris, C. Cirakoglu, G. D. Schellenberg, N. H. Gerber, and C. E. Furlong, *Arch. Biochem. Biophys.* **161**, 64 (1974).

seem more difficult to deal with than are the ammonium sulfate precipitation and the first DEAE passage steps described above.

Comments. The L-arabinose- and D-galactose-binding proteins isolated by the above procedures are stable indefinitely (at least 4 years) when stored lyophilized at 0°. In solution, at high protein concentrations, the arabinose-binding protein has a tendency to dimerize; however, this process can be reversed by addition of guanidine-HCl (2 *M*) and dialysis of diluted solutions against buffer containing dithiothreitol or 2-mercaptoethanol.

Both proteins recognize and bind their respective ligands over an extended range of temperature, pH, and ionic strength.[6]

The primary sequence of the arabinose-binding protein has been determined[7] and the structure solved to 2.8 Å resolution.[12] *In vitro* synthesis has established the presence of a precursor form and the sequence of the 23-residue signal peptide has been determined.[13] The primary sequence of the galactose-binding protein has also been determined[14] and the structure resolved to 4.1 Å resolution[15] at time of writing.

[12] F. A. Quiocho, G. L. Gilliland, and G. N. Phillips, *J. Biol. Chem.* **252**, 5142 (1977).
[13] V. G. Wilson and R. W. Hogg, *J. Biol. Chem.* **255**, 6745 (1980).
[14] W. Mahoney, R. W. Hogg, and M. A. Hermodson, *J. Biol. Chem.* **256**, 4350 (1981).
[15] F. A. Quiocho and J. W. Pflugrath, *J. Biol. Chem.* **255**, 559 (1980).

[77] Ribose-Binding Protein from *Escherichia coli* [1]

By CLEMENT E. FURLONG

The ribose-binding protein of *Escherichia coli* is required for the high affinity transport of ribose, and for chemotaxis toward this sugar.[2]

Assay Methods

Principle. Ribose-binding protein does not chemically alter ribose, it simply binds the sugar in a reversible interaction. Most assays that will measure the reversible interaction of protein with ligand can be used to

[1] Supported by Grant GM 15253 and Career Development Award K04 GM 00452 from the U.S. Public Health Service.
[2] D. R. Galloway and C. E. Furlong, *Arch. Biochem. Biophys.* **184**, 496 (1977).

measure this association.[3] The two most commonly used procedures are equilibrium dialysis[4] and the nitrocellulose filter binding assay.[5]

Reagents

Tris-HCl, 1 M, pH 7.5 (23°)

[^{14}C]Ribose (approximately 50 μm, 50 μCi/μmol)

Procedure for Membrane Filter Assay. The most convenient assay to use during purification procedures is the nitrocellulose filter assay. A small sample of column fraction (5–100 μl or more) is brought to a final concentration of 5–10 μM [^{14}C]ribose and filtered under vacuum through a piece (square or round) of wetted Schleicher & Schuell B-6 membrane (other nitrocellulose membrane filters are also satisfactory; however, some have reduced binding capacity). The membrane filter is washed under vacuum with 0.25 ml of 50 mM Tris-HCl, pH 7.5. The filters are dried and counted by scintillation spectrometry. One to three milliliters of scintillation solution usually suffice. This assay does not work well for crude extracts owing to competition for protein binding sites on the filter.

Procedure for Equilibrium Dialysis Assay. More quantitative measurements may be made by equilibrium dialysis in sacs or microdialysis chambers.[4,6] The latter are more convenient to use, but may give higher apparent K_D values due to dilution of the ligand specific activity by protein-donated unlabeled ligand.

For this assay 0.05–1.0 ml of protein solution in 50 mM Tris, pH 7.5, is introduced to one side of the membrane of a microdialysis cell and the same volume of [^{14}C]ribose-containing buffer to the opposite side. For measuring the total ribose binding activity, the final free ribose concentration (determined by counting a sample from the nonprotein side of the membrane after equilibration) should be 5–10 μM.[4] For assaying column fractions, the initial ribose concentration should be 0.2–0.5 μM. For K_d determinations, the final free ribose concentration is varied between 10^{-8} and 5 × 10^{-6} M. The bound ribose is determined by counting samples of equal volume from each side of the membrane after equilibration, and subtracting the calculated ribose concentration of the nonprotein side from that of the protein side. This value gives the amount of ribose bound by the protein contained in the removed sample.[4]

[3] D. L. Oxender and S. C. Quay, *Methods Membr. Biol.* **6**, 183 (1976).

[4] C. E. Furlong, R. G. Morris, M. Kandrach, and B. P. Rosen, *Anal. Biochem.* **47**, 514 (1972).

[5] C. E. Furlong and J. H. Weiner, *Biochem. Biophys. Res. Commun.* **38**, 1076 (1970).

[6] R. C. Willis, R. G. Morris, C. Cirakoglu, G. D. Schellenberg, N. Gerber, and C. E. Furlong, *Arch. Biochem. Biophys.* **161**, 64 (1974).

Purification Procedure

Choice of Strain

Ribose-binding protein can be purified from ribose-positive strains of wild-type *Escherichia coli* (e.g., K12) grown on ribose as a carbon source. However, owing to the relatively high cost of ribose, it is more convenient to use strains that constitutively express the ribose operon. The glycerol growth procedure described below is designed for constitutive strains. Such strains may be isolated by the procedure of Hayashi *et al.*[7] Basically, overnight tryptone agar plates that each contain approximately 600–1200 mutagenized[8] colonies are sprayed with a solution of 100 mM ribose, 0.2% chloramphenicol, taking care not to spray in such a manner as to wash away the colonies. After 2 hr, the plates are sprayed with a 1%, solution of triphenyltetrazolium dye. The ribose constitutive colonies turn red while the wild-type colonies remain white. The resulting red colonies should be purified[8] and tested for full constitutivity by comparing the ribose transport[9] of mutant cells grown in ribose with the transport of cells grown in glycerol. The medium described below is used for growth of the cells with 2% carbon sources.[2]

Growth of Cells

Prepare the following medium (per 1000 ml, pH 6.5): K_2HPO_4, 13.9 g; KH_2PO_4, 4.1 g; NH_4Cl, 5.35 g; $MgSO_4$, 0.5 g; H_2O, 900 ml. Autoclave, then add 100 ml of 50% sterile glycerol[10] and 1 ml of sterile trace elements.[11,12]

Procedure. Cells may be grown to high densities in special media,[6,12] including the one described here,[6] which provides up to 30 g or more of wet cell paste per liter of culture under ideal conditions. The following

[7] S. Hayashi, J. P. Koch, and E. C. C. Lin, *J. Biol. Chem.* **239**, 3098 (1964).

[8] J. Miller, "Experiments in Molecular Genetics." Cold Spring Harbor Lab., Cold Spring Harbor, New York, 1972.

[9] R. C. Willis and C. E. Furlong, *J. Biol. Chem.* **249**, 6926 (1974).

[10] Glycerol may be used as a carbon source for strains that constitutively express the ribose operon. Strains that are inducible for ribose utilization should be grown with ribose as sole carbon source. The culture should be started with 2% ribose, and additional ribose (20% sterile stock solution) should be added as required for optimal growth.

[11] Trace elements are prepared by heating a 100-ml solution containing 0.05 g of $CaCl_2$, 0.018 g of $ZnSO_4 \cdot 7 H_2O$, 0.016 g of $CuSO_4 \cdot 5 H_2O$, 0.018 g of $CoCl_2 \cdot 6 H_2O$, 2.0 g of disodium EDTA (ethylenediaminetetraacetic acid) until the components dissolve. After cooling, 2 g of $FeCl_2 \cdot 6 H_2O$ and 100 μl of 1 M HCl are added. This mixture is then sterilized by filtration and frozen in aliquots.

[12] A. D. Kelmers, C. W. Hancher, E. F. Phares, and G. D. Novelli, this series, Vol. 20, p. 3.

growth procedure is convenient for the usual laboratory schedule. A 10-liter fermenter is inoculated with the desired strain and grown to an A_{600} of 2–3. The cells are chilled overnight, then restarted early the next day. At this point, 4 drops of antifoam (Dow Corning FG10) are added, along with an additional 50 ml of sterile 50% glycerol. Agitation and aeration will need to be increased throughout the run. The pH is maintained at or above 6.5 by adding concentrated NH_4OH. This is most easily done with a pump coupled to a pH controller; however, it may be done manually. At an A_{600} of 10–12, 10 ml of sterile 2 M $MgSO_4$ and 200 ml of sterile 50% glycerol are added. Antifoam is added as required. (A fermenter equipped with mechanical foam breaker allows for the minimum use of antifoam.) An additional 200 ml of 50% glycerol are added at an A_{600} of 15 and again at an A_{600} of 25. The NH_4OH added to control the pH also serves to provide additional nitrogen to the culture. The cell growth will slow drastically when the A_{600} reaches 20–35 depending on the strain used and the rates of aeration and stirring. As soon as growth ceases, the cells are prepared for the osmotic shock procedure.

Osmotic Shock Procedure

Principle. Ribose-binding protein is located in the periplasmic space. The osmotic shock procedure developed by Leon Heppel and co-workers[6,13,14] selectively releases the periplasmic proteins from cells without releasing cytoplasmic proteins. Since the periplasmic proteins account for only 3–4% of total cell protein, the osmotic shock procedure provides an initial 30-fold purification of ribose-binding protein.

Release of Periplasmic Proteins. Just prior to harvest, the culture is brought to a final concentration of 0.03 M Tris-HCl (pH 7.3, 23°), 0.03 M NaCl by addition of the appropriate amount of 1 M stock solutions. The cells are then harvested (most conveniently in a 4° continuous-flow centrifuge), weighed, and resuspended (in a large blender controlled by a variable voltage transformer) in 7.5–10 volumes (w/v) of 0.033 M Tris-HCl (pH 7.3). An equal volume of 0.033 M Tris-HCl (pH 7.3) containing 40% sucrose (w/v) and 2 mM EDTA is added slowly with stirring. The cells are again harvested (most conveniently by continuous-flow centrifugation at 4°). The plasmolyzed pellet is rapidly resuspended in 20 volumes of cold deionized water by mixing for 45 sec in large blender. To avoid cell lysis, concentrated $MgCl_2$ solution is then immediately added to give a final concentration of 1 mM. After an additional 15 sec of blending, the cells are removed from the osmotic shock fluid by centrifugation (continu-

[13] H. C. Neu and L. A. Heppel, *J. Biol. Chem.* **240,** 3685 (1965).
[14] N. G. Nossal and L. A. Heppel, *J. Biol. Chem.* **241,** 3055 (1966).

ous flow, 4°), and the supernatant solution (osmotic shock fluid) is saved. The ribose-binding protein is purified from the clarified osmotic shock fluid as described below. The centrifuge flow rates should be adjusted to provide clear supernatant solutions in each of the above steps. If a conventional centrifuge is used, at least 10 min at 10,000 g will be required to sediment the cells. Longer times (or slower flow rates) will be required to remove cells from the sucrose solution. The remaining steps including buffer calibrations are carried out at ambient temperature unless indicated otherwise.

Purification of the Ribose-Binding Proteins

The pH of the osmotic shock fluid is adjusted to 4.5 with 7.5% acetic acid. After stirring for 30 min, the precipitate is removed by centrifugation. The supernatant solution is applied at the ratio of 7 mg of protein per milliliter of exchanger bed volume to a high-flow-rate CM-cellulose column (Whatman CM-32 prepared according to the manufacturer's instructions) equilibrated with 10 mM sodium acetate buffer, pH 4.5. The high flow rate is achieved by rigorously removing the fines from the exchanger before packing the column and using a height-to-diameter ratio of 2–5 : 1. A sintered-glass funnel attached to a large vacuum flask works well for this procedure. The column is eluted with 3–4 column volumes of 0.3 M NaCl in 10 mM sodium acetate, pH 5.0. Fractions of half a column volume are collected and assayed. The first fraction usually contains very little if any binding activity. Those fractions containing significant ribose-binding activity are pooled, and the pH is adjusted to 8.0 with concentrated NH$_4$OH. This step should result in approximately 80% recovery of the binding activity.

The resulting pool is dialyzed against a solution that contains sufficient ammonium sulfate to bring the entire system (volume inside sac plus volume outside) to a final concentration of 242.3 g/liter. The contents of the sac are occasionally agitated to assure a uniform concentration of ammonium sulfate. After equilibration (6–10 hr), the insoluble protein is removed by centrifugation and the supernatant solution is returned to the rinsed dialysis sac. Sufficient ammonium sulfate is added to the solution outside the sac to saturate the entire system (approximately 525 g per liter of original total system volume), and ammonium carbonate buffer (pH 8) sufficient to provide a concentration of 50 mM for the entire system. The dialysis is continued for an additional 6–10 hr with occasional mixing of the sac contents. At equilibrium, ammonium sulfate crystals should be present outside the sac. This step results in an additional concentration of the sac contents and produces a pellet that is easily collected by centrifugation. Before collecting the pellet, the contents of the sac are mixed to resuspend sedimented protein precipitate. Any precipitate remaining in the sac may be rinsed out with some of the saturated ammonium sulfate solution. The precipitated protein is collected by centrifugation (20 min, 10,000 g), resuspended in 1/200 the original volume, and dialyzed against several changes of 10 mM Tris-HCl, pH 8.3(4°). The dialysis sac should be tied to allow for a 50% increase in volume. Thorough dialysis is required for the ribose-binding protein to interact with the DEAE column.

DEAE-cellulose (Whatman DE-23) is prepared according to the manufacturer's instructions. After rigorously removing the fine particles from the prepared exchanger, a column with a height-to-diameter ratio of 30 : 1 is packed and equilibrated with 10 mM Tris-HCl, pH 8.3. The dialyzed material is loaded on the column at a ratio of 7 mg of protein per milliliter of bed volume. The column is washed with 3 column volumes of column equilibration buffer and eluted with a gradient of 0 to 50 mM NaCl in 10 mM Tris-HCl, pH 8.3. The yield for this step should be 50–90%.

Ribose-binding activity is measured by either of the assays or by antibody diffusion[6] if antiserum is available. The fractions containing ribose-binding activity are analyzed by sodium dodecyl sulfate gel electrophoresis,[15] acid urea gel electrophoresis,[2] or slab gel isoelectric focusing (e.g., on LKB gels). The fractions containing pure ribose-binding protein are pooled, dialyzed against deionized water, and lyophilized.

Variations of this purification scheme have been published.[9,16] If the DEAE column does not provide protein of adequate purity, SP-Sephadex[9] or high-pressure liquid chromatography with gel filtration through a Waters I-125 column[16] should be used as a final step in the procedure. In the latter procedure, ribose-binding protein elutes after approximately 6 min. The yield of protein depends on the strain used. Typically, 60–300 mg should be obtainable from 1 kg of wet cell paste.

Properties of Ribose-Binding Protein

The pure protein should exhibit a single band with a molecular weight of approximately 29,500 when analyzed by SDS gel electrophoresis, or a single band with a pI of approximately 6.6 when analyzed by slab gel electrofocusing. The dissociation constant for ribose is approximately 0.14 μM when determined by equilibrium dialysis under conditions where no effort is made to remove unlabeled bound ligand before the determination. The protein binds 1 mol of ribose per mole of protein.[9]

[15] U. K. Laemmli, *Nature* (*London*) **227**, 680 (1970).
[16] F. T. Robb and C. E. Furlong, *J. Supramol. Struct.* **13**, 183 (1980).

[78] D-Xylose-Binding Protein (Periplasmic) from *Escherichia coli*

By A. Stephen Dahms, William Huisman, Gerald Neslund, and Clarence Ahlem

This protein is a periplasmic component that functions in osmotic shock-sensitive D-xylose transport in *E. coli*.[1]

Assay Method

Principle. The xylose-binding protein was assayed by equilibrium dialysis.

Procedure. For routine binding analysis, samples (0.4 ml) are dialyzed against 10 ml of 0.01 M potassium phosphate buffer, pH 6.9, containing D-[U-^{14}C]xylose (2.92 or 4.7 μCi/μmol) at a concentration of 1 μM. Dialysis is carried out at 4° for 12–18 hr; isotopic equilibrium is attained after 3 hr in the absence of the xylose-binding protein. At the end of the dialysis period, the dialysis sacs are removed and cut open; 100-μl samples are withdrawn for protein and ^{14}C determination. Identical samples of dialyzate are taken simultaneously to determine the amount of unbound [^{14}C]xylose. Nanomoles of xylose bound are computed by subtraction of the disintegrations per minute per milliliter (dpm/ml) in the dialyzate from the dpm/ml inside the dialysis sac. Dialysis is also carried out in rotating microdialysis chambers.[2] Dialysis tubing (0.25 inch, Arthur H. Thomas Co., Philadelphia, Pennsylvania) is prepared by boiling 1 mM EDTA (pH 7) for 5 min and rinsing with distilled water prior to boiling for another 5-min period in 95% ethanol and rinsing with 100 ml of distilled water per foot of tubing. The prepared tubing is stored at 4° in 1 mM EDTA (pH 7).

Definition of Unit and Specific Activity. One unit of xylose-binding activity is defined as the amount that binds 1 nmol of [^{14}C]xylose. Specific activity is defined as the number of units per milligram of protein. Protein was determined by absorbance at 595 nm with Coomassie Blue G-250 according to the method of Bradford[3] with bovine serum albumin as the standard.

---

[1] C. N. Ahlem, G. Neslund, and A. S. Dahms, *Fed. Proc., Fed. Am. Soc. Exp. Biol.* (in press).

[2] C. E. Furlong, C. Cirakoglu, R. C. Willis, and P. A. Santy, *Anal. Biochem.* **51**, 597 (1973).

[3] M. M. Bradford, *Anal. Biochem.* **72**, 248 (1976).

METHODS IN ENZYMOLOGY, VOL. 90

PURIFICATION OF D-XYLOSE-BINDING PROTEIN

Fraction	Volume (ml)	Units[a]	Protein (mg/ml)	Specific activity (units/mg protein)		Recovery (%)
				1 μM D-xylose	Saturating D-xylose	
Osmotic shock supernatant	150	187	0.50	1.5	2.5	100
AffiGel Blue	135	169	0.24	3.15	5.25	90
DEAE-cellulose	12	112	0.275	20.7	33.9	60

[a] Calculated at saturating ligand concentrations.

Purification Procedure[1]

Growth of the Organism. The organism (*E. coli* K12) was supplied by Dr. G. Bertani of the Karolinska Institute and grown aerobically at 37° in a xylose mineral salts medium containing 7 g of K_2HPO_4, 3 g of KH_2PO_4, 1 g of $(NH_4)_2SO_4$, 0.5 g of Na_3 citrate · 3 H_2O, 0.1 g $MgSO_4$ · 7 H_2O, and 5 g D-xylose per liter (pH 7.0). The xylose and mineral salts are autoclaved separately. Stirred Erlenmeyer flasks or sparged carboys are inoculated with 0.01 volume of a late log phase culture grown in the same medium. The cells are harvested by centrifugation at 4000 g after 18–24 hr and processed immediately.

Preparation of Periplasmic Fluid. All operations are carried out at 0–4° unless otherwise indicated. Cold osmotic shock is carried out by the procedure of Neu and Heppel.[4] Cells are washed in 10 mM Tris–30 mM NaCl, pH 7.3. Pellets are resuspended in 33 mM Tris–0.1 mM EDTA–20% sucrose in proportions of 80 ml per gram wet weight of cells and allowed to stand at room temperature for 20 min. The suspension was centrifuged at 8000 g for 30 min. The resulting pellet is rapidly suspended in cold 0.5 mM $MgCl_2$ and stirred for 10 min before centrifugation at 12,000 g for 20 min. The supernatant is carefully decanted under strong light and recentrifuged for 10 min to remove cell debris. The resultant supernatant (referred to as the crude shockate) is concentrated by pressure dialysis (PM-10 membrane, Amicon Corp., Lexington, Massachusetts) to protein concentrations of 0.5–2 mg/ml.

Negative Adsorption on AffiGel Blue. AffiGel Blue (100–200 mesh, Bio-Rad, Richmond, California) is equilibrated with 10 mM potassium phosphate (pH 6.9) at 0–4°. Negative adsorption is carried out batchwise. AffiGel Blue is gently mixed with 2–3 volumes of concentrated shockate

[4] H. C. Neu and L. A. Heppel, *J. Biol. Chem.* **240**, 3685 (1965).

for 10–15 min. The solution is centrifuged at 10,000 *g* for 10 min, and the supernatant is carefully removed.

DEAE-Cellulose Chromatography. The supernatant from the preceding step is applied to a column (2 × 7.5 cm, 22°) of DEAE-cellulose (Whatman DE-23) that has been equilibrated with 10 m*M* potassium phosphate, pH 6.9. The protein is eluted with a 250-ml linear gradient of 0 to 0.15 *M* NaCl in the above buffer. Fractions (4 ml) are collected at the rate of 1 ml/min. Fractions containing xylose binding activity are stored at −20°. The protein is purified 15-fold with an overall recovery of 60%.

A typical purification is summarized in the table.

Properties

Substrate Specificity. Substrate specificity was determined by competition of unlabeled carbohydrates with [14C]xylose. The specificity for D-xylose is apparently absolute. The binding kinetics are monophasic; the K_d was found to be 0.63 μM at 4°. Carbohydrates not bound (at concentrations up to 100-fold the D-xylose concentration) were D-galactose, L-arabinose, L-fucose, L-galactose, L-rhamnose, D-lyxose, D-xylonate, D-ribose, D-arabinose, xylitol, L-xylose, D-glucose, *o*-nitrophenyl-β-D-xylopyranoside, *p*-nitrophenyl-α-D-xylopyranoside, and methyl α- and methyl β-D-xylopyranosides. Methyl α-D-xylofuranoside was bound with an K_d of 33 μM, suggesting that the furanose configuration may be the preferred isomer. Neither tetraiodofluorescein (50 μM) nor NAD$^+$ (1 m*M*) diminished D-xylose binding.

Stability. Crude shockate or purified binding protein was stable for over 2 months at −20° and for over 1 week at 4°. Binding activity was diminished by heating at 80° (half-life, 10 m*M* potassium phosphate buffer, pH 6.9, 5 min).

Molecular Properties. Purified D-xylose-binding protein migrates electrophoretically as a single component upon SDS–polyacrylamide 9% gels with a molecular weight of 37,500 ± 1000 corresponding to 34 nmol of xylose bound per milligram of binding protein at saturation. The protein was free of xylose dehydrogenase,[5] xylokinase,[6] xylose isomerase,[7] and xylose reductase[8] activities and was found only in xylose-grown cells. Gel filtration on Sephadex G-200 yielded a molecular weight of 28,000 ± 2000. The binding protein contains no cysteine and two tryptophans per mole; intrinsic fluorescence (λ_{ex} 284, λ_{em} 345) is increased 32% in the presence of

[5] A. S. Dahms, *Biochem. Biophys. Res. Commun.* **60**, 1433 (1974).
[6] F. J. Simpson, this series, Vol. 9 [81].
[7] K. Yamanka, this series, Vol. 9 [104].
[8] C. Chiang and S. G. Knight, this series, Vol. 9 [37].

D-xylose and saturates at 0.5 μM D-xylose, suggesting the presence of a tryptophan in the area of the xylose binding site. Xylose-binding protein binds to tetraiodofluorescein–agarose (with elution by 1 mM NAD$^+$) but does not bind to Cibacron Blue F3G-A agarose; no tetraiodofluorescein interaction can be detected spectroscopically in contrast to the L-arabinose-binding protein.[9] The purified protein has a 280 : 260 ratio of 1.09 and an $A_{1\,cm}^{1\%}$ of 12.9. Oxidation of one of the two tryptophan residues with N-bromosuccinimide[10] resulted in a complete loss of binding activity.

pH Optimum. Xylose-binding activity was constant over the 5–8 pH range. Binding activity was reduced to 50% at pH 3.8 and 9.5, respectively.

Cofactors. Binding was not affected by NEM, iodoacetic acid, 2-mercaptoethanol, MgCl$_2$, or EDTA at 1 mM or by high ionic strength (0.4 M NaCl).

[9] R. W. Hogg, M. A. Hermodson, F. A. Quiocho, G. N. Philips, G. L. Gilliland, and M. F. Newcomer, *J. Supramol. Biol., Suppl.* **1**, 602 (1977).
[10] T. F. Spande and B. Witkep, this series, Vol. 11 [61].

Section VIII

Procedures Yielding Several Enzymes

[79] Purification of All Glycolytic Enzymes from One Muscle Extract

By ROBERT K. SCOPES and ANN STOTER

Muscle tissue cytoplasm is a very convenient source of the glycolytic enzymes; approximately 75% of the soluble protein can be accounted for by the enzymes from phosphorylase to lactate dehydrogenase,[1] plus the two other enzymes intimately involved in adenylate charge control, creatine kinase and adenylate kinase. The tissue is easily extracted by homogenization, and contains only low amounts of proteases that might modify the enzymes. Many of these enzymes are very useful for enzymic analytical procedures, and over the years this laboratory has developed a variety of multienzyme purification procedures to satisfy our analytical requirements. In addition we have needed all the enzymes for experiments on reconstitution of the muscle glycolytic system.[2-6] This chapter describes a scheme whereby it is possible to purify all the glycolytic enzymes (except hexokinase), creatine kinase, adenylate kinase, and phosphorylase *b* kinase from the one extract of rabbit muscle. Over 20 years ago Czok and Bücher[1] described a similar process in which eight of these enzymes were crystallized, using only ammonium sulfate and selective heat denaturation as fractionation procedures. We also use these, but with the development of ion-exchange chromatography since then, and of affinity elution procedures,[7,8] more flexibility is introduced. Although the present total scheme may prove to be impractical because of limitations on equipment and personnel, it is possible to carry it out with yields for individual enzymes varying from 20 to 80%. However, for one particular enzyme it is often possible to omit several intermediate steps; thus the details described below do not necessarily represent the easiest way of purifying a particular enzyme.

Enzyme Assays

The measurements of activity of each of the 17 enzymes described are by well-documented continuous spectrophotometric methods that it is

[1] R. Czok and T. Bücher, *Adv. Protein Chem.* **15**, 315 (1960).
[2] R. K. Scopes, *Biochem. J.* **134**, 197 (1973).
[3] R. K. Scopes, *Biochem. J.* **138**, 119 (1974).
[4] R. K. Scopes, *Biochem. J.* **142**, 79 (1974).
[5] G. R. Eagle and R. K. Scopes, *Arch. Biochem. Biophys.* **208**, 593 (1981).
[6] G. R. Eagle and R. K. Scopes, *Arch. Biochem. Biophys.* **210**, 540 (1981).
[7] R. K. Scopes, *Biochem. J.* **161**, 253 (1977).
[8] R. K. Scopes, *Biochem. J.* **161**, 265 (1977).

unnecessary to repeat in detail here.[2,7,9] Buffers were chosen to give near optimum activity. These were 30 mM Tris, triethanolamine, imidazole, pH 8.0, 7.5, or 7.0 adjusted to pH with HCl; in most cases they included 50 mM KCl and 3 mM magnesium acetate, whether or not these (cations) are necessary. Reaction mixtures are made up containing all components except one substrate; approximately 0.2 mg/ml of bovine serum albumin is present in each mixture. One milliliter of the mixture is measured into a 10-mm pathlength cuvette of 1.2-ml capacity, and 10–50 μl of an appropriate stock solution of the missing substrate are added. Any blank rate due to contamination of the coupling enzymes with the enzyme being assayed is noted, then a sample of the solution being tested is added (e.g., 0.1 to 10 μl). If necessary the test solution is diluted 10- or 100-fold with imidazole buffer, pH 7.0, containing 1% bovine serum albumin, so that a more accurate sample can be taken using a larger-capacity microliter syringe. Measurements are made at room temperature (21–28°) and corrected to 25° using a factor of 1.06 per degree, which is a good average value for these enzymes. Activities are expressed in units of micromoles of product formed per minute.

Extraction Procedure

Chilled muscle tissue from the back and hind legs of two rabbits, totaling 800–900 g, is diced, and connective tissue is removed as far as possible, before homogenization with 2.5 ml/g of extraction buffer, generally 30 mM potassium phosphate, pH 7.0, containing 1 mM EDTA and 10 mM 2-mercaptoethanol. However, if it is desired to purify both phosphorylase b kinase and phosphofructokinase, this buffer is not ideal, because, in the presence of endogenous ATP in fresh muscle, phosphate extracts a portion of the actomyosin. The actomyosin in the extract precipitates in the first step, along with these two kinases, which can adversely affect their further purification. A better extractant for these is of lower ionic strength and contains fluoride, which not only prevents extraction of actomyosin, but also activates phosphofructokinase[10] and blocks glycolytic acidification during the preparation of the extract. But at this lower ionic strength some other enzymes, notably aldolase, may adsorb to the insoluble material and so be obtained in lower yield. If phosphorylase b kinase is not required, the actomyosin-containing precipitate from the phosphate extract can be worked up for phosphofructokinase using the

[9] H.-U. Bergmeyer, "Methods in Enzymatic Analysis," 2nd Ed. Academic Press, New York, 1970.
[10] K.-H. Ling, F. Marcus, and H. A. Lardy, *J. Biol. Chem.* **240**, 1893 (1965).

isopropanol heat treatment described by Ling *et al.*[10] Because of the better overall yields, the phosphate extractant is used in the procedure described.

After homogenization with cold buffer for approximately 1 min, the homogenate is stirred gently at room temperature for 30 min, keeping the pH at or above 7.0 by addition of 1 M Tris base. It is then centrifuged in the cold at 5000 g for 30 min, and the extract is poured through Miracloth (Chicopee Mills Inc., Milltown, New Jersey) to remove fat particles.

Primary Fractionation

Keeping the extract at about 5°, the pH is lowered to 5.8 using 1 M acetic acid; after 10–15 min, the precipitate that forms is removed by centrifugation as above. The supernatant should now be a clear pink and is ready for the ammonium sulfate fractionation. The precipitate, which contains both phosphorylase *b* kinase and phosphofructokinase, is suspended in 20 mM Tris-HCl buffer, pH 8.0, containing 0.2 mM EDTA and 10 mM 2-mercaptoethanol, to a final volume of about one-tenth the original extract volume, and the pH is adjusted to 7.5. This is called fraction P.

To the supernatant, 275 g per liter of ammonium sulfate is added with steady stirring over a period of 10 min, and the stirring is continued for a further 20–30 min. This process can be carried out at room temperature; the temperature of the cold extract does not rise above 15°. The precipitate is collected by centrifugation (at 0–5°) at 5000 g for 25 min, and the volume of the supernatant is measured before adding the next batch of ammonium sulfate. The precipitate is taken up in a minimum volume of Tris-HCl buffer as above and transferred to dialysis tubing; this is fraction A (0–1.81 M ammonium sulfate).

Fraction B (1.81–2.44 M ammonium sulfate) is collected in the same way after dissolving 100 g of ammonium sulfate per liter of supernatant from A, stirring, and centrifuging as above. The precipitate is dissolved to a volume of about 0.15 times the original extract volume, and the pH is adjusted to 7.0 using 1 M Tris base.

Fraction C (2.44–3.02 M ammonium sulfate) is collected in the same way, after dissolving 100 g of ammonium sulfate per liter of supernatant from B, stirring, centrifuging and dissolving as above. After the precipitate is dissolved in buffer, the pH is adjusted to 7.0.

To the supernatant from fraction C, a further 35 g of ammonium sulfate per liter are dissolved, and the pH of the solution, as measured directly, is adjusted to 7.5 using concentrated ammonia solution. The container is covered and left in the cold overnight, or for up to 48 hr, before collecting

TABLE I
APPROXIMATE PERCENTAGE OF DISTRIBUTION OF THE ENZYMES AFTER PRIMARY FRACTIONATION

Enzyme	EC number	Abbre- viation	P	A	B	C	D	S n fr
Phosphorylase	2.4.1.1	PH	—	*90*	5	—	—	
Phosphorylase *b* kinase	2.7.1.38	PH*b*K	*80*	10	—	—	—	
Amylo-1,6-glucosidase	3.2.1.33	A1,6G	20	*60*	—	—	—	
Phosphoglucomutase	2.7.5.1	PGM	—	10	*50*	40	—	
Phosphoglucose isomerase	5.3.1.9	PGI	—	15	*60*	25	—	
Phosphofructokinase	2.7.1.11	PFK	*60*	10	20	—	—	
Aldolase	4.1.2.13	ALD	—	15	*85*	—	—	
Triosephosphate isomerase	5.3.1.1	TIM	—	—	5	*40*	30	
Glycerol-phosphate dehydrogenase	1.1.1.8	GOPDH	—	—	*80*	10	—	
Glyceraldehyde-phosphate dehydrogenase	1.2.1.12	GAPDH	—	—	—	20	*60*	
Phosphoglycerate kinase	2.7.2.3	PGK	—	—	15	*60*	10	
Phosphoglycerate mutase	2.7.5.3	PGAM	—	10	*60*	20	—	
Enolase	4.2.1.11	EN	—	—	20	*60*	10	
Pyruvate kinase	2.7.1.40	PK	—	5	*65*	15	—	
Lactate dehydrogenase	1.1.1.27	LDH	—	15	*70*	15	—	
Creatine kinase	2.7.3.2	CK	—	5	25	*60*	10	
Adenylate kinase	2.7.4.3	AMPK	—	10	25	*60*	5	

[a] The italic numbers indicate the fraction used for further purification of each enzyme. Or extract = 100%.

the precipitate by centrifugation. This precipitate is taken up in a small amount of 20 m*M* EDTA, pH 8.0 (NaOH), containing 10 m*M* 2-mercaptoethanol. This is fraction D.

The approximate distribution of activities of the enzymes in these fractions is shown in Table I. Note that in some cases enzymes are distributed over several fractions, and the one chosen is a compromise; different ammonium cuts may be more appropriate if only one enzyme is being purified.

Secondary Fractionation Procedures

Fraction P

Fraction P, if obtained from a fluoride extract, can be worked up for phosphorylase *b* kinase and phosphofructokinase. To the cloudy fraction at pH 7.5, 0.25 volume of room temperature-saturated (4.05 *M*) ammonium sulfate is mixed in, and the mixture is clarified by centrifugation for at least 10^6 g × min (e.g., 35,000 g for 30 min). The supernatant is carefully decanted from the soft precipitate through Miracloth to remove lipid particles. To every milliliter of this supernatant a further 0.15 ml of saturated ammonium sulfate is added; after stirring in the cold for 20 min, the mixture is centrifuged (20,000 g for 15 min). Most of the phosphorylase b kinase is in the precipitate; after dissolving it in a small amount of buffer, it is purified on a gel filtration column (Sepharose 6B or Sephacryl S-300, 5-ml sample applied to 100 cm^3 column), collecting the first major peak eluted. This overall procedure for phosphorylase b kinase is essentially the same as that described by de Lange et al.[11] To the supernatant from the phosphorylase b kinase precipitation, a further 0.7 ml of saturated ammonium sulfate per milliliter is added, and the precipitate is collected as above. Phosphofructokinase is mainly in this fraction, and, after dissolving it in Tris-phosphate buffer, pH 8.0 (0.1 M Tris to pH 8 with H_3PO_4), the fraction is desalted through a column of Sephadex G-25, at least 5 times the volume of sample, then applied to DEAE-cellulose in the Tris-phosphate buffer, using 0.2 ml of adsorbant per milligram of protein. The phosphofructokinase is eluted with 0.3 M Tris-phosphate buffer.[10]

If the extract were in phosphate buffer, the above fractionation scheme would not be so successful because of contamination by actomyosin.

Fraction A

Fraction A contains phosphorylase and a lot of high-molecular-weight proteins as contaminants. For phosphorylase b preparation it is worked up exactly as described by Fischer and Krebs[12] with modifications described by de Lange et al.[11] using repeated recrystallization as a purification procedure. For preparation of amylo-1,6-glucosidase, fraction A is first refractionated with ammonium sulfate, after carefully estimating the amount of ammonium sulfate in the fraction by noting the increase in volume over that of the buffer used to dissolve the precipitate, collecting the 0.8–1.15 M fraction for amylo-1,6-glucosidase and 1.15–1.4 M fraction for phosphorylase b. The phosphorylase b is worked up as above, and amylo-1,6-glucosidase is treated with urea and purified on DEAE-cellulose as described by Watts and Nelson.[13] Amylo-1,6-glucosidase can also be isolated from the first phosphorylase b crystal supernatant if the ammonium sulfate refractionation is not carried out.

[11] R. J. de Lange, R. G. Kemp, W. D. Riley, R. A. Cooper, and E. G. Krebs, *J. Biol. Chem.* **243**, 2200 (1968).
[12] E. H. Fischer and E. G. Krebs, *J. Biol. Chem.* **231**, 65 (1958).
[13] T. E. Watts and T. E. Nelson, *Anal. Biochem.* **49**, 479 (1972).

Fraction B

As can be seen from Table I, 7 of the 17 enzymes are concentrated in fraction B. The procedure for isolating these involves three ion-exchange columns, using affinity elution from two of them. Unless otherwise stated, all ion-exchange and gel-filtration columns described below are run at room temperature.

First a DEAE-cellulose column (Whatman DE-52) is prepared in 20 mM triethanolamine buffer adjusted to pH 7.5 with MOPS.[14] The MOPS should be recrystallized from 80% ethanol before use; it is used as the counterion on DEAE-cellulose, so that the nonadsorbed fraction can be run on to CM-cellulose (Whatman CM-52) without having to change the buffer other than to lower its pH. Fraction B is desalted on a large Sephadex G-25 column preequilibrated in 0.2 mM EDTA, pH 8.0. The total volume of the column should be as much as the volume of the original extract if complete desalting is to be achieved. Thus, from 2 liters of extract, the sample (B) should be in not more than 300 ml, and a column of 1500–2000 ml used. After desalting, triethanolamine is added to 20 mM and the pH is lowered to 7.5 with MOPS; the volume is adjusted to 500 ml with buffer and run onto the DEAE-cellulose column (15–20 cm^2 in cross section, 6–8 cm tall) at 300 ml/hr. Only phosphoglucomutase and glycerol-phosphate dehydrogenase adsorb; all the other enzymes pass through. The column is washed with 150 ml of buffer before commencing a linear gradient using 250 ml of 0.25 M triethanolamine acetate, pH 7.5, mixing into 250 ml of starting buffer. The first two main peaks contain phosphoglucomutase (dephospho and phospho forms), followed by a peak containing glycerol-phosphate dehydrogenase. The phosphoglucomutase peaks are combined, salted out with 50 g of ammonium sulfate per 100 ml, and stored for future purification on CM-cellulose.[8] Glycerol-phosphate dehydrogenase is salted out with 60 g of ammonium sulfate per 100 ml, dissolved in a small volume of 30 mM, pH 6.0, succinate buffer, and saturated ammonium sulfate is added until turbidity increases. This precipitate is removed by centrifugation, and a few drops more of ammonium sulfate are added. Glycerol-phosphate dehydrogenase crystallizes under these conditions.

The fraction that passes through the DEAE-cellulose column is collected, and its pH is lowered to 6.8 using 1 M MOPS. The total volume is doubled by adding water; the buffer now consists of 10 mM triethanolamine (as a cation)–MOPS buffer, pH 6.8, and the same buffer is used to preequilibrate a CM-cellulose column (cross section 30 cm^2,

---

[14] Abbreviations: MOPS, (N-morpholino)propanesulfonic acid; MES, (N-morpholino)-ethanesulfonic acid; TES, N-tris(hydroxymethyl)methyl-2-aminoethanesulfonic acid.

height 15 cm). The sample is run onto the column at 600 ml/hr and washed in with a little pH 6.8 buffer. All the pyruvate kinase, lactate dehydrogenase, and aldolase is adsorbed; phosphoglucose isomerase and phosphoglycerate mutase pass through. The three adsorbed enzymes are all conveniently affinity-eluted at pH 7.5. First 300–400 ml of potassium-TES[14] buffer (20 mM KOH adjusted to pH 7.5 with TES) are washed into the column, followed by 400 ml of the same buffer containing 0.5 mM phosphoenolpyruvate. Pyruvate kinase is affinity-eluted, most of the activity emerging with the first 250 ml of phosphoenolpyruvate buffer. The next 400 ml of buffer contain 0.2 mM NADH, which affinity-elutes lactate dehydrogenase, concentrated in the first 200 ml of NADH buffer. Finally, aldolase is eluted by 400 ml of buffer containing 0.5 mM fructose 1,6-bisphosphate. Each of these three enzymes is salted out (3.4 M ammonium sulfate) and, after being redissolved in pH 7.5 buffer, can be crystallized by adding saturated ammonium sulfate until a turbidity develops. The protein concentration at crystallization should be 10–20 mg/ml. In order to remove trace amounts of potentially interfering enzymes, gel filtration is preferable to crystallization (see Tertiary Purification).

The nonadsorbed fraction containing phosphoglucose isomerase and phosphoglycerate mutase is readjusted to pH 6.8 if necessary and run onto a column of phosphocellulose (Sigma C2258; ~15 cm^2 × 6 cm) in the same triethanolamine–MOPS buffer, at about 400 ml/hr. The sample is washed in, then phosphoglucose isomerase is eluted by inclusion of 0.5 mM glucose 6-phosphate in 150 ml of the same buffer; this is followed by 150 ml of buffer containing both 0.1 mM 2,3-bisphosphoglycerate and 0.2 mM 3-phosphoglycerate, which elutes the phosphoglycerate mutase. Each of these preparations should be sufficiently pure to crystallize after salting out with ammonium sulfate (3.4 M), redissolving in pH 6.8 buffer, and adding saturated ammonium sulfate to turbidity.

Fraction C

Fraction C is used for five enzymes, all being adsorbed, and then eluted from CM-cellulose. Because of the instability of some enzymes at lower pH values, the following three-column method is used. The fraction is desalted as described above for fraction B and made to pH 6.5 using a 10× concentrated stock buffer to give a final composition of 10 mM KOH, 0.5 mM magnesium acetate, 0.1 mM EDTA, pH 6.5, with MES[14] (magnesium acetate is included to stabilize enolase). The pH is carefully checked before running the sample onto a CM-cellulose column (15–20 cm^2 × 10 cm) at 300 ml/hr. Creatine kinase and triosephosphate isomerase pass through; adenylate kinase, enolase, and phosphoglycerate kinase adsorb. After the column is washed with buffer, 0.2 mM ADP is included in 150 ml of pH 6.5 buffer, eluting the adenylate kinase. The column is then washed with potassium-Tricine buffer (10 mM KOH, 0.5 mM magnesium acetate to pH 8.0 with Tricine), and a major peak consisting principally of enolase is eluted as the pH rises. After a red band has left the column, phosphoglycerate kinase is eluted using 150 ml of Tricine buffer containing 0.5 mM 3-phosphoglycerate. Each of these enzymes benefits by a tertiary purification procedure of gel filtration.

The fraction that was not adsorbed on the column at pH 6.5 is lowered to pH 6.0 with 1 M MES. A second CM-cellulose column (15–20 cm^2 × 5 cm) is equilibrated in K–MES buffer, pH 6.0, and run in the cold room. Creatine kinase is liable to be lost by denaturation at room temperature at pH 6.0 on a CM-cellulose column, but in the cold it adsorbs and can be eluted with good recovery. After passing the fraction through the CM-cellulose column, the column is washed with 300 ml of K–MES pH 6.2, followed by 200 ml of the same buffer containing 0.2 mM ATP; the creatine kinase is eluted by the ATP. The solution that passes through the column at pH 6.0 is now lowered to pH 5.5 using 1 M picolinic acid, and the CM-cellulose column used for the creatine kinase preparation is reequilibrated with potassium picolinate, pH 5.5. This buffer consists of 10 mM KOH adjusted to pH 5.5 with picolinic acid. To ensure that the pH of the CM-cellulose is lowered to 5.5, a preliminary wash with 10× more concentrated buffer should be used. Alternatively, the adsorbent can be removed from the column and its pH lowered directly. The sample is run on, and the nonadsorbed fraction is discarded. The column is then washed with K–MES buffer, pH 6.5, and triosephosphate isomerase is eluted as the pH rises. Both creatine kinase and triosephosphate isomerase should be subjected to gel filtration before final storage.

Fraction D

Fraction D when dissolved in the right amount of buffer, starts to crystallize (glyceraldehyde phosphate dehydrogenase) in a few minutes. If too much buffer has been used, a little solid ammonium sulfate should be added until a slight turbidity develops. If too little buffer has been used and the fraction is thickly turbid, slow addition of more buffer should allow the crystals to form. When the crystals are developing well, recognized as a beautiful sheen in the red solution, up to 10% of the volume of saturated ammonium sulfate can be added to ensure that most of the enzyme crystallizes out. After a few hours or overnight, the crystals are collected by centrifugation and are redissolved for recrystallization. Recrystallization can be repeated several times; alternatively, gel filtration will remove traces of myoglobin, phosphoglycerate kinase, and adenylate kinase, which are detrimental to certain analytical uses of the enzyme. The first crystal supernatant contains much triosephosphate isomerase, which can be purified on CM-cellulose at pH 5.5 as described above for fraction C.

Tertiary "Cleaning-Up" Procedures

An enzyme is described as "pure" when it meets certain criteria, which may not be adequate for all purposes. If it shows a single band on electrophoresis, it may be "99% pure," but the 1% of impurity (perhaps spread over several components) may be completely unacceptable for some analytical purposes. When an enzyme is being used in a coupling system in a spectrophotometric enzyme assay, the contaminating level of the enzyme being measured should not exceed 0.1% and preferably should not exceed 0.01% in terms of enzyme units (which may be an even smaller percentage by weight if the contaminant has the higher specific

TABLE II

SPECIFIC ACTIVITIES (AT 25°) OF THE FINAL ENZYME PREPARATIONS AND THE
APPROXIMATE PERCENTAGE RECOVERY OF EACH[a]

Enzyme	Specific activity in original extract (units/mg)	Specific activity of final preparation (units/mg)	Recovery (%)
Phosphorylase	1.1	22	60
Phosphorylase b kinase	ND[b]	ND	ND
Amylo-1,6-glucosidase	ca. 0.02	3	40
Phosphoglucomutase	8	750	25
Phosphoglucose isomerase	6	750	40
Phosphofructokinase	2.5	180	35
Aldolase	2.3	18	50
Triosephosphate isomerase	140	6000	30
Glycerol-phosphate dehydrogenase	0.9	220	60
Glyceraldehyde-phosphate dehydrogenase	20	120	50
Phosphoglycerate kinase	13	700	45
Phosphoglycerate mutase	19	1000	40
Enolase	4.5	80	35
Pyruvate kinase	11	340	50
Lactate dehydrogenase	20	600	40
Creatine kinase	5	100	50
Adenylate kinase	6	1500	50

[a] Also see Scopes.[7,8]
[b] ND, not determined.

activity). On the other hand, noninvolved impurities can be tolerated to a much higher level in analytical uses. Many of the preparations described above can be crystallized after concentration by salting out, and the crystalline preparation meets the condition of being "99% pure."

To remove final traces of undesirable contaminants, repeated crystallization may work but results in substantial losses; it is usually preferable to carry out gel filtration. Fortunately most undesirable contaminants in these preparations have different molecular weights. Gel filtration on Sephacryl S-200 has proved to be the most successful method. The buffer used consists of 50 mM NaCl, 3 mM MgSO$_4$, 10 mM Tris-HCl, 5 mM sodium azide, pH 7.5. Columns are maintained and run at room temperature at speeds of 15–20 cm/hr. Column sizes of 100 ml to 1500 ml are used (length 20–30 times the diameter) to deal with samples (10–20 mg of protein per milliliter) of volumes of 3–50 ml, i.e., columns of volume at least 30 times the volume of the applied sample. Specific activities and approximate percentage recoveries of the final preparations are listed in Table II. Table III lists some examples of the uses of the enzymes and the contaminating enzymes removed by gel filtration or crystallization steps.

TABLE III

SOME ANALYTICAL USES OF THE ENZYMES PURIFIED FROM MUSCLE EXTRACT[a]

Enzyme	Uses		Undesirable trace impurities likely to be present before tertiary processing	Gel filtration needed
	Metabolite assays	Enzyme assays		
PH	Glycogen, P$_i$	—	—	No
PGM	Glycogen, P$_i$, glucose-1-P	PH	—	No[b]
PGI	Fructose-6-P	Fructose-1,6-P$_2$ase	—	Yes
ALD	Fructose-1,6-P$_2$	PFK, GOPDH	LDH, PFK	No[b]
TIM	Fructose-1,6-P$_2$	PFK, ALD	—	Yes
GOPDH	Fructose-1,6-P$_2$, triose-P, glycerol-P	PFK, ALD, TIM	ALD, TIM	No[b]
GAPDH	ATP, 3-P-glycerate, P$_i$	PGK	PGK, AMPK	Yes
PGK	ATP, 3-P-glycerate, P$_i$	GAPDH	GAPDH, AMPK	Yes
EN	2-P-glycerate	PGAM	PGAM, PK	Yes
PK	2-P-glycerate, PEP, ADP, AMP	PGAM, EN, kinases	PGAM, EN, kinases	Yes
LDH	2-P-glycerate, PEP, ADP, AMP, pyruvate, lactate	PGAM, EN, PK, kinases	PGAM, EN, PK, kinases	Yes[b]
CK	Creatine, P-creatine	—	AMPK	Yes
AMPK	AMP	—	CK	Yes

[a] For abbreviations, see Table I.
[b] Recrystallization procedure may be more appropriate.

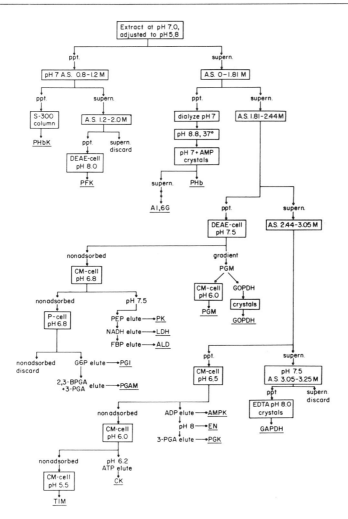

FIG. 1. Flow diagram for primary and secondary fractionation steps. Abbrevations used are as follows: A.S., ammonium sulfate fractionation; ppt., precipitate after centrifugation; supern., supernatant after centrifugation; -cell, -cellulose ion-exchange chromatography. For abbreviations of enzyme and metabolite names, see Table I and text.

In summary, we have been able to develop a fractionation procedure that allows purification of 17 enzymes associated with the conversion of glycogen to lactate in muscle tissue. A flow diagram outlining the process is shown in Fig. 1. Although rabbit muscle was the source, many of the enzymes from muscle of pig, chicken, and some other species have been purified by similar methods. Some notable differences in behavior make these species preferable sources for some of the enzymes. The whole procedure could theoretically be completed in about 3 days if limitless apparatus and personnel were available. However, storage of the redissolved ammonium sulfate precipitates for a few days at 0°, or longer at −30°, allows the whole process to be comfortably extended over a couple of weeks. The enzymes are stored as crystalline suspension in ammonium sulfate at 5°, or frozen concentrated solutions without salt, and mostly retain their activity for many months or years.

[80] Isolation of Several Abundant Muscle Enzymes[1]

By JAMES K. PETELL, ROBERT B. KILLION, and HERBERT G. LEBHERZ

A number of abundant enzymes derived from skeletal muscles are among the best characterized proteins of animal origin. In the present work, we describe rapid, reproducible procedures for the routine isolation of several of these abundant proteins from chicken breast muscle. This tissue is a rich source of the enzymes and is readily obtained commercially. The procedures are applicable for the isolation of these enzymes from large (greater than 100 g) or small (less than 0.5 g) amounts of tissue, and the enzyme purifications can be completed within 1–2 days. In addition, several "shortcuts" are given in which several of the enzymes can be isolated from crude muscle extracts in essentially single steps.

Enzymes. The chicken enzymes isolated were phosphorylase (EC 2.4.1.1), enolase (EC 4.2.1.11), creatine kinase (EC 2.7.3.2), fructose-bisphosphate aldolase (EC 4.1.2.13), glyceraldehyde-3-phosphate dehydrogenase (EC 1.2.1.13), phosphoglyceromutase (EC 2.7.5.3), and triosephosphate isomerase (EC 5.3.1.1). In addition, serum albumin, shown to be identical to muscle β-actinin,[1a] is isolated from muscle specimens during the purification procedures.

Enzyme Assays

The continuous spectrophotometric assays described by others are used to measure the activities of phosphorylase,[2] enolase,[3] aldolase,[4]

[1] This work was supported in part by research grants from the National Institutes of Health (GM-23045) and the Muscular Dystrophy Association and by an American Heart Association Established Investigatorship awarded to H. G. L.

[1a] C. W. Heizmann, G. Muller, E. Jenny, K. J. Wilson, F. Landon, and A. Olomucki, *Proc. Natl. Acad. Sci. U.S.A.* **78**, 74 (1981).
[2] J. Mendicino, H. Abou-Issa, R. Medicus, and N. Katowich, this series, Vol. 42, p. 375.
[3] F. Wold, this series, Vol. 42, p. 329.
[4] H. G. Lebherz and W. J. Rutter, this series, Vol. 42, p. 249.

glyceraldehyde-3-phosphate   dehydrogenase,[5]   phosphoglyceromutase,[6] triosephosphate isomerase,[7] and creatine kinase.[8] All enzyme assays are performed in a final volume of 1 ml at 25° using a Gilford Model 250 spectrophotometer equipped with a temperature-controlled cuvette holder. Assays of creatine kinase are performed both in the presence and in the absence of substrate to correct for "background" activity. All units of enzyme activity are expressed as micromoles of substrate converted to product per minute, and specific activities are defined as units per milligram of protein. Protein concentrations are estimated by the $A_{280} : A_{260}$ method of Warburg and Christian[9] or by reaction with Coomassie Blue.[10]

Preparation of Gel Filtration and Ion-Exchange Columns

Sephadex G-25 (Pharmacia) used for equilibration of samples in appropriate buffers is allowed to "swell" in buffer at room temperature for at least 1 day before preparation of columns. Phosphocellulose (Schleicher & Schuell), carboxymethyl cellulose (Whatman), and DEAE-cellulose (Whatman) are prepared for chromatography as follows: The substituted celluloses are suspended in 2–3 volumes of 0.5 N NaOH; after 15 min, the slurries are filtered through filter paper in Büchner funnels under vacuum. The packed cakes of cellulose are washed under suction 3 or 4 times with 500 ml of 0.5 N NaOH. Then, the celluloses are washed several times with water. The washed cakes are suspended in about 10 volumes of water, and the pH of the slurries is adjusted to the appropriate value by the dropwise addition of 6 N HCl. The suspensions are allowed to settle, and fine particles are removed by decanting the partially settled material. Removal of fines is repeated at least twice, and then the celluloses are equilibrated with the desired buffer before preparation of the columns. After use, the columns are washed with 1 M NaCl to remove residual protein and the celluloses are recycled using the washing procedure described above. Recycled celluloses have the same chromatographic properties as new material.

Preparation of Muscle Homogenates

Fresh or frozen chicken breast muscle is minced into small pieces with scissors. The minces are homogenized in 10 volumes of cold 50 mM

[5] W. S. Allison and N. O. Kaplan, *J. Biol. Chem.* **239**, 2140 (1964).
[6] S. Grisolia and J. Carreras, this series, Vol. 42, p. 435.
[7] P. M. Burton and S. G. Waley, *Biochem. J.* **100**, 702 (1966).
[8] H. M. Eppenberger, D. M. Dawson, and N. O. Kaplan, *J. Biol. Chem.* **242**, 204 (1967).
[9] O. Warburg and W. Christian, *Biochem. Z.* **310**, 384 (1942).
[10] M. M. Bradford, *Anal. Biochem.* **72**, 248 (1976).

potassium phosphate, 1 mM EDTA, 1 mM magnesium acetate, 1 mM 2-mercaptoethanol, pH 7.5. For small muscle samples a glass–glass hand homogenizer may be used for tissue disruption. For muscle samples greater than 3 g, homogenization in a Sorvall Omnimixer or other type of mechanical blender is more convenient. It is important that homogenization buffers contain phosphate or salt (0.2 M) to ensure complete solubilization of the enzymes from particulate muscle material. It is well known that some muscle enzymes interact with myofibrillar proteins[11]; therefore, extraction conditions that result in complete solubilization of the enzymes must be chosen when attempting to quantitate the levels of these proteins in muscle tissues. More than 95% of the total amount of the enzymes considered here is solubilized from myofibrils with the 50 mM phosphate buffer.[12]

The soluble fraction of muscle homogenates, containing the enzymes of interest, is obtained by centrifugation of the crude extracts at 12,000 g for 10 min in a Sorvall or other appropriate centrifuge.

Isolation of the Enzymes

Five grams of frozen chicken breast muscle is used as starting material.

Fractionation by Ammonium Sulfate Precipitation. The enzymes of interest can be separated into three different fractions by ammonium sulfate precipitation. The soluble fraction of muscle homogenates is adjusted to 40% saturation by the slow addition of solid ammonium sulfate (Schwarz-Mann ultrapure, 24.2 g per 100 ml of sample). After stirring for 30 min in the cold, the precipitated protein is collected by centrifugation for 10 min at 12,000 g. The precipitated material contains phosphorylase. The supernatant fraction is titrated to pH 8.4 by the dropwise addition of concentrated ammonium hydroxide, and the sample is adjusted to 70% saturation by the slow addition of solid ammonium sulfate (20.5 g per 100 ml of sample). After stirring for 30 min, the precipitated protein is collected by centrifugation as before. The precipitated material contains serum albumin (β-actinin), creatine kinase, enolase, aldolase, glyceraldehyde-3-phosphate dehydrogenase, and phosphoglyceromutase. The supernatant fraction is adjusted to 90% saturation by the slow addition of solid ammonium sulfate (14.6 g per 100 ml of sample); after stirring for 30 min, the precipitated protein is collected by centrifugation as described above. This fraction contains triosephosphate isomerase.

[11] H. G. Lebherz, M. J. Sardo, J. K. Petell, and J. E. Shackleford, *in* "Protein Turnover and Lysosome Function" (H. Segal and D. Doyle, eds.), p. 655. Academic Press, New York, 1978.

[12] J. K. Petell, M. J. Sardo, and H. G. Lebherz, *Prep. Biochem.* **11**, 69 (1981).

The ammonium sulfate precipitates are dissolved in appropriate buffers (see below) and are used for further purification of the enzymes.

Isolation of Phosphorylase by Ion Exchange on DEAE-Cellulose. Phosphorylase is the major component present in the protein fraction precipitated by 40% saturated ammonium sulfate. Final purification of this enzyme is accomplished by ion-exchange chromatography on DEAE-cellulose essentially as described by Cohen *et al.*[13] The protein precipitate is dissolved in 5 ml of 5 mM β-glycerol phosphate, 0.5 mM EDTA, 1 mM 2-mercaptoethanol, pH 7.0. The preparation is centrifuged to remove any insoluble material, and the supernatant is equilibrated in this buffer by gel filtration on a 2.2- $\times$ 25-cm Sephadex G-25 column or by dialysis. The equilibrated sample is applied to a 1.8- $\times$ 12-cm DEAE-cellulose column, and the column is washed with this buffer until the A_{280} of the eluate is less than 0.05. Then phosphorylase is specifically eluted from the column with 75 mM β-glycerol phosphate, 0.5 mM EDTA, 1 mM 2-mercaptoethanol, pH 7, and fractions containing phosphorylase activity are pooled.

Purification of Serum Albumin (β-Actinin), Phosphoglyceromutase, Creatine Kinase, Enolase, Aldolase, and Glyceraldehyde-3-phosphate Dehydrogenase by Chromatography on Phosphocellulose. The protein preparation that was precipitated between 40 and 70% ammonium sulfate saturation at pH 8.4 is dissolved in 5 ml of 1 mM 2-(N-morpholino)ethanesulfonic acid (MES), 1 mM EDTA, 1 mM magnesium acetate, 1 mM 2-mercaptoethanol, which was titrated to pH 6.0 with 1 N NaOH. Any insoluble material is removed by centrifugation, and the sample is equilibrated in this buffer by gel filtration on a Sephadex G-25 column (2.2 $\times$ 25 cm) or by dialysis. Then, the sample is applied to a 1.8- $\times$ 12-cm phosphocellulose column equilibrated in the above buffer. Under these conditions, serum albumin (β-actinin) is the only protein not retained on the column (Fig. 1). This protein is identical to serum albumin on the basis of subunit molecular weight, mobility during electrophoresis in the presence and the absence of 8 M urea, chromatographic properties, and immunological properties.[12] After collection of the albumin fractions, the column is washed with buffer until the A_{280} of the eluate is less than 0.05. Then the column is washed with buffer containing 50 mM potassium phosphate at pH 6.0.

Under these conditions, phosphoglyceromutase is eluted first from the column immediately followed by creatine kinase (Fig. 1). Fractions (4 ml each) containing these two enzymes are pooled separately. Then, enolase, together with some contaminating proteins, is eluted by washing the column with 100 mM Tris-HCl, 1 mM EDTA, 1 mM magnesium acetate, 1 mM 2-mercaptoethanol, pH 7.8. During this washing procedure, the pH

[13] P. Cohen, T. Duewer, and E. H. Fischer, *Biochemistry* **10**, 2683 (1971).

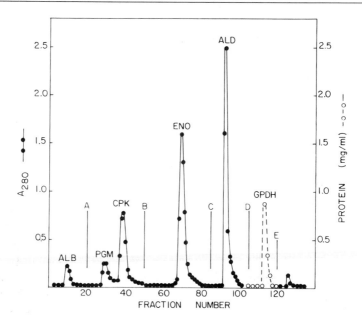

FIG. 1. Purification of serum albumin, phosphoglyceromutase, creatine kinase, enolase, aldolase, and glyceraldehyde-3-phosphate dehydrogenase by chromatography on phosphocellulose. The protein fraction precipitated between 40 and 70% saturation of ammonium sulfate at pH 8.4 was dissolved in and equilibrated in 1 mM 2-(N-morpholino)ethanesulfonic acid, 1 mM EDTA, 1 mM magnesium acetate, 1 mM 2-mercaptoethanol, pH 6.0, and was applied to a 1.8- $\times$ 12-cm phosphocellulose column equilibrated in the same buffer. Albumin (ALB) was the only protein not retained on the column. Phosphoglyceromutase (PGM) followed immediately by creatine kinase (CPK) were eluted from the column with buffer containing 50 mM phosphate (A). Then, enolase (ENO) was eluted with 100 mM Tris-HCl, 1 mM EDTA, 1 mM magnesium acetate, 1 mM 2-mercaptoethanol, pH 7.8 (B). Aldolase (ALD) was eluted with the Tris buffer containing 1 mM fructose-bisphosphate (C). Then, glyceraldehyde-3-phosphate dehydrogenase (GPDH) was eluted with Tris buffer containing 2 mM ATP (D). Remaining protein was eluted with buffer containing 1 M NaCl (E). The elution profile of the column was followed by absorbance at 280 nm or, in the case of GPDH, by assaying fractions with Coomassie Blue. See text for details.

of the eluate rises sharply from 6.0 to 7.8, and enolase is eluted from the column during this pH transition (Fig. 1). Final purification of enolase is accomplished as described below. Then, aldolase is specifically eluted from the column with this 100 mM Tris-HCl buffer, which contains 1 mM fructose bisphosphate (Sigma, sodium salt, grade II) (Fig. 1). Finally, glyceraldehyde-3-phosphate dehydrogenase is eluted from the column with buffer containing 2 mM ATP (Sigma, sodium salt, grade II) (Fig. 1). Note: Because ATP absorbs highly at 280 nm, the elution profile of this dehydrogenase is followed by enzymic activity measurements or by protein determinations by reacting fractions with Coomassie Blue[10] (Fig. 1).

Final purification of enolase is accomplished as follows. The protein sample eluted from the phosphocellulose column with 100 mM Tris-HCl is equilibrated in 10 mM Tris-HCl, 1 mM EDTA, 1 mM magnesium acetate, 1 mM 2-mercaptoethanol, pH 7.5, by gel filtration on a Sephadex G-25 column or by dialysis. Then, the sample is applied to a 1.1- × 10-cm carboxymethyl cellulose column equilibrated in this same buffer. Under these conditions, enolase is not retained on the column whereas contaminating proteins are. Fractions containing enolase activity are pooled.

Isolation of Triosephosphate Isomerase. The protein material that precipitated between 70% and 90% saturated ammonium sulfate is dissolved in 2 ml of the phosphate buffer used for tissue homogenization, which also contained 40% saturated ammonium sulfate. The solution is titrated to pH 5.4 with 1 M acetic acid. The sample is stirred at room temperature for 1 hr; during this time any creatine kinase present in the preparation is precipitated from solution. The precipitated protein is collected by centrifugation and discarded. The supernatant is equilibrated in 1 mM MES, 1 mM magnesium acetate, 1 mM EDTA, 1 mM 2-mercaptoethanol, pH 6.0, by gel filtration on a Sephadex G-25 column or by dialysis. After centrifugation at 12,000 g to remove any insoluble material, the supernatant is applied to a 1.1- × 9-cm phosphocellulose column equilibrated in the same buffer. The column is washed with this buffer until the A_{280} of the eluate is less than 0.05. Then, triosephosphate isomerase is specifically eluted from the column with this buffer, which also contains 50 mM phosphate, pH 6.0. Fractions containing triosephosphate isomerase are pooled.

Summary of the Purification Procedure and Purity of the Isolated Enzymes. The purity of the proteins isolated from chicken breast muscle are routinely tested by electrophoretic analysis in 9% polyacrylamide gels containing 0.1% sodium dodecyl sulfate. The gel reagents and buffer systems are those suggested by Laemmli,[14] and the gels are stained for protein with Coomassie Blue. Densitometric analysis of the stained gels shows that all isolated proteins are greater than 95% pure. Summaries of the recoveries of catalytic activity and specific activities of the purified enzymes are presented in the table. These values are for the isolation of the enzymes from a frozen muscle specimen. The use of fresh muscle as starting material results in considerably higher specific activities of some of the enzymes and also, in some cases, improves yield.[12]

"Shortcuts"

During the course of developing the purification procedures described above, we noticed that several "shortcuts" could be taken if there was

[14] U. K. Laemmli, *Nature (London)* **277**, 680 (1970).

SPECIFIC ACTIVITIES AND YIELDS OF ENZYMES ISOLATED
FROM FROZEN CHICKEN BREAST MUSCLE

Enzyme	Specific activity (IU/mg protein)	Yield (%)
Phosphorylase	11	41
Enolase	93	45
Creatine kinase	253	49
Aldolase	13	60
Glyceraldehyde-3-phosphate dehydrogenase	41	53
Phosphoglyceromutase	295	19
Triosephosphate isomerase	2576	11

interest in isolating only one or several of the enzymes under consideration. For example, aldolase and glyceraldehyde-3-phosphate dehydrogenase can be isolated in single steps as follows: Muscle tissue is homogenized in 5 volumes of 10 mM Tris-HCl, 0.15 M NaCl, 1 mM 2-mercaptoethanol, 1 mM EDTA, pH 7.3. The homogenate is centrifuged and the supernatant is diluted with an equal volume of water to reduce the ionic strength of the preparation. Then, the sample is applied directly to a phosphocellulose column, equilibrated in the above Tris buffer minus NaCl. The column is washed with 100 mM Tris-HCl, 1 mM EDTA, 1 mM 2-mercaptoethanol, pH 7.8, until the pH of the eluate is greater than 7.6 and the A_{280} of the eluate is less than 0.05. Aldolase is then specifically eluted from the column with buffer containing 1 mM fructose bisphosphate followed by elution of glyceraldehyde-3-phosphate dehydrogenase with buffer containing 2 mM ATP. The two enzymes obtained in this fashion are essentially homogeneous as judged by electrophoretic analysis. Also, creatine kinase can be isolated from crude muscle extracts in a single step[12] as follows: The soluble fraction of muscle homogenates is adjusted to 40% saturation by the slow addition of solid ammonium sulfate. After stirring for 30 min in the cold, precipitated proteins are collected by centrifugation and discarded. The supernatant fraction is titrated to pH 5.4 by the dropwise addition of 7 N acetic acid. The sample is stirred at room temperature for at least 1 hr. During this time, creatine kinase specifically precipitates from solution. The kinase is collected by centrifugation and is dissolved in an appropriate buffer. Electrophoretic analysis shows that creatine kinase isolated in this fashion is more than 97% pure.[12]

Comments

Many of the present procedures, with minor modification, may be used for the isolation of these enzymes from other species. For example, the specific elution of aldolase from phosphocellulose with low levels of fructose bisphosphate has been used to isolate this enzyme from a number of animal and plant sources.[15] The ability of ATP to effect the specific elution of glyceraldehyde-3-phosphate dehydrogenase from phosphocellulose presumably reflects the fact that this nucleotide is a competitive inhibitor of the enzyme.[16] The close homology of glyceraldehyde-3-phosphate dehydrogenase derived from diverse biological sources suggests that elution of the enzyme from phosphocellulose with low levels of ATP may become a general method for isolation of this enzyme. The specific precipitation of creatine kinase from 40% saturated ammonium sulfate at pH 5.4 can be used to isolate this enzyme from a number of vertebrate sources.[17] Interestingly, we have been able to isolate arginine kinase, the invertebrate counterpart of creatine kinase, from lobster tail muscle using an identical procedure.[17] The similar solubility behavior of creatine kinase and arginine kinase under these conditions presumably reflects the close evolutionary relationship between these two enzymes.

The versatility and reproducibility of the present procedures allow for the routine isolation of muscle enzymes in "regulation studies" as well as in obtaining large amounts of these proteins for structural analysis.

[15] J. A. Heil and H. G. Lebherz, *J. Biol. Chem.* **253**, 6599 (1978).
[16] S. T. Yang and W. C. Deal, *Biochemistry* **8**, 2806 (1969).
[17] J. K. Petell, R. B. Killion, and H. G. Lebherz, unpublished observation (1981).

[81] Simultaneous Purification of Glyceraldehyde-3-phosphate Dehydrogenase, 3-Phosphoglycerate Kinase, and Phosphoglycerate Mutase from Pig Liver and Muscle

By Klaus D. Kulbe, Harald Foellmer, and Joachim Fuchs

$$GAP + NAD^+ + P_i \underset{}{\overset{GAPDH}{\rightleftharpoons}} 1,3\text{-BPG} + NADH + H^+$$

$$1,3\text{-BPG} + MgADP \underset{}{\overset{PGK}{\rightleftharpoons}} 3\text{-PG} + MgATP$$

$$3\text{-PG} + 2,3\text{-BPG} \underset{}{\overset{PGM}{\rightleftharpoons}} 2\text{-PG} + 2,3\text{-BPG}$$

$$\overline{GAP + NAD^+ + P_i + MgADP \rightleftharpoons 2\text{-PG} + NADH + MgATP + H^+}$$

The three enzymes D-glyceraldehyde-3-phosphate : NAD^+ oxidoreductase (phosphorylating) (EC 1.2.1.12), ATP : 3-phospho-D-glycerate 1-phosphotransferase (EC 2.7.2.3), and 2,3-bisphospho-D-glycerate : 2-phospho-D-glycerate phosphotransferase (EC 2.7.5.3) catalyze consecutive steps in the glycolytic and also in the gluconeogenetic pathways. In order to detect tissue-characteristic forms or isoenzymes in functionally distinct but metabolically related tissues such as liver and muscle,[1] the three enzymes have been prepared from liver and muscle tissue of the same species by applying similar purification procedures including affinity chromatography.[2] For reasons of economy each set of three enzymes have been isolated from the same extract of liver or muscle tissue, respectively. Preparations were carried out from bovine,[3-5] pig,[6,7] and human[8] tissues in a similar way and also from baker's yeast.[9] In this contribution we describe a simultaneous purification procedure for GAPDH, PGK, and PGM from pig (*Sus domesticus*) tissues.[6,7] On the other hand an alternative sequence of isolation steps for the three enzymes from bovine muscle (*Bos taurus*) will be given briefly.[10] D'Alessio

[1] K. D. Kulbe, Habilitationsschrift, Medical University, Hannover, Germany (1975).
[2] K. D. Kulbe, M. Bojanovski, H. Foellmer, J. Fuchs, and R. Schuer, *Colloq.—Inst. Natl. Sante Rech. Med.* **86**, 444 (1979).
[3] F. Heinz and K. D. Kulbe, *Hoppe-Seyler's Z. Physiol. Chem.* **351**, 249 (1970).
[4] M. Bojanovski, K. D. Kulbe, and W. Lamprecht, *Eur. J. Biochem.* **45**, 321 (1974).
[5] J. Fuchs, Ph.D. Thesis, Technical University, Hannover, Germany (1978).
[6] H. Foellmer, Ph.D. Thesis, Technical University, Hannover, Germany (1976).
[7] H. Foellmer, M.-R. Kula, and K. D. Kulbe, *Hoppe-Seyler's Z. Physiol. Chem.* **358**, 232 (1977).
[8] K. D. Kulbe, unpublished results (1978).
[9] K. D. Kulbe and R. Schuer, *Anal. Biochem.* **93**, 46 (1979).
[10] J. Fuchs and K. D. Kulbe, *Hoppe-Seyler's Z. Physiol. Chem.* **360**, 1146 (1979).

and Josse[11,12] reported the preparation of GAPDH, PGK, and PGM from *Escherichia coli* using a common purification scheme.

Assay Method

Principles. The activities of the three enzymes are determined by using coupled enzymic reactions leading to a consumption of stoichiometric amounts of NADH.[13] Measurements are carried out at 360 nm and 25° in 1-cm light path cuvettes. The assay principle and procedure for 3-phosphoglycerate kinase is as given elsewhere in this volume.[14] Glyceraldehyde-3-phosphate dehydrogenase is determined in the same way, but in the presence of an excess of yeast phosphoglycerate kinase (10 μg, 5 units).

Phosphoglycerate mutase activity is assayed according to the method of Czok and Eckert[15] by coupling 2-phosphoglycerate production with the reactions catalyzed by enolase, pyruvate kinase, and the NADH/NAD$^+$-dependent lactate dehydrogenase (forward test).

Procedure. The concentrations of the reaction participants in the final volume of 1 ml of 50 mM triethanolamine-HCl buffer, pH 7.5, are as follows: 5.25 mM 3-phospho-D-glycerate, tricyclohexylammonium salt; 0.15 mM 2,3-bisphospho-D-glycerate, pentacyclohexylammonium salt; 0.80 mM ADP, Na$_2$; 0.32 mM NADH, Na$_2$; 4 mM MgSO$_4$; 100 μg of enolase (4 units); 20 μg of pyruvate kinase (4 units); and 50 μg of lactate dehydrogenase (15 units). The reaction is started by addition of 10 μl of suitably diluted 3-phosphoglycerate mutase solution.

Units. One unit is defined as the amount of enzyme required for the formation of 1 μmol of product per minute at 25°. Determinations of protein or ammonium sulfate concentration, conductivity measurements, and centrifugation conditions were as described previously.[14]

Enzyme Purification from Pig Liver and Pig Muscle

General. Well-reproducible procedures for the purification of the three consecutive enzymes glyceraldehyde-3-phosphate dehydrogenase, phosphoglycerate kinase, and phosphoglycerate mutase from the same extract of either pig liver or pig muscle are described. The first four steps (extrac-

[11] G. D'Alesio and J. Josse, *J. Biol. Chem.* **246**, 4319 (1971).
[12] G. D'Alesio and J. Josse, this series, Vol. 42, p. 139.
[13] O. Warburg and W. Christian, *Biochem. Z.* **314**, 149 (1943).
[14] K. D. Kulbe and M. Bojanovski, this volume [20].
[15] R. Czok and L. Eckert, *in* "Methods of Enzymatic Analysis" (H. U. Bergmeyer, ed.), p. 229. Academic Press, New York, 1965.

tion, ammonium sulfate fractionation, Sephadex G-100 gel filtration, DEAE-Sephadex A-50 ion-exchange chromatography; see Fig. 1) are the same for all of the enzymes. Finally, the individual enzymes are obtained in a homogeneous state by applying differently substituted Sepharose 4B matrices for affinity chromatography as follows: (a) NAD$^+$–Sepharose for liver (200 mg/kg tissue; specific activity 75 units/mg; yield 38%) and muscle glyceraldehyde-3-phosphate dehydrogenase (780 mg/kg; 80 units/mg; 48%); (b) ATP–Sepharose for liver phosphoglycerate kinase (65 mg/kg; 210 units/mg; 23%); and (c) C$_6$-Sepharose for liver phosphoglycerate mutase (36 mg/kg; 308 units/mg; 21%). In contrast, phosphoglycerate kinase (284 mg/kg; 235 units/mg; 38%) and phosphoglycerate mutase (160 mg/kg; 340 units/mg; 28%) from pig muscle could be separated only by gradient elution from hydroxyapatite.

Liver and muscle tissue of the same animal (pig, *Sus domesticus*) are obtained fresh from the local slaughterhouse and used immediately or stored frozen without effects on yield. All steps are carried out at about 4°.

Extraction. Liver tissue of a freshly killed pig is first thoroughly freed of blood by extensive rinsing with cold water with a tube fitted to a water supply. For better entrance into the blood vessels, a pipette tip is fixed at the end of the tube. Washing is stopped when the color of the liver tissue has changed from red to light yellow-brown. One kilogram of liver or muscle tissue is cut into small pieces and homogenized in a commercial blender with 1500 ml of 20 mM sodium phosphate–20 mM EDTA–0.2% 2-mercaptoethanol buffer, pH 7.5.

After blending, the mixture is allowed to stand for 30 min. Then debris are removed by centrifugation at 13,700 g for 40 min. Extraction is repeated twice, 1000 ml of the same buffer being used each time. The combined extracts are filtered through quartz wool to remove particles of fat. Liver extracts are cleared by ultracentrifugation at 100,000 g for 1 hr.

Ammonium Sulfate Fractionation. To the supernatant, solid ammonium sulfate is added slowly to a concentration of 1.4 M. After 1 hr of stirring, the precipitate is removed by centrifugation for 40 min and discarded. The clear supernatant is concentrated to about 10% of its volume by hollow-fiber ultrafiltration (Romikon GM 80, Rohm & Haas, Darmstadt, Germany).

Sephadex G-100 Gel Filtration. Concentrated protein solutions from the preceding step are pumped onto a 11- × 120-cm column of Sephadex G-100 and equilibrated with the 20 mM sodium phosphate buffer used for tissue extraction; activity is eluted with the same buffer. Fractions of 75 ml are collected at a flow rate of 112 ml/hr. Two pools are combined separately (fractions 27–33 for GADPH; fractions 36–47 for PGK/PGM) and concentrated to about 100 ml each by hollow-fiber ultrafiltration (HXL 50, Amicon Corp., Lexington, Massachusetts).

DEAE-Sephadex A-50 Ion-Exchange Chromatography. During the concentration procedure in the preceding step, the buffer composition is changed to 10 mM potassium phosphate–1 mM EDTA–0.2% 2-mercaptoethanol, pH 7.5. Chromatography is carried out on separate 5- × 60-cm columns previously equilibrated with the same buffer. The GAPDH from both pig muscle and pig liver and PGK and PGM from pig muscle could be eluted from these columns with equilibration buffer, pH 7.5. In the case of PGK and PGM from pig liver, PGK is also eluted under starting conditions. After that, the column is eluted with 500 ml of the same buffer containing 120 mM NaCl. Then a linear gradient to 300 mM NaCl in equilibration buffer, pH 7.5, is applied to elute liver PGM. Fractions (12 ml) containing more than 10 units/ml are pooled.

Affinity Chromatography on NAD$^+$–p-Aminobenzamidohexyl-Sepharose 4B. The final purification of the GAPDH enzymes from pig liver and muscle is carried out using the derivatized Sepharose synthesized from CNBr-activated Sepharose.[16] By reaction with 1,6-diaminohexane, coupling of the resulting aminohexyl-Sepharose to p-nitrobenzoylazide,[17] and binding of NAD$^+$ according to Barry and O'Carra.[18] The affinity gel is stored in 5 mM sodium acetate buffer, pH 5.0, containing 0.02% sodium azide. Enzyme solution (25 ml) is applied to the top of a 2.5- × 8-cm column of NAD$^+$–p-aminobenzamidohexyl-Sepharose 4B previously equilibrated with 20 mM potassium phosphate–1 mM EDTA–0.2% 2-mercaptoethanol buffer, pH 7.5. Essentially all the GAPDH activity is adsorbed to the column material. Initially, elution is carried out with 2 column volumes of this buffer containing 250 mM KCl. After washing with 1 volume of the salt-free equilibration buffer, final elution of the enzyme is performed with 10 mM NAD$^+$ (1 volume) at a flow rate of 120 ml/hr, and fractions of 6 ml are collected. Fractions containing GADPH are pooled and concentrated either by dialysis against 20% polyethylene glycol 20,000 in the above buffer or by ultrafiltration.

Affinity Chromatography of Pig Liver Phosphoglycerate Kinase on ATP–Sepharose. ATP–Sepharose[18,19] (25 ml) is poured onto a 2.5- × 5-cm column and equilibrated with 10 mM potassium phosphate buffer, pH 7.5, containing 0.2% 2-mercaptoethanol and 1 mM EDTA. A 100-ml sample of liver PGK (45 mg of protein, 2500 units) resulting from the ion-exchange step is applied, and elution with equilibration buffer is carried out until nonbound protein is removed. After that, elution is continued first by 1 mM MgATP^{2-} and finally by 200 mM KCl in the starting buffer, pH 7.5.

[16] S. C. March, I. Parikh, and P. Cuatrecasas, *Anal. Biochem.* **60,** 149 (1974).
[17] P. Cuatrecasas, *J. Biol. Chem.* **245,** 3059 (1970).
[18] S. Barry and P. O'Carra, *Biochem. J.* **135,** 595 (1973).
[19] R. Lamed, Y. Levin, and M. Wilchek, *Biochim. Biophys. Acta* **304,** 231 (1973).

Hydrophobic Chromatography of Pig Liver PGM on C_6-Sepharose. Preparative-scale experiments are carried out on a 2.6- × 14-cm column of C_6-Sepharose[20] equilibrated with 10 M potassium phosphate–1 mM EDTA–0.2% 2-mercaptoethanol, pH 7.5. A sample obtained by DEAE-Sephadex chromatography (455 mg, 18,000 units) is applied; after washing with 10 mM NaCl in the equilibration buffer, PGM activity is eluted by applying a linear gradient to 100 mM NaCl in this buffer.

Separation of Pig Muscle PGK from PGM on Hydroxyapatite (HTP-BioGel). A 2.5- × 30-cm column is filled with HTP-BioGel and equilibrated with 55 mM potassium phosphate–0.2% 2-mercaptoethanol buffer, pH 6.5. The PGK activity is adsorbed and then eluted with a linear gradient of 55 to 155 mM potassium phosphate in the equilibrating buffer (1.5 liters each, pH 6.5).

The quantitative results of a typical purification of GAPDH, PGK, and PGM from pig liver and pig muscle, respectively, are summarized in Table I and Table II.

Comments on Simultaneous Enzyme Preparation. As shown in Fig. 1 the purification procedure for the liver (A) and muscle (B) enzymes is similar during the first three steps (extraction,[21] ammonium sulfate fractionation, Sephadex G-100 gel filtration). To start with maximum activity for all three enzymes and simultaneously purify the enzymes as far as possible, it is necessary to use the supernatant of the 1.4 M ammonium sulfate step for the column chromatographic experiments. This solution is concentrated by hollow-fiber ultrafiltration to about 300 ml during a 5-hr period. The following chromatographic step yielded two fractions containing GAPDH and PGK/PGM, respectively. The chromatographic behavior of the muscle and liver enzymes during the molecular sieve chromatography was identical. GAPDH from liver and muscle tissue behaved identically on DEAE-Sephadex A-50 under the experimental conditions applied. Whereas phosphoglycerate kinase and phosphoglycerate mutase from pig liver could be separated on DEAE-Sephadex columns by applying a salt gradient, the corresponding enzymes from muscle tissue eluted together in a broad peak. They were, however, well separated from the GAPDH activity remaining from the preceding gel filtration step. In this case the final separation of PGK and PGM was achieved by applying potassium phosphate gradient elution on hydroxyapatite.

Essentially all GAPDH activity could be adsorbed to NAD$^+$–p-amino-benzamidohexyl-Sepharose. The bulk of the foreign proteins was eluted

---

[20] Z. Er-el, Y. Zaidenzaig, and S. Shaltiel, *Biochem. Biophys. Res. Commun.* **49**, 383 (1972).

[21] By ultracentrifugation at 100,000 g, nearly the whole glycogen content of the liver extract could be separated; in turn, it was possible to centrifuge the ammonium sulfate precipitates quantitatively, too. This treatment was especially needed for liver tissue obtained on Tuesday to Friday from the slaughterhouse, but was unnecessary for muscle extracts.

TABLE I

SCHEME FOR SIMULTANEOUS PURIFICATION OF D-GLYCERALDEHYDE-3-PHOSPHATE DEHYDROGENASE (GAPDH), PHOSPHOGLYCERATE KINASE (PGK), AND PHOSPHOGLYCERATE MUTASE (PGM) FROM PIG LIVER[a]

Step	Protein (mg)	GADPH		PGK		PGM	
		Specific activity (units/mg protein)	Activity (units)	Specific activity (units/mg protein)	Activity (units)	Specific activity (units/mg protein)	Activity (units)
Extraction	63,655	0.5	38,520	0.92	59,000	0.82	52,000
Ammonium sulfate, 1.4 M	20,160	1.4	28,000	1.9	38,000	2.1	43,000
Sephadex G-100	4,800	4.6	22,000	—	—	—	—
DEAE-Sephadex A-50	2,600	—	—	12.2	29,000	8.1	21,000
	390	40.0	15,500	—	—	—	—
	350	—	—	56.3	19,700	—	—
	250	—	—	—	—	54.8	13,700
NAD$^+$–Sepharose 4B	195	75.0	14,630	—	—	—	—
ATP–Sepharose 4B	65	—	—	210.0	13,650	—	—
C$_6$-Sepharose 4B	36	—	—	—	—	308.0	11,100
Yield:	—		38%		23%		21%

[a] The starting material was 1 kg of liver tissue.

TABLE II

Scheme for Simultaneous Purification of d-Glyceraldehyde-3-phosphate Dehydrogenase (GAPDH), Phosphoglycerate Kinase (PGK), and Phosphoglycerate Mutase (PGM) from Pig Muscle[a]

Step	Protein (mg)	GAPDH		PGK		PGM	
		Specific activity (units/mg protein)	Activity (units)	Specific activity (units/mg protein)	Activity (units)	Specific activity (units/mg protein)	Activity (units)
Extraction	62,000	2.1	131,000	2.9	176,800	3.2	196,000
Ammonium sulfate, 1.4 M	36,400	3.7	144,480	4.7	172,200	5.3	193,280
Sephadex G-100	6,080	16.8	102,320	—	—	—	—
DEAE-Sephadex A-50	11,620	—	—	12.3	142,400	7.8	90,400
	1,360	66.2	90,000	—	—	—	—
	2,960	—	—	32.8	97,200	24.7	73,510
NAD+-Sepharose 4B	780	80	62,400	—	—	—	—
Hydroxyapatite	284	—	—	235	66,700	—	—
	160	—	—	—	—	340	54,400
Yield	—	48%		38%		28%	

[a] The starting material was 1 kg of muscle tissue.

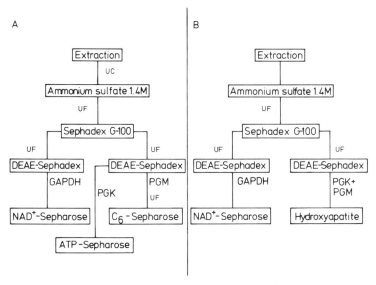

FIG. 1. Flow scheme for the simultaneous purification of D-glyceraldehyde-3-phosphate dehydrogenase (GAPDH), phosphoglycerate kinase (PGK), and phosphoglycerate mutase (PGM) from either pig liver (A) or pig muscle (B) extracts. UC, ultracentrifugation; UF, ultrafiltration.

with wash solution containing KCl. GAPDH activity was not eluted in a significant amount until NAD^+ was included in the eluting buffer. The concentrated fractions normally contained about 75–85% of the activity applied and showed specific activities between 65 and 90 units/mg. The biospecificity of this affinity gel was examined by comparison to the chromatographic behavior of pig GAPDH on AMP--p-aminobenzamido-hexyl-Sepharose and p-aminobenzamidohexyl-Sepharose. The NAD^+–p-aminobenzamidohexyl-Sepharose gel could be used at least up to 10 days without loss of binding capacity and specificity (storage at 2°, pH 5.6, sodium azide). As suggested by Barry and O'Carra[18] the gel material is altered as the result of instability of NAD^+ and should therefore not be used after this time. This disadvantage is more than compensated for by the high degree of biospecificity and the excellent capacity (50–70 mg of specifically bound protein per 100 ml of gel), qualifying it for preparative-scale purifications. Compared to chromatography on both ε-aminocaproyl-NAD^+–Sepharose[22] and NAD^+–adipoyl-Sepharose,[19] biospecificity of binding, capacity, and labor for preparation of NAD^+–p-aminobenzamidohexyl-Sepharose is superior. Their capacity was only

[22] P. O. Larsson and K. Mosbach, *Biotechnol. Bioeng.* **13**, 393 (1971).

7 mg/100 ml and 5.4 mg/100 ml of the affinity gel. GAPDH activity bio-specifically bound to NAD^+–adipoyl-Sepharose 4B was eluted at a NAD^+ concentration of 0.16 mM, whereas the much stronger interaction between GAPDH and the ϵ-aminocaproyl-NAD^+–Sepharose is demonstrated by the fact that about 10 mM NAD^+ is required for removal of this dehydrogenase.

Pig liver phosphoglycerate kinase was finally purified by affinity chromatography on ATP-substituted Sepharose 4B.[19] Only a small portion of the PGK activity applied was eluted biospecifically by 1 mM MgATP^{2-}; the main portion of the enzyme was removed from the column by 200 mM potassium chloride. The enzyme was homogeneous on SDS gel electrophoresis. The capacity of the hydrazide–ATP–Sepharose 4B for liver PGK was determined to be 100–120 mg per 100 ml of gel. Enzyme elution by applying a KCl gradient was not successful. Because of the good flow properties, high stability, and ease of synthesis, this material was chosen for preparative enzyme purification.

In 1975 Thompson et al.[23] showed that enzymes possessing a nucleotide fold in their tertiary structure were specifically bound to Blue Dextran–Sepharose. Our observations indicate that the results obtained by Thompson from microscale experiments could not be applied directly to preparative-scale work. Specifically, none of the PGM activity and 85% of the PGK activity were bound to Blue Dextran–Sepharose, and only about 50% of the PGK applied was eluted biospecifically with 1 mM MgATP^{2-}. However, the same concentration of the structure analog NAD^+ does not remove any PGK activity. Phosphoglycerate kinase from a bovine liver extract could be purified to homogeneity on Blue Dextran–Sepharose 4B by biospecific elution with 1 mM Mg ATP^{2-} in 0.1 M Tris at pH 7.5 after preliminary ammonium sulfate fractionation, heat treatment, and Sephadex G-75 gel filtration. In this case, the yield of the affinity step was about 90%.[1]

Later, a gradient procedure for purification of pig liver PGK on Blue Dextran–Sepharose was established. In this step 94% of the PGK activity was recovered at about 35 mM KCl. However, the binding capacity of this gel was minimal for pig liver PGK (3.3 mg/100 ml) and small for the bovine liver (11.5 mg/100 ml) and the yeast enzyme. This is in contrast to the findings of Stellwagen et al.[23,24] On the other hand, we were able to bind more than 1 g of yeast PGK to 100 ml of Cibacron Blue 3G-A–Sepharose 4B and to remove the enzyme nearly quantitatively at about 0.33 M NaCl.[9,14] Since the dye components of Blue Dextran– and Ciba-

[23] S. T. Thompson, K. H. Cass, and E. Stellwagen, Proc. Natl. Acad. Sci. U.S.A. 72, 669 (1975).

[24] E. Stellwagen, Acc. Chem. Res. 10, 92 (1977).

cron Blue–Sepharose are very similar in structure, the differences observed in binding and elution properties to PGK enzymes of various origin may be attributed to different matrix–dye linkages.[25]

Although PGK and PGM from pig muscle, which copurify on Sephadex G-100 and DEAE-Sephadex, could be separated on hydroxyapatite, the corresponding liver enzymes behave differently. After this separation, liver PGK was purified preparatively on ATP–Sepharose and liver PGM by hydrophobic chromatography on n-hexyl-Sepharose 4B. As shown in microscale experiments at low and high ionic strength, pig liver PGM is bound more and more strongly with increasing chain length. Quantitative desorption was possible up to the hexyl length, while PGM adsorption to the C_7-Sepharose was irreversible at 200 mM NaCl, pH 7.5. Therefore, C_6-Sepharose was used for the preparative scale purification of liver PGM. The recovery for this step amounts to 74%, and the enrichment was about eightfold. The capacity of the C_6-Sepharose for pig liver PGM was determined to be about 800 mg per 100 ml of gel.

The data obtained for the purification of the six enzymes are summarized in Tables I and II. Activity yields range between 21 and 38% for the three liver enzymes and between 28 and 48% for the corresponding muscle proteins. Since purification takes no longer than 5 days for each set of three enzymes, high specific activities could be achieved. Last, but not least, this was made possible by avoiding time-consuming steps such as dialysis and complete ammonium sulfate fractionation. The purified proteins were stored either in 15% solutions of polyethylene glycol (type 6000) or suspended in 2.8 M ammonium sulfate solution, pH 7.5.

Crystallization. A new method for the crystallization of proteins from polyethylene glycol (PEG) has been developed. Ten milliliters of the ultrafiltrate with 20 mg of GAPDH were dialyzed for 6 hr at 4° against the same buffer containing 10% polyethylene glycol (type 6000). During the next 12 hr, four portions of 5 g of PEG each were added to the dialysis solution until an opalescent turbidity appeared. The solution was kept overnight at 4° without further stirring. After centrifugation, the precipitate was redissolved in the above 10 mM phosphate buffer and recrystallized as described. The second crystal suspension was gently centrifuged and could be stored for 4 months without loss of specific activity. This method is based on the finding of Hönig and Kula[26] of a linear relationship between the log of molecular weight of proteins and the concentration of the PEG concentration required for their precipitation. All enzymes could be obtained in homogeneous form as demonstrated for the three liver enzymes by SDS–polyacrylamide gel electrophoresis. Disc electro-

[25] E. Bollin, K. Vastola, D. Oleszek, and E. Sulkowski, *Prep. Biochem.* **8**, 259 (1978).
[26] W. Hönig and M.-R. Kula, *Anal. Biochem.* **72**, 502 (1976).

TABLE III
Isoelectric Points of Glyceraldehyde-3-phosphate
Dehydrogenase (GAPDH), Phosphoglycerate Kinase (PGK),
and Phosphoglycerate Mutase (PGM) from Pig Liver and Pig
Muscle as Obtained by Two Different Methods of
Isoelectric Focusing

Tissue	Method	GAPDH pH	PGK pH	PGM pH
Liver	Polyacrylamide gel	8.5	5.1	6.2
	Sucrose density gradient	8.52	4.9	5.95
Muscle	Polyacrylamide gel	8.5	8.0	9.0
	Sucrose density gradient	8.5	7.7	8.85

phoresis both at pH 4.3 and pH 8.9 of native pig liver PGM leads to a three-band pattern indicating a microheterogeneity of this protein under the conditions applied. Similar results were obtained for bovine liver PGM.[5] This may reflect the relative instability of the liver enzymes observed also by some other workers, especially under extreme conditions of pH.

Properties

Comparative studies were undertaken to see how triosephosphate metabolizing enzymes from liver and muscle tissue of the same species may differ in their molecular properties. Whereas the two dehydrogenases were found to be very similar in most of their properties,[6,27] the phosphoglycerate kinases and the phosphoglycerate mutases from liver and muscle of the same animal were shown to possess largely different properties.[5,6,28,29] Differences observed in their isoelectric points were reflected in their chromatographic and electrophoretic behavior (Table III).

Glyceraldehyde-3-phosphate Dehydrogenase. The specific activity for both enzymes varies between 70 and 85 units/mg. Pig liver and pig muscle GAPDH were shown to be homogeneous by sedimentation analysis, by the appearance of a single protein band in SDS–polyacrylamide gel electrophoresis, and by end-group determination. Molecular weight (144,000) and isoelectric properties (pH 8.5) were found to be very similar. The

[27] H. Foellmer, K. D. Kulbe, W. Lamprecht, E. Rieke, and M.-R. Kule, *Eur. J. Biochem.* (submitted for publication).

[28] H. Foellmer, J. Fuchs, and K. D. Kulbe, *Eur. J. Biochem.* (submitted for publication).

[29] K. D. Kulbe, J. Fuchs, and H. Foellmer, *Hoppe-Seyler's Z. Physiol. Chem.* **360**, 1168 (1979).

results of amino acid analysis and thiol group titration have confirmed that all GAPDH enzymes from mammalian liver and muscle possess 4 cysteine residues per subunit. Kinetic analysis showed that about 3.5 out of the 16 SH groups per tetrameric enzyme are "fast-reacting" thiol groups. As for the muscle type of GAPDH, fluorescence quenching experiments with pig liver enzyme exhibit a negative cooperativity in coenzyme (NAD^+) binding.

Amino acid composition data and peptide maps of tryptic digests reveal an extended similarity (but not identity) of the two dehydrogenases from liver and muscle of the same animal. The 28 N-terminal amino residues within the primary sequence of pig liver GAPDH are identical to the pig muscle protein and to other mammalian enzymes of the same catalytic function.[30,31] Bovine liver GAPDH has the same amino acid sequence for the residues 1–22[30]; this also holds for this sequence stretch of rabbit muscle GAPDH.[1]

Phosphoglycerate Kinase. Detailed results on pig liver and pig muscle phosphoglycerate kinase will be published elsewhere.[28] In general, clear differences were detected between the liver and muscle kinases, e.g., in kinetic data, molecular weight (liver enzyme 47,000, muscle enzyme 42,000), amino acid composition, and peptide maps.[29,31]

Phosphoglycerate Mutases. The mutases from pig liver and pig muscle and also from bovine liver and bovine muscle are so markedly different in their amino acid composition that, together with the data for the isoelectric points, number and reactivity of sulfhydryl groups, and tryptic peptide maps, the existence of tissue-characteristic molecular forms (isozymes) can be postulated.[5,6,10,28,29]

Simultaneous Purification of Aldolase, Glyceraldehyde-3-phosphate
Dehydrogenase, Phosphoglycerate Kinase, and Phosphoglycerate
Mutase from Bovine Muscle (*Bos taurus*)

An alternative scheme has been developed from the simultaneous purification of aldolase (EC 4.1.2.13), GAPDH, PGK, and PGM from bovine muscle (see Fig. 2).[5,10] Most important for this procedure is the fact that ion-exchange chromatography on DEAE-cellulose at pH 7.5 was used only to remove impurities from the common extract of the four enzymes, not to separate them.

That means that all fractions containing sufficient activity of any one of the four enzymes are combined for gradient elution on a hydroxyapatite

[30] K. D. Kulbe, K. W. Jackson, and J. Tang, *Biochem. Biophys. Res. Commun.* **67**, 35 (1975).
[31] K. D. Kulbe, J. N. Tang, and K. W. Jackson, unpublished results (1975).

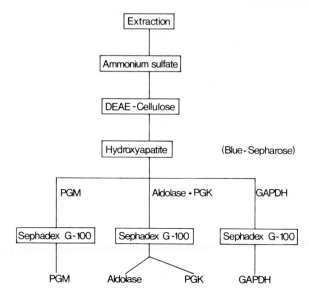

Fig. 2. Flow scheme for the simultaneous purification of aldolase, D-glyceraldehyde-3-phosphate dehydrogenase (GAPDH), phosphoglycerate kinase (PGK), and phosphoglycerate mutase (PGM) from bovine muscle.

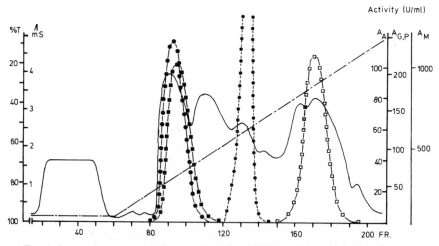

Fig. 3. Hydroxyapatite (BioGel HTP) chromatography of a bovine muscle extract previously treated by ammonium sulfate precipitation and DEAE-cellulose chromatography (see Fig. 2). A column (5 × 30 cm) equilibrated with 10 mM sodium phosphate–5 mM 2-mercaptoethanol, pH 6.5, was used. The four enzymes were eluted by applying a linear gradient to 500 mM sodium phosphate, pH 6.5 (2 × 800 ml): ●——●, phosphoglycerate kinase; ●--●, phosphoglycerate mutase; □——□, D-glyceraldehyde-3-phosphate dehydrogenase; ■——■, aldolase; -----, sodium phosphate gradient; ——, % transmission (or protein).

column at pH 6.5 (Fig. 3). Finally, Sephadex G-100 gel filtration was used to separate aldolase (160,000 M_r) and phosphoglycerate kinase (45,000 M_r), which cochromatographed on hydroxyapatite, and to prepare homogeneous glyceraldehyde-3-phosphate dehydrogenase (145,000 M_r) and phosphoglycerate mutase (58,000 M_r) individually. Yields for the four proteins ranged from 30 to 43% (217–1223 mg of enzyme per kilogram of muscle).

Some properties of the bovine liver and bovine muscle mutases have already been published.[29]

Section IX

Carboxylases and Decarboxylases

[82] Ribulose-1,5-bisphosphate Carboxylase/Oxygenase from Spinach, Tomato, or Tobacco Leaves

By STEPHEN D. MCCURRY, ROBERT GEE, and N. E. TOLBERT

$$\text{Ribulose 1,5-bisphosphate} + CO_2 \rightarrow 2 \text{ 3-phosphoglycerate}$$
$$\text{Ribulose 1,5-bisphosphate} + O_2 \rightarrow \text{2-phosphoglycolate} + \text{3-phosphoglycerate}$$

Ribulose-1,5-bisphosphate carboxylase/oxygenase (EC 4.1.1.39) is present in all photosynthetic organisms and comprises up to 50% of the soluble protein in the leaf of higher plants.[1] Numerous methods have been described for purification of this enzyme,[2-9] but preparations, except from spinach leaves, have had low specific activities and none were stable during prolonged storage. The following methods are currently in use in our laboratory for purification and long-term storage of the enzyme from spinach and for purification and crystallization of the enzyme from tobacco leaves with higher specific activities. The theoretical maximum specific activity of the carboxylase from spinach leaves is 2.8 μmol/min per milligram of protein,[10] but nearly all previous research has been done with the enzyme prepared from spinach leaves with specific activities of 1.0 to 2.0. Assay procedures are described elsewhere.[11]

Purification with Ammonium Sulfate and Chromatography

Reagents

Saturated ammonium sulfate at 4°, pH adjusted after 100-fold dilution to 7.4 with concentrated NH$_4$OH
Bicine buffer, 25 m*M*, pH 7.8–8.0, at 4° containing 10 m*M* 2-mercaptoethanol and 1 m*M* EDTA (grinding buffer)
Polyvinylpolypyrrolidone

[1] R. G. Jensen and J. T. Bahr, *Annu. Rev. Plant Physiol.* **28**, 379 (1977).
[2] J. M. Paulsen and M. D. Lane, *Biochemistry* **5**, 2350 (1966).
[3] J. Goldthwaite and L. Bogard, this series, Vol. 42, p. 481.
[4] P. H. Chan, K. Sakano, S. Singh, and S. G. Wildman, *Science* **176**, 1145 (1972).
[5] T. J. Andrews, G. H. Lorimer, and N. E. Tolbert, *Biochemistry* **12**, 11 (1973).
[6] M. I. Siegel and M. D. Lane, this series, Vol. 62, p. 472.
[7] G. H. Lorimer, M. R. Badger, and T. J. Andrews, *Biochemistry* **15**, 529 (1976).
[8] J. V. Schloss, E. F. Phares, M. V. Long, I. L. Norton, C. D. Stringer, and F. C. Hartman, *J. Bacteriol.* **137**, 490 (1979).
[9] F. J. Ryan and N. E. Tolbert, *J. Biol. Chem.* **250**, 4234 (1975).
[10] N. P. Hall, J. Pierce, and N. E. Tolbert, *Arch. Biochem. Biophys.* **212**, 115–119 (1981).
[11] J. W. Pierce, S. D. McCurry, R. M. Mulligan, and N. E. Tolbert, this series, Vol. 89 [9].

Procedure. Large preparations of enzyme can be made once a year, when fresh spinach leaves from the greenhouse or field are available, and stored at $-80°$ as an $(NH_4)_2SO_4$ suspension for use throughout the year. Enzyme prepared from store-bought spinach leaves has lower specific activity. Each batch of 250 g of deveined fresh spinach leaves is homogenized in a 4-liter Waring blender, and 2–4 batches are combined for a preparation. Larger preparations have not had as high a specific activity, and consequently several preparations have to be run separately through the whole procedure. The leaves are homogenized for 40–50 sec at low speed in Bicine grinding buffer with 2% polyvinylpolypyrrolidone (w/v), in the ratio of 2 ml of buffer per gram of tissue. This and all subsequent steps are performed at $4°$ as rapidly as possible. Polyvinylpolypyrrolidone (2%) is added partially to absorb phenolic compounds, whose oxidation products appear to lower the specific activity of the enzyme. Nevertheless, variations in total and specific activity have been attributed in part to the level of polyphenol oxidase in leaf homogenates that is not entirely eliminated by the polyvinylpolypyrrolidone treatment.[12] The ground slurry is filtered through 6–8 layers of cheesecloth and then through 2 layers of Miracloth (Chicopee Mills, Milltown, New Jersey) and centrifuged for 30–40 min at 10,000 to $11,000\,g$. The supernatant from this centrifugation is filtered through cheesecloth or glass wool on a funnel, further to remove suspended material. The volume is recorded, and the supernatant is made 37% saturated with saturated ammonium sulfate [volume (ml) $\div 1.7 =$ ml of saturated $(NH_4)_2SO_4)$]. The sample is stirred slowly for about 15 min and allowed to stand for 30 min before centrifugation at 10,000 to $11,000\,g$ for 80 min. The supernatant is made 50% saturated with the saturated ammonium sulfate, stirred, and allowed to stand before it is centrifuged again. The supernatant is discarded, and the precipitated enzyme is gently resuspended in a minimal volume of the Bicine buffer. From 500 g of spinach leaves, the volume should be 20–25 ml at this stage. The resuspended protein is clarified by centrifugation at 16,000 g for 20 min. This is an essential step to remove particles that otherwise would clog up the subsequent Sepharose column.

The enzyme solution is loaded onto a Sepharose 4B column, 2.5 or 2.6×75 or 90 cm, and eluted with the Bicine buffer. The maximum sample size for a column of these dimensions is 25–30 ml of very concentrated protein. Normally all steps up to this point are performed in 1 day and the Sepharose column is run overnight in the cold room with an automatic fraction collector. If it is not possible to start this column the same day, the enzyme can be stored overnight at $4°$ as a 50% saturated

---

[12] P. J. Koivuniemi, N. E. Tolbert, and P. S. Carlson, *Plant Physiol.* **65,** 828 (1980).

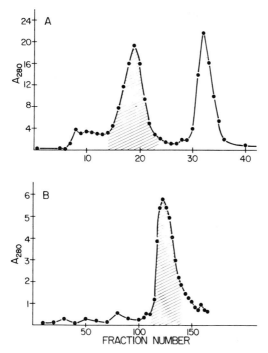

FIG. 1. Separation of ribulose-bisphosphate carboxylase by sequential chromatography using (A) Sepharose 4B and (B) DEAE-cellulose. The hatched peak indicates fractions with carboxylase activity used in the subsequent step.

$(NH_4)_2SO_4$ suspension without loss of activity (see section on routine storage, below). For instance, when the resuspended volume is too large to be chromatographed on one column, part of the sample can be precipitated and used later. Column fractions of 8–14 ml are collected and monitored by absorbance at 280 nm. A sample profile is shown in Fig. 1A. The first fractions are composed mostly of nucleic acids, the second peak (fractions 18 to 22) is the carboxylase, and the third peak is composed of proteins of lower molecular weight. If the Sepharose column has been overloaded, or if the preparation has stood too long as a crude extract, these peaks will be inadequately resolved. The pooled carboxylase peak from this column is loaded onto a DEAE-cellulose (Whatman DE-52) column (2.5–6 × 40–75 cm) that has been equilibrated with the Bicine buffer. The loaded column is washed with 600–1000 ml of buffer, and elution is effected with a 2-liter linear gradient of 0.0 to 0.4 M $NaHCO_3$ in Bicine buffer. It can also be eluted with KCl, but then the enzyme is not stable. Fractions are monitored for protein as before (Fig. 1B), the peak fractions are pooled, and the protein is precipitated by 50% saturated ammonium sulfate for concentration and storage as described below.

In solution, the purified protein ranges from being nearly colorless to yellow. the color is apparently caused by contaminants, such as polyphenols, bound to the protein and seems to depend on several variables, e.g., age and freshness of the spinach leaves, length of time spent in the crude extract, and the presence of reducing agents like 2-mercaptoethanol in the preparation. The absorption spectrum of the yellow color is a tail from a single peak in the 280 nm protein region of the spectrum, and it has no absorption characteristics of flavins or phenols.

Enzyme purified from fresh spinach by this method has had specific activities of 2.0–2.3 μmol/min per milligram of protein and was usually more than 99% pure. Nitrate reductase and phosphoglycolate phosphatase (M. Mulligan, unpublished) activities have been noted in these preparations, but these impurities can be removed by repeating the chromatographic steps. The yield from 500 g of leaves is 0.5–1.0 g of pure protein, yet this represents only a 10% overall recovery. Protein concentration may be estimated by its absorbance at 280 nm. In a cuvette with a 1-cm path, the A_{280} of purified carboxylase × 0.61 for the enzyme from spinach leaves equals the concentration of carboxylase in milligrams per milliliter. The coefficient for the carboxylase from tobacco leaves is 0.7. Protein may also be determined by a modified Lowry procedure with a bovine serum albumin standard, which gives nearly same calibration curve as the carboxylase.[10,13]

Preparation with Polyethylene Glycol[14]

Initial steps of this procedure for preparing the homogenate are the same as for the ammonium sulfate procedure. Using a 60% (w/v) solution of polyethylene glycol (PEG 4000), the supernatant is made to 18% PEG 4000 and stirred for 30 min. The resulting suspension is clarified by centrifugation, which removes precipitated nucleic acids and some protein. The supernatant is then made 20 mM in MgCl$_2$ by the addition of 2 M MgCl$_2$ and stirred. The combination of PEG 4000 and MgCl$_2$ precipitates the carboxylase, which is collected by centrifugation at 16,000 g for 30 min. The precipitate is dissolved in 100 ml of buffer and centrifuged at 27,000 g for 30 min for clarification. The enzyme solution may be further purified by a DEAE-cellulose column as in the ammonium sulfate procedure.

[13] A. Bensadoun and D. Weinstein, *Anal. Biochem.* **70**, 241 (1976).
[14] N. P. Hall and N. E. Tolbert, *FEBS Lett.* **96**, 167 (1978).

The advantage of this procedure is that it is possible to go through the PEG 4000 plus MgCl$_2$ step within 12–15 hr to obtain fairly pure enzyme of specific activity greater than 1.5 μmol/min per milligram and at yields of 0.5–1.0 g from 400 g of starting material. This preparation is at least 95% pure and has not been precipitated by (NH$_4$)$_2$SO$_4$. The small difference in purity makes the ammonium sulfate procedure necessary when very pure enzyme is required, as the DEAE-cellulose chromatography does not completely remove all the contamination in the enzyme precipitated by PEG 4000 and MgCl$_2$. The PEG 4000 procedure has been used successfully with leaves from different plants, but as yet there is not a large body of literature on the carboxylase prepared in this manner.

Routine Storage and Preparation of the Enzyme for Assay [15]

The suspension of precipitated enzyme from 50% saturated (NH$_4$)$_2$SO$_4$ is poured into a graduated cylinder, and the precipitate is allowed to settle overnight at 4°. The clarified supernatant is decanted, and the precipitated enzyme suspension is referred to as the ammonium sulfate slurry. Purified carboxylase has been generally stored as this slurry at 4°. Under these conditions, the enzyme is stable for 3–4 weeks, but after that time it looses activity rapidly. This loss of activity can be restored in preparations after a few months of storage by incubation with 50 mM dithiothreitol (DTT) as described below.

The preferred method of storage is by rapid freezing in liquid nitrogen and storage at −80°. The 50% ammonium sulfate slurry obtained at the end of the procedures is slowly dripped from a separatory funnel into a large flask of liquid nitrogen. The amount of liquid nitrogen and the rate of dripping must be slow enough to allow each drop of the enzyme solution to freeze rapidly into small beads. The beads are stored in closed jars in a −80° refrigerator, and later may be weighed out like a solid reagent. Enzyme stored in this manner has been stable for over a year. Loss of some activity in a few prepartions has been noted, and treatment with DTT (next section) restored activity. A desalted sample in 50–100 mM Bicine has also been frozen, but it was less stable than the frozen 50% saturation (NH$_4$)$_2$SO$_4$ slurry.

A sample of the frozen preparation is taken by weight or volume, thawed to 0° on ice, and centrifuged to collect the precipitated protein. The pellet is dissolved (10–40 mg/ml) in 50 mM Bicine buffer, with 50 mM DTT, and let stand at least 1 hr at room temperature. It is then dialyzed overnight against 50 mM Bicine buffer at room temperature to remove the

[15] N. P. Hall, S. D. McCurry, and N. E. Tolbert, *Plant Physiol.* **67**, 1220 (1981).

DTT and centrifuged to clarify; the A_{280} is read to calculate final protein concentration. The preparation is then activated and assayed as described elsewhere in this series.[11]

Reactivation by Dithiothreitol [15]

Carboxylase that is stored as a 50% saturated ammonium sulfate slurry at 4° loses almost all activity after 2 months. The only mechanism known to restore the activity partially is to allow the preparation to stand with high concentrations (50 mM) of dithiothreitol (DTT) for several hours or overnight at room temperature. Incubation of enzyme after storage at $-80°$ (with 25 mM DTT) has restored full activity. High concentrations of DTT have no effect on freshly prepared carboxylase. Generally enzyme treated with DTT is dialyzed before use to reduce the DTT level. Complete dialysis is mandatory for oxygenase assays in an oxygen electrode as the DTT is slowly oxidized. Removal of DTT by dialysis is also necessary before ultraviolet spectroscopy, as in protein determination, since oxidized DTT has a strong absorption at 283 nm.[16] Enzyme preparations treated with DTT will retain activity for several days.

Crystalline Tobacco Enzyme of High Specific Activity [17]

This procedure is modified from published methods for isolating and crystallizing the enzyme from tobacco leaves[18] by keeping CO_2 and $MgCl_2$ present as much of the time as possible so that the enzyme is in the apparently more stable and active state.

The preparation from tobacco leaves is essentially the same as from spinach leaves (see ammonium sulfate procedure above) through preparation of the homogenate and removal of the material precipitated by 37% saturated $(NH_4)_2SO_2$. Four milliliters of grinding buffer (100 mM Bicine at pH 8.0 with 10 mM mercaptoethanol and 1 mM EDTA) per gram of leaves are used. Before precipitating the enzyme with 50% saturated $(NH_4)_2SO_4$, the supernatant is made to 25.2 mM $MgCl_2$ and 25.2 mM $NaHCO_3$, pH 8, by adding 2 M solutions. Upon addition of more saturated $(NH_4)_2SO_4$ solution to reach 50% saturation, the concentration of $MgCl_2$ and $NaHCO_3$ will drop to 20 mM. The pellet of the enzyme obtained at 50% ammonium sulfate saturation is resuspended in a minimal volume (about 15 ml) of the buffer containing at 25 mM Bicine at pH 8.0, 20 mM $MgCl_2$, 20 mM $NaHCO_3$, 1 mM EDTA, and 10 mM 2-mercaptoethanol. The enzyme is then chromatographed on a Sepharose 4B column that had been equili-

[16] W. W. Cleland, *Biochemistry* **3**, 480 (1964).
[17] R. Gee, G. T. Santora, and N. E. Tolbert, unpublished.
[18] S. D. Kung, K. Sakano, and S. G. Wildman, *Biochem. Biophys. Acta* **365**, 138 (1974).

brated with this buffer and eluted with the same buffer while monitoring for absorbance at 280 nm. The fractions containing the ribulosebisphosphate carboxylase/oxygenase are concentrated by adding $(NH_4)_2SO_4$ to 50% saturation. The protein is centrifuged down and resuspended in a buffer of 25 mM Bicine at pH 8.0, 200 mM NaCl, 3 mM MgCl$_2$, 20 mM NaHCO$_3$, and 10 mM mercaptoethanol. The enzyme preparation is then dialyzed in Schleicher & Schuell (S&S) collodion tubes against four changes of the suspending buffer over a period of 4 hr to reduce the salt concentration. Crystallization is accomplished by further dialysis for 48 hr against four changes of 25 mM Bicine at pH 8.0 with 1 mM EDTA. The crystals are dissolved in a 25 mM Bicine buffer at pH 8.0, 1 mM EDTA, 20 mM MgCl$_2$, 20 mM NaHCO$_3$, and 10 mM mercaptoethanol. The enzyme is recrystallized by dialyzing it again in the S&S collodion bags overnight with one change in the buffer (25 mM Bicine at pH 8.0 with 1 mm EDTA).

Preparation of Ribulose-P$_2$ Carboxylase from Tomato Leaves[17]

Previous preparations of the enzyme from tomato leaves by the $(NH_4)_2SO_4$ and chromatographic procedure have had specific activities of 0.1 to 0.5 with the major loss occurring after Sepharose 4B chromatography. By incorporating a high concentration of NaHCO$_3$ and MgCl$_2$ into the buffers after clarification of the homogenate by adding $(NH_4)_2SO_4$ to 37% saturation, the specific activity during the remaining steps of the purification procedure is maintained between 1 and 1.5. Otherwise the procedure is a modification of that described in the preceding section for purification of the enzyme from tobacco leaves.

The volume of the supernatant after the 37% saturated $(NH_4)_2SO_4$ precipitation is measured and made 25.2 mM NaHCO$_3$ and 25.2 mM MgCl$_2$ by the addition of 2 M solutions. The ribulose-P$_2$ carboxylase protein is then precipitated by adding saturated $(NH_4)_2SO_4$ to 50% saturation. The precipitate is resuspended in a small volume of the 25 mM Bicine buffer at pH 8.0 containing 1 mM EDTA, 20 mM NaHCO$_3$, 20 mM MgCl$_2$, and 10 mM 2-mercaptoethanol. This preparation is applied to the Sepharose 4B column that has been equilibrated with the same buffer and eluted with this buffer. The protein peak with carboxylase is pooled and the enzyme is precipitated by 50% saturated $(NH_4)_2SO_4$ and resuspended in about 15 ml of a buffer containing 25 mM Bicine at pH 8.0, 1 mM EDTA, 10 mM NaHCO$_3$, and 10 mM MgCl$_2$. The preparation was dialyzed against 20 volumes of this buffer for 3 hr with two changes of the buffer to reduce the $(NH_4)_2SO_4$ concentration. Finally, the enzyme is put on a DEAE-cellulose column that has been preequilibrated with the latter buffer. The column is washed with 200 ml of the buffer and then eluted with a gradient of 10 mM to 500 mM NaHCO$_3$ in the same buffer.

[83] Ribulosebisphosphate Carboxylase/Oxygenase from *Rhodospirillum rubrum*[1,2]

By John V. Schloss,[3] E. F. Phares, Mary V. Long, I. Lucile Norton, Claude D. Stringer, and Fred C. Hartman

$$\text{D-Ribulose 1,5-bisphosphate} + CO_2 + H_2O \xrightarrow{\text{Mg}^{2+}} \text{2 D-3-phosphoglycerate}$$

$$\text{D-Ribulose 1,5-bisphosphate} + O_2 \xrightarrow{\text{Mg}^{2+}} \text{D-3-phosphoglycerate} + \text{2-phosphoglycolate}$$

Assay Method

Principle. Ribulosebisphosphate carboxylase activity is determined spectrophotometrically at 340 nm.[4] Oxidation of NADH is monitored during the conversion of 3-phosphoglycerate, the product of the carboxylation reaction, to glycerol 3-phosphate as catalyzed in concert by 3-phosphoglycerate phosphokinase, glyceraldehyde-3-phosphate dehydrogenase, triosephosphate isomerase, and glycerophosphate dehydrogenase. Four moles of NADH are oxidized per mole of ribulosebisphosphate cleaved. Carboxylase activity may also be determined directly with [14]CO_2 as substrate[5]; the enzyme's oxygenase activity can be measured with an oxygen electrode.[5] By use of isotopically labeled ribulose bisphosphate and CO_2, carboxylase and oxygenase activities can be determined simultaneously.[6]

Stock Reagents

Tris hydrochloride, 333 mM, pH 7.8
MgCl$_2$, 100 mM
Glutathione (reduced form), 100 mM
KHCO$_3$, 1.32 M
ATP (tetrasodium salt) (Sigma Chemical Co., St. Louis, Missouri)

[1] Research from the authors' laboratory was sponsored by the Office of Health and Environmental Research, U.S. Department of Energy, under contract W-7405-eng-26 with the Union Carbide Corporation.
[2] Ribulosebisphosphate carboxylase, EC 4.1.1.39.
[3] Supported by Grant GM 1974 from the National Institutes of General Medical Sciences, National Institutes of Health, while a predoctoral fellow at the University of Tennessee–Oak Ridge Graduate School of Biomedical Sciences.
[4] E. Racker, this series, Vol. 5 [29a].
[5] G. H. Lorimer, M. R. Badger, and T. J. Andrews, *Anal. Biochem.* **78**, 66 (1977).
[6] D. B. Jordan and W. L. Ogren, *Plant Physiol.* **67**, 237 (1981).

Glyceraldehyde-3-phosphate dehydrogenase, 17.5 mg/ml, 70 units/mg (ammonium sulfate suspension, Sigma)

3-Phosphoglycerate phosphokinase, 10.6 mg/ml, 2640 units/mg (ammonium sulfate suspension, Sigma)

Glycerophosphate dehydrogenase/triosephosphate isomerase, 12 mg/ml, 75 units of dehydrogenase/mg, 750 units of isomerase/mg (ammonium sulfate suspension, Sigma)

Assay Solutions. The solutions are stored frozen at $-20°$.

Buffer: Mix 6 ml of Tris-HCl, 1 ml of MgCl$_2$, 1 ml of KHCO$_3$, 1 ml of glutathione, and 1 ml of H$_2$O. As needed, the solution is thawed and 55 mg of ATP are added.

Coupling enzyme: Combine 400 μl of glyceraldehyde-3-phosphate dehydrogenase, 80 μl of 3-phosphoglycerate phosphokinase, and 100 μl of glycerophosphate dehydrogenase–triosephosphate isomerase and dialyze the resulting mixture exhaustively at 4° against 50 mM Bicine–5 mM glutathione–0.1 mM EDTA–20% (v/v) glycerol (pH 8.0). The dialyzed solution is divided into four equal portions for storage. As needed, a thawed portion is diluted with H$_2$O to 1.0 ml.

Ribulose bisphosphate (free acid), 2.5 mM. An aqueous suspension of the barium salt[7] is solubilized with Dowex 50 (H$^+$), which is subsequently removed by filtration.

NADH (Sigma), 2 mM

Procedure. The assay is carried out at 25°. To a 1-ml quartz cuvette with a 1-cm path length are added 500 μl of the buffer–ATP mixture, 100 μl of NADH, 100 μl of coupling enzymes, 200 μl of ribulose bisphosphate, and 0–95 μl of H$_2$O. (With a final volume of 1.0 ml, concentrations of assay constituents are as follows: 100 mM Tris, 66 mM KHCO$_3$, 0.5 mM ribulose bisphosphate, 5 mM MgCl$_2$, 5 mM ATP, 0.2 mM NADH, 5 mM glutathione, 20 μg/ml of phosphoglycerate phosphokinase, 175 μg/ml of glyceraldehyde-3-phosphate dehydrogenase, and 30 μg/ml of glycerophosphate dehydrogenase–triosephosphate isomerase.) The contents are thoroughly mixed, and the cuvette is placed into a recording spectrophotometer. The absorbance (A) range at 340 nm of the spectrophotometer is set between 0.8 and 1.0, and 5–100 μl (to give a final volume of 1.0 ml) of the solution to be assayed are added with rapid mixing. The decrease in A is linear with respect to time for several minutes provided the ΔA/min does not exceed 0.2 and provided the sample of carboxylase is preincubated with Mg^{2+} and bicarbonate.[5]

Definition of Enzyme Unit and Specific Activity. One unit of activity is defined as the conversion of 1 μmol of ribulose bisphosphate into 2 μmol

[7] B. L. Horecker, J. Hurwitz, and A. Weissbach, *Biochem. Prep.* **6**, 83 (1958).

of 3-phosphoglycerate per minute and represents a decrease in A at 340 nm of 24.88 absorbance units per minute. Specific activity is defined as units per milligram of protein. Purified carboxylase has an $\epsilon_{1cm}^{1\%}$ of 12.0.[8]

Purification Procedure

Growth of Cells

An initial culture of $R.$ $rubrum$ (strain S-1) was a gift from Professor Gary A. Sojka of Indiana University. Cells are maintained heterotrophically on the synthetic malate medium of Ormerod et $al.$[9] with 0.1% $(NH_4)_2SO_4$ as the nitrogen source. These cells serve as inocula for autotrophic cultures grown on the Ormerod medium (lacking malate) and CO_2 (2% CO_2 in H_2) as the sole carbon source. A stock potassium phosphate solution (200× concentrate) is autoclaved separately from the other salts of the Ormerod medium and after cooling is added to the remaining sterile medium components; this minimizes precipitation of insoluble phosphates. Two-liter autotrophic cultures are serially transferred, 10% of the culture being used to inoculate the next culture. After several passages, these cells, which have been grown autotrophically for many generations, exhibit very high concentrations of ribulosebisphosphate carboxylase. Of the total extractable soluble protein, about 45% is the carboxylase; this compares favorably to green plants in which the carboxylase comprises up to 50% of the soluble leaf protein.[10] The quantity of the carboxylase found in $R.$ $rubrum$ grown on malate is only 0.1% of the soluble protein.[11] Two-liter cultures, adapted to 2% CO_2, are used as inocula for 40-liter cultures. Cultures are maintained at pH 6.5 by the addition of sterile 10 N NaOH. A New Brunswick MicroFerm laboratory fermentor is used to maintain the temperature of the cultures between 28° and 30° and to stir them continuously at 60 rpm. Two sides of the glass carboys (30 × 30 × 60 cm) are lined with 12 Lumiline lamps (60 W, General Electric Co.). One 200-W medium flood (Westinghouse) and two 150-W projector spotlights (Sylvania) are used to illuminate the front, back, and bottom, respectively, of each carboy. A 40-liter culture is harvested 14 days after inoculation by continuous-flow centrifugation with a Sharples (T-1) centrifuge; 250 g of cell paste are obtained. Cells are frozen in liquid nitrogen and stored at $-80°$.

[8] C. D. Stringer, I. L. Norton, and F. C. Hartman, $Arch.$ $Biochem.$ $Biophys.$ **208,** 495 (1981).
[9] J. G. Ormerod, K. S. Ormerod, and H. Gest, $Arch.$ $Biochem.$ $Biophys.$ **94,** 449 (1961).
[10] N. Kawashima and S. G. Wildman, $Annu.$ $Rev.$ $Plant$ $Physiol.$ **21,** 325 (1970).
[11] L. E. Anderson and R. C. Fuller, $Plant$ $Physiol.$ **42,** 497 (1967).

PURIFICATION OF *Rhodospirillum rubrum* RIBULOSEBISPHOSPHATE
CARBOXYLASE/OXYGENASE

Step and fraction	Volume (ml)	Total protein (mg)	Total activity (units)	Specific activity (units/mg protein)	Yield (%)
1. Extract	70	2119[a]	5596	2.6	100
2. DEAE, pH 8.0 (condensed and dialyzed)	29	772[b]	3948	5.1	71
3. DEAE, pH 7.6					
Pool 1 (condensed and dialyzed)	7	456[b]	2753	6.0 ⎫	
Pool 2 (condensed and dialyzed)	2.5	116[b]	630	5.4 ⎭	60

[a] Based on Lowry determinations with bovine serum albumin as standard.
[b] Based on absorbance at 280 nm using $\epsilon_{1cm}^{1\%}$ of 12.0.[8]

Enzyme Isolation

The procedure given here is a modification of that published by Schloss *et al.*[12] Two other detailed methods for the isolation of the enzyme appeared earlier.[13,14]

A summary of the purification is given in the table. All operations are carried out at 4°, and aqueous solutions are prepared with glass-distilled water. The pH values reported for all the buffers used are those observed at 25°.

Step 1. A cell suspension is prepared from 56 g of frozen cell paste and 56 ml of 20 mM Tris-HCl–50 mM NaHCO$_3$–10 mM MgCl$_2$–5 mM 2-mercaptoethanol–1 mM EDTA–1 mM phenylmethylsulfonyl fluoride (pH 8.0). Cells are ruptured by ten 2-min periods of sonic treatment with a Braunsonic 1510 sonifier, and the resulting homogenate is centrifuged at 44,000 rpm for 2 hr in a Beckman 60 Ti rotor.

Step 2. The supernatant from step 1 is applied to a DEAE-cellulose (Whatman DE-52) column (2.5 × 55 cm) that has been equilibrated with the cell-suspension buffer (lacking phenylmethylsulfonyl fluoride). After the column is washed with 1 liter of equilibration buffer (flow rate = 60 ml/hr), it is eluted with a 4-liter linear gradient of 0 to 0.3 M NaCl in the equilibration buffer (flow rate = 30 ml/hr). Fractions of the carboxylase

[12] J. V. Schloss, E. F. Phares, M. V. Long, I. L. Norton, C. D. Stringer, and F. C. Hartman, *J. Bacteriol.* **137**, 490 (1979).
[13] L. E. Anderson, this series, Vol. 42 [68].
[14] F. R. Tabita and B. A. McFadden, *J. Biol. Chem.* **249**, 3453 (1974).

(centered at 0.13 M NaCl) are pooled and dialyzed against 50 mM NaHCO$_3$–5 mM MgCl$_2$–1 mM EDTA–10 mM 2-mercaptoethanol (pH 8.0), which contains 56.1 g of (NH$_4$)$_2$SO$_4$/100 ml of the summed volumes of dialysis buffer and pooled fractions (the pH is readjusted to 8.0 with concentrated NH$_4$OH after the (NH$_4$)$_2$SO$_4$ has been added). Precipitated protein is collected by centrifugation and dissolved in about 20 ml of 20 mM Tris citrate–10 mM MgCl$_2$–5 mM NaHCO$_3$–0.1 mM EDTA–10 mM 2-mercaptoethanol (pH 7.6). [For the preparation of citrate buffers, the pH of a solution of the desired concentration of citric acid (e.g., 20 mM in the buffer just described) is adjusted with solid Tris (free base), and after the other components have been added, the pH is readjusted with Tris or HCl as necessary.] The protein solution is then dialyzed against the same Tris citrate buffer to remove the remaining (NH$_4$)$_2$SO$_4$.

Step 3. The dialyzed sample is applied to a second DEAE-cellulose column (2.5 × 11.5 cm) equilibrated with the 20 mM Tris citrate buffer described in step 2. After the column has been washed with 110 ml of equilibration buffer (flow rate = 30 ml/hr), it is eluted with 30 mM Tris citrate–10 mM MgCl$_2$–5 mM NaHCO$_3$–0.1 mM EDTA–10 mM 2-mercaptoethanol (pH 7.6) (flow rate = 30 ml/hr). The carboxylase emerges between 50 and 200 ml after beginning elution of the column with 30 mM Tris citrate. Fractions from 60 to 120 ml and 121 to 150 ml are pooled separately and dialyzed against buffered (NH$_4$)$_2$SO$_4$ as described for step 2. The precipitated protein from each pool is collected by centrifugation, dissolved in 50 mM Bicine–10 mM MgCl$_2$–66 mM NaHCO$_3$–1 mM EDTA–10 mM 2-mercaptoethanol (pH 8.0) at a protein concentration of 30–70 mg/ml, and dialyzed against the dissolution buffer containing 20% (v/v) glycerol. Based on polyacrylamide gel electrophoresis, the carboxylase from the first pool (representing 80% of the units recovered) is homogeneous and that from the second pool is about 90% pure. The enzyme may be stored in the glycerol-containing buffer at $-80°$ for at least a year without noticeable decline in specific activity.

Properties[15]

Molecular Properties. The pure enzyme sediments with a $s_{20,w}$ of 6.2 S[12,16]; its molecular weight as determined by light scattering is 114,000,[16] and its subunit molecular weight as judged by SDS–polyacrylamide gel electrophoresis is 56,000.[12,16] Based on partial sequence information,[8,12]

[15] For extensive reviews, see R. G. Jensen and J. T. Bahr, *Annu. Rev. Plant Physiol.* **28**, 379 (1977); B. A. McFadden, *Acc. Chem. Res.* **13**, 394 (1980); G. H. Lorimer, *Annu. Rev. Plant Physiol.* **32**, 349 (1981).

[16] F. R. Tabita and B. A. McFadden, *J. Biol. Chem.* **249**, 3459 (1974).

the enzyme's two subunits are identical and show little homology with the large subunit of the maize enzyme whose primary structure was deduced from the sequence of its gene.[17] In contrast to ribulosebisphosphate carboxylase from plants, the amino-terminal residue of the *R. rubrum* enzyme is not blocked.

Catalytic Properties.[12,16,18,19] Kinetic parameters of the purified enzyme are as follows: V_{max} for carboxylation with ribulose bisphosphate as variable substrate at 66 mM NaHCO$_3$ = 6.3 units/mg (the rate for oxygenation is generally about 15% of that for carboxylation); K_m of ribulose bisphosphate = 34 μM; K_m of NaHCO$_3$ = 60 mM; K_m of Mg^{2+} = 210 μM. Mn^{2+} and Fe^{2+} can substitute for Mg^{2+} in the carboxylase reaction; Mn^{2+} and Co^{2+} can support oxygenase activity. Some variability among kinetic constants reported by several laboratories reflect in part differences in assay conditions but, more important, complications arising from the enzyme's requirement for CO$_2$ and divalent metal ions as essential activators and from competitive inhibition by O$_2$ and CO$_2$ of the carboxylation and oxygenation reactions, respectively.

Activators and Inhibitors. Like all ribulosebisphosphate carboxylase/oxygenases, the *R. rubrum* enzyme is devoid of both activities in the absence of divalent metal ions and CO$_2$.[12,14,20,21] The activation process involves carbamate formation between CO$_2$ and a lysyl ϵ-amino group at a site distinct from the binding site for substrate CO$_2$.[22–25] Mg^{2+} stabilizes the carbamate and perhaps bridges activator CO$_2$ and substrate CO$_2$.[26,27] Many phosphate esters and other di- or multivalent anions competitively inhibit the *R. rubrum* carboxylase; however, the inhibition is generally weaker than that observed for the corresponding plant enzymes.[12] For example, the K_i for 6-phosphogluconate is 74 μM with the spinach enzyme but 1.7 mM with the *R. rubrum* enzyme.[12,28] 2-Carboxyribitol 1,5-bisphosphate and 2-carboxyhexitol 1,6-bisphosphate are quite good competitive inhibitors having K_i values of 2 μM (C. S. Herndon and F. C.

[17] L. McIntosh, C. Poulsen, and L. Bogorad, *Nature (London)* **288**, 556 (1980).

[18] P. D. Robinson, M. N. Martin, and F. R. Tabita, *Biochemistry* **18**, 4453 (1979).

[19] J. T. Christeller, *Biochem. J.* **193**, 839 (1981).

[20] J. T. Christeller and W. A. Laing, *Biochem. J.* **173**, 467 (1978).

[21] W. B. Whitman, M. N. Martin, and F. R. Tabita, *J. Biol. Chem.* **254**, 10184 (1979).

[22] M. H. O'Leary, R. J. Jaworski, and F. C. Hartman, *Proc. Natl. Acad. Sci. U.S.A.* **76**, 673 (1979).

[23] H. M. Miziorko and R. C. Sealy, *Biochemistry* **19**, 1167 (1980).

[24] G. H. Lorimer and H. M. Miziorko, *Biochemistry* **19**, 5321 (1980).

[25] G. H. Lorimer, *Biochemistry* **20**, 1236 (1981).

[26] J. Pierce, N. E. Tolbert, and R. Barker, *Biochemistry* **19**, 934 (1980).

[27] G. H. Lorimer, *Annu. Rev. Plant Physiol.* **32**, 349 (1981).

[28] F. R. Tabita and B. A. McFadden, *Biochem. Biophys. Res. Commun.* **48**, 1153 (1972).

Hartman, unpublished data) and 60 μM,[29] respectively. Although the interaction has not been thoroughly studied, the transition-state analog 2-carboxyarabinitol 1,5-bisphosphate irreversibly inhibits the *R. rubrum* enzyme (F. C. Hartman, unpublished data) similarly to spinach carboxylase.[26] A lysyl residue at the active site of *R. rubrum* carboxylase is selectively modified by pyridoxal phosphate,[30,31] and an active-site methionine has been identified with the affinity label 2-bromoacetyl-aminopentitol 1,5-bisphosphate.[32]

[29] G. L. R. Gordon, V. B. Lawlis, Jr., and B. A. McFadden, *Arch. Biochem. Biophys.* **199**, 400 (1980).
[30] P. D. Robison, W. B. Whitman, F. Waddill, A. F. Riggs, and F. R. Tabita, *Biochemistry* **19**, 4848 (1980).
[31] C. S. Herndon, I. L. Norton, and F. C. Hartman, *Biochemistry* **21**, 1380 (1982).
[32] B. Fraij and F. C. Hartman, *J. Biol. Chem.* **257**, 3501 (1982).

[84] Pyruvate Decarboxylase from Sweet Potato Roots

By KAZUKO ÔBA and IKUZO URITANI

$$\text{Pyruvate} \rightleftharpoons \text{acetaldehyde} + CO_2$$

Assay Method

Principle. Pyruvate decarboxylase [2-oxo-acid carboxy-lase, EC 4.1.1.1.] is assayed by a method in which the decarboxylase is coupled with alcohol dehydrogenase [EC 1.1.1.1][1] in the presence of thiamine pyrophosphate and NADH. The rate of disappearance of NADH is followed spectrophotometrically at 340 nm.

Reagents

Histidine-HCl buffer, 0.25 M, pH 6.5 at 25°
Thiamine pyrophosphate (TPP), 14 mM, stored at $-20°$
Yeast alcohol dehydrogenase, 5 mg/ml, stored at $-20°$
MgCl$_2$, 10 mM
Pyruvate, sodium salt, 0.67 M, stored at $-20°$
NADH, 6.7 mM, prepared fresh daily

Procedure. In a reference quartz cell (0.7 ml, 1-cm light path), are placed 0.08 ml of buffer, 0.02 ml of pyruvate, 0.005 ml of enzyme, and

[1] J. Ullrich, J. H. Wittorf, and C. J. Gubler, *Biochim. Biophys. Acta* **113**, 595 (1966).

water to make the final volume of 0.4 ml. In a test quartz cell (0.7 ml, 1-cm light path), are placed 0.08 ml of buffer, 0.01 ml of TPP, 0.003 ml of alcohol dehydrogenase, 0.014 ml of $MgCl_2$, 0.01 ml of NADH, 0.005 ml of enzyme, and water to make the final volume of 0.38 ml. When the decrease in absorbance at 340 nm ceases, the reaction is initiated by the addition of 0.02 ml of pyruvate. The rate of decrease in absorbance at 340 nm is followed in a recording spectrophotometer. Since the crude extract and some preparations from the early fractionation steps contain low but significant NADH oxidase and L-lactate dehydrogenase, a correction must be made for these activities. A corrected value for pyruvate decarboxylase is obtained by subtracting the reaction rate in the solution containing NADH, pyruvate and enzyme from the reaction rate in the complete reaction medium.

Definition of Enzyme Unit and Specific Activity. One unit of enzyme activity is defined as the amount of enzyme required to produce 1 nmol of acetaldehyde per minute under the above assay condition based on the fact that 1 nmol of acetaldehyde yields 1 nmol of NAD^+ by oxidation of NADH in coupled assay. Specific activity is defined as the number of enzyme units per milligram of protein. Protein content is calculated by multiplying the nitrogen content by 6.25. Nitrogen content is determined with Nessler's reagent[2] after digestion of the trichloroacetic acid-insoluble fraction with 10 N H_2SO_4 containing 0.002% $CuSeO_3$.

Purification Procedure

Sampling of Sweet Potato Root Tissue. Sweet potato (*Ipomoea batatas* Lam, *cv* Norin No. 1) roots are harvested in the autumn and stored at 10–14° until used. Roots are dipped in a solution of 0.1% sodium hypochlorite for 20 min for sterilization and washed with water for 20 min, and tissue is taken from the parenchymatous tissue.

Crude Extracts. All steps below are carried out at 0–4°. Sweet potato root tissue (100 g) is mixed with 300 ml of chilled (4°) 20 mM potassium phosphate buffer (pH 7.5) containing 0.7 M mannitol, 1 mM EDTA, 3 g of potassium isoascorbate, and 20 g of Polyclar AT (GAF Corp., New York) and homogenized twice (20 sec and 30 sec) in a blender (Sakuma Co. Ltd) at maximum speed. The homogenate is squeezed through four layers of cotton gauze and centrifuged at 20,000 g for 30 min at 4°.

Gel Filtration, Sephadex G-25. To remove polyphenol compounds, the supernatant solution is passed through a Sephadex G-25 column (6 × 60 cm) previously equilibrated with the above medium not containing isoascorbate.

[2] M. J. Johnson, *J. Biol. Chem.* **137**, 575 (1941).

Ammonium Sulfate Precipitation. To the protein fraction obtained from the effluent of a Sephadex G-25 column is added solid ammonium sulfate (57.0 g/100 ml) at pH 7.0. The precipitate obtained by centrifugation at 20,000 g for 30 min is suspended in 20 ml of 10 mM potassium phosphate buffer (pH 7.0) containing 1 mM MgCl$_2$ and 0.5 mM DTT, and the suspension is dialyzed overnight against 2.5 liters of the buffer (with two buffer changes).

DE-52 Cellulose Column Chromatography, pH 7.0. The dialyzed solution (about 30 ml) is centrifuged to remove insoluble material, diluted with 50 ml of the buffer, and applied to a DE-52 cellulose column (2.6 × 14 cm) previously equilibrated with 10 mM potassium phosphate buffer (pH 7.0) containing 10% (w/v) sucrose, 1 mM MgCl$_2$, and 0.5 mM DTT. The column is washed with 100 ml of the equilibration buffer, and the enzyme is then eluted with a linear gradient of 0 to 0.4 M KCl in the same buffer (400 ml) at a flow rate of 1 ml/min. To the active protein fractions is added solid ammonium sulfate (57.0 g/100 ml) at pH 7.0. The resulting precipitate is collected by centrifugation at 15,000 g for 20 min and suspended in 10 mM potassium phosphate buffer (pH 6.5) containing 1 mM MgCl$_2$ and 0.5 mM DTT. The suspension is dialyzed against the same buffer for 3 hr at 4°.

Gel Filtration, Sephadex G-200. An 8-ml aliquot of the enzyme solution is then applied to a Sephadex G-200 column (2.5 × 65 cm) previously equilibrated with 10 mM potassium phosphate buffer (pH 6.5) containing 1 mM MgCl$_2$ and 0.5 mM DTT. The proteins are eluted with the same buffer. The enzyme activity is detected in the void volume, which is determined using Blue Dextran 2000. The active protein fractions are pooled, concentrated in a collodion bag (SM 13200, Sartorius GmbH, Göttingen, F.R.G.) with dialysis against the same buffer. The collodion bag is mounted on the bottom of a glass tube in a rubber stopper and set in a side arm filter flask. The flask is filled with buffer and stirred with a

PURIFICATION OF PYRUVATE DECARBOXYLASE FROM SWEET POTATO ROOTS

Fraction	Total protein (mg)	Total activity (units)	Specific activity (units/mg protein)	Purification (fold)
Sephadex G-25	311	5,930	17.9	1
Ammonium sulfate	223	14,300	64.3	3.6
DE-52 cellulose	90.4	12,800	141	7.9
Sephadex G-200	2.29	8,490	3,710	207
Sephadex G-200[a]	0.59	7,260	12,300	—

[a] Data from the separate experiment.

magnetic pellet. Then a vacuum is applied with a water aspirator. The concentrate is stored at $-20°$. A plot of enzyme activity as a function of volume of effluent from the Sephadex G-200 column yielded a single symmetrical peak.

A summary of the purification procedure for pyruvate decarboxylase is shown in the table.

Affinity chromatography has been used for the purification of L-lactate dehydrogenase from potato tubers[3] and lettuce leaves.[4]

Properties

Stability. Little or no loss of activity occurred during storage at $-20°$ for several months.

Purity and Molecular Weight. The enzyme protein which has a specific activity of 12,300 gives almost a single band on polyacrylamide gel electrophoresis[5] at pH 8.9,[6] and the pyruvate decarboxylase activities of the thinly sliced gel segments coincided with the protein band. The molecular weight of the enzyme is determined to be 240,000 by the method of Hedrick and Smith.[7] Subunits of the enzyme are separated by polyacrylamide gel electrophoresis in the presence of SDS, and the molecular weight of the monomer subunit is determined to be 60,000 by the method of Weber and Osborn.[8] It is likely that the sweet potato pyruvate decarboxylase is tetramer.

pH Optimum and Buffer Effect. The enzyme shows activity in histidine-HCl buffer from pH 5.70 to 7.50 with an optimum between 6.1 and 6.6. The activity is inhibited in phosphate buffer between pH 5.87 and 7.65, and phosphate is found to be competitive inhibitor with regard to pyruvate.[6]

Kinetic Properties. Lineweaver–Burk double-reciprocal plots of pyruvate deviated markedly from a straight line as the pyruvate concentration decreased. The Hill coefficient (n) is around 2 when the concentration of pyruvate is less than about 1 mM, and around 1 when higher concentrations of pyruvate are used. In the latter case, the K_m value for pyruvate is about 0.6 mM when histidine-HCl buffer is used.

Inhibitors and Activators. A number of mononucleotides, organic acids, metal ions, and EDTA were tested for effects on the enzyme activ-

---

[3] E. Poerio and D. D. Davies, *Biochem. J.* **191**, 341 (1980).
[4] T. Betsche, *Biochem. J.* **195**, 615 (1981).
[5] B. J. Davis, *Ann. N.Y. Acad. Sci.* **121**, 404 (1964).
[6] K. Ôba and I. Uritani, *J. Biochem.* (*Tokyo*) **77**, 1205 (1975).
[7] J. L. Hedrick and A. J. Smith, *Arch. Biochem. Biophys.* **126**, 155 (1968).
[8] K. Weber and M. Osborn, *J. Biol. Chem.* **244**, 4406 (1969).

ity.[6] Pyruvate decarboxylase activity from sweet potato root tissue is not affected by mononucleotides and metabolites such as citrate, α-ketoglutarate, and phosphoenolpyruvate, at a concentration of 1.25 mM. Metal ions activate pyruvate decarboxylase activity at a concentration of 0.5 mM in the following order: $Mg^{2+} > Zn^{2+} > Mn^{2+} > Ca^{2+}$. Ammonium ion is not effective at 1.0 mM. The enzyme activity in the presence of Mg^{2+} (0.5 mM) is reduced to 44% by addition of EDTA (10 mM). Phosphate is a competitive inhibitor for pyruvate.

Intracellular Localization. The enzyme activity is found only in the supernatant fraction centrifuged at 105,000 g for 2 hr.

Section X

Miscellaneous Enzymes

[85] Glyoxalase I from Human Erythrocytes

By BENGT MANNERVIK, ANNE-CHARLOTTE ARONSSON, and
GUDRUN TIBBELIN

$$R—CO—CHO + GSH \rightarrow R—CHOH—CO—SG$$

Glyoxalase I (EC 4.4.1.5, lactoyl-glutathione lyase) catalyzes the formation of S-2-hydroxyacylglutathione from 2-oxoaldehydes, such as methylglyoxal (R = $—CH_3$), and glutathione (GSH). A second enzyme, glyoxalase II (EC 3.1.2.6, hydroxyacylglutathione hydrolase), catalyzes the hydrolysis of the thiolester to yield a 2-hydroxycarboxylate, D-lactate when R = $—CH_3$, and regenerate GSH. Thus, the combination of the two enzymes and GSH constitutes the glyoxalase system, which effects the biotransformation of 2-oxoaldehydes to corresponding 2-hydroxy acids.[1] Before the system had been resolved into its constituent components, it was referred to as a single enzyme, "glyoxalase."[2,3] Glutathione and 2-oxoaldehydes spontaneously combine to a hemimercaptal adduct, and the non-Michaelian steady-state kinetics demonstrate that at least two of the three chemical species affect the reaction velocity. A branching reaction scheme and/or nonlinear competitive inhibition by GSH were proposed,[cf. 4] and the claim that the hemimercaptal adduct was the true substrate[5] could not easily be corroborated. However, the correspondence between the rates of formation of hemimercaptal and thiolester under various conditions[6,7] suggests that the hemimercaptal is indeed the substrate of glyoxalase I. The natural substrate of the enzyme is unknown, but methylglyoxal and some additional 2-oxoaldehydes can be formed *in vivo* (see Mannervik[8] for a review). Human glyoxalase I exists in the form of three isozymes that appear identical in their functional properties.[8] In blood from heterozygous individuals as well as in material combined from several individuals, all three enzyme forms are expected to be present.

[1] E. Racker, *J. Biol. Chem.* **190**, 685 (1951).
[2] C. Neuberg, *Biochem. Z.* **49**, 502 (1913).
[3] H. D. Dakin and H. W. Dudley, *J. Biol. Chem.* **14**, 423 (1913).
[4] B. Mannervik, B. Górna-Hall, and T. Bartfai, *Eur. J. Biochem.* **37**, 270 (1973).
[5] M. Jowett and J. H. Quastel, *Biochem. J.* **27**, 486 (1933).
[6] D. L. Vander Jagt, E. Daub, J. A. Krohn, and L.-P. B. Han, *Biochemistry* **14**, 3669 (1975).
[7] E. Marmstal and B. Mannervik, *FEBS Lett.* **131**, 301 (1981).
[8] B. Mannervik, *in* "Enzymatic Basis of Detoxication" (W. B. Jakoby, ed.), Vol. 2, p. 263. Academic Press, New York, 1980.

Assay Method

Principle. The increase in absorbance at 240 nm due to thiolester formation from GSH and methylglyoxal is monitored spectrophotometrically.[1]

Reagents

Sodium phosphate buffer, 0.1 M, pH 7.0
Methylglyoxal, 40 mM
Glutathione, 50 mM

Procedure. Add 500 μl of buffer, 50 μl of methylglyoxal, 20 μl of GSH, and deionized water to a final volume of 1 ml (after addition of enzyme) to a 1-ml quartz cuvette (1-cm light path). The components are allowed to equilibrate for 2 min at 30°, and the enzymic reaction is then started by addition of glyoxalase I. The absorption coefficient for the product, S-D-lactoylglutathione, is 3.37 mM^{-1} cm^{-1}.[1]

Definition of Unit and Specific Activity. A unit of glyoxalase I is defined as the quantity of enzyme that catalyzes the formation of 1 μmol of S-D-lactoylglutathione per minute at 30°. The specific activity is expressed as units per milligram of protein. The protein determinations used for the data recorded here were made by use of a microbiuret method[9] involving bovine serum albumin as a standard protein.

Purification Procedure

All operations should be performed at about 5° unless otherwise stated.

Source of Enzyme. Erythrocytes are obtained from human blood; outdated material from blood banks or cells discarded in the preparation of plasma proteins can be used. For small quantities of blood, the erythrocytes may be collected by centrifugation and washed with 0.9% (w/v) NaCl. In the processing of larger quantities (as described below) the erythrocytes are usually not washed. A convenient procedure to recover the erythrocytes is to fill large plastic tubings with blood and hang the tubings vertically in a cold room. After sedimentation of the erythrocytes, the tubings are frozen; the plasma fraction can then be cut off in the frozen state.

Solutions

Buffer A: Tris-HCl, 10 mM, pH 7.8
Buffer B: Tris-TES, pH 6.8 (obtained by titrating 10 mM TES,

---

[9] J. Goa, *Scand. J. Clin. Lab. Invest.* **5**, 218 (1953).

N-tris(hydroxymethyl)methyl-2-aminoethanesulfonic acid, with 10
mM Tris base), containing 2 mM dithioerythritol
Ammonia, concentrated solution, 13.3 M
Phenylmethanesulfonyl fluoride (PMSF), 0.1 M in isobutanol, freshly
  prepared

Affinity Matrix. S-Hexylglutathione, which is used both as an immobilized ligand of the affinity matrix and as a soluble ligand in the elution
of glyoxalase I, can be synthesized according to Method A of Vince *et al.* [10]
The synthesis is described in this series. [11] The ligand is coupled to
epoxy-activated Sepharose 6B (Pharmacia Fine Chemicals). [12] A more detailed description of the procedure is available. [11]

Step 1. Denaturation of Hemoglobin. The procedure described here
involves 60 liters of human erythrocytes, which are processed in four
batches of 15 liters. After step 3, the batches are combined for the remaining purification steps. The cold suspension of erythrocytes (15 liters) is
adjusted to pH 8.3 with ammonia, and the protease inhibitor PMSF is
added to a final concentration of 50 μM. A cold ($-20°$) mixture of 2.25
liters of n-butanol and 0.9 liter of chloroform is added with continuous
mechanical stirring. After this addition, the mixture is stirred for 60 min at
room temperature (20–22°). A heavy brick-red precipitate of denatured
hemoglobin develops and is allowed to sediment at room temperature
overnight (12–18 hr). The next day, the supernatant fraction is cleared by
centrifugation at 7000 g for 45 min. The sediment is discarded. In steps 1
and 2 the centrifugations may be performed at room temperature.

Step 2. Acetone Fractionation. To the supernatant fraction of step 1 is
added 1.5 volumes of acetone at room temperature. The mixture is stirred
manually for 10 min (to avoid the fire hazard of sparks from an electric
motor) and the precipitate is allowed to sediment at room temperature for
about 30 min. The clear part of the supernatant fraction is siphoned off and
discarded, and the precipitate is then collected by centrifugation at 7000 g
for 5 min. The precipitate is suspended in 2 liters of 0.1 M Tris-HCl, pH
7.8, containing 50 μM PMSF and 0.1 mM dithioerythritol (DTE), and
stirred at 5° overnight. The next day the suspension is centrifuged at 19,000g
for 2 hr, and the supernatant fraction is saved. The remaining precipitate
is stirred with 2 liters of the same buffer in order to extract additional
enzyme. After 4 hr the suspension is centrifuged as above. The supernatant fraction is combined with the first centrifugate. The pooled fractions
are dialyzed overnight at 5° against a total of 30 liters of buffer A contain-

[10] R. Vince, S. Daluge, and W. B. Wadd, *J. Med. Chem.* **14**, 402 (1971).
[11] B. Mannervik and C. Guthenberg, this series, Vol. 77, p. 231.
[12] A.-C. Aronsson, E. Marmstål, and B. Mannervik, *Biochem. Biophys. Res. Commun.* **81**, 1235 (1978).

ing 20 μM PMSF and 0.1 mM DTE. Preferably, a method involving continuous change of buffer should be used. From this stage on, the enzyme is kept at 5°.

Step 3. Ammonium Sulfate Fractionation. Solid ammonium sulfate (280 g/liter) is added with stirring to the dialyzed solution of step 2. The pH value of the solution is monitored and maintained at pH 7.8 by addition of ammonia, if necessary. After 60 min the precipitate formed is collected by centrifugation at 7000 g for 60 min and discarded. Glyoxalase I is precipitated by further addition of ammonium sulfate (176 g/liter) and 60 min later collected by centrifugation as described above. The precipitate is dissolved in 3 liters of buffer A. At this stage the material from the four batches are combined.

Step 4. Affinity Chromatography (I) on S-Hexylglutathione–Sepharose 6B. The combined material from step 3 is applied to a column (4 × 8.5 cm) of S-hexylglutathione–Sepharose 6B, previously equilibrated with buffer A containing 0.2 M NaCl. After application of the sample, at a rate of 150 ml/hr, the column is washed with 300 ml of the buffer used for equilibration followed by 300 ml of buffer A. The enzyme is eluted with 300 ml of buffer A containing 10 mM GSH and 6 mM S-hexylglutathione. During elution the flow rate is decreased to 50 ml/hr. The enzyme activity is recovered in approximately 100 ml of the effluent.

Step 5. Chromatography on Sephadex G-75. The pooled fractions from step 4 are concentrated to about 10 ml by use of Millipore immersible-CX ultrafilters. The concentrate is chromatographed on a column (4 × 30 cm) of Sephadex G-75 (Fine), previously equilibrated with buffer A. The elution is carried out with the same buffer at a flow rate of 200 ml/hr. Enzyme-containing fractions of the effluent are pooled.

Step 6. Affinity Chromatography II. To the pooled effluent from step 5 is added solid NaCl to a final concentration of 0.2 M before the second affinity chromatography on S-hexylglutathione–Sepharose 6B. (The enzyme binds more tightly at high than at low ionic strength.) The enzyme sample is applied to a column (4 × 4.5 cm) equilibrated as in step 4, and the matrix is also washed in the same manner as before. The enzyme is eluted from the column with 100 ml of buffer A containing 10 mM GSH and 6 mM S-hexylglutathione (flow rate, 50 ml/hr).

Step 7. Chromatography on DEAE-Sepharose. The active fractions from step 6 are pooled and treated with 10 mM DTE for 1 hr before application to a column (2 × 8 cm) of DEAE-Sepharose CL-6B, previously equilibrated with buffer B. The bed is washed with 50 ml of buffer B after application of the enzyme, and elution is effected by a linear concentration gradient of 0 to 0.1 M NaCl in the same buffer (total volume, 1000

TABLE I
PURIFICATION OF GLYOXALASE I FROM HUMAN ERYTHROCYTES[a]

Fraction	Volume (ml)	Total activity (kilounits)	Specific activity (units/mg protein)	Yield (%)
Erythrocytes	60,000	1490	0.067	(100)
Butanol–CHCl₃ denaturation	23,600	981	1.27	66
Acetone fractionation	4,170	564	1.26	38
Ammonium sulfate fractionation	3,000	338	1.39	22
Affinity chromatography I	112	232	647	16
Sephadex G-75 chromatography	122	225	546	15
Affinity chromatography II	42	220	1066	15
DEAE-Sepharose chromatography	285	169	1150	11

[a] The first four fractions represent material from four batches, which were subsequently pooled and processed together in the last four purification steps.

ml; flow rate, 50 ml/hr). The active fractions can be concentrated to about 2 ml by ultrafiltration without decrease of specific activity.

A summary of the results of a representative purification of glyoxalase I from human erythrocytes is given in Table I. The purification can be carried out on a smaller scale, but step 1 gives a lower recovery of active enzyme when the volume of erythrocytes is of the order of a few liters or less. It should also be noted that step 1 does not work well when applied to the purification of glyoxalase I from porcine erythrocytes.[13]

Separation of the Isozymes of Glyoxalase I. The procedure summarized in Table I yields a mixture of the three isozymes of glyoxalase I. In the absence of DTE or other thiol-containing reagents, more than three forms of the enzyme appear. The additional components are believed to be oxidized forms of the protein. The isozymes can be separated and isolated, according to step 7, if a smaller column and smaller load of protein are used.[14]

A sample from step 7 containing 10–25 mg of protein is equilibrated with buffer B by gel filtration or dialysis. Material from step 6 can be used without this equilibration. After treatment with 10 mM DTE for 1 hr, the sample is applied to a column (1 × 5 cm) of DEAE-Sepharose CL-6B, equilibrated with buffer B containing 2 mM DTE. The three isozymes of glyoxalase I are separated and eluted with a linear concentration gradient of NaCl, exactly as described for step 7. The eluted fractions should

[13] A.-C. Aronsson and B. Mannervik, *Biochem. J.* **165**, 503 (1977).
[14] A.-C. Aronsson, G. Tibbelin, and B. Mannervik, *Anal. Biochem.* **92**, 390 (1979).

TABLE II

MOLECULAR AND KINETIC PROPERTIES OF GLYOXALASE I
FROM HUMAN ERYTHROCYTES[a]

Property	Value
Molecular weight	46,000
Number of subunits	2
Stokes' radius	2.8 nm
$s_{20,w}$	4.0 S
Isoelectric point at 4°	pH 4.8[b]
Essential metal (1 per subunit)	Zn
pH optimum of activity	Constant between pH 6.5 and 7.5
Specific activity	1150 μmol/min/mg protein
Apparent steady-state kinetic parameters (at 2 mM free GSH)	
K_m for methylglyoxal	0.13 mM
k_{cat} for methylglyoxal	68,000 min^{-1}
K_m for phenylglyoxal	0.04 mM
k_{cat} for phenylglyoxal	64,000 min^{-1}

[a] Data compiled from Aronsson et al.[14] and Marmstål et al.[15]

[b] Minor differences between isozymes.

preferably be analyzed by means of electrophoresis in gels to avoid cross-contamination of the isozymes in the pooling of the eluted fractions.

Properties

Purity. The enzyme obtained by the procedure summarized in Table I appears homogeneous as judged by various analytical procedures, except for the presence of the three isozymes. Antibodies to human glyoxalase I (raised in rabbits) do not reveal any heterogeneity of the enzyme in immunoelectrophoretic analyses. The isozymes appear to be identical in all respects except for the electrophoretic and chromatographic properties that allow their separation. Even their amino acid compositions show no significant differences.[15] Antibodies raised against any of the three isozymes are fully reactive with the other two isozymes; the antibodies against human glyoxalase I cross-react to some extent with the enzyme from rat liver and from porcine erythrocytes, but not with enzyme from yeast (*Saccharomyces cerevisiae*).[16]

[15] E. Marmstål, A.-C. Aronsson, and B. Mannervik, *Biochem. J.* **183**, 23 (1979).
[16] K. Larsen, Ph.D. Dissertation, Univ. of Stockholm, Stockholm, 1981.

Molecular and Kinetic Properties. Some properties of the enzyme are listed in Table II. The enzyme is composed of two subunits of equal weight[14] and contains 1 Zn per subunit that is essential to catalytic activity.[12] The metal can be replaced by Co, Mn, and Mg to give enzyme with high catalytic activity; the substitutions do not significantly affect the apparent K_m value of the enzyme.[17] *S-p*-Bromobenzylglutathione is a strong reversible inhibitor ($K_i = 0.08 \ \mu M$), which can be used to titrate the active site of the enzyme.[17] The substrate specificity of the human enzyme has not been investigated extensively; methylglyoxal and phenylglyoxal can be used as the 2-oxaldehydes (cf. Table II). Interestingly, 3-phosphohydroxypyruvaldehyde is not active with human enzyme[18] in spite of being a good substrate for the enzyme from yeast. Mannervik[8] summarizes the known facts about the substrate specificity of glyoxalase I from various sources; the human enzyme is believed to have properties in common with the other mammalian enzymes investigated.

Acknowledgments

The work from the authors' laboratory has been supported by the Swedish Natural Science Research Council and the Swedish Cancer Society.

[17] A.-C. Aronsson, S. Sellin, G. Tibbelin, and B. Mannervik, *Biochem. J.* **197**, 67 (1981).
[18] P. Christen, personal communication.

[86] Glyoxalase I[1] from Mouse Liver

By BEDII ORAY and SCOTT J. NORTON

$$CH_3\overset{\displaystyle O}{\overset{\|}{C}}-\overset{\displaystyle O}{\overset{\|}{C}}-H \;+\; GSH \;\rightleftharpoons\; CH_3\overset{\displaystyle O}{\overset{\|}{C}}-\underset{\displaystyle OH}{\overset{\displaystyle |}{C}}H-SG$$

methylglyoxal + glutathione   hemimercaptal

$$CH_3\overset{\displaystyle O}{\overset{\|}{C}}-\underset{\displaystyle OH}{\overset{\displaystyle |}{C}}H-SG \;\xrightarrow{\text{glyoxalase I}}\; CH_3-\underset{\displaystyle OH}{\overset{\displaystyle |}{C}}H-\overset{\displaystyle O}{\overset{\|}{C}}-SG$$

hemimercaptal   S-D-lactoylglutathione

Assay Method

Principle. The method is based upon the increase in absorbance at 240 nm due to the formation of S-D-lactoylglutathione. A modification of the procedure of Racker[2] is used.

Reagents

Imidazole · HCl, 200 mM, pH 7.0; MgSO$_4$, 16 mM
Methylglyoxal 100 mM; standardized according to the method of Friedmann[3]
Glutathione 100 mM; prepared fresh daily
Enzyme solution

Procedure.[4] Add 237 μl of 100 mM methylglyoxal and 30 μl of 100 mM glutathione to 2733 μl of imidazole buffer. The reaction mixture is allowed to stand for at least 2 min at room temperature to ensure equilibration. The hemimercaptal concentration at equilibrium is calculated to be 0.7 mM using K_{eq} = 3.1 mM.[5] The enzymic production of S-D-lactoylglutathione (E_{240} = 3.37 mM^{-1} cm^{-1}) is followed by measuring the increase at 240 nm for 2 min at 25° on a recording spectrophotometer. The reaction is initiated by the addition of a rate-limiting volume of enzyme preparation (1–20 μl) to 3.0 ml of reaction mixture, and initial rates

[1] S-D-Lactoyl-glutathione methylglyoxal-lyase (isomerizing); EC 4.4.1.5
[2] E. Racker, *J. Biol. Chem.* **190**, 685 (1951).
[3] T. E. Friedmann, *J. Biol. Chem.* **73**, 331 (1927).
[4] B. Oray and S. J. Norton, *Biochim. Biophys. Acta* **483**, 203 (1977).
[5] R. Vince, S. Daluge, and W. B. Wadd, *J. Med. Chem.* **14**, 402 (1971).

are determined by the slope of the linear portion of a change in absorbance vs time plot. The reference cell contains all reaction mixture components with the exception of the enzyme preparation.

Definition of Enzyme Units. One unit of glyoxalase I activity is defined as the amount of the enzyme catalyzing the formation of 1 μmol of S-D-lactoylglutathione per minute in the routine enzyme assay system. Specific activity is expressed as units per milligram of protein, the protein being measured by the colorimetric Coomassie Blue method of Bradford.[6]

Purification Procedure

Column Material Preparation. The preparations of column materials are based on the method of Cuatrecasas.[7] To 100 ml of washed Sepharose 4B, 25 g of finely divided cyanogen bromide are added with stirring in a well-ventilated hood. The pH is maintained at 10.5 by the addition of 5 N NaOH. The temperature is maintained between 18° and 20° by the addition of ice to the reaction mixture. When all cyanogen bromide is dissolved (12–15 min), the activated Sepharose is quickly washed with minimum of 10 volumes of cold 0.1 M sodium carbonate (pH 10.2). The step of washing the activated Sepharose is accomplished in less than 90 sec. The appropriate ligand (3.6 mmol of either ethylamine or S-octylglutathione) dissolved in 100 ml of the same wash solution, is then added with gentle stirring to the activated Sepharose. The mixture is stirred overnight at 5°. It is then washed extensively with distilled water and placed in the buffer used for chromatography. The degree of substitution of the ligands is not determined.

S-Octylglutathione is prepared by allowing 1-bromooctane to react with reduced glutathione employing Method A of Vince *et al.*[5]

Crude Preparation. Swiss mice (24–27 g) are purchased from Timco, Houston, Texas. The mice are sacrificed by asphyxiation in CO_2. The livers are removed immediately and homogenized (0° for 45–60 sec at medium speed with a VirTis homogenizer) in three volumes of a solution of 1 mM potassium phosphate (pH 7.0), 1 mM $MgSO_4$, and 20% glycerol. The homogenate is centrifuged at 100,000 g for 1 hr. The supernatant fraction thus obtained (approximately 225 ml per 100 g of liver) is designated as the crude preparation of glyoxalase I and contains 25–30 mg of protein per milliliter. When the livers are not used immediately, they are frozen in solid CO_2 and then stored at −30°. All purification steps are conducted at 4°, and enzyme preparations are stored at −30°.

[6] M. M. Bradford, *Anal. Biochem.* **72**, 248 (1976).
[7] P. Cuatrecasas, *Annu. Rev. Biochem.* **40**, 259 (1971).

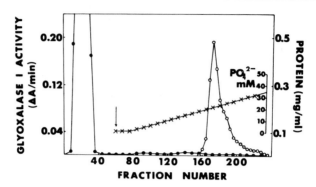

FIG. 1. Ethylamine–Sepharose hydrophobic chromatography. Experimental details are given in the text. Crude preparation (220 ml) was applied to a 2.5- × 45-cm preequilibrated column. Fraction volumes were 20 ml. The arrow marks the application of the phosphate gradient (1 to 50 mM, pH 7.0). ○——○, Glyoxalase I activity, ΔA/min (20 μl of each fraction were used in the assay); ●——●, protein concentration, mg/ml; ×——×, phosphate concentration, mM. From Oray and Norton.[4]

Ethylamine–Sepharose 4B Hydrophobic Chromatography. Sufficient crude preparation to give 6.2 g of protein is placed on a 2.5- × 45-cm ethylamine–Sepharose (hydrophobic) column. The column material was previously equilibrated with a solution of 1 mM potassium phosphate (pH 7.0), 1 mM MgSO$_4$, and 20% glycerol. After loading, the column is

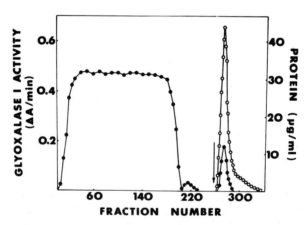

FIG. 2. S-Octylglutathione–Sepharose affinity chromatography. Experimental details are given in the text. Pooled fractions (1280 ml) from the hydrophobic chromatography column were applied to a 1.5- × 20-cm preequilibrated column. Fraction volumes were 6.9 ml. The arrow marks the application of glutathione (20 mM). ○——○, Glyoxalase I activity, ΔA/min (10 μl of each fraction were used in the assay); ●——●, protein concentration, μg/ml. From Oray and Norton.[4]

PURIFICATION OF GLYOXALASE I[a]

Fraction	Volume (ml)	Total activity (units)	Specific activity (units/mg protein)	Yield, step (%)	Purification Step	Purification Overall
Crude preparation	220	4883	0.81	—	1.0	1.0
Ethylamine	1280	2789	16.4	57	20.2	20.2
S-Octylglutathione	545	2571	943.6	92	57.6	1162

[a] From Oray and Norton.[4]

washed with the equilibration solution until the protein being eluted is at a constant, low level (monitored at 280 nm). A phosphate gradient (1 to 50 mM, pH 7.0) in 5 mM MgSO$_4$ and 20% glycerol is used to eluate glyoxalase I. Figure 1 shows typical protein concentration, glyoxalase I activity, and phosphate concentration profiles obtained. The phosphate concentrations are determined by the method of Chen et al.[8]

S-Octylglutathione–Sepharose 4B Affinity Chromatography. Pooled active fractions from the ethylamine–Sepharose column are added to a 1.5- × 20-cm S-octylglutathione–Sepharose affinity chromatography column that has been equilibrated with a solution of 10 mM potassium phosphate (pH 7.0), 5 mM MgSO$_4$, and 20% glycerol. After application of the sample, the column is washed with the equilibration solution until no protein can be detected in the washings. The column is then treated with a 50 mM imidazole · HCl (pH 7.2) solution containing 5 mM MgSO$_4$ and 20% glycerol. Homogeneous glyoxalase I is then eluted by the addition of the above imidazole buffer medium containing 20 mM GSH (pH 7.2). The protein concentration and activity profiles are shown in Fig. 2. Volume reduction of the pooled active fractions from the affinity column was achieved by ultrafiltration.

The results of a typical purification are presented in the table. Purifications approaching 1200-fold with an overall yield of greater than 50% were obtained.

Properties

General. Glyoxalase I from mouse liver has a molecular weight of 43,000[4,9]; it is a dimer and is apparently composed of identical subunits of molecular weights approximating 21,500. Homogeneity was established

[8] P. S. Chen, Jr., T. Y. Toribara, and H. Warner, *Anal. Chem.* **28**, 1756 (1956).
[9] M. V. Kester and S. J. Norton, *Biochim. Biophys. Acta* **391**, 212 (1975).

by multiple electrophoretic determinations and by sedimentation equilibrium centrifugation. The purified enzyme exhibited a specific activity of 944 IU per milligram of protein employing the routine enzyme assay system.

Stability. When the enzyme was stored $-30°$ in the presence of glutathione, no significant loss of activity was observed for at least 6 months.

Inhibitors. There have been numerous reports in the literature on the inhibition of glyoxalase I by different glutathione derivatives or substrate analogs[5,10-13]; these compounds are typically competitive inhibitors. There have been reports on some glyoxalase I inhibitors that are not substrate analogs[14-16]; both competitive and noncompetitive inhibitions were observed.

[10] W. O. Kermack and N. A. Matheson, *Biochem. J.* **65**, 48 (1957).
[11] M. V. Kester, J. A. Reese, and S. J. Norton, *J. Med. Chem.* **17**, 413 (1974).
[12] G. W. Phillips and S. J. Norton, *J. Med. Chem.* **18**, 482 (1975).
[13] R. Vince, M. Wolf, and C. Sanford, *J. Med. Chem.* **16**, 951 (1973).
[14] S. Kurasawa, H. Naganawa, T. Takeuchi, and H. Umezawa, *Agric. Biol. Chem.* **39**, 2009 (1975).
[15] S. Kutasawa, T. Takeuchi, and H. Umezawa, *Agric. Biol. Chem.* **40**, 559 (1976).
[16] K. T. Douglas and I. N. Nadvi, *FEBS Lett.* **106**, 393 (1979).

[87] Glyoxalase II[1] from Mouse Liver

By Bedii Oray and Scott J. Norton

$$CH_3-CH-\overset{\displaystyle O}{\overset{\displaystyle \|}{C}}-SG \xrightarrow[H_2O]{\text{glyoxalase II}} CH_3-CH-COOH + GSH$$

S-D-Lactoylglutathione   D-lactic acid   + glutathione

(with OH groups on the CH of both the substrate and product)

Assay Method

Principle. The method is based upon the decrease in absorbance at 240 nm due to the disappearance of S-D-lactoylglutathione.

Reagents

Potassium phosphate buffer, 100 mM, pH 7.0
S-D-lactoylglutathione, 12 mM, prepared and purified by the method of Uotila[2]
Enzyme solution

Procedure.[3] Add 125 μl of 12 mM S-D-lactoylglutathione to 2875 μl of 100 mM potassium phosphate buffer (pH 7.0). The enzymic disappearance of S-D-lactoylglutathione (E_{240} = 3.37 mM^{-1} cm^{-1}) is followed at 240 nm for 5 min at 25° on a recording spectrophotometer. The reaction is initiated by the addition of a rate-limiting volume of enzyme preparation (5–75 μl) to 3.0 ml of reaction mixture, and the initial rates are determined by the slope of the linear portion of a change in absorbance vs time plot.

Definition of Enzyme Units. One unit of glyoxalase II activity is defined as the amount of the enzyme catalyzing the conversion of 1 μmol of S-D-lactoylglutathione per minute in the routine enzyme assay system. Specific activity is expressed as units per milligram of protein, the protein being measured by the colorimetric Coomassie Blue method of Bradford.[4]

Purification Procedure

Affinity Chromatographic Material Preparation. The preparation of column materials is based on the method of Cuatrecasas.[5] To 100 ml of

[1] S-2-Hydroxyacylglutathione hydrolase; EC 3.1.2.6.
[2] L. Uotila, *Biochemistry* **12**, 3938 (1973).
[3] B. Oray and S. J. Norton, *Biochim. Biophys. Acta* **611**, 168 (1980).
[4] M. M. Bradford, *Anal. Biochem.* **72**, 248 (1976).
[5] P. Cuatrecasas, *Annu. Rev. Biochem.* **40**, 259 (1971).

washed Sepharose 4B, 25 g of finely divided cyanogen bromide is added with stirring; use a well-ventilated hood. The pH is maintained at 10.5 by the addition of 5 N NaOH. The temperature is maintained between 18° and 20° by the addition of ice to the reaction mixture. When all the cyanogen bromide is dissolved (12–15 min), the activated Sepharose 4B is quickly washed with at least 10 volumes of cold 0.1 M sodium carbonate (pH 10.2). The step of washing the activated Sepharose is accomplished in less than 90 sec. Oxidized glutathione (3.6 nmol), dissolved in 100 ml of 0.1 M sodium carbonate, is then added with gentle stirring to the activated Sepharose. The mixture is stirred overnight at 5°. It is then washed extensively with distilled water and placed in the buffer used for chromatography. The degree of substitution with oxidized glutathione is not determined.

Crude Preparation. Swiss mice (24–27 g) are purchased from Timco, Houston, Texas. The mice are sacrificed by cervical dislocation. The livers are removed immediately and homogenized (0° for 2 min at medium speed with a VirTis homogenizer) in 2 volumes of a solution of 10 mM potassium phosphate (pH 7.0) and 20% glycerol. The homogenate is centrifuged at 100,000 g for 1 hr. The supernatant fraction thus obtained is designated as the crude preparation of glyoxalase II. When the livers are not used immediately, they are frozen in solid CO_2 and then stored at −30°. All purification steps are conducted at 4°, and enzyme preparations are stored at −30°.

First GSSG–Sepharose 4B Affinity Chromatography. A sufficient volume of the crude preparation to give 7.0 g of protein is placed on a 2.6- × 70-cm GSSG–Sepharose affinity chromatographic column. The column material was previously equilibrated with a solution of 10 mM potassium phosphate (pH 7.0) and 20% glycerol. After loading, the column is washed with the equilibration solution until the bulk of the protein is eluted. The column is then washed with a solution of 50 mM potassium phosphate (pH 7.0) and 20% glycerol, and then again with the equilibration solution. A gradient of a competitive inhibitor of glyoxalase II, S-octylglutathione[6] (1 to 5 mM, a total of 2 liters), prepared in the equilibration solution, is used to elute glyoxalase II. Figure 1 shows typical protein concentration and glyoxalase II activity profiles obtained. All of the glyoxalase I activity was eluted with the early fractions containing the bulk of the protein. The GSSG–Sepharose affinity material can be used repeatedly. After each use, the column materials are washed with the equilibration solution containing 1 M NaCl, and then again with the equilibration solution before the reuse.

[6] R. Vince, S. Daluge, and W. B. Wadd, *J. Med. Chem.* **14**, 402 (1971).

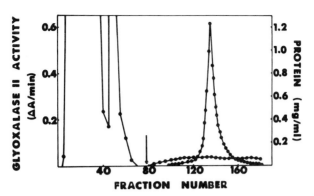

FIG. 1. First GSSG–Sepharose affinity chromatography. Experimental details are given in the text. Crude preparation (380 ml) was applied to a 2.6- × 70-cm preequilibrated column. Fraction volumes were 21.0 ml. The arrow marks the application of the S-octylglutathione gradient (1 to 5 mM). O——O, Glyoxalase II activity, ΔA/min (75 μl of each fraction was used in the assay); ●——●, protein concentration, mg/ml. From Oray and Norton.[3]

Second GSSG–Sepharose 4B Affinity Chromatography. Active fractions from the first column are combined and diluted twofold with the equilibration solution to decrease the S-octylglutathione concentration. The primary reason for the dilution of the pooled active fractions is to allow binding of the enzyme to the second affinity column, since binding is prevented by 2 mM S-octylglutathione. S-Octylglutathione apparently stabilizes glyoxalase II, and complete removal of this inhibitor ($K_i = 1.5$ mM) by dialysis results in substantial loss of activity.

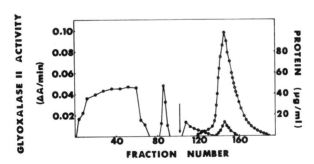

FIG. 2. Second GSSG–Sepharose affinity chromatography. Experimental details are given in the text. Pooled and diluted fractions (920 ml) from the first column were applied to a 2.6- × 40-cm preequilibrated column. Fraction volumes were 18.4 ml. The arrow marks the application of S-octylglutathione (2 mM). O——O, Glyoxalase II activity, ΔA/min (75 μl of each fraction was used in the assay); ●——●, protein concentration, μg/ml. From Oray and Norton.[3]

PURIFICATION OF GLYOXALASE II[a]

Fraction	Volume (ml)	Total activity (units)[b]	Specific activity (units/mg protein)	Yield, step (%)	Purification Step	Purification Overall
Crude preparation	380	1743	0.248	100	1.0	1.0
First GSSG column	460	1656	60.3	95	243.0	243.0
Second GSSG column	650	1407	614.7	85	10.2	2479.0

[a] From Oray and Norton.[3]

[b] The routine assay mixture contained 0.5 M S-D-lactoylglutathione, a concentration that is only approximately twice the value of K_m (0.27 mM). When the enzyme preparations were assayed at near-saturating levels of substrate, the activity was increased by a factor of 1.5. Thus, the correct specific activity of the homogeneous enzyme is approximately 920 IU/mg.

The combined, diluted fractions from the first column are then added to a second GSSG–Sepharose affinity chromatographic column (2.6 × 40 cm) that has been equilibrated with a solution of 10 mM potassium phosphate (pH 7.0) containing 20% glycerol. After loading, the column is again washed with the same equilibrated solution. Homogeneous glyoxalase II is then eluted by addition of a solution of 2.0 mM S-octylglutathione in the equilibration solution. Typical protein concentration and activity profiles are shown in Fig. 2. Volume reduction of the pooled active fractions from this column was achieved by ultrafiltration.

The results of a typical purification are presented in the table. An approximate 2500-fold purification with an overall yield of 85% was obtained.

Properties

General. Glyoxalase II from mouse liver has a molecular weight of 29,500 as estimated by SDS–polyacrylamide gel electrophoresis. Homogeneity was established by multiple disc gel electrophoretic determinations. The purified enzyme exhibited a specific activity of 920 IU per milligram of protein. The enzyme is a basic protein with a pI of approximately 8.1.

Stability. When the enzyme was stored at −30° in the presence of 1 mM S-octylglutathione, no significant loss of activity was observed for at least 6 months.

Kinetic Properties and Inhibitors. The K_m for the substrate, S-D-lactoylglutathione, is 0.27 mM; Uotila found the K_m to be 0.19 mM for

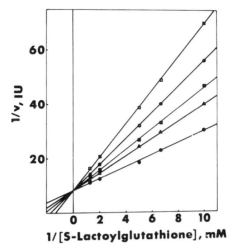

F<small>IG</small>. 3. Inhibition of glyoxalase II by the hemimercaptal of glutathione and methylglyoxal. The calculated concentrations of the hemimercaptal were 0 mM (●), 0.05 mM (▲), 0.1 mM (■), 0.2 mM (○), 0.5 mM (□). Experimental details are given in the text. From Oray and Norton.[3]

the enzyme from human liver.[7] Both methylglyoxal and glutathione have been shown to be weak inhibitors of human liver glyoxalase II; however, combinations of methylglyoxal and glutathione show a greater than additive inhibitory activity, presumably due to a greater inhibition by the hemimercaptal adduct of methylglyoxal and glutathione (the substrate of glyoxalase I).[7] Inhibition studies by the hemimercaptal were conducted on the homogeneous mouse liver glyoxalase II, and the inhibition was found to be competitive with a K_i of 0.3 mM (Fig. 3).

[7] L. Uotila, *Biochemistry* **12**, 3944 (1973).

[88] Galactose-1-phosphate Uridylyltransferase from *Entamoeba histolytica*

By PATRICIA LOBELLE-RICH and RICHARD E. REEVES

$$\text{Galactose 1-phosphate} + \text{MgUTP} \rightleftharpoons \text{UDPgalactose} + \text{MgPP}_i$$

Galactose-1-phosphate uridylyltransferase (EC 2.7.7.10) is a UTP-utilizing transferase involved in the metabolism of galactose. In *Entamoeba histolytica* this enzyme is constitutive and can utilize glucose 1-phosphate as an alternative carbohydrate substrate. The organism possesses a great excess of glucose-1-phosphate uridylyltransferase activity (EC 2.7.7.9), but lacks the common UDPglucose–hexose-1-phosphate uridylyltransferase (EC 2.7.7.12). The title enzyme activity was discovered in *Saccharomyces*[1] and was found in *Streptococcus faecalis*[2] and in *Phaseolus aureus*.[3] It is present in minor quantities in mammalian systems[4] and provides a method other than the Leloir pathway for the metabolism of galactose 1-phosphate in adult galactosemics. The prior authors[1-4] did not mention the carbohydrate specificity of the enzyme, but Lee *et al.*[5,6] isolated from *Bifidobacterium bifidum* a homogeneous enzyme that exhibited the same activity toward UDPgalactose and UDPglucose in the pyrophosphorylase direction.

The production of pyrophosphate by this enzyme allows it to be monitored spectrophotometrically by the use of a linking enzyme system.[7]

Assay Method

Principle. The assay method given below monitors the rate of formation of pyrophosphate by galactose-1-phosphate uridylyltransferase (or any other pyrophosphate generating system such as glucose-1-phosphate uridylyltransferase). The assay is linear with the amount of enzyme up to 10 milliunits of enzyme per milliliter in the cuvette.

Reagents

Imidazole-HCl buffer, 0.2 M, pH 7
UTP, 80 mM

[1] H. M. Kalckar, B. Braganca, and A. Munch-Petersen, *Nature (London)* **172**, 1038 (1953).
[2] J. H. Pazur and J. S. Anderson, *J. Biol. Chem.* **238**, 3155 (1963).
[3] R. B. Frydman, E. F. Neufeld, and W. Z. Hassid, *Biochim. Biophys. Acta* **77**, 332 (1963).
[4] K. J. Isselbacher, *J. Biol. Chem.* **232**, 429 (1958).
[5] L. Lee, A. Kimura, and T. Tochikura, *Biochim. Biophys. Acta* **527**, 301 (1978).
[6] L. Lee, A. Kimura, and T. Tochikura, *J. Biochem (Tokyo)* **86**, 923 (1979).
[7] See this volume [17].

MgCl$_2$, 40 mM

Galactose 1-phosphate, dipotassium salt, 100 mM

Fructose 6-phosphate, disodium salt, 60 mM

NADH, 10 mM

Assay enzymes

A solution containing sulfate-free rabbit muscle fructose-bisphosphate aldolase, 50 units/ml; sulfate-free rabbit muscle glycerol-3-phosphate dehydrogenase, 150 units/ml; and triosephosphate isomerase 1250 units/ml; in 1 mM NaEDTA, pH 7.

A solution of 6-phosphofructokinase (pyrophosphate) in 20 mM imidazole-HCl, pH 7, and 25% glycerol, 4 units/ml[7]

Procedure. To a quartz cuvette of 1-cm light path, the following components are added: 100 μl of imidazole buffer, 5 μl of UTP, 10 μl of MgCl$_2$, 20 μl of galactose 1-phosphate, 10 μl of fructose 6-phosphate, 5 μl of NADH, 10 μl of each of the assay enzyme solutions, and water to a final volume of 0.39 ml. The cuvettes are equilibrated at 30° for 5 min. The reaction is initiated by the addition of up to 3 milliunits of enzyme and is monitored spectrophotometrically by the decrease in absorbance at 340 nm. A control lacking substrate is subtracted from the assay values. This correction becomes negligible after the Sephacryl S-200 fractionation. When glucose 1-phosphate is substituted for galactose 1-phosphate as substrate a second control must be run to correct for possible contamination of the amoebal phosphofructokinase by glucose-1-phosphate uridylyltransferase. This is done by monitoring the cuvette in the presence of glucose 1-phosphate prior to the addition of enzyme.

Definition of Unit and Specific Activity. One unit of enzyme activity is defined as the galactose 1-phosphate-dependent formation of 1 μmol of pyrophosphate per minute and is evidenced by the oxidation of 2 μmol of NADH per minute at 30° under above standard assay conditions. The molar extinction coefficient of NADH is taken to be 6220 M^{-1} cm^{-1}. Specific activity is defined as units per milligram of protein. Protein concentrations are determined by the method of Lowry *et al.*[8]

Purification Procedures

The K-9 strain of *Entamoeba histolytica* (ATCC 30015) is grown and treated through the first ammonium sulfate precipitation as described elsewhere in this volume.[7] The buffer employed is 20 mM imidazole-HCl, pH 7, and all steps are done at 4°.

---

[8] O. H. Lowry, N. J. Rosebrough, A. L. Farr, and R. J. Randall, *J. Biol. Chem.* **193,** 265 (1951).

PURIFICATION OF GALACTOSE-1-PHOSPHATE URIDYLYLTRANSFERASE FROM LYOPHILIZED
AMOEBAS REPRESENTING 3.1 g OF FRESH *Entamoeba histolytica* CELLS

Treatment	Enzyme (units)	Protein (mg)	Specific activity (units/mg protein)	Recovery (%)
Step 1. Crude enzyme	8.6	50	0.17	(100)
Step 2. Sephacryl S-200	7.1	13	0.55	83
Step 3. DEAE-cellulose	5.1	1.3	4	59
Step 4. Blue Sepharose	2.7	0.02	135	31

Step 1. Crude Enzyme. The ammonium sulfate pellet reserved for the preparation of galactose-1-phosphate uridylyltransferase[7] is redissolved in about 0.6 ml of buffer per milliliter of fresh cells. This solution constitutes the crude enzyme (see Comment 1 below).

Step 2. Sephacryl S-200 Column Fractionation. A Sephacryl S-200 column (51 cm $\times$ 5.73 cm^2) is equipped with a flow adaptor and equilibrated with buffer. The crude enzyme is applied to it and eluted with buffer. Ten-milliliter fractions are collected. The six fractions with a ratio of galactose-1-phosphate uridylyltransferase to glucose-1-phosphate uridylyltransferase activity greater than one are pooled.

Step 3. DEAE-Cellulose Fractionation. A column (4.1 cm $\times$ 3.77 cm^2) of DEAE-cellulose, which had been pretreated according to a published method,[9] is prepared and equilibrated with buffer. The combined fractions from step 2 are applied. The enzyme is eluted with buffer. Ten milliliter fractions are collected. Enzyme from the five peak fractions is pooled. (See Comment 2 below.)

Step 4. Blue Sepharose Fractionation. A Blue Sepharose column (10.3 cm $\times$ 4.65 cm^2) is prepared and equilibrated with buffer. The pooled fractions from step 3 are applied to the column. It is washed with buffer until protein can no longer be detected in the effluent by monitoring the 280:260 absorbance ratio. The enzyme is then eluted with 1 mM MgUTP in 20 mM imidazole-HCl, pH 7. Five-milliliter samples are collected. Enzyme pooled from the three peak fractions is stored at $-20°$ in 20% glycerol. (See Comment 3 below.)

The results of a typical purification are summarized in the table.

Comments about the Procedure

1. The crude enzyme solution contains 65–85% of the galactose-1-phosphate uridylyltransferase originally present in the cells and 10–20%

[9] D. J. South and R. E. Reeves, this series, Vol. 42 [31].

of the original glucose-1-phosphate uridylyltransferase activity. *Entamoeba histolytica* has about 8 times as much of the latter as of the former.

2. Under the conditions given, the remaining glucose-1-phosphate uridylyltransferase is retained by the DEAE-cellulose column.

3. This purified enzyme preparation was used to obtain kinetic data for the carbohydrate substrates. The determination of the K_m for UTP employed enzyme from step 2.

Properties

Stability. Purified galactose-1-phosphate uridylyltransferase lost activity when frozen in 20 mM imidazole-HCl, pH 7, but in the presence of added glycerol (20% v/v) it retained 70% of its activity over 6 weeks at $-20°$.

Molecular Weight. The molecular weight of the enzyme, as determined by chromatography on a Sephadex G-100 column, was found to be 40,000.

Specificity. Amoebic galactose-1-phosphate uridylyltransferase can use either galactose 1-phosphate or glucose 1-phosphate as a carbohydrate substrate. Galactose 1-phosphate gives a maximum velocity of 1.35 times that of glucose 1-phosphate. UTP is the only uridylyl group donor utilized; ATP, ITP, CTP, and GTP are not substrates for this enzyme. Both Mg^{2+} and Mn^{2+} stimulate the enzyme reaction.

Kinetic Values for Substrates. Kinetic studies were performed using the spectrophotometric assay described, and the data were fitted to Eq. (1) using the FORTRAN programs of Cleland,[10]

$$v = VA/(K + A) \tag{1}$$

where A is the substrate concentration.

The K_m value for galactose 1-phosphate is 415 μM; for UTP, 34 μM; and for glucose 1-phosphate, 230 μM.

---

[10] W. W. Cleland, this series, Vol. 63 [103].

[89] Sucrose-6-phosphate Hydrolase from *Streptococcus mutans*

By Bruce M. Chassy and E. Victoria Porter

Sucrose 6-phosphate → D-glucose 6-phosphate + D-fructose

Streptococci that transport sucrose by the action of a phosphoenol-pyruvate (PEP)-dependent phosphotransferase system (PTS) produce intracellular sucrose 6-phosphate that is hydrolyzed by a specific low K_m sucrose-6-phosphate hydrolase. The isolation of this enzyme from sucrose-grown cells of *Streptococcus mutans* DR0001 (ATCC 33534) is described.[1]

Assay Method

Principle. The continuous spectrophotometric assay is based upon the following sequence of reactions:

$$\text{Sucrose 6-phosphate} \xrightarrow{\text{S6P hydrolase}} \text{glucose 6-phosphate} + \text{fructose}$$

$$\text{NADP}^+ + \text{glucose 6-phosphate} \xrightarrow{\text{G6P dehydrogenase}} \text{6-phosphogluconic acid} + \text{NADPH} + \text{H}^+$$

When glucose-6-phosphate dehydrogenase is present in excess, the rate of sucrose 6-phosphate hydrolysis is proportional to the rate of NADP reduction, which is measured as an increase in absorbance at 340 nm.[1]

Reagents

2-(*N*-Morpholino)ethanesulfonic acid (MES) buffer, 1 *M*, pH 7.1
Sucrose 6-phosphate, 20 m*M* [2]
NADP$^+$, 10 m*M*, pH 7.1
Glucose-6-phosphate dehydrogenase (EC 1.1.1.49), type XV Sigma (salt free, lyophilized powder)

Procedure. The following reagents are added to a semi-microcuvette (0.4 × 1 × 3 cm, internal dimensions) with a 1-cm light path: 0.1 ml of MES buffer, 0.05 ml of sucrose 6-phosphate, 0.05 ml of NADP$^+$, 0.01 ml of glucose-6-phosphate dehydrogenase (3–5 IU), sucrose-6-phosphate hydrolase, and water to a final volume of 0.5 ml. Increases in absorbance

---

[1] B. M. Chassy and E. V. Porter, *Biochem. Biophys. Res. Commun.* **89**, 307 (1979).

[2] Prepared synthetically as described by Chassy and Porter[1] or by F. Kunst, M. Pascal, J. Lepesant, J. Walle, and R. Dedonder [*Eur. J. Biochem.* **42**, 611 (1974)] and E. J. St. Martin and C. E. Wittenberger [*Infect. Immun.* **26**, 487 (1979)].

at 340 nm can be measured in a Gilford multiple-sample absorbance spectrophotometer. Assays are conducted at ambient temperature (ca 22°).

Evaluation of the Assay. As long as glucose-6-phosphate dehydrogenase is present in excess, the reaction rate, as measured by $NADP^+$ reduction, is proportional to sucrose 6-phosphate hydrolysis. Crude extracts of *S. mutans* lack 6-phosphogluconate dehydrogenase activity, so that the reduction of $NADP^+$ is stoichiometric with glucose 6-phosphate formed by sucrose 6-phosphate hydrolysis.

Definition of Unit and Specific Activity. A unit of sucrose-6-phosphate hydrolase is defined as the amount of enzyme that catalyzes the hydrolysis of 1 μmol of sucrose 6-phosphate in 1 min in the assay described above. Specific activity is expressed as units per milligram of protein. Protein is determined by the Coomassie Blue dye-binding assay described by Bradford.[3]

Purification Procedure

Growth of the Organism. *Streptococcus mutans* DR0001 is maintained by monthly transfer in NIH fluid thioglycolate broth (fluid thioglycolate, Baltimore Biological Laboratories, supplemented with 200 ml of beef infusion and 50 g of $CaCO_3$ per liter). The growth medium contains the following (amounts per liter): 5 g of yeast extract, 5 g of trypticase, 5 g of $K_2HPO_4 \cdot 3\ H_2O$, 50 mg of Na_2CO_3, 0.5 ml of salt solution (containing 800 mg of $MgSO_4 \cdot 7\ H_2O$, 40 mg of $FeSO_4 \cdot 7\ H_2O$, and 19 mg of $MnSO_4 \cdot 4\ H_2O$ in 100 ml of water), and carbohydrate (glucose or sucrose). Seed cultures are prepared by overnight growth in media containing 5 mM glucose. Batch cultures are grown in media containing 30 mM sucrose. Bottles are inoculated with seed cultures (5%, v/v) and incubated overnight at 37°. The cells are harvested by centrifugation (13,000 g for 20 min) and washed twice by resuspension in and centrifugation from 0.1 M potassium phosphate buffer, pH 7.5. Cell harvests and all the following procedures are performed at 0–4°.

Preparation of Cell Extracts. Fresh (or previously frozen) cell pastes (11 g wet weight resulting from 2 liters of batch culture) are suspended to make 50 ml of an approximately 20% (w/v) suspension in 0.1 M N-2-hydroxyethylpiperazine-N'-2-ethanesulfonic acid (HEPES) buffer, pH 7.5, containing 5 mM dithiothreitol (DTT). The cell suspensions are subjected to ultrasonic disruption for 9 min at full power using a Branson W-350 sonifier with continuous cooling. The suspensions are clarified by centrifugation at 43,000 g for 30 min, and the supernatant fluid is

---

[3] M. M. Bradford, *Anal. Biochem.* **72**, 248 (1976).

PURIFICATION OF SUCROSE-6-PHOSPHATE HYDROLASE FROM *Streptococcus mutans* DR0001

Preparation	Volume (ml)	Activity (units/ml)	Total units	Protein (mg/ml)	Specific activity (units/mg protein)	Yield (%)	Purification (fold)
Crude extract	43	1.14	49	3.15	0.36	100	1.00
DEAE-cellulose	4	8.25	33	2.10	3.93	67	11
Ultrogel AcA-54	10	2.75	27.5	0.10	25.56	56	71

dialyzed overnight against 4 liters of 5 mM HEPES buffer, pH 7.5, containing 5 mM DTT.

DEAE-Cellulose Chromatography. The dialyzed crude extract (43 ml) is applied (flow rate 1 ml/min) to a 2.5- × 40-cm column of DEAE-cellulose (DE-52, Whatman) previously equilibrated with 0.1 M HEPES buffer, pH 7.5, containing 5 mM DTT. A linear gradient from 0.1 M HEPES buffer, pH 7.5, containing 5 mM DTT to 0.4 M KCl in the same buffer is applied over a 16-hr elution period. Sucrose-6-phosphate hydrolase activity elutes at 0.34 M KCl (fractions 134 through 144; 7.5 ml/fraction). These fractions are combined and pressure-concentrated (PM-10 membrane; Amicon Corp., Lexington, Massachusetts) to about 8–10 ml. The concentrate is transferred to 8-mm dialysis sacks, and the volume is further reduced to about 4 ml with Aquacide III (Calbiochem Corp., La Jolla, California).

Ultrogel AcA-54 Chromatography. The pooled concentrate from DEAE-cellulose chromatography is applied at a flow rate of 0.2 ml/min to a 1.4- × 60-cm column of Ultrogel AcA-54 equilibrated with 0.1 M HEPES buffer, pH 7.5, containing 5 mM DTT. Sucrose-6-phosphate hydrolase activity is present in fractions 54 through 66, with the peak activity in fraction 60 (1 ml/fraction). Fractions containing activity are pooled, concentrated (as described above under DEAE-cellulose chromatography), and adjusted to 30% (v/v) glycerol prior to freezing. The total enzyme activity in the concentrate corresponds to a 56% recovery; the specific activity indicates that a 71-fold purification was obtained.

A typical purification is summarized in the table.

Notes on the Purification Procedure

1. For larger batches of cells, excellent cell breakage and enzyme yield has been obtained with a Bead-Beater (Biospec Products, Bartlesville, Oklahoma).

2. Larger batches (2–3×) may be processed by the same DEAE-cellulose chromatography step with little loss in separation efficiency. DEAE-Sephacel may be substituted for DE-52.

3. Substitution of Sephacryl S-200 for Ultrogel AcA-54 resulted in nonspecific losses of the enzyme. This step can be easily scaled up; for larger batches, a 2.5- × 90-cm column may be used.

4. As judged by disc-gel electrophoresis, the final product is essentially homogeneous when this strain of *S. mutans* is used. However, the purity of preparations isolated from other strains ranged from 25 to 80%. The two-step purification described here results in a homogeneous preparation primarily owing to the higher levels of activity and starting specific activity found in *S. mutans* DR0001.

Properties

Effect of Growth Substrate. Sucrose 6-phosphate hydrolase is found in *S. mutans* cells cultured on a number of sugars; however, the best yield is obtained from sucrose-grown cells.

Substrate Specificity. Sucrose 6-phosphate (K_m = 0.21 mM) and sucrose (K_m = 40–120 mM, depending on the phosphate concentration) are the only known substrates of the enzyme. Sucrose is a competitive inhibitor of sucrose 6-phosphate hydrolysis (K_i = 8.12 mM). Invertase activity cochromatographs and coelectrophoreses with sucrose-6-phosphate hydrolase. Thus, it appears that a single protein catalyzes both enzymic activities in *S. mutans*. Sucrose 6'-phosphate (D-glucopyranosyl-1-$\alpha \rightarrow$ β-2-D-fructofuranoside 6'-phosphate) is not a substrate for the enzyme.

Metal Ion Requirements. Activity in both crude extracts and purified preparations is independent of added metals ions or cofactors. A number of divalent and monovalent cations, including Mg^{2+}, Mn^{2+}, Ca^{2+}, Zn^{2+}, NH^+, K^+, and Na^+ were without appreciable effect on enzyme activity. The enzyme was not inhibited by 10 mM EDTA.

Buffer and pH Optimum. Among the buffers tested, MES was superior to HEPES and Tris. The pH optimum was 7.1.

Cofactors and Activators. ATP, ADP, AMP, cAMP, FDP, F6P, PEP, pyruvate, 3-PGA (1 and 10 mM), and phosphate (10 or 100 mM) were without marked effect on activity.

Molecular Weight Estimation. Based upon its peak elution fraction from an Ultrogel AcA-54 column (previously calibrated with proteins of known molecular weight) sucrose-6-phosphate hydrolase isolated from *S. mutans* DR0001 had an apparent M_r of 42,000. Similar values have been reported for the enzyme isolated from a number of *S. mutans* strains.[4]

Stability. The purified enzyme is stable for days at room temperature, weeks at 4°, and indefinitely when frozen in 30% glycerol at −20°.

[4] E. V. Porter and B. M. Chassy, *Abstr. 81st Ann. Meet. Am. Soc. Microbiol.* K134, p. 159 (1981).

Author Index

Numbers in parentheses are footnote reference numbers and indicate that an author's work is referred to although his name is not cited in the text.

Subject Index

A

Tumor, Ehrlich ascites,
  phosphofructokinase, 35–38
Turbatrix aceti, aldolase, 258, 262

V

Veillonela alcalescens, acetate kinase,
  179

W

Wheat, chloroplast, sedoheptulose-1,7-
  bisphosphatase, 392–396

X

ᴅ-Xylose-binding protein, *Escherichia
  coli*, 473–476
  cofactors, 476
  enzyme unit, 473
  molecular properties, 475, 476
  molecular weight, 475
  pH optimum, 476
  protein determination, 473

purification, 474, 475
specific activity, 473
stability, 475
substrate specificity, 475

Y

Yeast
  brewers', extract, preparation, for
    isolation of phosphoglycerate
    kinase, 111
  galactokinase, 30–35
  glyoxalase I, 540, 541
  phosphofructokinase, 49–60
  phosphoglycerate kinase, 107,
    110–114, 115–120
  pitching, 51, 52
  transketolase, 209–217

Z

Zaocys dhumnades, muscle, fructose-1,6-
  bisphosphatase, 349–351